Practical Skills in
Biology

Visit the *Practical Skills in Biology, fourth edition* Companion Website at **www.pearsoned.co.uk/practicalskills** to find valuable **student** learning material including:

- Self-test questions to test your knowledge

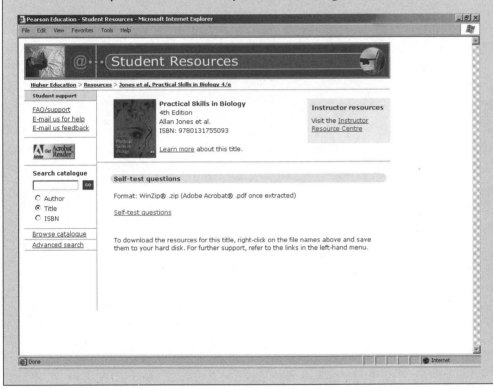

Practical Skills in
Biology

ALLAN JONES
ROB REED
JONATHAN WEYERS

Fourth edition

PEARSON
Benjamin
Cummings

Harlow, England • London • New York • Boston • San Francisco • Toronto
Sydney • Tokyo • Singapore • Hong Kong • Seoul • Taipei • New Delhi
Cape Town • Madrid • Mexico City • Amsterdam • Munich • Paris • Milan

Pearson Education Limited
Edinburgh Gate
Harlow
Essex CM20 2JE
England

and Associated Companies throughout the world

Visit us on the World Wide Web at:
www.pearsoned.co.uk

First published 1994
Second edition 1998
Third edition 2003
Fourth edition 2007

ISBN: 978-0-13-175509-3

British Library Cataloguing-in-Publication Data
A catalogue record for this book is
available from the British Library

10 9 8 7 6 5 4 3 2 1
11 10 09 08 07

Typeset in 10/12pt Times by 73
Printed by Ashford Colour Press Ltd, Gosport

The publisher's policy is to use paper manufactured from sustainable forests.

Contents

Contents

Supporting resources

Visit **www.pearsoned.co.uk/practicalskills** to find
valuable online resources

Companion Website for students
- Self-test questions to test your knowledge

For instructors
- Downloadable PowerPoint slides of all figures from the book

For more information please contact your local Pearson Education sales
representative or visit **www.pearsoned.co.uk/practicalskills**

List of boxes

List of boxes

Preface to the fourth edition

All knowledge and theory in biology has originated from observation and experiment. As a result, laboratory and field work are important components of undergraduate training, and successful students develop a number of practical skills, ranging from those required to observe, measure and record accurately to those associated with operating up-to-date analytical equipment, alongside broader skills involved in teamwork and effective study. In creating the fourth edition of *Practical Skills in Biology*, we have maintained the approach of the previous editions, with the aim of supporting students (and lecturers) over a wide range of practical topics. As before, this is provided in a concise but user-friendly manner, with key points and definitions, illustrations and worked examples, tips and hints, 'how to' boxes and checklists where appropriate. Safety issues are now emphasised through the use of a safety icon ⚠.

There are over 20 improved or new figures. We have made more effective use of the margin, with over 90 new tips, examples and definitions. Naturally, we have also used the opportunity of this new edition to update the content and add fresh material. Topics newly emphasised or covered for the first time include personal development planning (PDP), curriculum choices, learning styles, thinking processes, avoiding plagiarism, copyright, e-learning, bioethics, digital imaging, sterile technique for eukaryote cell culture, ELISA assays, and the PCR. Some areas move faster than others, and in particular the chapters on online resources, photography and imaging, and molecular biology have seen many changes. There is a new box on writing styles as some students would welcome guidance in this area. There are also new boxes on the effective use of presentation software for preparing talks and posters. Mindful of the value of spreadsheets in the biosciences, we have added several boxes explaining how to use these to produce and amend graphs and tables, to carry out statistical analyses and to calibrate instruments. We have updated all the references, and checked on the availability of online sources, and have added over 80 new ones. Noting the influential Quality Assurance Agency *Subject Benchmark Statement for the Biosciences* in the UK, first published in 2002, we have provided up-to-date material on all of the generic skills that this highlights, along with practical aspects of many of the subject-specific topics within plant and animal biology, together with microbiology. We should like to take this opportunity to thank our wives and families for their continued support, and to recognise the following colleagues and friends who have provided assistance, comment and food for thought at various points during the production of all four editions: Richard A'Brook, Margaret Adamson, Gail Alexander, Steve Atkins, Janet Aucock, Chris Baldwin, Abdellah Barekate, Gary Black, Geoff Bosson, Olivia Bragg, Sally Brown, Eldridge Buultjens, Richard Campbell, Cathy Caudwell, Bob Cherry, Mirela Cuculescu, John Dean, Charlie Dixon, Jackie Eager, Brian Eddy, Neil Fleming, Jennifer Gallacher, Karen Gowlett-Holmes, Alan Grant, Margaret Gruver, Mhairi Haggerty, Bryan Harrison, Rod Herbert, Dave Holmes, Helen Hooper, David Hopkins, Steve Hubbard, Jane Illés, Hugh Ingram, Wendy James, Andy Johnston, Alan Jones, Lorraine Kay, Ian Kill, Rhonda Knox, Lisa Lee-Jones, Phil Manning, Pete Maskrey, Fiona McKie-Bell, Kathleen McMillan, Steve Millam, Kirsty Millar, Stephen Moore, Rachel Morris, Fiona O'Donnell, Roy Oliver, Neil Paterson, John Raven, Steve Reed, Pete Rowell, David Sillars, J. Andrew C. Smith, Philip Smith, Susan Smith, Peter Sprent, Rob Sunley, Bill Tomlinson, Ruth Valentine, Lorraine Walsh, Dave Wealleans, Will Whitfield, Ian Winship, Bob Young and Hilary-Kay Young. We should also like to thank the staff of Pearson Education for their friendly support over the years, and would wish to acknowledge Pauline Gillet, Owen Knight, Simon Lake and Alex Seabrook, for their encouragement and commitment to the *Practical Skills* series. Our thanks are also extended to Tim Parker, David Hemsley, Ron Hawkins and Tony Clappison, for their excellent work at the proofreading stage. As with the previous editions, we should be grateful to hear of any errors you might notice, so that these can be put right at the earliest opportunity.

ALLAN JONES (a.m.jones@dundee.ac.uk)
ROB REED (rob.reed@unn.ac.uk)
JONATHAN WEYERS (j.d.b.weyers@dundee.ac.uk)

Guided tour

Tips and Hints provide useful hints and practical advice, and are highlighted in the text margin

Key Points highlight critical features of methodology

Examples are included in the margin to illustrate important points without interrupting the flow of the main text

Figures are used to illustrate key points, techniques and equipment

Safety Notes highlight specific hazards and appropriate practical steps to minimise risk

Worked Examples and 'How To' boxes set out essential procedures in a step-by-step manner

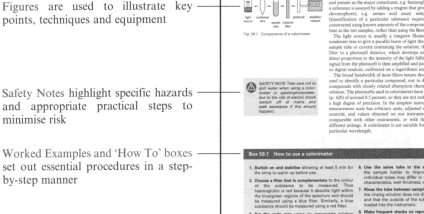

Definitions of key terms and concepts are highlighted in the text margin

Sources for Further Study and Text References every chapter is supported by a section giving additional printed and electronic sources

Study Exercises are included in every chapter to reinforce learning with problems and practical exercises

For the student

This book aims to provide guidance and support over the broad range of your undergraduate course, including laboratory classes, project work, lectures, tutorials, seminars and examinations, as outlined below.

Chapters 1–7 cover general skills

These include a number of transferable skills that you will develop during your course: for example, self-evaluation; time management; teamwork; preparing for examinations; creating a CV.

Chapters 8–18 deal with IT, library resources and communicating information

These chapters will help you get the most out of the resources and information available in your library, and on the World Wide Web, as well as providing helpful guidance on the use of software packages for data analysis, preparing assignments, essays and laboratory reports, etc. The ability to evaluate information is an increasingly important skill in contemporary society, and practical advice is provided here.

Chapters 19–60 cover a wide range of specific practical skills required in biology

These are based on the authors' experience of the questions students often ask in practical classes, and the support that is needed in order to get the most out of particular exercises. The text includes tips, hints, definitions, worked examples and 'how to' boxes that set out the key procedures in a step-by-step manner, with appropriate comments on safe working practice. The material ranges from basic laboratory procedures, such as preparing solutions, through specimen collection, identification and manipulation to the more advanced practical procedures that you might use during a final year project, e.g. radioisotope work.

Chapters 61–66 explain data analysis and presentation

This will be an important component of your course and you will find that these chapters guide you through the skills and techniques required, ranging from the presentation of results as graphs or tables through to the application of statistical tests. Worked examples are used to reinforce the numerical aspects wherever possible.

Study exercises and problems

We added these as a new feature in the third edition, following comments from students and staff at UK universities, who felt that it would provide a useful opportunity to practise some of the skills covered in the book and a check on the understanding of the material. We hope that the exercises will continue be useful both to learners and to their tutors.

Most of the problems assume students are working on their own, using the information supplied; however, tutors might wish to provide alternative starting material (e.g. a set of data from a practical class). We have also assumed that students will have access to a scientific calculator and, sometimes, to a networked PC with typical 'office' programs (especially word processor and spreadsheet), plus Internet access via a modem and browser. Where a library is mentioned, this is assumed to include access to standard reference works and a selection of scientific journals.

Answers are given at the back of the book for all exercises. For numerical problems, the working out is also shown along with the final answer, while the non-numerical exercises also have 'answers' in the form of tips, general guidance or illustrative examples, etc. We recommend that students work together for some exercises – this is a valuable means of learning and, where there is no single correct answer to a problem, teamwork provides a mechanism for checking and discussing your answers.

We hope that you will find this book a helpful guide throughout your course, and beyond.

Acknowledgements

We are grateful to the following for permission to reproduce copyright material:

Figure 11.2 from screen shot of Microsoft Excel 2002, http://www.microsoft.com/office/excel/evaluation/tour/page2.asp, copyright © 2001 Microsoft Corporation. Screen shot reprinted by permission from Microsoft Corporation; Figure 22.1 from Monograph No. 8559 Sodium Chloride from *The Merck Index: An Encyclopedia of Chemicals, Drugs and Biologicals*, Fourteenth Edition, Maryadele J. O'Neil, Patricia E. Heckelman, Cherie B. Koch, Kristin J. Roman, eds. (Merck & Co., Inc., Whitehouse Station, NJ, USA, 2006). Reproduced with permission from The Merck Index, Fourteenth edition, copyright © 2006 by Merck & Co., Inc., Whitehouse Station, NJ, USA. All rights reserved; Figure 38.2 adapted from Figure from *New Flora of the British Isles*, 2nd edn, Cambridge University Press, (Stace, C.A., 1997); Figure 39.4 property of bioMérieux S.A. Andrea Bannuscher; Figure 43.2 from photograph of *Olympus Model CX41* binocular microscope, supplied by Olympus Microscopes, Olympus Optical Company (UK) Ltd, published courtesy of Olympus Optical Company (UK) Ltd; Figure 43.2 from photograph of *Olympus Model SZX12* dissecting microscope, supplied by Olympus Microscopes, Olympus Optical Company (UK) Ltd, published courtesy of Olympus Optical Company (UK) Ltd; Figures 55.1 and 55.3 from figures from *Methods for physical and chemical analysis of fresh waters*, 2nd edn, International Biological Programme, No. 8, Blackwell Scientific, (Golterman, H.L., Clymo, R.S., and Ohnstad, M.A.M., 1978), reproduced by kind permission of Blackwell Publishing Ltd; Figure 58.1 and Table 58.1 from diagram and table of properties of low-speed centrifuge *Model MSE Centaur 2*, supplied by Fisher Scientific UK Ltd, reproduced by kind permission of Fisher Scientific UK Ltd.

In some instances we have been unable to trace the owners of copyright material, and we should appreciate any information that would enable us to do so.

List of abbreviations

A	absorbance
ACDP	Advisory Committee on Dangerous Pathogens
ADP	adenosine diphosphate
ANOVA	analysis of variance
ATP	adenosine triphosphate
BSA	bovine serum albumin
CCCP	carbonylcyanide m-chlorophenylhydrazine
CFU	colony-forming unit
COSHH	Control of Substances Hazardous to Health
CoV	coefficient of variance
DNA	deoxyribonucleic acid
D.P.M	disintegrations per minute
EDTA	ethylenediaminetetraacetic acid
EIA	enzyme immunoassay
ELISA	enzyme-linked immunosorbent assay
EM	electron microscopy
EMR	electromagnetic radiation
F	Faraday constant
FTP	file transfer protocol
g	acceleration due to gravity
GC	gas chromatography
HEPES	N-[2-hydroxyethyl]piperazine-N$'$-[2-ethanesulphonic acid]
HPLC	high-performance liquid chromatography
Ig	immunoglobulin
IR	infra-red (radiation)
IRMA	immunoradiometric assay
K_m	Michaelis constant
K_w	ionisation constant of water
LAN	local area network
LM	light microscopy
LSD	least significant difference
M	molar (mol l^{-1})
MPN	most probable number
M_r	relative molecular mass
MS	mass spectrometry
NAD$^+$	nicotinamide adenine dinucleotide (oxidised form)
NADH	nicotinamide adenine dinucleotide (reduced form)
NADP$^+$	nicotinamide adenine dinucleotide phosphate (oxidised form)
NADPH	nicotinamide adenine dinucleotide phosphate (reduced form)
NH	null hypothesis
OD	optical density
PAR	photosynthetically active radiation
PCR	polymerase chain reaction
PDP	personal development planning
PFD	photon flux density
PFU	plaque-forming unit
pH	$-\log_{10}$ proton concentration (activity), in mol m^{-1}
PI	photosynthetic irradiance
pK_a	$\log_{10}$ acid dissociation constant
PPFD	photosynthetic photon flux density

R	universal gas constant
RCF	relative centrifugal field
rDNA	recombinant deoxyribonucleic acid
R_F	relative frontal mobility
RIA	radioimmunoassay
RID	radioimmunodiffusion
RNA	ribonucleic acid
RP-HPLC	reverse phase high-performance liquid chromatography
r.p.m.	revolutions per minute
SD	standard deviation
SE	standard error (of the sample mean)
SEM	scanning electron microscopy
SI	Système Internationale d'Unités
STP	standard temperature and pressure
T	Absolute temperature (in Kelvin)
TEM	transmission electron microscopy
TLC	thin-layer chromatography
TRIS	tris(hydroxymethyl)aminomethane *or* 2-amino-2-hydroxymethyl-1, 3-propanediol
URL	uniform resource locator
UV	ultraviolet (radiation)
V_{max}	maximum velocity
WWW	World Wide Web
WYSIWYG	what you see is what you get

Study and examination skills

1 The importance of transferable skills

Skills terminology – *different phrases may be used to describe transferable skills, depending on place or context. These include: 'personal transferable skills' (PTS), 'key skills', 'core skills' and 'competences'.*

This chapter outlines the range of transferable skills and their significance to biologists. It also indicates where practical skills fit into this scheme. Having a good understanding of this topic will help you place your work at university in a wider context. You will also gain an insight into the qualities that employers expect you to have developed by the time you graduate. Awareness of these matters will be useful when carrying out personal development planning (PDP) as part of your studies.

The range of transferable skills

Table 1.1 provides a comprehensive listing of university-level transferable skills under six skill categories. There are many possible classifications – and a different one may be used in your institution or field of study. Note particularly that 'study skills', while important, and rightly emphasised at the start of many courses, constitute only a subset of the skills acquired by most university students.

The phrase '*Practical Skills*' in the title of this book indicates that there is a special subset of transferable skills related to work in the laboratory or field. However, although this text deals primarily with skills and techniques required for laboratory practicals, fieldwork and associated studies, a broader range of material is included. This is because the skills concerned are important, not only in the biosciences but also in the wider world. Examples include time management, evaluating information and communicating effectively.

Using course materials – *study your course handbook and the schedules for each practical session to find out what skills you are expected to develop at each point in the curriculum. Usually the learning objectives/outcomes (p. 22) will describe the skills involved.*

 KEY POINT *Biology is essentially a practical subject, and therefore involves highly developed laboratory and field skills. The importance that your lecturers place on practical skills will probably be evident from the large proportion of curriculum time you will spend on practical work in your course.*

The word 'skill' implies much more than the robotic learning of, for example, a laboratory routine. Of course, some of the tasks you will be asked to carry out in practical classes *will* be repetitive. Certain techniques require manual dexterity and attention to detail if accuracy and precision are to be attained, and the necessary competence often requires practice to make perfect. However, a deeper understanding of the context of a technique is important if the skill is to be appreciated fully and then transferred to a new situation. That is why this text is not simply a 'recipe book' of methods and why it includes background information, tips and worked examples, as well as study exercises to test your understanding.

Example The skills involved in teamwork cannot be developed without a deeper understanding of the interrelationships involved in successful groups. The context will be different for every group and a flexible approach will always be required, according to the individuals involved and the nature of the task.

Transferability of skills

Transferable skills are those that allow someone with knowledge, understanding or ability gained in one situation to adapt or extend this for application in a different context. In some cases, the transfer of a skill is immediately obvious. Take, for example, the ability to use a spreadsheet to summarise biological data and create a graph to illustrate results. Once the key concepts and commands are learned (Chapter 11),

The importance of transferable skills

Table 1.1 Transferable skills identified as important in the biosciences. The list has been compiled from several sources, including the UK Quality Assurance Agency for Higher Education *Subject Benchmark Statement for the Biosciences* (QAAHE, 2002). Particularly relevant chapters are shown for the skills covered by this book (numbers in **bold** green text indicate a deeper, or more extensive, treatment).

Skill category	Examples of skills and competences	Relevant chapters in this textbook
Generic skills for biologists	Having an appreciation of the complexity and diversity of life and life processes	34 35 36 **37** 38 39 53 54
	Reading and evaluating biological literature with a full and critical understanding	4 8 **9**
	Ability to communicate a clear and accurate account of a biological topic, both verbally and in writing	**13 14 15** 16 17 18
	Applying critical and analytical skills to evaluate evidence regarding theories and hypotheses in biology	9 31
	Using a variety of methods for studying biology	34 35 36 37 38 39 40 41 42 43 44 45 46 47 48 49 50 51 52 53 54 55 56 57 58 59 60
	Having the ability to think independently, set personal tasks and solve problems	33 **64**
Intellectual skills	Recognising and applying biological theories, concepts and principles	9 31
	Analysing, synthesising and summarising information critically	61 62 63 64 65 66
	Obtaining evidence to formulate and test hypotheses; applying knowledge to address familiar and unfamiliar problems	25 26 27 28 29 30 **31** 32 33
	Recognising and explaining moral, ethical and legal issues in biological research	**19 20** 29 36 40
Experimental and observational skills	Carrying out basic laboratory and field techniques and understanding the principles that underlie them	19 20 21 22 23 24 25 26 27 28 29 32 38 39 41 42 43
	Working in lab or field safely, responsibly and legally, with due attention to ethical aspects	**19 20** 29 33 34 36 40
	Designing, planning, conducting and reporting on biological investigations and data arising from them	14 17 29 31 33
	Obtaining, recording, collating and analysing data in the field and laboratory	25 27 29 30 31 34 35 36 37 38 39 **49** 61 62 63 64 65 66
Numeracy, communication and IT skills	Understanding and using data in several forms (e.g. numerical, textual, verbal and graphical)	4 9 27 61 62 63 64
	Communicating in written, verbal, graphical and visual forms	**13 14 15** 16 17 18 **62 63 64**
	Citing and referencing the work of others in an appropriate manner	8
	Obtaining data, including the concepts behind sampling and sampling errors, calibration and types of error	27 **30** 31 **32 50** 61 64 65 66
	Processing, interpreting and presenting data, and applying appropriate statistical methods for summarising and analysing data	61 62 63 64 **65** 66
	Solving problems with calculators and computers, including the use of tools such as spreadsheets	10 **11** 19 **64**
	Using computer technology to communicate and as a source of information in biology	10 11 12
Interpersonal and teamwork skills	Working individually or in teams as appropriate; identifying individual and group goals and acting responsibly and appropriately to achieve them	3
	Recognising and respecting the views and opinions of others	3
	Evaluating your own performance and that of others	3 7
	Appreciating the interdisciplinary nature of contemporary biology	1 18
Self-management and professional development skills	Working independently, managing time and organising activities	2 29 31 33
	Identifying and working towards targets for personal, academic and career development	1 7
	Developing an adaptable and effective approach to study and work (including revision and exam technique)	**4 5 6**

they can be applied to many instances outside the biosciences where this type of output is used. This is not only true for similar data sets, but also in unrelated situations, such as making up a financial balance sheet and creating a pie chart to show sources of expenditure. Similarly, knowing the requirements for good graph drawing and tabulation (Chapters 62 and 63), perhaps practised by hand in earlier work, might help you use spreadsheet commands to make the output suit your needs.

Other cases may be less clear but equally valid. For example, towards the end of your undergraduate studies you may be involved in designing experiments as part of your project work. This task will draw on several skills gained at earlier stages in your course, such as preparing solutions (Chapters 21–24), deciding about numbers of replicates and experimental layout (Chapters 30 and 31) and perhaps carrying out some particular method of observation, measurement or analysis (Chapters 40–60). How and when might you transfer this complex set of skills? In the workplace, it is unlikely that you would be asked to repeat the same process, but in critically evaluating a problem or in planning a complex project for a new employer, you will need to use many of the time-management, organisational and analytical skills developed when designing and carrying out experiments. The same applies to information retrieval and evaluation and writing essays and dissertations, when transferred to the task of analysing or writing a business report.

> *Opportunities to develop and practise skills in your private or social life* – you could, for example, practise spreadsheet skills by organising personal or club finances using Microsoft Excel, or teamwork skills within any university clubs or societies you may join (see Chapter 7).

Personal development planning

Many universities have schemes for personal development planning (PDP), which may go under slightly different names such as progress file or professional development plan. You will usually be expected to create a portfolio of evidence on your progress, then reflect on this, and subsequently set yourself plans for the future, including targets and action points. Analysis of your transferable skills profile will probably form part of your PDP. Other aspects commonly included are:

> *Types of PDP portfolio and their benefits* – some PDP schemes are centred on academic and learning skills, while others are more focused on career planning. Some are carried out independently and others in tandem with a personal tutor or advisory system. Some PDP schemes involve creating an online portfolio, while others are primarily paper-based. Each method has specific goals and advantages, but whichever way your scheme operates, maximum benefit will be gained from being fully involved with the process.

- your aspirations, goals, interests and motivations;
- your learning style or preference (see p. 20);
- your assessment transcript or academic profile information (e.g. record of grades in your modules);
- your developing CV (see p. 38).

Taking part in PDP can help focus your thoughts about your university studies and future career. This is important in biology, because most biological sciences degrees do not lead only to a specific occupation. The PDP process will introduce you to some new terms and will help you to describe your personality and abilities. This will be useful when constructing your CV and when applying for jobs.

Definition

> *Employability* – the 'combination of in-depth subject knowledge, work awareness, subject-specific, generic and career management skills, and personal attributes and attitudes that enable a student to secure suitable employment and perform excellently throughout a career spanning a range of employers and occupations.' (Anon, 2007: Higher Education Academy Centre for Bioscience, *Define Employability in the Context of Teaching Bioscience*).

What your future employer will be looking for

At the end of your course, which may seem some time away, you will aim to get a job and start on your chosen career path. You will need to sell yourself to your future employer, firstly in your application form and curriculum vitae (Chapter 7), and perhaps later at interview. Companies rarely employ bioscience graduates simply because they know how to carry out a particular lab routine or because they can remember specific facts about their chosen degree subject. Instead, employers tend to look for a range of qualities and transferable skills that together define an attribute known as 'graduateness'. This encompasses, for example, the ability to work in a team, to speak effectively and write clearly about your work, to understand complex data and to manage a project to completion. All of these skills can be developed at different stages during your university studies.

 KEY POINT *Factual knowledge can be important in degrees with a strong vocational element, but understanding how to find and evaluate information is usually rated more highly by employers than the ability to memorise facts.*

Most likely, your future employer(s) will seek someone with an organised yet flexible mind, capable of demonstrating a logical approach to problems – someone who has a range of skills and who can transfer these skills to new situations. Many competing applicants will probably have similar qualifications. If you want the job, you will have to show that your additional skills place you above the other candidates.

Text references

Anon. *Define Employability in the Context of Teaching Bioscience*. Available: http://www.bioscience.heacademy.ac.uk/issues/employability/forum/define.htm
Last accessed: 09/04/07
[Part of HE Academy Centre for Bioscience website.]

QAAHE (2002) *Subject Benchmark Statement for the Biosciences*. Quality Assurance Agency for Higher Education, Gloucester.

Sources for further study

Drew, S. and Bingham, R. (1997) *The Student Skills Guide,* 2nd edn. Gower Publishing Ltd, Aldershot.

McMillan, K. and Weyers, J.D.B. (2006) *The Smarter Student: Study Skills and Strategies for Success at University*. Pearson Education, Harlow.

Race, P. (1999) *How to Get a Good Degree: Making the Most of Your Time at University*. Open University Press, Buckingham.

Study exercises

1.1 Evaluate your skills. Examine the list of skill topics shown in Table 1.1. Now create a new table with two columns, like the one on the right-hand side. The first half of this table should indicate *five* skills you feel confident about in column 1 and show where you demonstrated this skill in column 2 (for example, 'working in a team' and 'in a first year group project on marine biology'). The second half of the table should show *five* skills you do not feel confident about, or you recognise need development (e.g. communicating in verbal form). List these in column 1 and in column 2 list ways in which you think the course material for your current modules will provide you with the opportunity to develop these skills.

Skills I feel confident about	Where demonstrated
1.	
2.	
3.	
4.	
5.	
Skills that I could develop	Opportunities for development
6.	
7.	
8.	
9.	
10.	

(continued)

Study exercises (continued)

1.2 **Find skills resources**. For at least one of the skills in the second half of the Table in 1.1, check your university's library database to see if there are any texts on that subject. Borrow an appropriate book and read the relevant sections. Alternatively, carry out a search for relevant websites (there are many); decide which are useful and 'bookmark' them for future use (see Chapter 10).

1.3 **Analyse your goals and aspirations**. Spend a little time thinking what you hope to gain from university. See if your friends have the same aspirations. Think about and/or discuss how these goals can be achieved, while keeping the necessary balance between university work, paid employment and your social life.

2 Managing your time

Time management – A system for controlling and using time as efficiently and as effectively as possible.

Advantages of time management *– these include:*

- *a much greater feeling of control over your activities;*
- *avoidance of stress;*
- *improved productivity – achieve more in a shorter period;*
- *improved performance levels – work to higher standards because you are in charge;*
- *an increase in time available for non-work matters – work hard, but play hard too.*

Example The objective 'to spend an extra hour each week on directed study in microbiology next term' fulfils the SMART criteria, in contrast to a general intention 'to study more'.

One of the most important activities that you can do is to organise your personal and working time effectively. There is a lot to do at university and a common complaint is that there just isn't enough time to accomplish everything. In fact, research shows that most people use up a lot of their time without realising it through ineffective study or activities such as extended coffee breaks. Developing your time-management skills will help you achieve more in work, rest and play, but it is important to remember that putting time-management techniques into practice is an individual matter, requiring a level of self-discipline not unlike that required for dieting. A new system won't always work perfectly straight away, but through time you can evolve a system that is effective for you. An inability to organise your time effectively, of course, results in feelings of failure, frustration, guilt and being out of control in your life.

Setting your goals

The first step is to identify clearly what you want to achieve, both in work and in your personal life. We all have a general idea of what we are aiming for, but to be effective, your goals must be clearly identified and priorities allocated. Clear, concise objectives can provide you with a framework in which to make these choices. Try using the 'SMART' approach, in which objectives should be:

- Specific – clear and unambiguous, including what, when, where, how and why.
- Measurable – having quantified targets and benefits to provide an understanding of progress.
- Achievable – being attainable within your resources.
- Realistic – being within your abilities and expectations.
- Timed – stating the time period for completion.

Having identified your goals, you can now move on to answer four very important questions:

1. Where does your time go?
2. Where should your time go?
3. What are your time-wasting activities?
4. What strategies can help you?

Analysing your current activities

The key to successful development of time management is a realistic knowledge of how you currently spend your time. Start by keeping a detailed time log for a typical week (Fig. 2.1), but you will need to be truthful in this process. Once you have completed the log, consider the following questions:

- How many hours do I work in total and how many hours do I use for 'relaxation'?
- What range of activities do I do?
- How long do I spend on each activity?

Time slots	Activity									Notes
7.00–7.15										
7.15–7.30										
7.30–7.45										
7.45–8.00										
8.00–8.15										
8.15–8.30										
8.30–8.45										
8.45–9.00										
9.00–9.15										

Fig. 2.1 Example of how to lay out a time log. Write activities along the top of the page, and divide the day into 15-minute segments as shown. Think beforehand how you will categorise the different things you do, from the mundane (laundry, having a shower, drinking coffee, etc.) to the well timetabled (tutorial meeting, sports club meeting), and add supplementary notes if required. At the end of each day, place a dot in the relevant column for each activity and sum the dots to give a total at the bottom of the page. You will need to keep a diary like this for at least a week before you see patterns emerging.

- What do I spend most of my time doing?
- What do I spend the least amount of my time doing?
- Are my allocations of time in proportion to the importance of my activities?
- How much of my time is ineffectively used, e.g. for uncontrolled socialising or interruptions?

If you wish, you could use a spreadsheet (Chapter 11) to produce graphical summaries of time allocations in different categories as an aid to analysis and management. Divide your time into:

- Committed time – timetabled activities involving your main objectives/goals.
- Maintenance time – time spent supporting your general life activities (shopping, cleaning, laundry, etc.).
- Discretionary time – time for you to use as you wish, e.g. recreation, sport, hobbies, socialising.

Quality in time management – avoid spending a lot of time doing unproductive studying e.g. reading a textbook without specific objectives for that reading.

Avoiding time-wasting activities

Look carefully at those tasks that could be identified as time-wasting activities. They include gossiping, over-long breaks, uninvited interruptions and even ineffective study periods. Try to reduce these to a minimum, but do not count on eliminating them entirely. Remember also that some relaxation *should* be programmed into your daily schedule.

Being assertive – if friends and colleagues continually interrupt you, find a way of controlling them, before they control you. Indicate clearly on your door that you do not wish to be disturbed and explain why. Otherwise, try to work away from disturbance.

Organising your tasks

Having analysed your time usage, you can now use this information, together with your objectives and prioritised goals, to organise your activities, on both a short-term and a long-term basis. Consider using a diary-based system (such as those produced by Filofax, TMI, and Day-timer) that will help you plan ahead and analyse your progress.

WEEKLY DIARY   Week beginning:

DATE	Sunday	Monday	Tuesday	Wednesday	Thursday	Friday	Saturday
7–8 am		Breakfast	Breakfast	Breakfast	Breakfast	Breakfast	
8–9		Preparation	Preparation	Preparation	Preparation	Preparation	Breakfast
9–10	Breakfast	PE112(L)	PE112(L)	PE112(L)	PE112(L)	BIOL(P)	Travel
10–11	FREE	CHEM(L)	CHEM(L)	CHEM(L)	CHEM(L)	BIOL(P)	WORK
11–12	STUDY	STUDY	STUDY	STUDY	STUDY	BIOL(P)	WORK
12–1 pm	STUDY	BIOL(L)	BIOL(L)	BIOL(L)	BIOL(L)	TUTORIAL	WORK
1–2	Lunch	Lunch	Lunch	Lunch	Lunch	Lunch	Lunch
2–3	(VOLLEY-	CHEM(P)	STUDY	SPORT	PE112(P)	STUDY	WORK
3–4	BALL	CHEM(P)	STUDY	(VOLLEY-	PE112(P)	STUDY	WORK
4–5	MATCH)	CHEM(P)	STUDY	BALL	PE112(P)	SHOPPING	WORK
5–6	FREE	STUDY	STUDY	CLUB)	STUDY	TEA ROTA	WORK
6–7	Tea	Tea	Tea	Tea	Tea	Tea	Tea
7–8	FREE*	STUDY	STUDY	FREE*	STUDY	FREE*	FREE
8–9	FREE*	STUDY	STUDY	FREE*	STUDY	FREE*	FREE
9–10	FREE*	FREE*	STUDY	FREE*	STUDY	FREE*	FREE

	Sunday	Monday	Tuesday	Wednesday	Thursday	Friday	Saturday
Study (h)	2	10	11	4	11	6	0
Other (h)	13	5	4	11	4	9	15

Total study time = 44 h

Fig. 2.2 A weekly diary with an example of entries for a first year science student with a Saturday job and active membership of a volleyball club. Note that 'free time' changes to 'study time', e.g. for periods when assessed work is to be produced or during revision for exams. Study time (including attendance at lectures, practicals and tutorials) thus represents between 42 and 50 per cent of the total time.

Divide your tasks into several categories, such as:

- Urgent – must be done as a top priority and at short notice (e.g. doctor's appointment).
- Routine – predictable and regular and therefore easily scheduled (e.g. preparation, lectures or playing sport).
- One-off activities – usually with rather shorter deadlines and which may be of high priority (e.g. a tutorial assignment or seeking advice).
- Long-term tasks – sometimes referred to as 'elephant tasks' that are too large to 'eat' in one go (e.g. learning a language). These are best managed by scheduling frequent small 'bites' to achieve the task over a longer timescale.

You should make a weekly plan (Fig. 2.2) for the routine activities, with gaps for less predictable tasks. This should be supplemented by individual daily checklists, preferably written at the end of the previous working day. Such plans and checklists should be flexible, forming the basis for most of your activities except when exceptional circumstances intervene. The planning must be kept brief, however, and should be scheduled into your activities. Box 2.1 provides tips for effective planning and working.

Matching your work to your body's rhythm – *everyone has times of day when they feel more alert and able to work. Decide when these times are for you and programme your work accordingly. Plan relaxation events for periods when you tend to be less alert.*

Use checklists as often as possible – *post your lists in places where they are easily and frequently visible, such as in front of your desk. Ticking things off as they are completed gives you a feeling of accomplishment and progress, increasing motivation.*

 KEY POINT *Review each day's plan at the end of the previous day, making such modifications as are required by circumstances, e.g. adding an uncompleted task from the previous day or a new and urgent task.*

Box 2.1 Tips for effective planning and working

- Set guidelines and review expectations regularly.
- Don't procrastinate: don't keep putting off doing things you know are important – they will not go away but they will increase to crisis point.
- Don't be a perfectionist – perfection is paralysing.
- Learn from past experience – review your management system regularly.
- Don't set yourself unrealistic goals and objectives – this will lead to procrastination and feelings of failure.
- Avoid recurring crises – they are telling you something is not working properly and needs to be changed.
- Learn to concentrate effectively and don't let yourself be distracted by casual interruptions.

- Learn to say 'no' firmly but graciously when appropriate.
- Know your own body rhythms: e.g. are you a morning person or an evening person?
- Learn to recognise the benefits of rest and relaxation at appropriate times.
- Take short but complete breaks from your tasks – come back feeling refreshed in mind and body.
- Work in suitable study areas and keep your own workspace organised.
- Avoid clutter (physical and mental).
- Learn to access and use information effectively (Chapter 9).
- Learn to read and write accurately and quickly (Chapters 4 and 15).

Sources for further study

Anon. *Day-Timer*. Available: http://www.daytimer.co.uk Last accessed 09/04/07.
[Website for products Day-Timers Europe Ltd, Chene Court, Poundwell Street, Modbury, Devon PL21 0QJ.]

Anon. *Filofax*. Available: http://www.filofax.co.uk Last accessed 09/04/07.
[Website for products of Filofax UK, Unit 3, Victoria Gardens, Burgess Hill, West Sussex, RH15 9NB.]

Anon. *TMI Website*. Available http://www.tmi.co.uk Last accessed 09/04/07.
[Website for products of TMI (Time Manager International A/S), 50 High Street, Henley-in-Arden, Solihull, West Midlands B95 5AN.]

Mayer, J.L. (1999) *Time Management for Dummies*, 2nd edn. IDG Books Worldwide, Inc., Foster City, Calif.

Study exercises

2.1 Evaluate your time usage. Compile a spreadsheet to keep a record of your daily activities in 15-minute segments for a week. Analyse this graphically and identify areas for improvement.

2.2 List your short-, medium- and long-term tasks and allocate priorities. Produce several lists, one for each of the three timescales and prioritise each item. Use this list to plan your time management by scheduling high-priority tasks and leave low-priority activities to 'fill in' the spare time that you may identify. This task should be done on a regular (monthly) basis to allow for changing situations.

2.3 Plan an 'elephant' task. Spend some time planning how to carry out a large or difficult task (learning a language or learning to use a complex computer program) by breaking it down into achievable segments ('bites').

3 Working with others

Peer assessment – this term applies to marking schemes in which all or a proportion of the marks for a teamwork exercise are allocated by the team members themselves. Read the instructions carefully before embarking on the exercise, so you know which aspects of your work your fellow team members will be assessing. When deciding what marks to allocate yourself, try to be as fair as possible with your marking.

Gaining confidence through experience – the more you take part in teamwork, the more you know how teams operate and how to make teamwork effective for you.

It is highly likely that you will be expected to work with fellow students during practicals and study exercises. This might take the form of sharing tasks or casual collaboration through discussion, or it might be formally directed teamwork such as problem-based learning (Box 6.1) or preparing a poster (Chapter 13). Interacting with others can be extremely rewarding and realistically represents the professional world, where teamworking is common. The advantages of working with others include:

- Teamworking is usually synergistic in effect – it often results in better ideas, produced by the interchange of views, and better output, due to the complementary skills of team members.
- Working in teams can provide support for individuals within the team.
- Levels of personal commitment can be enhanced through concern about letting the team down.
- Responsibilities for tasks can be shared.

However, you can also feel both threatened and exposed if teamwork is not managed properly. Some of the main reasons for negative feelings towards working in groups include:

- Reservations about working with strangers – not knowing whether you will be able to form a friendly and productive relationship.
- Worries over rejection – a perception of being unpopular or being chosen last by the group.
- Levels of personal commitment – these can be enhanced through a desire to perform well, so the team as a whole achieves its target.
- Fear of being held back by others – especially for those who have been successful in individual work already.
- Lack of previous experience – worries about the kinds of personal interactions likely to occur and the team role likely to suit you best.
- Concerns about the outcomes of peer assessment – in particular, whether others will give you a fair mark for your efforts.

Teamwork skills

Some of the key skills you will need to develop to maximise the success of your teamworking activities include:

- Interpersonal skills. How do you react to new people? Are you able to both listen and communicate easily with them? How do you deal with conflicts and disagreements?
- Delegation/sharing of tasks. The primary advantage of teamwork is the sharing of effort and responsibility. Are you willing/able to do this? It involves trusting your team members. How will you deal with those group members who don't contribute fully?
- Effective listening. Successful listening is a skill that usually needs developing, e.g. during the exchange of ideas within a group.
- Speaking clearly and concisely. Effective communication is a vital part of teamwork, both between team members and when presenting team outcomes to others. Try to develop your communication skills through learning and practice (see Chapter 14).

- Providing constructive criticism. It is all too easy to be negative, but only constructive criticism will have a positive effect on interactions with others.

Collaboration for learning

Much collaboration is informal and consists of pairs or groups of individuals getting together to exchange materials and ideas while studying. It may consist of a 'brainstorming' session for a topic or piece of work, or sharing efforts to research a topic. This has much to commend it and is generally encouraged. However, it is vital that this collaborative learning is distinguished from the collaborative writing of assessed documents: the latter is not usually acceptable and, in its most extreme form, is plagiarism, usually with a heavy potential punishment in university assessment systems. Make sure you know what plagiarism is, what unacceptable collaboration is, and how they are treated within your institution (see p. 51).

 KEY POINT *Collaboration is inappropriate during the final phase of an assessed piece of work unless you have been directed to produce a group report. Collaboration is encouraged during research and learning activities but the final write-up must normally be your own work. The extreme of producing copycat write-ups is regarded as plagiarism (p. 51) and will be punished accordingly.*

The dynamics of teamworking

It is important that team activities are properly structured so that each member knows what is expected of them. Allocation of responsibilities usually requires the clear identification of a leader. Several studies of groups have identified different team roles that derive from differences in personality. You should be aware of such categorisations, both in terms of your own predispositions and those of your fellow team members, as it will help the group to interact more productively. Belbin (1993) identified eight such roles, recently extended to nine, as shown in Table 3.1.

In formal team situations, your course organiser should deal with these issues; even if they do not, it is important that you are aware of these roles and their potential impact on the success or failure of teamwork. You should try to identify your own 'natural' role: if asked to form a team, bear the different roles in mind during your selection of colleagues and your interactions with them. The ideal team should contain members capable of adopting most of these roles. However, you should also note the following points:

- People will probably fit one of these roles naturally as a function of their personality and skills.
- Group members may be suited to more than one of these roles.
- In some circumstances, team members may be required to adapt and take a different role from the one that they feel suits them.
- No one role is 'better' than any other. For good teamwork, the group should have a balance of personality types present.
- People may have to adopt multiple roles, especially if the team size is small.

Studying with others – *teaming up with someone else on your course for revision ('study buddying') is a potentially valuable activity and may especially suit some types of learners (p. 20). It can help keep your morale high when things get tough. You might consider:*

- *sharing notes, textbooks and other information;*
- *going through past papers together, dissecting the questions and planning answers;*
- *talking to each other about a topic (good for aural learners: see Box 5.1);*
- *giving tutorials to each other about parts of the course that have not been fully grasped.*

Web-based resources and support for brainstorming – *websites such as http://www.brainstorming.co.uk give further information and practical advice for teamworking.*

Recording group discussions – *make sure you structure meetings (including writing agendas) and note their outcomes (taking minutes and noting action points).*

Table 3.1 A summary of the team roles described by Belbin (1993). No one role should be considered 'better' than any other, and a good team requires members who are able to undertake appropriate roles at different times. Each role provides important strengths to a team, and its compensatory weaknesses should be accepted within the group framework.

Team role	Personality characteristics	Typical function in a team	Strengths	Allowable weaknesses
Coordinator	Self-confident, calm and controlled	Leading: causing others to work towards goals	Good at spotting others' talents and delegating activities	Often less creative or intellectual than others in the group
Shaper	Strong need for achievement; outgoing; dynamic; highly strung	Leading: generating action within a team; imposing shape and pattern to work	Providing drive and realism to group activities	Can be headstrong, emotional and less patient than others
Innovator[1]	Individualistic, serious-minded; often unorthodox	Generating new proposals and solving problems	Creative, innovative and knowledgeable	Tendency to work in isolation; ideas may not always be practical
Monitor–evaluator	Sober, unemotional and prudent	Analysing problems and evaluating ideas	Shrewd judgement	May work slowly; not usually a good motivator
Implementer	Well organised and self-disciplined, with practical common sense	Doing what needs to be done	Organising abilities and common sense	Lack of flexibility and tendency to resist new ideas
Teamworker	Sociable, mild and sensitive	Being supportive, perceptive and diplomatic; keeping the team going	Good listeners; reliable and flexible; promote team spirit	Not comfortable when leading; may be indecisive
Resource investigator	Extrovert, enthusiastic, curious and communicative	Exploiting opportunities; finding resources; external relations	Quick thinking; good at developing others' ideas	May lose interest rapidly
Completer–finisher	Introvert and anxious; painstaking, orderly and conscientious	Ensuring completion of activity to high standard	Good focus on fulfilling objectives and goals	Obsessive about details; may wish to do all the work to control quality
Specialist	Professional, self-motivated and dedicated	Providing essential skills	Commitment and technical knowledge	Contribute on a narrow aspect of project; tend to be single-minded

[1] May also be called 'plant' in some texts.

 KEY POINT *In formal teamwork situations, be clear as to how individual contributions are to be identified and recognised. This might require discussion with the course organiser. Make sure that recognition, including assessment marks, is truly reflective of effort. Failure to ensure that this is the case can lead to disputes and feelings of unfairness.*

Your lab partner

Many laboratory sessions in the life sciences involve working in pairs. In some cases, you may work with the same partner for a series of practicals or for a complete module. The relationship you develop as a team is important to your progress, and can enhance your understanding of the material and the grades you obtain. Tips for building a constructive partnership include:

- Introduce yourselves at the first session and take a continuing interest in each other's interests and progress at university.
- At appropriate points, discuss the practical (both theory and tasks) and your understanding of what is expected of you.

- Work jointly to complete the practical effectively, avoiding the situation where one partner dominates the activities and gains most from the practical experience.
- Share tasks according to your strengths, but do this in such a way that one partner can learn new skills and knowledge from the other.
- Make sure you ask questions of each other and communicate any doubts about what you have to do.
- Discuss other aspects of your course, e.g. by comparing notes from lectures or ideas about in-course assessments.
- Consider meeting up outside the practical sessions to study, revise and discuss exams.

Text reference

Belbin, R.M. (1993) *Team Roles at Work*. Butterworth-Heinemann, Oxford.

Source for further study

Belbin, R.M. *The Belbin Website*. Available http://www.belbin.com/
Last accessed 09/04/07.

Study exercises

3.1 **Evaluate your 'natural' team role(s).** Using Table 3.1 as a source, decide which team role best fits your personality.

3.2 **Keep a journal during a group activity.** Record your feelings and observations about experiences of working with other students. After the event, review the journal, then draw up a strategy for developing aspects where you feel you might have done better.

3.3 **Reflect upon your teamwork abilities.** Draw up a list of your reactions to previous efforts at collaboration or teamwork and analyse your strengths and weaknesses. How could these interactions have been improved or supported more effectively?

Choose note-taking methods appropriately – *the method you choose to take notes might depend on the subject; the lecturer and their style of delivery; and your own preference.*

Note-taking is an essential skill that you will require in many different situations, such as:

- listening to lectures and seminars;
- attending meetings and tutorials;
- reading texts and research papers;
- finding information on the World Wide Web.

 KEY POINT *Good performance in assessments and exams is built on effective learning and revision (Chapters 5 and 6). However, both ultimately depend on the quality of your notes.*

Taking notes from lectures

Compare lecture notes with a colleague – *looking at your notes for the same lecture may reveal interesting differences in approach, depth and detail.*

Taking legible and meaningful lecture notes is essential if you are to make sense of them later, but many students find it difficult when starting their university studies. Begin by noting the date, course, topic and lecturer on the first page of each day's notes. Number every page in case they get mixed up later. The most popular way of taking notes is to write in a linear sequence down the page, emphasising the underlying structure via headings, as in Fig. 15.3. However, the 'pattern' and 'Mind Map' methods (Figs. 4.1 and 4.2) have their advocates: experiment, to see which method you prefer.

Adjusting to the different styles of your lecturers – *recognise that different approaches to lecture delivery demand different approaches to note-taking. For example, if a lecturer seems to tell lots of anecdotes or spend much of the time on examples during a lecture, do not switch off – you still need to be listening carefully to recognise the key take-home messages. Similarly, if a lecture includes a section consisting mainly of images, you should still try to take notes – names of organisms, locations, key features, even quick sketches. These will help prompt your memory when revising. Do not be deterred by lecturers' idiosyncrasies; in every case you still need to focus and take useful notes.*

Whatever technique you use, don't try to take down all the lecturer's words, except when an important definition or example is being given, or when the lecturer has made it clear that he/she is dictating. Listen first, then write. Your goal should be to take down the structure and reasoning behind the lecturer's approach in as few words and phrases as possible. At this stage, follow the lecturer's sequence of delivery. Use headings and leave plenty of space, but don't worry too much about being tidy – it is more important that you get down the appropriate information in a readable form. Use abbreviations to save time. Recognise that you may need to alter your note-taking technique to suit different lecturers' styles.

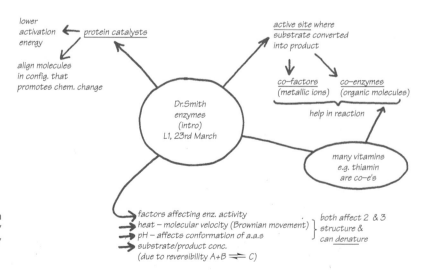

Fig. 4.1 An example of 'pattern' notes, an alternative to the more commonly used 'linear' format. Note the similarity to the 'spider diagram' method of brainstorming ideas (Fig. 15.2).

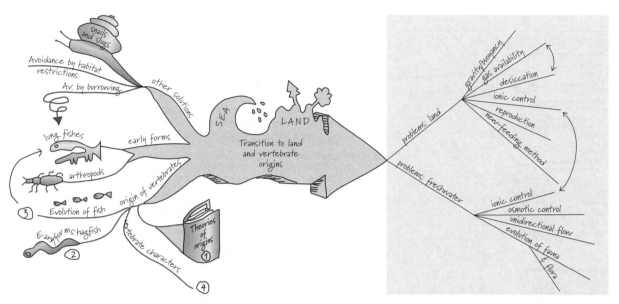

Fig. 4.2 Example of the 'Mind Map' approach to note-taking and 'brainstorming'. Start at the centre with the overall topic title, adding branches and sub-branches for themes and subsidiary topics. 'Basic' maps consist of a branched hierarchy overwritten with key words (e.g. shaded portion). Connections should be indicated with arrows; numbering and abbreviations are encouraged. To aid recall and creativity, Buzan and Buzan (2006) recommend use of colour, different fonts, 3-dimensional doodles and other forms of emphasis (e.g. non-shaded portion).

Example Commonly used abbreviations include:

$\exists$	there are, there exist(s)
$\therefore$	therefore
$\because$	because
$\propto$	is proportional to
$\rightarrow$	leads to, into
$\leftarrow$	comes from, from
$\rightarrowtail$	involves several processes in a sequence
$1°, 2°$	primary, secondary (etc.)
$\approx, \cong$	approximately, roughly equal to
$=, \neq$	equals, not equal to
$\equiv, \not\equiv$	equivalent, not equivalent to
$<, >$	smaller than, bigger than
$\gg$	much bigger than
$[X]$	concentration of X
$\sum$	sum
Δ	change
f	function
$\#$	number
∞	infinity, infinite

You should also make up your own abbreviations relevant to the context, e.g. if a lecturer is talking about photsynthesis, you could write 'PS' instead.

Printing PowerPoint slides – use the 'Black and White' option on the Print menu to avoid wasting ink on printing coloured backgrounds. If you wish to use colour, remember that slides can be difficult to read if printed in small format. Always print a sample page before printing the whole lecture.

Make sure you note down references to texts and take special care to ensure accuracy of definitions and numerical examples. If the lecturer repeats or otherwise emphasises a point, highlight (e.g. by underlining) or make a margin note of this – it could come in useful when revising. If there is something you don't understand, ask at the end of the lecture, and make an appointment to discuss the matter if there isn't time to deal with it then. Tutorials may provide an additional forum for discussing course topics.

Lectures delivered by PowerPoint or similar presentation programs

Some students make the mistake of thinking that lectures delivered as computer-based presentations with an accompanying handout or Web resource require little or no effort by way of note-taking. While it is true that you may be freed from the need to copy out large diagrams and the basic text may provide structure, you will still need to adapt and add to the lecturer's points. Much of the important detail and crucial emphasis will still be delivered verbally. Furthermore, if you simply listen passively to the lecture, or worse, try to work from the handout alone, it will be far more difficult to understand and remember the content.

If you are not supplied with handouts, you may be able to print out the presentation beforehand, perhaps in the '3 slides per page' format that allows space for notes alongside each slide (Fig. 4.3). Scan through this before the lecture if you can; then, during the presentation, focus on listening to what the lecturer has to say. Note down any extra details, points of emphasis and examples. After lectures, you could also add notes from supplementary reading. The text in presentations can be converted to word processor format if you have access to the electronic file. In PowerPoint, this can be achieved from the *Outline View* option on the *View* menu. You can copy and paste text between programs in the normal fashion, then modify font size and colour as appropriate.

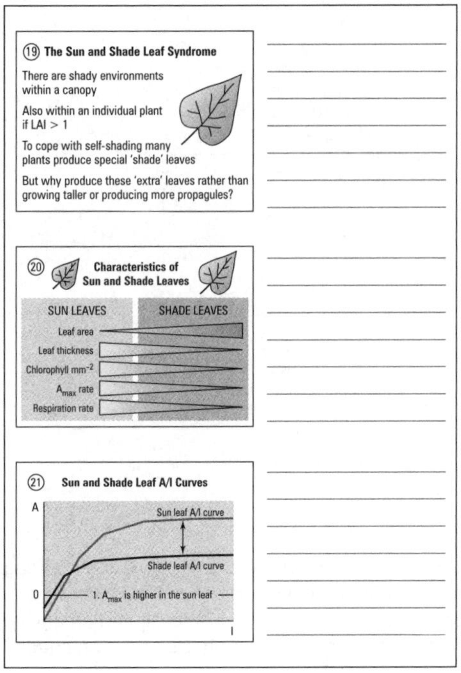

Fig. 4.3 An example of a printout from PowerPoint in 'Handouts (3 slides per page)' format

Scanning effectively – *you need to stay focused on your key words, otherwise you may be distracted by apparently interesting but irrelevant material.*

Spotting sequences – *writers often number their points (firstly, secondly, thirdly, etc.) and looking for these words in the text can help you skim it quickly.*

Making sure you have all the details – *when taking notes from a text or journal paper: (a) always take full details of the source (Chapter 8); (b) if copying word-for-word make sure you indicate this using quotes and take special care to ensure you do not alter the original.*

'Making up' your notes

As soon as possible after each lecture, work through your notes, tidying them up and adding detail where necessary. Add emphasis to any headings you have made, so that the structure is clearer. If you feel it would be more logical for your purposes, change the order. Compare your notes with material in a textbook and correct any inconsistencies. Make notes from, or photocopy, any useful material you see in text-books, ready for revision.

Taking notes from books and journal papers

Scanning and skimming are useful techniques to find sections to read in detail and to make notes from.

Scanning

This involves searching for relevant content. Useful techniques are to:

- decide on key words relevant to your search;
- check that the book or journal title indicates relevance;
- look through the contents page (either paper titles in a journal volume, or chapter titles in a text);
- look at the index if present.

Skimming

This is a valuable way to gain the maximum amount of information in the minimum amount of time, by reading as little of a text as is required. Essentially, the technique (also termed 'surveying') requires you to look at the structure of the text, rather than the detail. In a sense, you are trying to see the writer's original plan and the purpose behind each part of the text. Look through the whole of the piece first, to gain an overview of its scope and structure. Headings provide an obvious clue to structure, if present. Next, look for the 'topic sentence' in each paragraph (p. 97), which is often the first. You might then decide that the paragraph contains a definition that is important to note, or it may contain examples, so may not be worth reading for your purpose.

When you have found relevant material, note-taking fulfils the vital purpose of helping you understand and remember the information. If you simply read it, either directly or from a photocopy, you are at risk of accomplishing neither. The act of paraphrasing (using different words to give the same meaning) makes you think about the meaning and express this for yourself. It is an important active learning technique. A popular method of describing skimming and note-taking is called the SQ3R technique (Box. 4.1).

KEY POINT *Obtaining information and understanding it are distinct parts of the process of learning. As discussed in Chapter 5 (Table 5.1), you must be able to do more than recall facts to succeed.*

Methods for finding and evaluating texts and articles are discussed further in Chapters 8 and 9.

Box 4.1   The SQ3R technique for skimming texts

Survey Get a quick overview of the contents of the book or chapter, perhaps by rapidly reading the contents page or headings.

Question Ask yourself what the material covers and how precisely it relates to your study objectives.

Read Now read the text, paying attention to the ways it addresses your key questions.

Recall Recite to yourself what has been stated every few paragraphs. Write notes of this if appropriate, paraphrasing the text rather than copying it.

Review Think about what you have read and/or review your notes as a whole. Consider where it all fits in.

Sources for further study

Buzan, T. (2006) *The Ultimate Book of Mind Maps*. Harper Thorsons, New York.

Buzan, T., Morris, S. and Smith, J. (1998) *Understanding Mind Maps . . . in a Week*. Hodder & Stoughton, London.

Anon. *Mind map*. Available: http://en.wikipedia.org/wiki/Mind_mapping. Last accessed 09/04/07
[An independent review of mind-mapping and its history.]

Study exercises

4.1 Experiment with a new note-taking technique. If you haven't tried the pattern or mind-mapping methods (Figs 4.1 and 4.2), carry out a trial to see how they might work for you. Research the methods first by consulting appropriate books or websites.

4.2 Carry out a 'spring clean' of your desk area and notes. Make a concerted effort to organise your notes and handouts, investing if necessary in files and folders. This will be especially valuable at the start of a revision period.

4.3 Try out the SQ3R technique. The next time you need to obtain information from a text, compare this method (Box 4.1) with others you may have adopted in the past. Is it faster, and does it aid your ability to recall the information?

5 Learning and revising

There are many different ways of learning and at university you have the freedom to choose which approach to study suits you best. You should tackle this responsibility with an open mind, and be prepared to consider new options. Understanding how you learn best and how you are expected to think about your discipline will help you to improve your approach to study and to understand your branch of biology at a deeper level. Adopting active methods of studying and revision that are suited to your personality can make a significant difference to your performance. Your department will publish material that can help too. Taking account of learning outcomes and marking/assessment criteria, for example, can help you focus your revision.

 KEY POINT *At university, you are expected to set your own agenda for learning. There will be timetabled activities, assessments and exam deadlines, but it is your responsibility to decide how you will study and learn, how you will manage your time, and, ultimately, what you will gain from the experience. You should be willing to challenge yourself academically to discover your full potential.*

Learning styles

Significance of learning styles *– no one learning style is 'better' than the others; each has its own strengths and weaknesses. However, since many university exams are conducted using 'reading and writing' modes of communication, you may need to find ways of expressing yourself appropriately using the written word (see Box 5.1).*

Learning styles and teaching styles *– there may be a mismatch between your preferred learning style and the corresponding 'teaching style' used by your lecturers, in which case you will need to adapt appropriately (see Box 5.1).*

We don't all learn in the same way. Your preferred learning style is simply the one that suits you best for receiving, communicating and understanding information. It therefore involves approaches that will help you learn and perform most effectively. There are many different ways of describing learning styles, and you may be introduced to specific schemes during your studies. Although methods and terminology may differ among these approaches, it is important to realise that the important thing is the *process* of analysing your learning style, together with the way you use the information to modify your approach to studying, rather than the specific type of learner that you may identify yourself to be.

A useful scheme for describing learning styles is the VARK system devised by Fleming (2001). By answering a short online questionnaire, you can 'diagnose' yourself as one of the types shown in Box 5.1, which also summarises important outcomes relating to how information and concepts can be assimilated, learned and expressed. People show different degrees of alignment with these categories, and research indicates that the majority of students are multi-modal learners – that is, falling into more than one category – rather than being only in one grouping. By carrying out an analysis like this, you can become more aware of your personal characteristics and think about whether the methods of studying you currently use are those that are best suited to your needs.

 KEY POINT *Having a particular learning preference or style does not mean that you are automatically skilled in using methods generally suited to that type of learner. You must work at developing your ability to take in information, study and cope with assessment.*

Box 5.1 How to diagnose your learning preferences using the VARK learning styles scheme

Visit www.vark-learn to carry out the online diagnostic test, reflect on whether it is a fair description of your preferences, and think about whether you might change the way you study to improve your performance. None of the outcomes should be regarded as prescriptive – you should mix techniques as you see fit and only use methods that you feel comfortable adopting. Adapted with permission from material produced by Fleming (2001).

Learning style *Description of learning preferences*	Outcomes for your learning, studying and exam technique		
	Advice for taking in information and understanding it	Best methods of studying for effective learning	Ways to cope with exams so you perform better
Visual: *You are interested in colour, layout and design. You probably prefer to learn from visual media or books with diagrams and charts. You tend to add doodles and use highlighters on lecture and revision notes and express ideas and concepts as images.*	Use media incorporating images, diagrams, flow-charts, etc. When constructing notes, employ underlining, different colours and highlighters. Use symbols as much as you can, rather than words. Leave plenty of white space in your notes. Experiment with the 'mind map' style of note-taking (p. 17).	Use similar methods to those described in column two. Reduce lecture notes to pictures. Try to construct your own images to aid understanding, then test your learning by redrawing these from memory.	Plan answers diagrammatically. Recall the images and doodles you used in your notes. Use diagrams in your answers (making sure they are numbered and fully labelled, p. 168). As part of your revision, turn images into words.
Aural: *You prefer discussing subjects and probably like to attend tutorials and listen to lecturers, rather than read textbooks. Your lecture notes may be poor because you would rather listen than take notes.*	Make sure you attend classes, discussions and tutorials. Note and remember the interesting examples, stories, jokes. Leave spaces in your notes for later recall and 'filling'. Discuss topics with 'study buddy'. Record lectures (with lecturer's permission).	Expand your notes by talking with others and making additional notes from the textbook. Ask others to 'hear' you talk about topics. Read your summarised notes aloud to yourself. Record your vocalised notes and listen to them later.	When writing answers, imagine you are talking to an unseen examiner. Speak your answers inside your head. Listen to your voice and write them down. Practise writing answers to old exam questions.
Read–Write: *You prefer using text in all formats. Your lecture notes are probably good. You tend to like lecturers who use words well and have lots of information in sentences and notes. In note-taking, you may convert diagrams to text and text to bullet points.*	Focus on note-taking. You may prefer the 'linear' style of note-taking (p. 16). Use the following in your notes: lists; headings; glossaries and lists of definitions. Expand your notes by adding further information from handouts, textbooks and library readings.	Reduce your notes to lists or headings. Write out and read the lists again and again (silently). Turn actions, diagrams, charts and flowcharts into words. Rewrite the ideas and principles into other words. Organise diagrams and graphs into statements, e.g. 'The trend is . . .'.	Plan and write out exam answers using remembered lists. Arrange your words into hierarchies and points.

(continued)

Box 5.1 (continued)

Kinesthetic: *You tend to recall by remembering real events and lecturers' 'stories'. You probably prefer field excursions and lab work to theory and like lecturers who give real-life examples. Your lecture notes may be weak because the topics did not seem 'concrete' or 'relevant'.*	Focus on examples that illustrate principles. Concentrate on applied aspects and hands-on approaches, but try to understand the theoretical principles that underpin them. When taking in information, use all your senses – sight, touch, taste, smell, hearing.	Put plenty of examples, pictures and photographs into your notes. Use case studies and applications to help with principles and concepts. Talk through your notes with others. Recall your experience of lectures, tutorials, experiments or field trips.	Write practice answers and paragraphs. Recall examples and things you did in the lab or field trip. Role-play the exam situation in your own room.
Multi-modal: *Your preferences fall into two or more of the above categories. You are able to use these different modes as appropriate.*	If you are diagnosed as having two dominant preferences or several equally dominant preferences, read the study strategies above that apply to each of these. You may find it necessary to use more than one strategy for learning and communicating, feeling less secure with only one.		

Example A set of learning objectives taken from an introductory lecture on *The Ocean as a Habitat.*
'After this lecture, you should be able to:

☐ Define the following:

✓ Benthos
✓ Chemolithotroph
✓ Photolithotroph
✓ Plankton
✓ Upper mixed layer
✓ Upwelling

☐ Draw a diagram illustrating changes in light, plant nutrients, oxygen, temperature and density with depth in the ocean

☐ Discuss the role of seasonal mixing and upwellings, and inputs from rivers and the atmosphere, in the supply of nutrients to the surface of the ocean

☐ Demonstrate knowledge of magnitudes of ocean area and marine primary productivity relative to land and freshwater habitats.'

(With acknowledgements to Professor J.A. Raven, University of Dundee)

Thinking about thinking

The thinking processes that students are expected to carry out can be presented in a sequence, starting with shallower thought processes and ending with deeper processes, each of which builds on the previous level (see Table 5.1). The first two categories in this ladder apply to gaining basic knowledge and understanding, important when you first encounter a topic. Processes three to six are those additionally carried out by high-performing university students, with the latter two being especially relevant to final-year students, researchers and professionals. Naturally, the tutors assessing you will want to reward the deepest thinking appropriate for your level of study. This is often signified by the words they use in assessment tasks and marking criteria (column four, Table 5.1, and p. 102), and while this is not an exact process, being more aware of this agenda can help you to gain more from your studies and appreciate what is being demanded of you.

 KEY POINT When considering assessment questions, look carefully at words used in the instructions. These cues can help you identify what depth is expected in your answer. Take special care in multi-part questions, because the first part may require lower-level thinking, while in later parts marks may be awarded for evidence of deeper thinking.

The role of assessment and feedback in your learning

Your starting point for assessment should be the learning outcomes or objectives for each module, topic or learning activity. You will usually find them in your module handbook. They state in clear terms what your tutors expect you to be able to accomplish after participating in each part and reading around the topic. Also of value will be marking/assessment

Table 5.1 A ladder of thinking processes, moving from shallower thought processes (top of table) to deeper levels of thinking (bottom of table). This table is derived from research by Benjamin Bloom et al. (1956). When considering the cue words in typical question instructions, bear in mind that the precise meaning will always depend on the context. For example, while 'describe' is often associated with relatively simple processes of recall, an instruction like 'describe how the human brain works' demands higher-level understanding. Note also that while a 'cue word' is often given at the start of a question/instruction, this is not universally so.

Thinking processes and description (in approximate order of increasing 'depth')	Example in life sciences	Example of typical question structure, with *cue word* highlighted	Other cue words used in question instructions
1. **Knowledge (knowing facts).** If you know information, you can *remember* or *recognise* it. This does not always mean you understand it at a higher level.	You might know the order of bases in a piece of DNA but not understand what this means.	*Describe* the main components of a biological membrane.	<ul><li>define</li><li>list</li><li>state</li><li>identify</li></ul>
2. **Comprehension.** If you comprehend a fact, you *understand* what it means.	You might know the order of bases in a piece of DNA and understand that they code in triplets for specific amino acids.	*Explain* how membrane components are involved in the accumulation of solutes within living cells.	<ul><li>contrast</li><li>compare</li><li>distinguish</li><li>interpret</li></ul>
3. **Application.** To apply a fact means that you can *put it to use* in a particular context.	You might be able to take the DNA base sequence and work out the amino acid sequence of the protein for which they code.	Using the Nernst equation, and realistic values for the membrane potential and solute concentrations, *demonstrate* how Na⁺ ions must be actively transported out of the cells of marine organisms.	<ul><li>calculate</li><li>illustrate</li><li>solve</li><li>show</li></ul>
4. **Analysis.** To analyse information means that you are able to *break it down into parts* and show how these components *fit together*.	You might be able to construct a three-dimensional model of a protein derived from the base sequence.	Drawing on information about membrane structure, *defend* the endosymbiotic theory of eukaryote evolution.	<ul><li>compare</li><li>explain</li><li>consider</li><li>infer</li></ul>
5. **Synthesis.** To synthesise, you need to be able to *extract relevant facts* from a body of knowledge and use these to *address an issue in a novel way* or *create something new*.	You might be able to work out the function of a protein for which you know the sequence of bases, based on a comparison with other like proteins.	*Devise* an experiment to test the hypothesis that a specific membrane fraction contains a functional ATPase involved in glucose transport.	<ul><li>design</li><li>integrate</li><li>test</li><li>create</li></ul>
6. **Evaluation.** If you evaluate information, you *arrive at a judgement* based on its importance relative to the topic being addressed.	You might be able to comment on theories about how a protein has evolved, by considering the structure of related proteins and relating this to the taxonomic position of their source species.	*Evaluate* the relative importance of passive and active transport in the accumulation of heavy metal salts by the main groups of soil fungi.	<ul><li>review</li><li>assess</li><li>consider</li><li>justify</li></ul>

criteria or grade descriptors, which state in general terms what level of attainment is required for your work to reach specific grades. These are more likely to be defined at faculty/college/school/department level and consequently published in appropriate handbooks or websites. Reading learning outcomes and grade descriptors will give you a good idea of what to expect and the level of performance required to reach your personal goals. Relate them to both the material covered (e.g. in lectures and practicals, or online) and past exam papers. Doing this as you study and revise will indicate whether further reading and independent studying is required, and of what type. You will also have a much clearer picture of how you are likely to be assessed.

KEY POINT *Use the learning objectives for your course (normally published in the handbook) as a fundamental part of your revision planning. These indicate what you will be expected to be able to do after taking part in the course, so exam questions are often based on them. Check this by reference to past papers.*

There are essentially two types of assessment – formative and summative, although the distinction may not always be clear-cut (see margin). The first way you can learn from formative assessment is to consider the grade you obtained in relation to the work you put in. If this is a disappointment to you, then there must be a mismatch between your understanding of the topic and the marking scheme and that of the marker, or a problem in the writing or presentation of your assignment. This element of feedback is also present in summative assessment.

The second way to learn from formative assessment is through the written feedback and notes on your work. These comments may be cryptic, or scribbled hastily, so if you don't understand or can't read them, ask the tutor who marked the work. Most tutors will be pleased to explain how you could have improved your mark. If you find that the same comments appear frequently, it may be a good idea to seek help from your university's academic support unit. Take along examples of your work and feedback comments so they can give you the best possible advice. Another suggestion is to ask to see the work of another student who obtained a good mark, and compare it with your own. This will help you judge the standard you should be aiming for.

Preparing for revision and examinations

Before you start revising, find out as much as you can about each exam, including:

- its format and duration;
- the date and location;
- the types of questions;
- whether any questions/sections are compulsory;
- whether the questions are internally or externally set or assessed;
- whether the exam is 'open book', and if so, which texts or notes are allowed.

Your course tutor is likely to give you details of exam structure and timing well beforehand, so that you can plan your revision; the course handbook and past papers (if available) can provide further useful details. Always check that the nature of the exam has not changed before you consult past papers.

Organising and using lecture notes, assignments and practical reports

Time management when revising – this is a vital to success and is best achieved by creating a revision timetable (Box 5.2).

Filing lecture notes – make sure your notes are kept neatly and in sequence by using a ring binder system. File the notes in lecture or practical sequence, adding any supplementary notes or photocopies alongside.

Given their importance as a source of material for revision, you should have sorted out any deficiencies or omissions in your lecture notes and practical reports at an early stage. For example, you may have missed a lecture or practical due to illness, etc., but the exam is likely to assume attendance throughout the year. Make sure you attend classes whenever possible and keep your notes up to date. Your practical reports and any assignment work will contain specific comments from the teaching staff,

Box 5.2 How to prepare and use a revision timetable

1. **Make up a grid showing the number of days until your exams are finished.** Divide each day into several sections. If you like revising in large blocks of time, use a.m., p.m. and evening slots, but if you prefer shorter periods, divide each of these in two, or use hourly divisions (see also the table in study exercise 5.1).

2. **Write in your non-revision commitments,** including any time off you plan to allocate and physical activity at frequent intervals. Try to have about one-third or a quarter of the time off in any one day. Plan this in relation to your best times for useful work – for example, some people work best in the mornings, while others prefer evenings. If you wish, use a system where your relaxation time is a bonus to be worked for; this may help you motivate yourself.

3. **Decide on how you wish to subdivide your subjects** for revision purposes. This might be among subjects, according to difficulty (with the hardest getting the most time), or within subjects, according to topics. Make sure there is an adequate balance of time among topics and especially that you do not avoid working on the subject(s) you find least interesting or most difficult.

4. **Allocate the work to the different slots available on your timetable.** You should work backwards from the exams, making sure that you cover every exam topic adequately in the period just before each exam. You may wish to colour-code the subjects.

5. **As you revise, mark off the slots completed** – this has a positive psychological effect and will boost your self-confidence.

6. **After the exams, revisit your timetable** and decide whether you would do anything differently next time.

Using tutors' feedback – it is always worth reading any comments on your work as soon as it is returned. If you don't understand the comments, or are unsure about why you might have lost marks in an assignment, ask for an explanation.

indicating where marks were lost, corrections, mistakes, inadequacies, etc. Most lecturers are quite happy to discuss such details with students on a one-to-one basis and this information may provide you with 'clues' to the expectations of individual lecturers that may be useful in exams set by the same members of staff. However, you should never 'fish' for specific information on possible exam questions, as this is likely to be counterproductive.

Revision

Begin early, to avoid last-minute panic. Start in earnest several weeks beforehand, and plan your work carefully:

Recognise when your concentration powers are dwindling – take a short break when this happens and return to work refreshed and ready to learn. Remember that 20 minutes is often quoted as a typical limit to full concentration effort.

- Prepare a revision timetable – an 'action plan' that gives details of specific topics to be covered (Box 5.2). Find out at an early stage when (and where) your examinations are to be held, and plan your revision around this. Try to keep to your timetable. Time management during this period is as important as keeping to time during the exam itself.
- Study the learning objectives/outcomes for each topic (usually published in the course handbook) to get an idea of what lecturers expect from you.
- Use past papers as a guide to the form of exam and the type of question likely to be asked (Box 5.3).
- Remember to have several short (five-minute) breaks during each hour of revision and a longer break every few hours. In any day, try to work for a maximum of three-quarters of the time.
- Include recreation within your schedule: there is little point in tiring yourself with too much revision, as this is unlikely to be profitable.

Box 5.3   How to use past exam papers in your revision

Past exam papers are a valuable resource for targeting your revision.

1. **Find out where the past exam papers are kept.** Copies may be lodged in your department or the library; or they may be accessible online.

2. **Locate and copy relevant papers for your module(s).** Check with your tutor or course handbook that the style of paper will not change for the next set of examinations.

3. **Analyse the design of the exam paper.** Taking into account the length in weeks of your module, and the different lecturers and/or topics for those weeks, note any patterns that emerge. For example, can you translate weeks of lectures/ practicals into numbers of questions or sections of the paper? Consider how this might affect your revision plans and exam tactics, taking into account (a) any choices or restrictions offered in the paper, and (b) the different types of questions asked (i.e. multiple choice, short-answer or essay).

4. **Examine carefully the style of questions.** Can you identify the expectations of your lecturers? Can you relate the questions to the learning objectives? How much extra reading do they seem to expect? Are the questions fact-based? Do they require a synthesis based on other knowledge?

Can you identify different styles for different lecturers? Consider how the answers to these questions might affect your revision effort and exam strategy.

5. **Practise answering questions.** Perhaps with friends, set up your own mock exam when you have done a fair amount of revision, but not too close to the exams. Use a relevant past exam paper; don't study it beforehand! You need not attempt all of the paper at one sitting. You'll need a quiet room in a place where you will not be interrupted (e.g. a library). Keep close track of time during the mock exam and try to do each question in the length of time you would normally assign to it (see p. 31) – this gives you a feel for the speed of thought and writing required and the scope of answer possible. Mark each other's papers and discuss how each of you interpreted the question and laid out your answers and your individual marking schemes.

6. **Practise writing answer plans and starting answers.** This can save time compared with the 'mock exam' approach. Practise in starting answers can help you get over stalling at the start and wasting valuable time. Writing essay plans gets you used to organising your thoughts quickly and putting your thoughts into a logical sequence.

- Make your revision as active and interesting as possible (see below): the least productive approach is simply to read and reread your notes.
- Ease back on the revision near the exam: plan your revision to avoid last-minute cramming and overload fatigue.

Active revision

The following techniques may prove useful in devising an active revision strategy:

- 'Distil' your lecture notes to show the main headings and examples. Prepare revision sheets with details for a particular topic on a single sheet of paper, arranged as a numbered checklist. Wall posters are another useful revision aid.
- Confirm that you know about the material by testing yourself – take a blank sheet of paper and write down all you know. Check your full notes to see if you missed anything out. If you did, go back immediately to a fresh blank sheet and redo the example. Repeat, as required.
- Memorise definitions and key phrases: definitions can be a useful starting point for many exam answers. Make up lists of relevant facts

Aiding recall through effective note-taking – the Mind Map technique (p. 17), when used to organise ideas, is claimed to enhance recall by connecting the material to visual images or linking it to the physical senses.

Question-spotting – *avoid adopting this risky strategy to reduce the amount of time you spend revising. Lecturers are aware that this approach may be taken and try to ask questions in an unpredictable manner. You may find that you are unable to answer on unexpected topics that you failed to revise. Moreover, if you have a preconceived idea about what will be asked, you may also fail to grasp the nuances of the exact question set, and provide a response lacking in relevance.*

Revision checks – *it is important to test yourself frequently during revision, to ensure that you have retained the information you are revising.*

Final preparations – *try to get a good night's sleep before an exam. Last-minute cramming will be counterproductive if you are too tired during the exam.*

or definitions associated with particular topics. Test yourself repeatedly on these, or get a friend to do this. Try to remember *how many* facts or definitions you need to know in each case – this will help you recall them all during the exam.

- Use mnemonics and acronyms to commit specific factual information to memory. Sometimes, the dafter they are, the better they seem to work.
- Use pattern diagrams or Mind Maps as a means of testing your powers of recall on a particular topic (pp. 16–17).
- Draw diagrams from memory: make sure you can label them fully.
- Try recitation as an alternative to written recall. Talk about your topic to another person, preferably someone in your class. Talk to yourself if necessary. Explaining something out loud is an excellent test of your understanding.
- Associate facts with images or journeys if you find this method works.
- Use a wide variety of approaches to avoid boredom during revision (e.g. record information on audio tape, use cartoons, or any other method, as long as it's not just reading).
- Form a revision group to share ideas and discuss topics with other students.
- Prepare answers to past papers, e.g. write essays or, if time is limited, write essay plans (see Box 5.3).
- If your subject involves numerical calculations, work through representative problems.
- Make up your own questions: the act of putting yourself in the examiner's mindset by inventing questions can help revision. However, you should not rely on 'question-spotting': this is a risky practice!

The evening before your exam should be spent in consolidating your material, and checking through summary lists and plans. Avoid introducing new material at this late stage: your aim should be to boost your confidence, putting yourself in the right frame of mind for the exam itself.

Text references

Bloom, B., Englehart, M., Furst, E., Hill, W. and Krathwohl, D. (1956) *Taxonomy of Educational Objectives: The Classification of Educational Goals. Handbook I: Cognitive Domain*. Longmans, Green, New York and Toronto.

Fleming, N.D. (2001) *Teaching and Learning Styles: VARK Strategies*. Neil Fleming, Christchurch.

Fleming, N.D. *VARK: A Guide to Learning Styles*. Available: http://www.vark-learn.com/ Last accessed 09/04/07.

Sources for further study

Burns, R. (1997) *The Student's Guide to Passing Exams*. Kogan Page, London.

Hamilton, D. (1999) *Passing Exams: A Guide for Maximum Stress and Minimum Stress*. Cassell, London.

Many universities host study skills websites; these can be found using 'study skills', 'revision' or 'exams' as key words in a search engine.

Study exercises

5.1 **Draw up a revision timetable**. Use the techniques discussed in Box 5.2 to create a revision timetable for your forthcoming exams. You may wish to use or adapt the arrangement below, either on paper or within a spreadsheet.

5.2 **Make use of past exam papers**. Use the techniques discussed in Box 5.3 to improve your revision strategy.

5.3 **Try out a new active revision technique**. Try any or all of the methods mentioned on pages 26–27. when revising. Compare notes with a colleague – which seems to be the most successful technique for you and for the topic you are revising?

A revison timetable planner

Date	Morning		Lunch	Afternoon		Tea/Dinner	Evening	
	Session 1	Session 2		Session 1	Session 2		Session 1	Session 2

6 Curriculum options, assessment and exams

Many universities adopt a modular system for their biosciences degree courses. This allows greater flexibility in subject choice and accommodates students studying on different degree paths. Modules also break a subject into discrete, easily assimilated elements. They have the advantage of spreading assessment over the academic year, but they can also tempt you to avoid certain difficult subjects or to feel that you can forget about a topic once the module is finished.

 KEY POINT *You should select your modules with care, mind-ful of potential degree options and how your transcript and CV will appear to a prospective employer. If you feel you need advice, consult your personal tutor or study adviser.*

As you move between levels of the university system, you will be expected to have passed a certain number of modules, as detailed in the progression criteria. These may be expressed using a credit-point system. Students are normally allowed two attempts at each module exam and the resits often take place at the end of the summer vacation. If a student does not pass at the second attempt, they may be asked to 'carry' the subject in a subsequent year, and in severe cases of multiple failure, they may be asked to retake the whole year or even leave the course. Consequently, it is worth finding out about these aspects of your degree. They are usually published in relevant handbooks.

You are unlikely to have reached this stage in your education without being exposed to the examination process. You may not enjoy being assessed, but you probably want to do well in your course. It is therefore important to understand why and how you are being tested. Identifying and improving the skills required for exam success will allow you to perform to the best of your ability.

Assessed coursework

There is a component of assessed coursework in many modules. This often tests specific skills, and may require you to demonstrate thinking at deeper levels (see Table 5.1). The common types of coursework assessment are covered at various points in this book:

- Practical exercises (throughout).
- Essays (Chapters 15 and 16).
- Numerical problems (Chapter 64).
- Data analysis (Chapters 61, 65 and 66).
- Poster and spoken presentations (Chapters 13 and 14).
- Fieldwork (Chapter 29).
- Literature surveys and reviews (Chapter 18).
- Project work (Chapters 17 and 33).
- Problem-based learning (Box 6.1).

At the start of each year or module, read the course handbook or module guide carefully to find out when any assessed work needs to be submitted. Note relevant dates in your diary, and use this information to plan your work. Take special note if deadlines for different modules clash, or if they coincide with social or sporting commitments.

Box 6.1 Problem-based learning (PBL)

In this relatively new teaching method, you are presented with a 'real world' problem or issue, often working within a team. As you tackle the problem, you will gain factual knowledge, develop skills and exercise critical thinking (Chapter 9). Because there is a direct and relevant context for your work, and because you have to employ active learning techniques, the knowledge and skills you gain are likely to be more readily assimilated and remembered. This approach also more closely mimics workplace practices. PBL usually proceeds as follows:

1. **You are presented with a problem** (e.g. a case study, a hypothetical patient, a topical issue).

2. **You consider what issues and topics you need to research,** by discussion with others if necessary. You may need to identify where relevant resources can be found (Chapters 8 and 10).

3. **You then need to rank the issues and topics in importance,** allocating tasks to group members, if appropriate.

4. **Having carried out the necessary research, you should review what information has been obtained.** As a result, new issues may need to be explored and, where appropriate, allocated to group members.

5. **You will be asked to produce an outcome, such as a report, diagnosis, seminar presentation or poster.** An outline structure will be required, and for groups, further allocation of tasks to accomplish this goal.

If asked to carry out PBL as part of your course, it is important to get off to a good start. At first, the problem may seem unfamiliar. However, once you become involved in the work, you will quickly gain confidence. If working as part of a group, make sure that your group meets as early as possible, that you attend all sessions and that you do the necessary background reading. When working in a team, a degree of self-awareness is necessary regarding your 'natural' role in group situations (Table 3.1). Various methods are used for assessing PBL, and the assessment may involve peer marking.

 KEY POINT If, for some valid reason (e.g. illness), you will be late with an assessment, speak to your tutors as soon as possible. They may be able to take extenuating circumstances into account by not applying a marking penalty. They will let you know what paperwork you may require to submit to support your claim.

Summative exams – general points

Summative exams (p. 24) normally involve you answering questions without being able to consult other students or your notes. Invigilators are present to ensure appropriate conduct, but departmental representatives may be present for some of the exam. Their role is to sort out any subject-related problems, so if you think something is wrong, ask at the earliest opportunity. It is not unknown for parts of questions to be omitted in error, or for double meanings to arise, for example.

Planning

When preparing for an exam, make a checklist of the items you'll need (see p. 35). On the day of the exam, give yourself sufficient time to arrive at the correct room, without the risk of being late. Double-check the times and places of your exams, both well before the exam and also on arrival. If you arrive at the exam venue early, you can always rectify a mistake if you find you've gone to the wrong place.

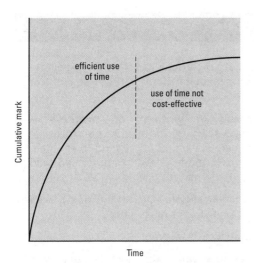

Fig. 6.1 Exam marks as a function of time. The marks awarded in a single answer will follow the law of diminishing returns – it will be far more difficult to achieve the final 25 per cent of the available marks than the initial 25 per cent. Do not spend too long on any one question.

Using the question paper – *unless this is specifically forbidden, you should write on the question paper to plan your strategy, keep to time and organise answers.*

Tackling the paper

Begin by reading the instructions at the top of the exam paper carefully, so that you do not make any errors based on lack of understanding of the exam structure. Make sure that you know:

- how many questions are set;
- how many must be answered;
- whether the paper is divided into sections;
- whether any parts are compulsory;
- what each question/section is worth, as a proportion of the total mark;
- whether different questions should be answered in different books.

Do not be tempted to spend too long on any one question or section: the return in terms of marks will not justify the loss of time from other questions (see Fig. 6.1). Take the first 10 minutes or so to read the paper and plan your strategy, before you begin writing. Do not be put off by those who begin immediately; it is almost certain they are producing unplanned work of a poor standard.

Underline the key phrases in the instructions, to reinforce their message. Next, read through the set of questions. If there is a choice, decide on those questions to be answered and decide on the order in which you will tackle them. Prepare a timetable which takes into account the amount of time required to complete each section and which reflects the allocation of marks – there is little point in spending one-quarter of the exam period on a question worth only 5 per cent of the total marks. Use the exam paper to mark the sequence in which the questions will be answered and write the finishing times alongside; refer to this timetable during the exam to keep yourself on course.

Reviewing your answers

At the end of the exam, you should allow at least 10 minutes to read through your script, to check for:

- errors of fact;
- missing information;
- grammatical and spelling errors;
- errors of scale and units;
- errors in calculations.

Make sure your name and/or ID number is on each exam book as required and on all other sheets of paper, including graph paper, even if securely attached to your script, as it is in your interest to ensure that your work does not go astray.

 KEY POINT Never leave any exam early. Most exams assess work carried out over several months in a time period of 2–3 hours and there is always something constructive you can do with the remaining time to improve your script.

Special considerations for different types of exam question

Essay questions

Essay questions let examiners test the depth of your comprehension and understanding as well as your recall of facts. Essay questions give you plenty of scope to show what you know. They suit those with a

---

Box 6.2   Writing under exam conditions

Always go into an exam with a strategy for managing the available time.

- **Allocate some time (say 5 per cent of the total) to consider which questions to answer and in which order.**

- **Share the rest of the time among the questions, according to the marks available.** Aim to optimise the marks obtained. A potentially good answer should be allocated slightly more time than one you don't feel so happy about. However, don't concentrate on any one answer (see Fig. 6.1).

- **For each question divide the time into planning, writing and revision phases** (see p. 101).

Employ time-saving techniques as much as possible.

- **Use spider diagrams** (Fig. 15.2) **or Mind Maps** (Fig. 4.2) to organise and plan your answer.

- **Use diagrams and tables** to save time in making difficult and lengthy explanations, but make sure you refer to each one in the text.

- **Use standard abbreviations** to save time repeating text but always explain them at the first point of use.

- **Consider speed of writing and neatness** – especially when selecting the type of pen to use – ballpoint pens are fastest, but they tend to smudge. You can only gain marks if the examiner can read your script.

- **Keep your answer simple and to the point**, with clear explanations of your reasoning.

Make sure your answer is relevant.

- **Don't include irrelevant facts** just because you memorised them during revision, as this may do you more harm than good. You must answer the specific question that has been set.

- **Time taken to write irrelevant material is time lost from another question.**

Adopting different tactics according to the exam – you should adjust your exam strategy (and revision methods) to allow for the differences in question types used in each exam paper.

Penalties for guessing – if there is a penalty for incorrect answers in a multiple-choice test, the best strategy is not to answer questions when you know your answer is a complete guess. Depending on the penalty, it may be beneficial to guess if you can narrow the choice down to two options (but beware false or irrelevant alternatives). However, if there are no such penalties, then you should provide an answer to all questions.

good grasp of principles but who perhaps have less ability to recall details.

Before you tackle a particular question, you must be sure of what is required in your answer. Ask yourself 'What is the examiner looking for in this particular question?' and then set about providing a *relevant* answer. Consider each individual word in the question and highlight, underline or circle the key words. Make sure you know the meaning of the terms given in Table 16.1 (p. 102) so that you can provide the appropriate information, where necessary. Spend some time planning your writing (see Chapter 15). Refer back to the question frequently as you write, to confirm that you are keeping to the subject matter. Box 6.2 gives advice on writing essays under exam conditions.

It is usually a good idea to begin with the question that you are most confident about. This will reassure you before tackling more difficult parts of the paper. If you run out of time, write in note form. Examiners are usually understanding, as long as the main components of the question have been addressed and the intended structure of the answer is clear. Common reasons for poor exam answers in essay-style questions are listed in Box 6.3.

Multiple-choice and short-answer questions

Multiple-choice questions (MCQs) and short-answer questions (SAQs) are generally used to test the breadth and detail of your knowledge. The various styles that can be encompassed within the SAQ format allow for

Box 6.3 Reasons for poor exam answers to essay-style questions

The following are reasons that lecturers cite when they give low marks for essay answers:

- **Not answering the exact question set.** Either failing to recognise the specialist terms used in the question, or failing to demonstrate an understanding of the terms by not providing definitions, or failing to carry out the precise instruction in a question, or failing to address all aspects of the question.

- **Running out of time.** Failing to match the time allocated to the extent of the answer. Frequently, this results in spending too long on one question and not enough on the others, or even failing to complete the paper.

- **Failing to answer all parts** of a multiple-part question, or to recognise that one part (perhaps involving more complex ideas) may carry more marks than another.

- **Failing to provide evidence** to support an answer. Forgetting to state the 'obvious' – either basic facts or definitions.

- **Failing to illustrate an answer appropriately**, either by not including a relevant diagram, or by providing a diagram that does not aid communication, or by not including examples.

- **Incomplete answer(s).** Failing to answer appropriately due to lack of knowledge.

- **Providing irrelevant evidence** to support an answer. 'Waffling' to fill space.

- **Illegible handwriting.**

- **Poor English**, such that facts and ideas are not expressed clearly.

- **Lack of logic** or structure to the answer.

- **Factual errors**, indicating poor note-taking or poor revision or poor recall.

- **Failing to correct obvious mistakes** by rereading an answer before submitting the script.

At higher levels, the following aspects are especially important:

- **Not providing enough in-depth information.**

- **Providing a descriptive rather than an analytical answer** – focusing on facts, rather than deeper aspects of a topic.

- **Not setting a problem in context**, or not demonstrating a wider understanding of the topic. (However, make sure you don't overdo this, or you may risk not answering the question set.)

- **Not giving enough evidence of reading around the subject.** This can be demonstrated, for example, by quoting relevant papers and reviews and by giving author names and dates of publication.

- **Not considering both sides of a topic/debate, or not arriving at a conclusion if you have done so.**

more demanding questions than MCQs, which may emphasise specific factual knowledge.

A good approach for MCQ papers is as follows:

1. First trawl: read through the questions fairly rapidly, noting the 'correct' answer in those you can attempt immediately, perhaps in pencil.
2. Second trawl: go through the paper again, checking your original answers and this time marking up the answer sheet properly.
3. Third trawl: now tackle the difficult questions and those that require longer to answer (e.g. those based on numerical problems).

One reason for adopting this three-phase approach is that you may be prompted to recall facts relevant to questions looked at earlier. You can also spend more time per question on the difficult ones.

When unsure of an answer, the first stage is to rule out options that are clearly absurd or have obviously been placed there to distract you. Next, looking at the remaining options, can you judge between contrasting pairs with alternative answers? Logically, both cannot be correct, so you should see if you can rule one of the pair out. Watch out, however, in case *both* may be irrelevant to the answer. If the question involves a

Answer the question as requested – This is true for all questions, but especially important for SAQs. If the question asks for a diagram, make sure you provide one; if it asks for n aspects of a topic, try to list this number of points; if there are two or more parts, provide appropriate answers to all aspects. This may seem obvious, but many marks are lost for not following instructions.

Examples These are the principal types of question you are likely to encounter in a practical or information-processing exam:

Manipulative exercises Often based on work carried out during your practical course. Tests dexterity, specific techniques (e.g. dissection; sterile technique).

'Spot' tests Short questions requiring an identification, or brief descriptive notes on a specific item (e.g. a prepared slide). Tests knowledge of seen material or ability to transfer this to a new example.

Calculations May include the preparation of aqueous solutions at particular concentrations (p. 131) and statistical exercises (p. 433). Tests numeracy.

Data analyses May include the preparation and interpretation of graphs (p. 390) and numerical information, from data either obtained during the exam or provided by the examiner. Tests problems-solving skills.

Drawing specimens Accurate representation and labelling will be important. Tests drawing and interpretation abilities.

Preparing specimens for examination with a microscope Tests staining technique and light microscopy technique.

Interpreting images Sometimes used when it is not possible to provide living specimens, e.g. in relation to fieldwork, or electron microscopy. Can test a variety of skills.

calculation, try to do this independently from the answers, so you are not influenced by them.

In SAQ papers, there may be a choice of questions. Choose your options carefully – it may be better to gain half marks for a correct answer to half a question, than to provide a largely irrelevant answer that apparently covers the whole question but lacks the necessary detail. For this form of question, few if any marks are given for writing style. Think in 'bullet point' mode and list the crucial points only. The time for answering SAQ questions may be tight, so get down to work fast, starting with answers that demand remembered facts. Stick to your timetable by moving on to the next question as soon as possible. Strategically, it is probably better to get part-marks for the full number of questions than good marks for only a few.

Practical and information-processing exams

The prospect of a practical or information-processing exam in biology may cause you more concern than a theory exam. This may be due to a limited experience of practical examinations, or to the fact that practical and observational skills are tested, as well as recall, description and analysis of factual information. Your first thoughts may be that it is not possible to prepare for such exams but, in fact, you can improve your performance by mastering the various practical techniques described in this book.

You may be allowed to take your laboratory reports and other texts into the practical exam. Don't assume that this is a soft option, or that revision is unnecessary: you will not have time to read large sections of your reports or to familiarise yourself with basic principles, etc. The main advantage of 'open book' exams is that you can check specific details of methodology, reducing your reliance on memory, provided you know your way around your practical manual. In all other respects, your revision and preparation for such exams should be similar to theory exams. Make sure you are familiar with all of the practical exercises, including any work carried out in class by your partner (since exams are assessed on individual performance). If necessary, check with the teaching staff to see whether you can be given access to the laboratory, to complete any exercises that you have missed.

At the outset of the practical exam, determine or decide on the order in which you will tackle the questions. A question in the latter half of the paper may need to be started early on in the exam period (e.g. an enzyme assay requiring 2-hour incubation in a 3-hour exam). Such questions are included to test your forward-planning and time-management skills. You may need to make additional decisions on the allocation of material, e.g. if you are given 30 sterile test tubes, there is little value in designing an experiment that uses 25 of these to answer question 1, only to find that you need at least 15 tubes for subsequent questions.

Make sure you explain your choice of apparatus and experimental design. Calculations should be set out in a stepwise manner, so that credit can be given, even if the final answer is incorrect (see p. 405). If there are any questions that rely on recall of factual information and you are unable to remember specific details, e.g. you cannot identify a particular specimen or slide, make sure that you describe the item fully, so that you gain credit for observational skills. Alternatively, leave a gap and return to the question at a later stage.

Oral exams and interviews

An oral interview is sometimes a part of final degree exams, representing a chance for the external examiner(s) to get to know the students personally and to test their abilities directly and interactively. In some departments, orals are used to validate the exam standard, or to test students on the borderline between exam grades. Sometimes an interview may form part of an assessment, as with project work or posters. This type of exam is often intimidating – many students say they don't know how to revise for an oral – and many candidates worry that they will be so nervous they won't be able to do themselves justice.

Preparation is just as important for orals as it is for written exams:

- Think about your earlier performances – if the oral follows written papers, it may be that you will be asked about questions you did not do so well on. These topics should be revised thoroughly. Be prepared to say how you would approach the questions if given a second chance.
- Read up a little about the examiner – he or she may focus their questions in their area of expertise.
- Get used to giving spoken answers – it is often difficult to transfer between written and spoken modes. Write down a few questions and get a friend to ask you them, possibly with unscripted follow-up queries.
- Research and think about topical issues in your subject area – some examiners will feel this reflects how interested you are in your subject.

Your conduct during the oral exam is important, too:

- Arrive promptly and wear reasonably smart clothing. Not to do either might be considered disrespectful by the examiner.
- Take your time before answering questions. Even if you think you know the answer immediately, take a while to check mentally whether you have considered all angles. A considered, logical approach will be more impressive than a quick but ill-considered response.
- Start answers with the basics, then develop into deeper aspects. There may be both surface and deeper aspects to a topic and more credit will be given to students who mention the latter.
- When your answer is finished, stop speaking. A short, crisp answer is better than a rambling one.
- If you don't know the answer, say so. To waffle and talk about irrelevant material is more damaging than admitting that you don't know.
- Make sure your answer is balanced. Talk about the evidence and opinions on both sides of a contentious issue.
- Don't disagree violently with the examiner. Politely put your point of view, detailing the evidence behind it. Examiners will be impressed by students who know their own mind and subject area. However, they will expect you to support a position at odds with the conventional viewpoint.
- Finally, be positive and enthusiastic about your topic.

Counteracting anxiety before and during exams

Adverse effects of anxiety need to be overcome by anticipation and preparation well in advance (Box 6.4). Exams, with their tight time limits,

---

Terminology – the oral exams are sometimes known simply as 'orals' or, borrowing Latin, as 'viva voces' (by or with the living voice) or 'vivas'.

Allow yourself to relax in oral exams – external examiners are experienced at putting students at ease. They will start by asking 'simple-to-answer' questions, such as what modules you did, how your project research went, and what your career aspirations are. Imagine the external examiner as a friend rather than a foe.

Creating an exam action list – knowing that you have prepared well, checked everything on your list and gathered together all you need for an exam will improve your confidence and reduce anxiety. Your list might include:

- Verify time, date and place of the exam
- Confirm travel arrangements to exam hall
- Double-check module handbooks and past papers for exam structure
- Think through use of time and exam strategy
- Identify a quiet place near the exam hall to carry out a last-minute check on key knowledge (e.g. formulae, definitions, diagram labels)
- Ensure you have all the items you wish to take to the exam, e.g.
  – pens, pencils (with sharpener and eraser);
  – ruler;
  – correction fluid;
  – calculator (allowable type);
  – sweets and drink, if allowed;
  – tissues
  – watch or clock;
  – ID card;
  – texts and/or notes, if an open-book exam;
  – mascot.
- Lay out clothes (if exam is early in the morning)
- Set alarm and/or ask a friend or family member to check you are awake on time

---

Box 6.4 Strategies for combating the symptoms of exam anxiety

Sleeplessness – this is commonplace and does little harm in the short term. Get up, have a snack, do some light reading or other activity, then return to bed. Avoid caffeine (e.g. tea, coffee and cola) for several hours before going to bed.

Lack of appetite – again commonplace. Eat what you can, but take sugary sweets into the exam to keep energy levels up in case you become tired.

Fear of the unknown – it can be a good idea to visit the exam room beforehand, so you can become familiar with the location. By working through the points given in the exam action list on p. 35 you will be confident that nothing has been left out.

Worries about timekeeping – get a reliable alarm clock or a new battery for an old one. Arrange for an alarm phone call. Ask a friend or relative to make sure you are awake on time. Make reliable travel arrangements, to arrive on time. If your exam is early in the morning, it may be a good idea to get up early for a few days beforehand.

Blind panic during an exam – explain how you feel to an invigilator. Ask to go for a supervised walk outside. Do some relaxation exercises (see below), then return to your work. If you are having problems with a specific question, it may be appropriate to speak to the departmental representative at the exam.

Feeling tense – shut your eyes, take several slow, deep breaths, do some stretching and relaxing muscle movements. During exams, it may be a good idea to do this between questions, and possibly to have a complete rest for a minute or so. Prior to exams, try some exercise activity or escape temporarily from your worries by watching TV or a movie.

Running out of time – don't panic when the invigilator says 'five minutes left'. It is amazing how much you can write in this time. Write note-style answers or state the areas you would have covered: you may get some credit.

---

are especially stressful for perfectionists. To counteract this tendency, focus on the following points during the exam:

- Don't expect to produce a perfect essay – this won't be possible in the time available.
- Don't spend too long planning your answer – once you have an outline plan, get started.
- Don't spend too much time on the initial parts of an answer, at the expense of the main message.
- Concentrate on getting all of the basic points across – markers are looking for the main points first, before allocating extra marks for the detail.
- Don't be obsessed with neatness, either in handwriting, or in the diagrams you draw, but make sure your answers are legible.
- Don't worry if you forget something. You can't be expected to know everything. Most marking schemes give a first class grade to work that misses out on up to 30 per cent of the marks available.

 KEY POINT *Everyone worries about exams. Anxiety is a perfectly natural feeling. It works to your advantage, as it helps provide motivation and the adrenaline that can help you 'raise your game' on the day.*

There is a lot to be said for treating exams as a game. After all, they are artificial situations contrived to ensure that large numbers of candidates can be assessed together, with little risk of cheating. They have conventions and rules, just like games. If you understand the rationale behind them and follow the rules, this will aid your performance.

After the exam – try to avoid becoming involved in prolonged analyses with other students over the 'ideal' answers to the questions; after all, it is too late to change anything at this stage. Go for a walk, watch TV for a while, or do something else that helps you relax, so that you are ready to face the next exam with confidence.

Sources for further study

Acres, D. (1998) *How to Pass Exams Without Anxiety: How to get Organised, Be Prepared and Feel Confident of Success.* How To Books, London.

Burns, R. (1997) *The Student's Guide to Passing Exams.* Kogan Page, London.

Hamilton, D. (1999) *Passing Exams: A Guide for Maximum Success and Minimum Stress.* Cassell, London.

Many universities host study skills websites; these can be found using 'study skills', 'revision' or 'exams' as key words in a search engine.

Study exercises

6.1 Analyse your past performances. Think back to past exams and any feedback you received from them. How might you improve your performance? Consider ways in which you might approach the forthcoming exam differently. If you have kept past papers and answers to continuous assessment exercises, look at any specific comments your lecturers may have made.

6.2 Share revision notes with other students. Make a revision plan (see pp. 10 and 25) and then allocate some time to discussing your revision notes with a colleague. Try to learn from his or her approach. Discuss any issues you do not agree upon.

6.3 Plan your exam tactics. Find out from your module handbook or past papers what the format of each paper will be. Confirm this if necessary with staff. Decide how you will tackle each paper, allocating time to each section and to each question within the sections (see p. 101). Write a personal checklist of requirements for the exam (see p. 35).

Many students only think about their curriculum vitae immediately before applying for a job. Reflecting this, chapters on preparing a CV are usually placed near the end of texts of this type. Putting the chapter near the beginning of this book emphasises the importance of focusing your thoughts on your CV at an early stage in your studies. There are four main reasons why this can be valuable:

1. Considering your CV and how it will look to a future employer will help you think more deeply about the direction and value of your academic studies.
2. Creating a draft CV will prompt you to assess your skills and personal qualities and how these fit into your career aspirations.
3. Your CV can be used as a record of all the relevant things you have done at university and then, later, will help you communicate these to a potential employer.
4. Your developing CV can be used when you apply for vacation or part-time employment.

 KEY POINT *Developing your skills and qualities needs to be treated as a long-term project. It makes sense to think early about your career aspirations so that you can make the most of opportunities to build up relevant experience. A good focus for such thoughts is your developing curriculum vitae, so it is useful to work on this from a very early stage.*

Skills and personal qualities

Skills (sometimes called competences) are generally what you have learned to do and have improved with practice. Table 1.1 summarises some important skills for biologists. This list might seem quite daunting, but your tutors will have designed your courses to give you plenty of opportunities for developing your expertise. Personal qualities, on the other hand, are predominately innate. Examples include honesty, determination and thoroughness (Table 7.1). These qualities need not remain static, however, and can be developed or changed according to your experiences. By consciously deciding to take on new challenges and responsibilities, not only can you develop your personal qualities, but you can also provide supporting evidence for your CV.

Personal qualities and skills and are interrelated because your personal qualities can influence the skills you gain. For example, you may become highly proficient at a skill requiring manual dexterity if you are particularly adept with your hands. Being able to transfer your skills is highly important (Chapter 1) – many employers take a long-term view and look for evidence of the adaptability that will allow you to be a flexible employee and one who will continue to develop skills.

Developing your curriculum vitae

The initial stage involves making an audit of the skills and qualities you already have, and thinking about those you might need to develop. Tables 7.1 and 1.1 could form a basis of this self-appraisal. Assessing

---

Definition

Curriculum vitae (or CV for short) – A Latin phrase that means 'the course your life has taken'.

Personal development planning (PDP) and your CV – *many PDP schemes (p. 5) also include an element of career planning that may involve creating a draft or generic CV. The PDP process can help you improve the structure and content of your CV, and the language you use within it.*

Understanding skills and qualities – *it may be helpful to think about how the skills and qualities in Tables 1.1 and 7.1 apply to particular activities during your studies, since this will give them a greater relevance.*

Focusing on evidence – *it is important to be able to provide specific concrete information that will back up the claims you make under the 'skills and personal qualities' and other sections of your CV. A potential employer will be interested in your level of competence (what you can actually do) and in situations where you have used a skill or demonstrated a particular quality. These aspects can also be mentioned in your covering letter or at interview.*

Table 7.1 Some positive personal qualities

Adaptability
Conscientiousness
Curiosity
Determination
Drive
Energy
Enthusiasm
Fitness and health
Flexible approach
Honesty
Innovation
Integrity
Leadership
Logical approach
Motivation
Patience
Performance under stress
Perseverance
Prudence
Quickness of thought
Seeing other's viewpoints
Self-confidence
Self-discipline
Sense of purpose
Shrewd judgement
Social skills (sociability)
Taking initiative
Tenacity
Tidiness
Thoroughness
Tolerance
Unemotional approach
Willingness to take on challenges

Seeing yourself as others see you – you may not recognise all of your personal qualities and you may need someone else to give you a frank appraisal. This could be anyone whose opinion you value: a friend, a member of your family, a tutor, or a careers adviser.

Setting your own agenda – you have the capability to widen your experience and to demonstrate relevant personal qualities through both curricular and extra-curricular activities.

Paying attention to the quality of your CV – your potential employer will regard your CV as an example of your very best work and will not be impressed if it is full of mistakes or badly presented, especially if you claim 'good written communication' as a skill!

your skills may be easier than critically analysing your personal characteristics. In judging your qualities, try to take a positive view and avoid being overly modest. It is important to think of your qualities in a specific context, e.g. 'I have shown that I am trustworthy, by acting as treasurer for the Biological Society', as this evidence will form a vital part of your CV and job applications.

If you can identify gaps in your skills, or qualities that you would like to develop, especially in relation to the needs of your intended career, the next step is to think about ways of improving them. This will be reasonably easy in some cases, but may require some creative thinking in others. A relatively simple example would be if you decided to learn a new language or to keep up with one you learned at school. There are likely to be many local college and university courses dealing with foreign languages at many different levels, so it would be a straightforward matter to join one of these. A rather more difficult case might be if you wished to demonstrate 'responsibility', because there are no courses available on this. One route to demonstrate this quality might be to put yourself up for election as an officer in a student society or club; another could be to take a leading role in a relevant activity within your community (e.g. voluntary work such as hospital radio). If you already take part in activities like these, your CV should relate them to this context.

Basic CV structures and their presentation

Box 7.1 illustrates the typical parts of a CV and explains the purpose of each part. Employers are more likely to take notice of a well-organised and -presented CV, in contrast to one that is difficult to read and assimilate. They will expect it to be concise, complete and accurate. There are many ways of presenting information in a CV, and you will be assessed partly on your choices.

- **Order.** There is some flexibility as to the order in which you can present the different parts (see Box 7.1). A chronological approach within sections helps employers gain a picture of your experience.
- **Personality and 'colour'.** Make your CV different by avoiding standard or dull phrasing. Try not to focus solely on academic aspects: you will probably have to work in a team, and the social aspects of teamwork will be enhanced by your outside interests. However, make sure that the reader does not get the impression that these interests dominate your life.
- **Style.** Your CV should reflect *your* personality, but not in such a way that it indicates too idiosyncratic an approach. It is probably better to be formal in both language and presentation, as flippant or chatty expressions will not be well received.
- **Neatness.** Producing a well-presented, word-processed CV is very important. Use a laser-quality printer and good-quality paper; avoid poor-quality photocopying at all costs.
- **Layout.** Use headings for different aspects, such as personal details, education, etc. Emphasise words (e.g. with capitals, bold, italics or underlining) sparingly and with the primary aim of making the structure clearer. Remember that careful use of white space is important in design.
- **Grammar and proofreading.** Look at your CV carefully before you submit it, as sloppy errors give a very poor impression. Even if you use a spellchecker, some errors may creep in. Ask someone whom you regard as a reliable proofreader to comment on it (many tutors will do this, if asked in advance).

Box 7.1 The structure and components of a typical CV and covering letter

There is no right or wrong way to write a CV, and no single format applies. It is probably best to avoid software templates and CV 'wizards' as they can create a bland, standardised result, rather than something that demonstrates your individuality.

You should include the following with appropriate sub-headings, generally in the order given below:

1. **Personal details.** This section *must* include your full name and date of birth, your address (both home and term-time, with dates, if appropriate) and a contact telephone number at each address. If you have an email account, you might also include this. You need only mention gender if your name could be either male or female.

2. **Education.** Choose either chronological order, or reverse chronological order and make sure you take the same approach in all other sections. Give educational institutions and dates (month, year) and provide more detail for your degree course than for your previous education. Remember to mention any prizes, scholarships or other academic achievements. Include your overall mark for the most recent year of your course, if it seems appropriate. Make sure you explain any gap years.

3. **Work experience.** Include all temporary, part-time, full-time or voluntary jobs. Details include dates, employer, job title and major duties involved.

4. **Skills and personal qualities.** Tables 1.1 and 7.1 give examples of the aspects you might include under this heading. Emphasise your strengths, and tailor this section to the specific requirements of the post (the 'job description'): for example, you might emphasise the practical skills you have gained during your degree studies if the post is a biological one, but concentrate on generic transferable skills and personal qualities for other jobs. Provide supporting evidence for your statements in all cases.

5. **Interests and activities.** This is an opportunity to bring out the positive aspects of your personality, and explain their relevance to the post you are applying for. Aim to keep this section short, or it may seem that your social life is more important than your education and work experience. Include up to four separate items, and provide sufficient detail to highlight the *positive* aspects of your interests (e.g. positions of responsibility, working with others, communication, etc.). Use sections 4 and 5 to demonstrate that you have the necessary attributes to fulfil the major requirements of the post.

6. **Referees.** Include the names (and titles), job descriptions, full postal addresses, contact telephone numbers and email addresses of two referees (rarely, some employers may ask for three). It is usual to include your personal tutor or course leader at university (who among other things will verify your marks), plus another person – perhaps a current or former employer, or someone who runs a club or society and who knows your personal interests and activities. Unless you have kept in touch with a particular teacher since starting university, it is probably best to choose current contacts, rather than those from your previous education.

Some other points to consider:

- Try to avoid jargon and over-complicated phrases in your CV: aim for direct, active words and phrases (see Box 15.1, p. 97).

- Most employers will expect your CV to be word-processed (and spellchecked). Errors in style, grammar and presentation will count against you, so be sure to check through your final version (and ask a reliable person to second-check it for you).

- Aim for a *maximum* length of two pages, printed single-sided on A4 paper, using a 'formal' font (e.g. Times Roman or Arial) of no less than 12 point for the main text. Always print on good-quality white paper. Avoid fussy use of colour, borders or fonts.

- Don't try to cram in too much detail. Use a clear and succinct approach with short sentences and lists to improve 'readability' and create structure. Remember that your aim is to catch the eye of your potential employer, who may have many applications to work through.

- It is polite to check that people are willing to act as a referee for you and to provide them with an up-to-date copy of your CV.

Your covering letter should have four major components:

1. **Letterhead.** Include your contact details, the recipient's name and title (if known) and address, plus any job reference number.

2. **Introductory paragraph.** Explain who you are and state the post you are applying for.

3. **Main message.** This is your opportunity to sell yourself to a potential employer, highlighting

(continued)

Box 7.1 (continued)

particular attributes and experience. Keep it to three or four sentences at the most and relate it to the particular skills and qualities demanded in the job or person specification.

4. **Concluding paragraph.** A brief statement that you look forward to hearing the outcome of your application is sufficient.

Finally, add either 'Yours sincerely' (where the recipient's name is known) or 'Yours faithfully' (in a letter beginning 'Dear Sir or Madam') and then end with your signature.

- **Relevance.** If you can, slant your CV towards the job description and the qualifications required (see below). Make sure you provide evidence to back up your assertions about skills, qualities and experience.
- **Accuracy and completeness.** Check that all your dates tally; otherwise, you will seem careless. It is better to be honest about your grades and (say) a period of unemployment, than to cover this up or omit details that an employer will want to know. They may be suspicious if you leave things out.

Adjusting your CV

Creating a generic CV – as you may apply for several jobs, it is useful to construct a CV in electronic format (e.g. as a Word file) which includes all information of potential relevance. This can then be modified to fit each post. Having a prepared CV on file will reduce the work each time you apply, while modifying this will help you focus on relevant skills and attributes for the particular job.

You should fine-tune your CV for each post. Employers frequently use a 'person specification' to define the skills and qualities demanded in a job, often under headings such as 'essential' and 'desirable'. This will help you decide whether to apply for a position and it assists the selection panel to filter the applicants. Highlight relevant qualifications as early in your CV as possible. Be selective – don't include every detail about yourself. Emphasise relevant parts and leave out irrelevant details, according to the job. Similarly, your letter of application is not merely a formal document but is also an opportunity for persuasion (Box 7.1). You can use it to state your ambitions and highlight particular qualifications and experience. However, don't go over the top – always keep the letter to a single page.

 KEY POINT *A well-constructed and relevant CV won't necessarily guarantee you a job, but it may well get you on to the short list for interview. A poor-quality CV is a sure route to failure.*

Sources for further study

Anon. (2000) *How to Write a Curriculum Vitae.* University of London Careers Service, London.

Anon. *Doctorjob.com Website.* Available: http://doctorjob.com
Last accessed 09/04/07.
[Created by GTI Specialist Publishers.]

Anon. *Graduate Prospects Website.* Available: http://www.prospects.ac.uk/cms/ShowPage/Home_page/p!eLaXi
Last accessed 09/04/07.
[Created and maintained by the Higher Education Careers Services Unit, contains good examples of typical CVs.]

Higher Education and Research Opportunities in the United Kingdom (HERO). *Applying for Jobs.* Available: http://www.hero.ac.uk/uk/studying/careers_and_lifelong_learning/applying_for_jobs285.cfm
Last accessed 09/04/07.
[Provides links to useful sources on job applications.]

Phillips, C. (1999) *Making Wizard Applications.* GTI Specialist Publishers, Wallingford.

Study exercises

7.1 Evaluate your personal attributes. Using Table 7.1 as a source, list *five* qualities that you feel that you would best use to describe yourself, and cite the evidence you might give to a potential employer to convince them that this was the case. List *five* attributes you could develop, then indicate how you might do this.

7.2 Create a generic CV. Drawing on your school record of achievement or similar, or any CV you may already have prepared, e.g. for a summer or part-time job, create a word-processed generic CV. Save the file in an appropriate (computer) folder and make a back-up copy. Print out a copy for filing. Periodically update the word-processed version. If necessary, keep or save different versions to be used in different contexts (e.g. when you apply for a vacation job).

7.3 Think about your future career and ask for advice. Make an appointment with one of the advisers in your university's careers service. Ask what the career options might be for people with your intended degree, or determine what qualifications or module options might be appropriate for occupations that attract you.

Information technology and library resources

The ability to find scientific information is a skill required for many exercises in your degree programme. You will need to research facts and published findings as part of writing essays, literature reviews and project introductions, and when amplifying your lecture notes and revising for exams. You must also learn how to follow scientific convention in citing source material as the authority for the statements you have made.

Sources of information

For essays and revision

You are unlikely to delve into the primary literature (p. 53) for these purposes – books and reviews are much more readable. If a lecturer or tutor specifies a particular book, then it should not be difficult to find out where it is shelved in your library, using the computerised index system. Library staff will generally be happy to assist with any queries. If you want to find out which books your library holds on a specified topic, use the system's subject index. You will also be able to search by author or by key words.

There are two main systems used by libraries to classify books: the Dewey Decimal system and the Library of Congress system. Libraries differ in the way they employ these systems, especially by adding further numbers and letters after the standard classification marks to signify e.g. shelving position or edition number. Enquire at your library for a full explanation of local usage.

The World Wide Web is an ever expanding resource for gathering both general and specific information (see Chapter 10). Sites fall into analogous categories to those in the printed literature: there are sites with original information, sites that review information and bibliographic sites. One considerable problem is that websites may be frequently updated, so information present when you first looked may be altered or even absent when the site is next consulted. Further, very little of the information on the WWW has been monitored or refereed. Another disadvantage is that the site information may not state the origin of the material, who wrote it or when it was written.

For literature surveys and project work

You will probably need to consult the primary literature. If you are starting a new research project or writing a report from scratch, you can build up a core of relevant papers by using the following methods:

- Asking around: supervisors or their postgraduate students will almost certainly be able to supply you with a reference or two that will start you off.
- Searching an online database: these cover very wide areas and are a convenient way to start a reference collection, although a charge is often made for access and sending out a listing of the papers selected (your library may or may not pass this on to you).
- Consulting the bibliography of other papers in your collection – an important way of finding the key papers in your field. In effect, you are taking advantage of the fact that another researcher has already done all the hard work.

Browsing in a library – this may turn up interesting material, but remember the books on the shelves are those not currently out on loan. Almost by definition, the latter may be more up to date and useful. To find out a library's full holding of books in any subject area, you need to search its catalogue (normally available as an online database).

Example The book *The Selfish Gene* by Richard Dawkins (1976; Oxford University Press) is likely to be classified as follows:

Dewey Decimal system: 591.51
where 591 refers to zoology
591.5 refers to ecology of animals
591.51 refers to habits and behaviour patterns

Library of Congress system: QL751
where Q refers to science
QL refers to zoology
QL75 refers to animal behaviour
QL751 refers to general works and treatises

Web resources – your university library will provide you with access to a range of web-based databases and information systems. The library web pages will list these and provide links, which may be worth bookmarking on your web browser. Resources especially useful to bioscientists include:

- ISI Web of Knowledge, *including the* Science Citation Index
- Ingentaconnect *(previously known as BIDS), including* Ingenta Medline
- CSA Illumina
- Dialog
- PubMed
- ScienceDirect
- Scopus
- Ovid, *including* Cinahl *and* Medline

Most of these electronic resources operate on a subscription basis and may require an 'Athens' username and password – for details of how to obtain these consult library staff or your library's website.

Journal/periodical/serial – any publication issued at regular intervals. In biosciences, usually containing papers (articles) describing original research findings and reviews of literature.

e-journal – a journal published online, consisting of articles structured in the same way as a paper-based journal. A valid username and password may be required for access (arranged via your library, if it subscribes to the e-journal).

The primary literature – this comprises original research papers, published in specialist scientific periodicals. Certain prestigious general journals (e.g. *Nature*) contain important new advances from a wide subject area.

Monograph – a specialised book covering a single topic.

e-book – a book published online in downloadable form.

ebrary – a commercial service offering e-books and other online resources.

HERON (Higher Education Resources ON demand) – a national service for UK higher education offering copyright clearance, digitisation and delivery of book extracts and journal articles.

Review – an article in which recent advances in a specific area are outlined and discussed.

Proceedings – volume compiling written versions of papers read at a scientific meeting on a specific topic.

Abstracts – shortened versions of papers, often those read at scientific meetings. These may later appear in the literature as full papers.

Bibliography – a summary of the published work in a defined subject area.

Alternative methods of receiving information – *RSS (really simple syndication) feeds and email updates from publishers are increasingly used to provide automated information services to academic clients, for example, by supplying links to relevant contents of new editions of online journals.*

- Referring to 'current awareness' online databases: these are useful for keeping you up to date with current research; they usually provide a monthly listing of article details (title, authors, source, author address) arranged by subject and cross-referenced by subject and author. Current awareness databases cover a wider range of primary literature than could ever be available in any one library.

  Examples relevant to the biosciences include: *Current Contents Connect* (ISI), the *Current Advances* series (Elsevier), *Biological Abstracts* and *Zoological Record Archive* (Thompson Scientific). Some online databases also offer a service whereby they will email registered users with updates based on saved search criteria. Consult library staff or your library website to see which of these databases and services are available to you.

- Using the *Science Citation Index* (SCI): this is a very valuable source for exploring the published literature in a given field, because it lets you see who has cited a given paper; in effect, SCI allows you to move forward through the literature from an existing reference. The Index is available online via ISI Web of Science.

For specialised information

You may need to consult reference works, such as encyclopedias, maps and books providing specialised information. Much of this is now available online (consult your library's information service or web pages). Three books worth noting are:

- The *Handbook of Chemistry and Physics* (Lide, 2005): the Chemical Rubber Company's publication (affectionately known as the 'Rubber Bible') giving all manner of physical constants, radioisotope half-lives, etc.
- *The Merck Index* (O'Neil et al., 2006), which gives useful information about organic chemicals, e.g. solubility, whether poisonous, etc.
- The *Geigy Scientific Tables* (8th edition), a series of six volumes (Lentner, 1981; 1984; 1986; 1990; 1992; Lentner et al., 1982) provides a wide range of information centred on biochemistry, e.g. buffer formulae, properties of constituents of living matter.

Obtaining and organising research papers

Obtaining a copy

It is usually more convenient to have personal copies of key research articles for direct consultation when working in a laboratory or writing. The simplest way of obtaining these is to photocopy the originals or download and/or print off copies online (e.g. as .pdf or HTML files). For academic purposes, this is normally acceptable within copyright law. If your library does not subscribe to the journal, it may be possible for them to borrow it from a nearby institute or obtain a copy via a national borrowing centre (an 'inter-library loan'). If the latter, you will have to fill in a form giving full bibliographic details of the paper and where it was cited, as well as signing a copyright clearance statement concerning your use of the copy.

Your department might be able to supply 'reprint request' postcards to be sent to the designated author of a paper. This is an unreliable method of obtaining a copy because it may take some time (allow at least 1–3 months) and some requests will not receive a reply. Taking into account

Copyright law – *In Europe, copyright regulations were harmonised in 1993 (Directive 93/98/EEC) to allow literary copyright for 70 years after the death of an author and typographical copyright for 25 years after publication. This was implemented in the UK in 1996, where, in addition, the Copyright, Designs and Patents Act (1988) allows the Copyright Licensing Agency to license institutions so that lecturers, students and researchers may take copies for teaching and personal research purposes – no more than a single article per journal issue, one chapter of a book, or extracts to a total of 5 per cent of a book.*

Storing research papers – *these can easily be kept in alphabetical order within filing boxes or drawers, but if your collection is likely to grow large, it will need to be refiled as it outgrows the storage space. You may therefore wish to add an 'accession number' to the record you keep in your database, and file the papers in sequence according to this as they accumulate. New filing space is only required at the 'end' and you can use the accession numbers to form the basis of a simple cross-referencing system.*

Using commercial bibliographic database software to organise your references – *for those with large numbers of references in their collection, and who may wish to produce lists of selected references in particular format, e.g. for inclusion in a project report or journal paper, systems like EndNote, Reference Manager or ProCite can reward the investment of time and money required to create a personal reference catalogue. Appropriate bibliographic data must first be entered into fields within a database (some versions assist you to search online databases and upload data from these). The database can then be searched and used to create customised lists of selected references in appropriate citation styles.*

the waste involved in postage and printing, it is probably best simply to photocopy or send for a copy via inter-library loan.

Organising papers

Although the number of papers you accumulate may be small to start with, it is worth putting some thought into their storage and indexing before your collection becomes disorganised and unmanageable. Few things are more frustrating than not being able to lay your hands on a vital piece of information, and this can seriously disrupt your flow when writing or revising.

Indexing your references

Whether you have obtained a printed copy, have stored downloaded files electronically, or have simply noted the bibliographic details of a reference, you will need to index each resource. This is valuable for the following reasons:

- You will probably need the bibliographic information for creating a reference list for an assignment or report.
- If the index also has database features, this can be useful, allowing you to search for key words or authors.
- If you include an 'accession number' and if you then file printed material sequentially according to this number, then it will help you to find the hard copy.
- Depending on the indexing system used, you can add comments about the reference that may be useful at a later time, e.g. when writing an introduction or conclusion.

The simplest way to create an index system is to put the details on reference cards, but database software can be more convenient and faster to sort, once the bibliographic information has been entered. If you do not feel that commercial software is appropriate for your needs, consider using a word processor or spreadsheet; their rudimentary database sorting functions (see Chapters 11 and 12) may be all that you require.

If you are likely to store lots of references and other electronic resources digitally, then you should consider carefully how this information is kept, for example by choosing file names that indicate what the file contains and that will facilitate sorting.

Making citations in text

There are two main ways of citing articles and creating a bibliography (also referred to as 'references' or 'literature cited').

The Harvard system

For each citation, the author name(s) and the date of publication are given at the relevant point in the text. The bibliography is organised alphabetically and by date of publication for papers with the same authors. Formats normally adopted are, for example, 'Smith and Jones (1983) stated that ...' or 'it has been shown that ... (Smith and Jones, 1983)'. Lists of references within parentheses are separated by semi-colons, e.g. '(Smith and Jones, 1983; Jones and Smith, 1985)', normally in order of date of publication. To avoid repetition within the same

Example Incorporating references in text – this sample shows how you might embed citations in text using the Harvard approach:
'... Brookes et al. (2001) proposed that protein A216 was involved in the degradation process. However, others have disputed this notion (Scott and Davis, 1997; Harley, 1998, 2000). Patel (1999a; 1999b) found that A216 is inactivated at pH values less than 5; while several authors (e.g. Hamilton, 1995; Drummond and Stewart, 2002) also report that its activity is strongly dependent on Ca^{2+} concentration ...'

Example Incorporating references in text – this sample shows how you might embed citations in text using the Vancouver approach:
'... Brookes et al. proposed that protein A216 was involved in the degradation process[1]. However, others have disputed this notion[2-4]. Patel[5,6] found that A216 is inactivated at pH values less than 5; while several authors[7,8] also report that its activity is strongly dependent on Ca^{2+} concentration ...'

Examples
Paper in journal:
Smith, A. B., Jones, C.D. and Professor, A. (1998). Innovative results concerning our research interest. *Journal of New Results*, 11, 234–5.
Book:
Smith, A. B. (1998). *Summary of My Life's Work.* Megadosh Publishing Corp., Bigcity. ISBN 0-123-45678-9.
Chapter in edited book:
Jones, C. D. and Smith, A. B. (1998). Earth-shattering research from our laboratory. In: *Research Compendium 1998* (ed. A. Professor), pp. 123–456. Bigbucks Press, Booktown.
Thesis:
Smith, A. B. (1995). *Investigations on my Favourite Topic.* PhD thesis, University of Life, Fulchester.

Note: if your references are handwritten, you should indicate italics by underlining text or numerals.

paragraph, an approach such as 'the investigations of Smith and Jones indicated that' could be used following an initial citation of the paper. Where there are more than two authors it is usual to write 'et al.'; this stands for the Latin *et alii* meaning 'and others'. If citing more than one paper with the same authors, put, for example, 'Smith and Jones (1987; 1990)' and if papers by a given set of authors appeared in the same year, letter them (e.g. Smith and Jones, 1989a; 1989b).

The numerical or Vancouver system

Papers are cited via a superscript or bracketed reference number inserted at the appropriate point. Normal format would be, for example: 'DNA sequences[4,5] have shown that ...' or 'Jones [55,82] has claimed that ...'. Repeated citations use the number from the first citation. In the true numerical method (e.g. as in *Nature*), numbers are allocated by order of citation in the text, but in the alpha-numerical method (e.g. the *Annual Review* series), the references are first ordered alphabetically in the Bibliography, then numbered, and it is this number that is used in the text. Note that with this latter method, adding or removing references is tedious, so the numbering should be done only when the text has been finalised.

 KEY POINT *The main advantages of the Harvard system are that the reader might recognise the paper being referred to and that it is easily expanded if extra references are added. The main advantages of the Vancouver system are that it aids text flow and reduces length.*

How to list your citations in a bibliography

Whichever citation method is used in the text, comprehensive details are required for the bibliography so that the reader has enough information to find the reference easily. Citations should be listed in alphabetical order with the priority: first author, subsequent author(s), date. Unfortunately, in terms of punctuation and layout, there are almost as many ways of citing papers as there are journals! Your department may specify an exact format for project work; if not, decide on a style and be consistent – if you do not pay attention to the details of citation you may lose marks. Take special care with the following aspects:

- Authors and editors: give details of *all* authors and editors in your bibliography, even if given as 'et al.' in the text.
- Abbreviations for journals: while there are standard abbreviations for the titles of journals (consult library staff), it is a good idea to give the whole title, if possible.
- Books: the edition should always be specified as contents may change between editions. Add, for example, '(5th edition)' after the title of the book. You may be asked to give the International Standard Book Number (ISBN), a unique reference number for each book published.
- Unsigned articles, e.g. unattributed newspaper articles and instruction manuals – refer to the author(s) in text and bibliography as 'Anon.'.
- Websites: there is no widely accepted format at present. You should follow departmental guidelines if these are provided, but if these are

not available, we suggest providing author name(s) and date in the text when using the Harvard system (e.g. Hacker, 2006), while in the bibliography giving the URL details in the following format: Hacker, A (2006) *University of Anytown Homepage on Aardvarks*. Available: http://www.myserver.ac.uk/homepage. Last accessed: 23/02/07. In this example, the web page was constructed in 2006, but accessed in February, 2007. If no author is identifiable, cite the sponsoring body (e.g. University of Anytown, 2006), and if there is no author or sponsoring body, write 'Anon.' for 'anonymous', e.g. Anon. (2006), and use Anon. as the 'author' in the bibliography. If the web pages are undated, *either* use the 'Last accessed' date for citation and put no date after the author name(s) in the reference list, *or* cite as 'no date' (e.g. Hacker, no date) and leave out a date after the author name(s) in the reference list – you should be consistent whichever option you choose.

- Unread articles: you may be forced to refer to a paper via another without having seen it. If possible, refer to another authority who has cited the paper, e.g. '... Jones (1980), cited in Smith (1990), claimed that ...'. Alternatively, you could denote such references in the bibliography by an asterisk and add a short note to explain at the start of the reference list.

- Personal communications: information received in a letter, seminar or conversation can be referred to in the text as, for example, '... (Smith, pers. comm.)'. These citations are not generally listed in the bibliography of papers, though in a thesis you could give a list of personal communicants and their addresses.

- Online material: some articles are published solely online and others online ahead of publication in printed form. The article may be given a DOI (digital object identifier), allowing it to be cited and potentially tracked before and after it is allocated to a printed issue (see http://www.doi.org/). DOIs allow for web page redirection by a central agency, and CrossRef (see http://www.crossref.org) is the official DOI registration organisation for scholarly and professional publications. DOIs can be used as 'live' hyperlinks in online articles, or cited in place of (and, when they become available, following) the volume and page numbers for the article, with the remainder of the details cited in the usual fashion: for example,

'Smith. A. and Jones B. (2006) Our latest important research in the form of a web-published article. *Online Biosciences* 8/2006 (p. 781). Published Online: 26 March 2006. DOI: 10.1083/mabi.200680019'.

Text references and sources for further study

Lentner, C. (ed.) (1981) *Geigy Scientific Tables, Vol. 1: Units of Measurement, Body Fluids, Composition of the Body. Nutrition*, 8th edn. Ciba-Geigy, Basel.

Lentner, C. (ed.) (1984) *Geigy Scientific Tables, Vol. 3: Physical Chemistry, Composition of Blood, Hematology, Somatometric Data*, 8th edn. Ciba-Geigy, Basel.

Lentner, C. (ed.) (1986) *Geigy Scientific Tables, Vol. 4: Biochemistry, Metabolism of Xenobiotics, Inborn Errors of Metabolism, Pharmacogenetics, Ecogenetics*, 8th edn. Ciba-Geigy, Basel.

Lentner, C. (ed.) (1990) *Geigy Scientific Tables, Vol. 5: Heart and Circulation*, 8th edn. Ciba-Geigy, Basel.

Lentner, C. (ed.) (1992) *Geigy Scientific Tables, Vol. 6: Bacteria, Fungi, Protozoa, Helminths*, 8th edn. Ciba-Geigy, Basel.

Lentner, C., Diem, K. and Seldrup, J. (eds) (1982) *Geigy Scientific Tables, Vol. 2: Introduction to Statistics, Statistical Tables, Mathematical Formulae,* 8th edn. Ciba-Geigy, Basel.

Lide, D.R. (ed.) (2005) *CRC Handbook of Chemistry and Physics,* 86th edn. CRC Press, Boca Raton, Florida.

McMillan, K.M. and Weyers, J.D.B. (2006) *The Smarter Student: Study Skills and Strategies for Success at University.* Pearson Education, Harlow.

O'Neil, M.J., Heckelman, P.E., Koch, C.B. and Roman, K.J. (2006) *Merck Index: An Encyclopedia of Chemicals, Drugs, and Biologicals,* 14th edn. Merck & Co., Inc., Whitehouse Station, New Jersey.

Pears, R. and Shield, G. *Cite Them Right! The Essential Guide to Referencing and Plagiarism.* Pear Tree Books, Newcastle upon Tyre.

Study exercises

8.1 Test your library skills. This exercise relies on the fact that most university-level libraries serving biosciences departments will take the scientific journal *Nature.* To help you answer these questions, it may be beneficial to attend a library induction session, if you haven't already. Alternatively, the library's help or enquiry desks may be able to assist you if you are having problems.

(a) First, find out and provide the name of the classification system that your university uses for cataloguing its books and periodicals.

(b) Using your library's cataloguing system (online, preferably) find out the appropriate local classification number for the journal *Nature.*

(c) Where precisely is *Nature* shelved in your library? (Your answer need refer only to most recent issues if some have been archived.)

(d) What is the exact title of the landmark papers in the following two volumes? (i) *Nature* **171,** 737–738 (1953); (ii) *Nature* **408,** 796–815 (2000).

8.2 Explore different methods of citing references. Go to your library and seek out the journal area for biology or life sciences. Choose *three* different journals in your subject area and from a recent edition write down how they would print a typical citation for a multi-author journal paper in the 'references' or 'literature cited' section. Where used, indicate italicised text with normal underline and bold text with wavy underline. Pay attention to punctuation. Compare these methods with each other, with the methods recommended on pp. 47–8 of this book and with the recommendation your department or your course handbook makes. Are they all the same?

8.3 Make website citations. Use a search engine (pp. 62–4) to find an informative website that covers each of the following:

(a) The use of SI units.

(b) Information about pollination of a specific flower by bats.

(c) The different types of blood cells.

Indicate how you would cite each website at the end of an essay (follow your department's guidelines or use those in this chapter).

8.4 Compare the Harvard and Vancouver methods of citation. Pair up with a partner in your class. Each person should then pick one of the two main methods of citation and consider its pros and cons independently. Meet together and compare your lists. Given the choice, which method would you choose for (a) a hand-written essay; (b) a word-processed review; (c) an article in an academic journal, and why?

9 Evaluating information

Example A Web search for the letters 'DNA' (e.g. using the search engine Google, p. 63), will reveal that this acronym appears in several hundred million websites. Not all of these deal with deoxyribonucleic acid – the listed websites include: *Doctors' Net Access*, a Web resource for American physicians; the website of the *Dermatology Nurses Association*; and that of *Delta Nu Alpha*, an international organisation for those working in the professions. These examples are easy to identify as irrelevant to research on 'DNA, the molecule', but considering the many websites that do actually cover this topic, which might contain the exact information you seek? How valid is the information? Is it biased towards one viewpoint or hypothesis? Does it represent current mainstream thinking on the topic? These are some of the issues that an evaluation of the information sources might deal with.

Checking the reliability of information, assessing the relative value of different ideas and thinking critically are skills essential to the scientific approach. You will need to develop your abilities to evaluate information in this way because:

- You will be faced with many sources of information, from which you will need to select the most appropriate material.
- You may come across conflicting sources of evidence and may have to decide which is the more reliable.
- The accuracy and validity of a specific fact may be vital to your work.
- You may doubt the quality of the information from a particular source.
- You may wish to check the original source because you are not sure whether someone else is quoting it correctly.

 KEY POINT *Evaluating information and thinking critically are regarded as higher-order academic skills. The ability to think deeply in this way is greatly valued in the biosciences and will consequently be assessed in coursework and exam questions (see Chapter 5).*

The process of evaluating and using information can be broken down into four stages:

1. Selecting and obtaining material. How to find sources is covered in Chapter 8. Printed books and journals are important, but if you identify a source of this kind there may be delays in borrowing it or obtaining a photocopy. If the book or journal is available online, then downloading or printing sections or papers will be more convenient and faster. The Internet is often a first port of call if you wish to find something out quickly. For many websites, however, it can be difficult to verify the authenticity of the information given (see Box 10.5).
2. Assessing the content. You will need to understand fully what has been written, including any technical terms and jargon used. Establish the relevance of the information to your needs and assure yourself that the data or conclusions have been presented in an unbiased way.
3. Modifying the information. In order to use the information, you may need to alter it to suit your needs. This may require you to make comparisons, interpret or summarise. Some sources may require translation. Some data may require mathematical transformation before they are useful. There is a chance of error in any of these processes and also a risk of plagiarism.
4. Analysis. This may be your own interpretation of the information presented, or an examination of the way the original author has used the information.

 KEY POINT *Advances in communications and information technology mean that we can now access almost limitless knowledge. Consequently, the ability to evaluate information has become an extremely important skill.*

Definition

Plagiarism – The unacknowledged use of another's work as if it were one's own. In this definition, the concept of 'work' includes ideas, writing, data or inventions and not simply words; and the notion of 'use' does not only mean copy 'word for word', but also 'in substance' (i.e. a copy of the ideas involved). Use of another's work is acceptable if you acknowledge the source (see Box 9.1).

Box 9.1 How to avoid plagiarism and copyright infringement

Plagiarism is defined on page 51. Examples of plagiarism include:

- Copying the work of a fellow student (past or present) and passing it off as your own.
- Using 'essay-writing services', such as those on offer on certain websites.
- Copying text or images from a source (book, journal article or website, for instance) and using this within your own work without acknowledgement.
- Quoting others' words without indicating who wrote or said them.
- Copying ideas and concepts from a source without acknowledgement, even if you paraphrase them.

Most students would accept that some of the above can only be described as cheating. However, many students, especially at the start of their studies, are unaware of the academic rule that they must *always* acknowledge the originators of information, ideas and concepts, and that not doing so is regarded a form of academic dishonesty. If you adopt the appropriate conventions that avoid such accusations, you will achieve higher marks for your work as it will fulfil the markers' expectations for academic writing.

Universities have a range of mechanisms for identifying plagiarism, from employing experienced and vigilant coursework markers and external examiners to analysing students' work using sophisticated software programs. Plagiarism is always punished severely when detected. Penalties may include awarding a mark of zero to all involved – both the copier(s) and the person whose work has been copied (who is regarded as complicit in the crime). Further disciplinary measures may be taken in some instances. In severe cases, such as copying substantive parts of another's work within a thesis, a student may be dismissed from the university.

If you wish to avoid being accused of plagiarism, the remedies are relatively simple:

1. **Make sure the work you present is always your own.** If you have been studying alongside a colleague, or have been discussing how to tackle a particular problem with your peers, make sure you write on your own when working on your assignments.

2. **Never be tempted to 'cut and paste'** from websites or online sources such as word-processed handouts. Read these carefully, decide what the important points are, express these *in your own words* and *provide literature citations to the original sources* (see Chapter 8). In some cases, further investigations may be required to find out details of the original sources. The lecturer's reading list or a book's references may help you here.

3. **Take care when note-taking.** If you decide to quote word for word, make sure you show this clearly in your notes with quotation marks. If you decide to make your own notes based on a source, make sure these are original and do not copy phrases from the text. In both cases, write down full details of the source at the appropriate point in your notes.

4. **Place appropriate citations throughout your text where required.** If you are unsure about when to do this, study reviews and articles in your subject area (see also Chapter 8).

5. **Show clearly where you are quoting directly from a source.** For short quotes, this may involve using quotation marks and identifying the source afterwards, as in the example '... as Samuel Butler (1877) wrote: "a hen is only an egg's way of making another egg". ' For a longer quotes (say 40 words or more), you should create a separate paragraph of quoted text, usually identified by inverted commas, indentation, italicisation or a combination of these. A citation must *always* be included, normally at the end. Your course handbook may specify a layout. Try not to rely too much on quotes in your work. If a large proportion of your work is made up from quotes, this will almost certainly be regarded as lacking in originality, scoring a poor mark.

Copyright issues are often associated with plagiarism, and refer to the right to publish (and hence copy) original material, such as text, images and music. Copyright material is indicated by a symbol © and a date (see, for example, page iv of this book). Literary copyright is the aspect most relevant to students in their academic studies. UK copyright law protects authors' rights for life and gives their estates rights for a further 70 years. Publishers have 'typographical copyright' that lasts for 25 years. This means that it is illegal to photocopy, scan or print out copyright material unless you have permission, or unless your copying is limited to an extent that could be considered 'fair dealing'. For educational purposes – private study or research – in a scientific context, this generally means:

- no more than 5 per cent in total of a work;
- one chapter of a book;
- one article per volume of an academic journal;
- 20 per cent of a short book;
- one separate illustration or map.

(continued)

Box 9.1 (continued)

You may only take one copy within the above limits, may not copy for others, and may not exceed these amounts *even if you own a copy of the original*. These rules also apply to Web-based materials, but sometimes you will find sites where the copyright is waived. Some copying may be licensed; you should consult your library's website or helpdesk to see whether it has access to licensed material. Up-to-date copyright information is generally posted close to library and departmental photocopiers.

Evaluating sources of information

One way of assessing the reliability of a piece of scientific information is to think about how it was obtained in the first place. Essentially, facts and ideas originate from someone's research or scholarship, whether they are numerical data, descriptions, concepts, or interpretations. Sources are divided into two main types:

1. Primary sources – those in which ideas and data are first communicated. The primary literature is generally published in the form of 'papers' (articles) in journals whether printed or online. These are usually refereed by experts in the academic peer group of the author, and they will check the accuracy and originality of the work and report their opinions back to the editors. This peer review system helps to maintain reliability, but it is not perfect. Books and, more rarely, websites and articles in magazines and newspapers, can also be primary sources but this depends on the nature of the information published rather than the medium. These sources are not formally refereed, although they may be read by editors and lawyers to check for errors and unsubstantiated or libellous allegations.

2. Secondary sources – those which quote, adapt, interpret, translate, develop or otherwise use information drawn from primary sources. It is the act of quoting or paraphrasing that makes the source secondary, rather than the medium. Reviews are examples of secondary scientific sources, and books and magazine articles are often of this type.

When information is modified for use in a secondary source, alterations are likely to occur, whether intentional or unintentional. Most authors do not deliberately set out to change the meaning of the primary source, but they may unwittingly do so, e.g. in changing text to avoid plagiarism or by oversimplification. Others may consciously or unconsciously exert bias in their reporting, for example, by quoting evidence that supports only one side of a debate. Therefore, the closer you can get to the primary source, the more reliable the information is likely to be. On the other hand, modification while creating a secondary source could involve correcting errors, or synthesising ideas and content from multiple sources.

Distinguishing between primary and secondary sources – try the 'IMRaD test'. Many primary sources contain information in the order: **I**ntroduction, **M**aterials and **M**ethods, **R**esults and **D**iscussion. If you see this format, and particularly if data from an experiment, study or observation are presented, then you are probably reading a primary source.

Example If a journalist wrote an article about a new 'flesh-eating bug' for the *New York Times* that was based on an article in the *British Medical Journal*, the *New York Times* article would be the secondary source, while the *British Medical Journal* article would be the primary source.

Taking account of the changing nature of websites and wikis – by their very nature, these sources may change. This means that it is important to quote accurately from them and to give a 'Last accessed' date when citing (see p. 49).

Authorship and provenance

Clearly, much depends on who is writing the source and on what basis (e.g. who paid them?). Consequently, an important way of assessing sources is to investigate the ownership and provenance of the work (who and where it originated from, and why).

Can you identify who wrote the information? If it is signed or there is a 'byline' showing who wrote it, you might be able to make a judgement on

Evaluating information

Finding out about authors and provenance – these pieces of information are easy to find in most printed sources and may even be presented just below the title, for convenience. In the case of the Web, it may not be so easy to find what you want. Relevant clues can be obtained from 'home page' links and the header, body and footer information. For example, the domain (p. 60) may be useful, while the use of the tilde symbol (~) in an address usually indicates a personal, rather than an institutional, website.

Assessing substance over presentation – just because the information is presented well (e.g. in a glossy magazine or particularly well-constructed website), this does not necessarily tell you much about its quality. Try to look below the surface, using the methods mentioned in this chapter.

Table 9.1 Checklist for assessing information in science. How reliable is the information you have been reading? The more 'yes' answers you can give below, the more trustworthy you can assume it to be.

Assessing sources

- Can you identify the author's name?
- Can you determine what relevant qualifications he/she holds?
- Can you say who employs the author?
- Do you know who paid for the work to be done?
- Is this a primary or secondary source?
- Is the content original or derived from another source?

Evaluating information

- Have you checked a range of sources?
- Is the information supported by relevant literature citation?
- Is the age of the source likely to be important regarding the accuracy of the information?
- Have you focused on the substance of the information presented rather than its packaging?
- Is the information fact or opinion?
- Have you checked for any logical fallacies in the arguments?
- Does the language used indicate anything about the status of the information?
- Have the errors associated with any numbers been taken into account?
- Have the data been analysed using appropriate statistics?
- Are any graphs constructed fairly?

the quality of what you are reading. This may be a simple decision, if you know or can assume that the writer is an authority in the area; otherwise a little research might help (for example, by putting the name into a search engine). Of course, just because Professor X thinks something does not make it true. However, if you know that this opinion is backed up by years of research and experience, then you might take it a little more seriously than the thoughts of a school pupil. If an author is not cited, effectively nobody is taking responsibility for the content. Could there be a reason for this?

Is the author's place of work cited? This might tell you whether the facts or opinions given are based on an academic study. Is there a company with a vested interest behind the content? If the author works for a public body, there may be publication rules to follow and they may even have to submit their work to a publications committee before it is disseminated. They are certainly more likely to get into trouble if they include controversial material.

Evaluating facts and ideas

However reliable the source of a piece of information seems to be, it is probably a good idea to retain a slight degree of scepticism about the facts or ideas involved. Even information from impeccable primary sources may not be perfect – different approaches can give different outcomes, and interpretations can change with time and with further advances in knowledge.

Table 9.1 provides a checklist to use when evaluating sources.

Critically examining facts and ideas is a complex task depending on the particular issues involved, and a number of different general approaches can be applied. You will need to decide which of the following general tips are useful in your specific case:

- Make cross-referencing checks – look at more than one source and compare what is said in each. The cross-referenced sources should be as independent as possible (for example, do not compare a primary source together with a secondary review based on it). If you find that all the sources give a similar picture, then you can be more confident about the reliability of the information.
- Look at the extent and quality of citations – if references are quoted, these indicate that a certain amount of research has been carried out beforehand, and that the ideas or results are based on genuine scholarship. If you are doubtful about the quality of the work, these references might be worth looking at. How up to date are they? Do they cite independent work, or is the author exclusively quoting their own work, or solely the work of one person?
- Consider the age of the source – the fact that a source is old is not necessarily a barrier to truth, but ideas and facts may have altered since the date of publication, and methods may have improved. Can you trace changes through time in the sources available to you? What key events or publications have forced any changes in the conclusions?
- Try to distinguish fact from opinion – to what extent has the author supported a given viewpoint? Have relevant facts been quoted, via literature citations or the author's own researches? Are numerical data used to substantiate the points used? Are these reliable and can you

verify the information, for example, by looking at the sources cited? Might the author have a reason for putting forward biased evidence to support a personal opinion?

- Analyse the language used – words and their use can be very revealing. Subjective wording might indicate a personal opinion rather than an objective conclusion. Propaganda and personal bias might be indicated by absolute terms, such as 'everyone knows...'; 'It can be guaranteed that ...', or a seemingly one-sided consideration of the evidence. How carefully has the author considered the topic? A less studious approach might be indicated by exaggeration, ambiguity, or the use of 'journalese' and slang. Always remember, however, that content should be judged above presentation.

- Look closely at any numbers – if the information you are looking at is numerical in form, have statistical errors been taken into consideration, and, where appropriate, quantified? If so, does this help you arrive at a conclusion about how genuine the differences are between important values?

- Think carefully about any hypothesis-testing statistics used – are the methods appropriate? Are the underlying hypotheses the right ones? Have the results of any tests been interpreted correctly in arriving at the conclusion? To deal with these matters, you will need at least a basic understanding of the 'statistical approach' and of commonly used techniques (see Chapters 65 and 66).

Critical thinking

Critical thinking involves the application of logic to a problem, issue or case study. It requires a wide range of skills. Key processes involved include: acquiring and processing information; creating appropriate hypotheses and formulating conclusions; and acting on the conclusions towards a specific objective.

 KEY POINT *Critical thinking needs reliable knowledge, but it requires you to use this appropriately to analyse a problem. It can be contrasted with rote learning – where you might memorise facts without an explicit purpose other than building your knowledge base.*

Critical thinking is particularly important in biology, because the subject deals with complex and dynamic systems. These can be difficult to understand for several reasons:

- they are often multi-faceted, involving many interactions;
- it can be difficult to alter one variable in an experiment without producing confounding variables (see p. 186);
- many variables may be unmeasured or unmeasurable;
- heterogeneity (variability) is encountered at all scales from the molecular scale to the ecosystem;
- perturbation of the system can lead to unexpected ('counter-intuitive') results.

As a result, conclusions in biological research are seldom clear-cut. Critical thinking allows you to arrive at the most probable conclusion from the results at hand; however, it also involves acknowledging that

Learning from examples – as your lecturers introduce you to case studies, you will see how biologists have applied critical thinking to understand the nature of cells, organisms and ecosystems. Some of your laboratory sessions may mimic the processes involved – observation, hypothesis, experimental design, data gathering and analysis and formulating a conclusion (see Chapter 31). These skills and approaches can be applied in your course, e.g. when writing about a biological issue or carrying out a research project.

'You can prove anything with statistics' – leaving aside the issue that statistical methods deal with probability, not certainty (Chapter 66), it is possible to analyse and present data in such a way that they support one chosen argument or hypothesis rather than another. Detecting a bias of this kind can be difficult, but the critical thinking skills involved are essential for all scientists (see e.g. Box 62.3).

other conclusions might be possible. It allows you to weigh up these possibilities and find a working hypothesis or explanation, but also to understand that your conclusions are essentially dynamic and might alter when new facts are known. Hypothesis-testing with statistics (Chapter 66) is an important adjunct to critical thinking because it demands the formulation of simple hypotheses and provides rational reasons for making conclusions.

Recognising fallacies in arguments is an important aspect of critical thinking. Philosophers and logicians recognise different forms of argument and many different fallacies in each form. Damer (2004) provides an overview of this wide-ranging and complex topic.

Interpreting data

Numerical data

Information presented in public, whether as a written publication or spoken presentation, is rarely in the same form as it was when first obtained. Chapters 50 and 61 deal with processes in which data are recorded, manipulated and transformed, while Chapter 65 describes the standard descriptive statistics used to 'encapsulate' large data sets. Chapter 64 covers some relevant mathematical techniques. Sampling (essentially, obtaining representative measurements) is at the heart of many observational and experimental approaches in biology (see Chapters 25, 30 and 31), and analysis of samples is a key component of hypothesis-testing statistics (Chapter 66). Understanding these topics and carrying out the associated study exercises will help you improve your ability to interpret numerical data.

Graphs

Frequently, understanding and analysis in science depends on your ability to interpret data presented in graphical form. Sometimes, graphs may mislead. This may be unwitting, as in an unconscious effort to favour a 'pet' hypothesis of the author. Graphs may be used to 'sell' a product, e.g. in advertising, or to favour a viewpoint as, perhaps, in politics. Experience in drawing and interpreting graphs will help you spot these flawed presentations, and understanding how graphs can be erroneously presented (Box 62.3) will help you avoid the same pitfalls.

Tables

Tables, especially large ones, can appear as a mass of numbers and thus be more daunting at first sight than graphs. In essence, however, most tables are simpler than most graphs. The construction of tables is dealt with in Chapter 63.

Analysing a graph – *this process can be split into six phases:*

1. *Considering the context and purpose of the graph.*
2. *Recognising the type of presentation and examining the axes.*
3. *Looking closely at the scale on each axis.*
4. *Examining the data presented (e.g. data points, symbols, curves).*
5. *Considering errors and statistics associated with the graph.*
6. *Reaching conclusions based on the above.*

These processes are amplified in Chapter 62

Analysing a table – *as with analysing a graph, this process can be split into six phases:*

1. *Considering the context and purpose.*
2. *Examining the sub-headings to see what information is contained in the rows and columns.*
3. *Considering the units used and checking any footnotes.*
4. *Comparing the data values across rows and/or down columns, looking for patterns, trends and unusual values.*
5. *Taking into account any statistics presented.*
6. *Reaching conclusions based on the above.*

Text reference

Damer, T.E. (2004) *Attacking Faulty Reasoning: A Practical Guide to Fallacy-Free Arguments*, 5th edn. Wadsworth, Belmont, California.

Sources for further study

Anon. *Critical Thinking*. Available: http://en.wikipedia.org/wiki/Critical_thinking
Last accessed: 09/04/07.
[A frequently updated review of critical thinking terminology and techniques, with additional links and references.]

Barnard, C.J., Gilbert, F.S. and MacGregor, P.K. (2001) *Asking Questions in Biology: Key Skills for Practical Assessments and Project Work*, 2nd edn. Prentice Hall, Harlow.

Smith, A. *Evaluation of Information Sources*. Available: http://www.vuw.ac.nz/staff/alastair_smith/evaln/evaln.htm
Last accessed: 09/04/07.
[Part of the Information Quality WWW Virtual Library.]

Van Gelder, T. *Critical Thinking on the Web*. Available: http://www.austhink.org/critical/
Last accessed: 09/04/07.
[A useful directory of Web resources on the topic of critical thinking, associated with *Austhink,* a group based in Melbourne, Australia, that specialises in complex reasoning and argument, including critical thinking research, training and consulting.]

Study exercises

9.1 **Distinguish between primary and secondary literature**. Based on their titles and any research you can do in your library, determine whether the following journals are primary or secondary sources.

   (a) *British Journal of Haematology*
   (b) *Proceedings of the National Academy of Sciences*
   (c) *Annual Review of Molecular Biology*
   (d) *Essays in Biochemistry*
   (e) *Microbiological Research*
   (f) *Trends in Biotechnology*
   (g) *The Lancet*
   (h) *Oecologia*
   (i) *Scientific American*
   (j) *Food Safety News*

9.2 **Consider a controversial issue from both sides**. Select a current biological topic being discussed in the newspapers or other media. Relatively controversial issues such as 'genetically modified crops' or 'stem cell research' would be good examples. Next, write out a statement that you might use for a motion in a debate, such as 'Genetic modification of crops is a good thing' or 'Creating chimeric organisms is justified if it results in medical advances'. The exercise consists of writing at least five points in support of either side of the argument, which you should organise in tabular form. If you can find more than five points, add these to your table, but for each point that you add to one side, you should add one to the other side.

9.3 **Analyse graphic presentations in the media**. Many newspapers provide graphic presentations related to current issues, and graphs are frequently used in television news reports. Practise critical thinking skills by determining whether the graphs presented are a fair representation of the facts.

Browser – A program to display web pages and other Internet resources.

FAQ – Frequently Asked Question; a file or Web page giving information on common queries, sometimes used as a file extension (.faq).

FTP – File Transfer Protocol; a mechanism of downloading files.

URL – Uniform Resource Locator; the 'address' for WWW resources.

Understanding the technology – you do not need to understand the workings of the Internet to use it – most of it is invisible to the user. To ensure you obtain the right facilities, you may need to know some jargon, such as terms for the speed of data transfer (megabits) and the nature of Internet addresses. Setting up a modem and/or local wireless network can be complex, but instructions are usually provided with the hardware. White and Downs (2005) and Gralla (2003) are useful texts if you wish to learn more about computing and the Internet.

Information and communication technology (ICT) is vital in the modern academic world and 'IT literacy' is a core skill for all bioscientists. This involves a wide range of computer-based skills, including:

- Accessing Web pages using a Web browser such as Microsoft Internet Explorer, Mozilla Firefox or Opera.
- Searching the Web for useful information and resources using a search engine such as Google, or a meta-search engine such as Dogpile.
- Finding what you need within online databases, such as library catalogues, or complex websites, such as your university's homepage.
- Downloading, storing and manipulating files.
- Communicating via the Internet.
- Using e-learning facilities effectively.
- Working with 'office'-type programs and other software (dealt with in detail in Chapters 11 and 12).

You will probably receive an introduction to your university's networked IT systems and will be required to follow rules and regulations that are important for the operation of these systems. Whatever your level of experience with PCs and the Internet, you should also follow the basic guidelines shown in Box 10.1. Reminding yourself of these from time to time will reduce your chances of losing data.

The Internet as a global resource

The Internet is a complex network of computer networks; it is loosely organised and no one group organises it or owns it. Instead, many private organisations, universities and government organisations fund and operate discrete parts of it.

The most popular application of the Internet is the WWW ('the Web'). It allows easy links to information and files which may be located on networked computers across the world. The WWW enables you to access millions of 'home pages' or 'websites' – the initial point of reference with many individuals, institutions, and companies. Besides text and images, these sites may contain 'hypertext links', highlighted words or phrases that take you to another location via a single mouse click.

You can gain access to the Internet either through a local area network (LAN) at your university, at most public libraries, at a commercial 'Internet cafe', or from home via a modem connected to a broadband or dial-up Internet service provider (e.g. Virgin Media, British Telecom or AOL).

 KEY POINT Most material on the Internet has not been subject to peer review or vetting. Information obtained from the WWW or posted on newsgroups may be inaccurate, biased or spoof; do not assume that everything you read is true or even legal.

Online communication

You will be allocated an email account by your university and should use this routinely for communicating with staff and fellow students, rather than using a personal account. You may be asked to use email to submit

Box 10.1 Important guidelines for using PCs and networks

Hardware

- Don't drink or smoke around the computer.
- Try not to turn the computer off more than is necessary.
- Never turn off the electricity supply to the machine while in use.
- Switch off the computer and monitor when not in use (saves energy and avoids dangers of 'hijacking').
- ⚠ Rest your eyes at frequent intervals if working for extended periods at a computer monitor. Consult Health and Safety Executive publications for up-to-date advice on working with display screens (http://www.hse.gov.uk/pubns/).
- Never try to reformat the hard disk without the help of an expert.

CDs and USB drives

- Protect CDs when not in use by keeping them in holders or boxes.
- Label USB (Universal Serial Bus) drives with your name and return details and consider adding these to a file stored on the drive.
- Try not to touch the surface of CDs, and if they need cleaning, do so carefully with a clean cloth, avoiding scratching. If floppy disks are used, keep these away from sources of magnetism (e.g. speakers).
- Keep disks and USB drives away from moisture, excess heat or cold.
- Don't use disks from others, unless you first check them for viruses.
- Don't insert or remove a disk or USB drive when it is operating (drive light on). Close all files before removing a USB drive and use the *Safely Remove Hardware* feature.
- Try not to leave a disk or USB drive in the drive when you switch the computer off.

File management

- Organise your files in an appropriate set of folders.
- Always use virus-checking programs on copied or imported files before running them.
- Make back-ups of all important files at frequent intervals (say, every ten minutes or half-hour), e.g. when using a word processor or spreadsheet.
- Periodically clear out redundant files.

Network rules

- Never attempt to 'hack' into other people's files.
- Do not give out any of your passwords to others. Change your password from time to time. Make sure it is not a common word, is longer than eight characters, and includes numerical characters and punctuation symbols, as well as upper- and lower-case letters.
- Never use network computers to access or provide financial or other personal information: spyware and Trojan programs may intercept your information.
- Never open email attachments without knowing where they came from; always virus-check attachments before opening.
- Remember to log out of the network when finished; others can access your files if you forget to log out.
- Be polite when sending email messages.
- Periodically reorganise your email folder(s). These rapidly become filled with acknowledgements and redundant messages that reduce server efficiency and take up your allocated filespace.
- Do not play games without approval – they can affect the operation of the system.
- If you are setting up your own network, e.g. in your flat, always install up-to-date firewall software, anti-spyware and anti-virus programs.

The Golden Rule – always make back-up copies of important files and store them well away from your working copies. Ensure that the same accident cannot happen to both copies.

Spam, junk mail and phishing – these should be relatively easy to identify, and should never be responded to or forwarded. Some may look 'official' and request personal or financial details (for example, they may pretend to come from e.g. your bank, and ask for account details). Never send these details by email or your identity may be used illegally.

work as an attachment, or you may be asked to use a 'digital drop-box' within the university's e-learning system (Box 10.2). When using email at university, follow these conventions, including etiquette, carefully:

- Check your email account regularly (daily). Your tutors may wish to send urgent messages to you in this way.
- Respond promptly to emails. Even if you are just acknowledging receipt, it is polite to indicate that you have received and understood a message.
- Be polite. Email messages can seem to be abrupt and impersonal. Take care to read your messages through before sending and if you are at all in doubt, do not send your message right away: reread at a later time and consider how others might view what you say.

Box 10.2 Getting to grips with e-learning

Some key aspects of tackling e-learning are outlined below.

1. **Develop your basic IT skills, if required.** e-learning requires only basic IT skills, such as: use of keyboard and mouse; word processing; file management; browsing and searching. If you feel weak on any of these, seek out additional courses offered by the IT administration or your department.

2. **Visit your e-learning modules regularly.** You should try to get into a routine of doing this on a daily basis at a time that suits you. Staff will present up-to-date information (e.g. lecture-room changes) via the 'announcements' section, may post information about assessments, or links to the assessments themselves, and you may wish to provide feedback or look at discussion boards and their threads.

3. **Participate.** e-learning requires an active approach.
   - At the start of each new course, spend some time getting to know what's been provided online to support your learning. As well as valuable resources, this may include crucial information such as learning objectives (pp. 22, 24), dates of submission for coursework and weighting of marks for different elements of the course.
   - If you are allowed to download lecture notes (e.g. in the form of PowerPoint presentations), do not think that simply reading through these will be an adequate substitute for attending lectures and making further notes (see page 17).
   - Do not be tempted to 'lurk' on discussion boards: take part. Ask questions; start new threads; answer points raised by others if you can.

- Try to gain as much as you can from formative online assessments (p. 24). If these include feedback on your answers, make sure you learn from this and if you do not understand it, consult your tutors.
- Learn from the critical descriptions that your lecturers provide of linked websites. These pointers may help you to evaluate such resources for yourself in future (pp. 53–55).
- Don't think that you will automatically assimilate information and concepts, just because you are viewing them online. The same principles apply as with printed media: you must apply active learning methods (p. 26).
- Help your lecturers by providing constructive feedback when they ask for it. You may find this easier to do when using the computer interface, and it may be more convenient than hurriedly filling out a feedback sheet at the end of a session.

4. **Organise files and Web links.** Take the time to create a meaningful folder- and file-naming system for downloaded material in tandem with your own coursework files and set up folders on your browser for bookmarked websites (*Favorites* in Internet Explorer).

5. **Take care when submitting coursework.** Make sure you keep a back-up of any file you email or submit online and check the version you are sending carefully. Follow instructions carefully, for example regarding file type, or how to use your system's 'digital drop-box'.

- Consider content carefully. Only send what you would be happy to hear being read out loud to classmates or family.
- When communicating with tutors, take care with language and names. Slang phrases and text message shorthand are unlikely to be understood. Overfamiliarity does not go down well.
- Use email for academic purposes – this includes discussing coursework with classmates, but not forwarding off-colour jokes, potentially offensive images, links to offensive websites, etc. In fact, doing so may break regulations and result in disciplinary action.
- Beware of spam, junk and 'phishing' via email.

Similar rules apply to discussion boards.

Newsgroups – these can be useful for getting answers to a specific problem: just post a query to the appropriate group and wait for someone to reply. Bear in mind that this may be the view of an individual person.

The Usenet Newsgroup service is an electronic discussion facility, and there are thousands of newsgroups representing different interests and topics. Any user can contribute to the discussion within a topic by posting their own message; it is like email, but without privacy, since your message becomes available to all other subscribers. To access a newsgroup, your system must be running, or have access to, a newsgroup server that has subscribed to the newsgroup of interest. Obtain a list of newsgroups available on your system from the IT administration service and search it for those of interest, then join them. Contact your network administrator if you wish to propose the addition of a specific newsgroup, but expect a large amount of information to be produced.

Internet tools

The specific programs you will use for accessing the Internet will depend on what has been installed locally, on the network you are using, and on your Internet service provider. The best way to learn the features of the programs is to try them out, making full use of whatever help services are available.

e-learning systems

Most university departments present their courses through a mixture of face-to-face sessions (e.g. lectures, tutorials, practicals) and online resources (e.g. lecture notes, websites, discussion boards, computerised tests and assessments). This constitutes 'blended learning' on your part, with the online component also being known as e-learning.

The e-learning element is usually delivered through an online module within a virtual learning environment (e.g. Blackboard, WebCT, Moodle). It is important not to neglect the e-learning aspects of your course just because it may not be as rigidly timetabled as your face-to-face sessions. This flexibility is to your advantage, as you can work when it suits you, but it requires discipline on your part. Box 10.2 provides tips for making the most of the e-learning components of your courses.

Internet browsers

These are software programs that interact with remote server computers around the world to carry out the tasks of requesting, retrieving and displaying the information you require. Many different browsers exist, but the most popular are Internet Explorer, Mozilla Firefox and Opera. These three browsers dominate the market and have plug-ins and add-on programs available that allow, for example, video sequences to be seen online. Many browsers incorporate email and newsgroup functions. The standard functions of browsers include:

- accessing Web documents;
- following links to other documents;
- printing the current document;
- maintaining a history of visited URLs (including 'bookmarks' for key sites);
- searching for a term in a document;
- viewing images and image maps.

Browsers provide access to millions of websites. Certain sites specialise in providing catalogued links to other sites; these are known as portals and can be of enormous help when searching within a particular area of interest. Your university's library website will almost certainly provide a useful portal to catalogues and search services, often arranged by subject area, and this is often the first port of call for electronic resources; get to know your way around this part of the website as early as possible during your course.

Examples Common domains and sub-domains include:
.ac   academic
.com   commercial
.co   commercial
.edu   education (USA mainly)
.gov   government (USA and UK)
.mil   military (USA only)
.net   Internet-based companies
.org   organisation
.uk   United Kingdom

When using a Web browser program to get to a particular page of information on the Web, all you require is the location of that page, i.e. the URL (uniform resource locator). Most Web-page URLs take the form http:// or https://, followed by the various terms (domains and sub-domains) that direct the system to the appropriate site. If you don't have a specific URL in mind but wish to explore appropriate sites, you will need to use a search tool with the browser.

WWW search tools

With the proliferation of information on the Web, one of the main problems is finding the exact information you require. There are a variety of information services that you can use to filter the material on the network. These include:

- search engines (Boxes 10.3 and 10.4);
- meta-search engines;
- subject directories;
- subject gateways (portals).

Box 10.3   Useful tips for using search engines

- **Keywords should be chosen with care.** Try to make them as specific as possible, e.g. search for 'archaebacteria', rather than 'bacteria' or 'microorganisms'.

- **Most search engines are case-insensitive.** Thus 'Nobel Prize' will return the same number of hits as 'nobel prize'. If in doubt, use lower case throughout.

- **Putting keyword phrases in double quotes (e.g. "dog violet") will result in a search for sites with the phrase as a whole** rather than sites with both (all) parts of the phrase as separate words (i.e. 'dog' and 'violet' at different places within a site). This feature allows you to include common words normally excluded in the search, such as 'the'.

- **Use multiple words/phrases plus similar words to improve your search,** for example 'terrestrial crustacean "land crab" Australia'. If you can, use scientific terms, as you are likely to find more relevant sites, e.g. search for the name of a particular species such as '*Coenobita compressus*'.

- **Adding words preceded with + or – will add or exclude sites with that word present** (e.g. 'blood diseases –leukaemia' will search for all blood diseases excluding leukaemia). This feature can also be used to include common words normally excluded by the search engine.

- **Check that your search terms have the correct spelling**, otherwise you may only find sites with the same mis-spelled word. In some cases, the search engine may prompt you with an alternative

(correct) spelling. If a word has an alternative US spelling (e.g. color, hemoglobin), then a search may only find hits from sites that use the spelling you specify.

- **Boolean operators (AND, OR, NOT) can be used with some search engines to specify combinations of keywords** to more precisely filter the sites identified (e.g. 'plant NOT engineering' will avoid sites about engineering plant and focus on botanical topics).

- **Some search engines allow 'wildcards' to be introduced with the symbol** *. For example, this will allow you to specify the root of a word and include all possible endings, as with anthropomorph*, which would find anthropomorphic, anthropomorphism, etc. If the search engine does not allow wildcards, then you will need to be especially careful with the key words used, including all possible words of relevance.

- **Numbers can be surprisingly useful in search engines.** For example, typing in EC 1.1.1.1 will find sites concerned with alcohol dehydrogenase, as this is its code number. If you know the phone number for a person, institute or company or the ISBN of the book, this can often help you find relevant pages quickly.

- **If you arrive at a large site and cannot find the point at which your searched word or phrase appears, press Control+F together** and a 'local' search window will appear, allowing you to find the point(s) where it is mentioned.

Box 10.4 Getting the most from Google searches

Google (http://www.google.com) has become the search engine of choice for millions of people, due to its simplicity and effectiveness. However, you may be able to improve your searches by understanding its default settings and how they can be changed.

- **Download the Google toolbar to your browser.** This is available from the Google homepage and will give you quick access to the Google search facility.

- **Understand how standard operators are used.** For combinations of keywords Google uses the minus operator '−' instead of NOT (exclude) and '+' instead of AND (include). Since Google usually ignores small words ('stop words' such as *in* or *the*), use '+' to include them in a search. Where no operator is specified, Google assumes that you are looking for both terms (i.e. '+' is default). If you want to search for alternative words, you can use 'OR' (e.g. *sulphur OR sulfur*). Google does not allow brackets and also ignores most punctuation marks.

- **While wildcard truncation of words using '*' is not allowed, you can use '*' to replace a whole word (or number).** For example, if you type the phrase *"a virus is approximately * nanometres"* your results will give you results for web pages where the wildcard is replaced by a number.

- **Search for exact wording.** By placing text in double inverted commas (''), you can ensure that only websites with this exact phrasing will appear at the head of your search results.

- **Search within your results to improve the outcome.** If your first search has produced a large number of results, use the *Search within results* option near the bottom of each page to type in a further word or phrase.

- **Search for words within the title of a web page.** Use the command *inititle:* to find a webpage, for example *intitle: "tissue culture"* returns web pages with this phrase in the title (note that phrases must always be in double speech marks, not single quotes).

- **Search within a website.** Use the *site:* command to locate words/phrases on a specific website, for example *site:unicef.org dysentery* returns only those results for this disease on the UNICEF website (unicef.org). Pressing *Control+F* when visiting a web page will give you a pop-up search window.

- **Locate definitions, synonyms and spellings.** The operator *define:* enables you to find the meaning of a word. If you are unsure as to the spelling of a word, try each possibility: Google will usually return more results for the correct spelling and will often also prompt you with the correct spelling (*Did you mean...?*).

- **Find similar Web pages.** Simply click the *Similar pages* option at the end of a Google search result to list other sites (note that these sites will not necessarily include the term(s) searched for).

- **If a Web link is unavailable, try the cached (stored) page.** Clicking on *Cached* at the end of a particular result should take you to the stored page, with the additional useful feature that the search term(s) will be highlighted.

- **Use the calculator functions.** Simply enter a calculation and press *Enter* to display the result, for example '*10+(2*4)*' returns 18. The calculator function can also carry out simple interconversion of units, e.g. '*2 feet 6 inches in metres*' returns 0.762 (see Box 26.1 for interconversion factors between SI and non-SI units).

- **Try out the advanced search features.** In addition to the standard operators these include the ability to specify the number of results per page (e.g. 50, to reduce the use of the *next* button), language (e.g. English), file format (e.g. for PDF files), recently updated Web pages (e.g. past three months), usage (e.g. free to use/share).

- **Find non-text material.** These include images, video and maps – always check that any material you use is not subject to copyright limitations (p. 52).

- **Use Google alerts to keep up to date.** This function (http://www.google.co.uk/alerts) enables you to received regular updated searches by email.

- **Use Google Scholar to find articles and papers.** Go to http://scholar.google.com/ and type in either the general topic or specific details for a particular article, e.g. author names or words from the title. Results show titles/authors of articles, with links to either the full article, abstract or citation. A useful feature is the *Cited by ...* link, taking you to those papers that have cited the article in their bibliography and enabling you to carry out forward citation searching to locate more recent papers. Also try out the advanced scholar search features to limit your search to a particular author, journal, date or subject area. However, you should note that Google Scholar provides only a basic search facility to easily accessible articles and should not be viewed as a replacement for your library's electronic journal holdings and searching software. For example, if you find the title of a paper *via* Google Scholar you may be able to locate the electronic version through your own library's databases, or request it via inter-library loan (p. 46). Another significant limitation is that older (more cited) references are typically listed first.

- **Use Google Earth to explore locations.** This allows you to zoom in on satellite images to find locations.

Search engines such as Google (http://www.google.com/), Altavista (http://uk.altavista.com/) and Lycos (http://www.lycos.com/) are tools designed to search, gather, index, and classify Web-based information. Searching is usually by keyword(s), although specific phrases can be defined. Many search engines offer advanced searching tools such as the use of Boolean operators to specify combinations of keywords to more precisely filter the sites. Box 10.3 provides tips for refining keyword searches while Box 10.4 provides tips for enhancing searches with Google.

It is important to realise that each search engine will cover at most about 40 per cent of the available sites; if you want to carry out an exhaustive search it is necessary to use several to cover as much of the Web as possible. Meta-search engines make this easier. These operate by combining collections of search engines. Examples include Mamma (http://www.mamma.com/), Dogpile (http://www.dogpile.com/index .gsp/) and Metacrawler (http://www.metacrawler.com/index.html/).

Some useful approaches to searching include the following:

- For a comprehensive search, use a variety of tools including search engines, meta-search engines and portals or directories.
- For a complex, finely specified search, employ Boolean operators and other tools to refine your keywords as fully as possible (Box 10.3). Some search engines allow you to include and exclude terms or restrict by date.
- Use 'cascading' searching when available – this is searching within the results of a previous search.
- Use advanced search facilities to limit your search, where possible, to the type of medium you are looking for (e.g. graphics, video), language, sites in a specific country (e.g. UK) or to a subject area (e.g. news only).

However well defined your search is, you will still need to evaluate the information obtained. Chapter 9 covers general aspects of this topic, while Box 10.5 provides specific advice on assessing the quality of information provided on websites.

Directories

A directory is a list of Web resources organised by subject. It can usually be browsed and may or may not have a search facility. Directories often contain better-quality information than the lists produced by search engines, as they have been evaluated, often by subject specialists or librarians. The BUBL information service life sciences directory of links, at: http://bubl.ac.uk/link/lif.html/ is a good example.

Using the Internet as a resource

A common way of finding information on the Web is by browsing or 'surfing'. However, this can be time-consuming; try to restrict yourself to sites known to be relevant to the topic of interest. Some of the most useful sites are those that provide hypertext links to other locations. Some other

'Dissecting' a Web address – if a URL is specified, you can often find out more about a site by progressively deleting sections of the address from the right-hand side. This will often take you to 'higher levels' of the site, or to the home page of the organisation or company involved.

Downloading files from the Internet and emails – read-only files are often available as 'pdf' files that can be viewed by Adobe reader software (available free from http://www.adobe.com), while other files may be presented as attachments to emails or as links from Web pages that can be opened by suitable software (e.g. Microsoft Word or 'paint' programs like Paint Shop Pro). Take great care in the latter cases as the transfer of files can result in the transfer of associated viruses. Always check new files for viruses (especially .exe files) before running them, and make sure your virus-detecting software is kept up to date.

Box 10.5   How to evaluate information on the World Wide Web

It is often said that 'you can find anything on the Web'. The two main disadvantages of this are, firstly, that you may need to sift through many sources before you find what you are looking for and, secondly, that the sources you find will vary in their quality and validity. *It is important to realise that evaluating sources is a key aspect of using the Internet for academic purposes, and one that you will need to develop during the course of your studies.* The ease with which you can 'point and click' to reach various sources should not make you complacent about evaluating their information content. The following questions can help you to assess the quality of a website – the more times you can answer 'yes', the more credible the source is likely to be, and vice versa.

Authority
- Is the author identified?
- Are the author's qualifications or credentials given?
- Is the owner, publisher or sponsoring organisation identified?
- Is an address given (postal and/or email)?

It is sometimes possible to get information on authority from the site's metadata (try the 'View' 'Source' option in Internet Explorer, or look at the URL to see if it gives any clues as to the organisation, e.g. does the domain name end in .ac, .edu, .gov or .org, rather than .co or .com).

Content
- Is there any evidence that the information has been peer-reviewed (p. 53), edited or otherwise validated, or is it based on such sources?
- Is the information factual or based on personal opinions?
- Is the factual data original (primary) or derived from other sources (secondary)?

- Are the sources of specific factual information detailed in full (p. 54)
- Is there any indication that the information is up to date, or that the site has been recently updated?
- What is the purpose of the site and who is it aimed at?
- Is the content relevant to the question you are trying to answer?
- Is there any evidence of a potential conflict of interest, or bias? (Is the information comprehensive and balanced, or narrowly focused?)
- Did you find the information via a subject-specific website (e.g. a bioscience gateway such as BIOME), or through a more general source, such as a search engine (e.g. Google)?

The above questions are similar to those that you would use in assessing the value of a printed resource (pp. 53–4), and similar criteria should be applied to Web-based information. You should be especially wary of sites containing unattributed factual information or data whose primary source is not given.

Presentation
- What is your overall impression of how well the site has been put together?
- Are there many grammatical or spelling mistakes?
- Are there links to other websites, to support statements and factual information?

The care with which a site has been constructed can give you an indication of the credibility of the author/organisation. However, while a poorly presented site may cause you to question the credibility of the information, the reverse is not always necessarily true: don't be taken in by a slick, well-presented website – authority and content are *always* more important than presentation.

The impermanence of the Web – the temporary nature of much of the material on the Web is a disadvantage for academic purposes because it may change or even disappear after you have cited it. You may also find it difficult or impossible to find out who authored the material (p. 54). A case in point are wikis, such as Wikipedia (www.wikipedia.org). This online encyclopedia has many potential authors and the content may change rapidly as a result of new submissions or edits; nevertheless, it can be a useful resource for up-to-date general information about a wide range of topics, though it is not necessarily regarded as the best approach for researching assignments.

resources you can use on the WWW are:

- Libraries, publishers and commercial organisations. Your university library is likely to subscribe to one or more databases providing access to scientific articles; these include BIDS (Bath Information and Database Services (http://www.bids.ac.uk), ISI Web of Science (http://wos.mimas.ac.uk/), and Science Direct (http://www.sciencedirect.com/). A password is usually required, especially for off-campus use; consult your library staff for further details. Some scientific database sites give free access, without subscription or password; these include

Using traditional sources – remember that using the Internet to find information is not a substitute for visiting your university library. Internet resources complement rather than replace CD-ROM and more traditional printed sources.

Remembering useful websites – create a 'bookmark' (= add a 'favourite') for the ones you find of value, to make revisiting easy. This can be done from the menu of your browser program. Make a copy of your bookmark file occasionally, to avoid loss of this information.

Examples Selected websites of biological interest:

American Society of Plant Biologists: http://www.aspb.org/resourcelinks/

Thomson Scientific Free Resources http://scientific.thomson.com/free/

Society for General Microbiology Web resources: http://www.sgm.ac.uk/links/

The Smithsonian (USA): http://www.si.edu/

HERO World-wide Library Resources: http://www.hero.ac.uk/uk/reference_and_subject_resources/resources/worldwide_library_resources3796.cfm/

World Wide Web Virtual Library (Biosciences): http://vlib.org/Biosciences

Note that URLs may change – make a keyword search using a search engine to find a particular site if the URL information you have does not lead you to an active page.

National Center for Biotechnology Information (USA) (http://www.ncbi.nlm.nih.gov/) and the Highwire Press (http://highwire.stanford.edu/). Others allow free searching, but require payment for certain articles, e.g. The Scientific World (http://www.thescientificworld.com/) and Infotrieve (http://www4.infotrieve.com/). The Natural History Book Service (http://www.nhbs.co.uk) provides useful information on relatively inaccessible literature such as government reports. Publishers such as Pearson and booksellers such as Amazon.com provide online catalogues and e-commerce sites that can be useful sources of information (see http://vig.pearsoned.co.uk/ and http://www.amazon.com).

- Online journals and e-books. A number of traditional journals have websites. You can keep up to date by visiting the websites of *Nature* (http://www.nature.com/), *New Scientist* (http://www.newscientist.com/) and *Scientific American* (http://www.sciam.com/), or the Elsevier Science Direct website (http://www.sciencedirect.com/science/journals). Some scientific societies make their journals and other publications available via their websites, e.g. the American Society for Microbiology, at: http://www.asm.org/. Journals solely published in electronic format are also available (e.g. *Molecular Vision*, http://www.molvis.org/molvis/) but some require a subscription password for access; check whether your institute is a subscriber.

- Data and images. Archives of text material, video clips and photographs can be accessed, and much of the material is readily available. The HEA Centre for Biosciences Image Bank (http://www.bioscience.heacademy.ac.uk/imagebank) is a good example. When downloading such material, you should (i) check that you are not breaching copyright and (ii) avoid potential plagiarism by giving a full citation of the source, if you use such images in an assignment (see p. 52).

- Biological institutions. Many museums, botanic gardens, zoos, culture collections, scientific societies and other biological institutions around the world are now online. Use their sites to obtain specific information about collections, resources, etc. They frequently provide lists of other relevant sites or topics, e.g. the Natural History Museum (http://www.nhm.ac.uk/) or the Society for General Microbiology (http://www.sgm.ac.uk/).

- Databases. In addition to those covering the scientific literature, others focus on specific topics (e.g. UK Countryside and Nature Conservation, http://www.naturenet.net/index.php, or academic employment, http://jobs.ac.uk/). Box 54.2 (p. 339) provides further guidance on using databases in the important field of bioinformatics.

Text references

Gralla, P. (2003) *How the Internet Works*, 7th edn. Pearson, Harlow.

White, R. and Downs, T. (2005) *How Computers Work*, 8th edn. Pearson, Harlow.

Sources for further study

Anon. *The Essentials of Google Search*. Available: http://www.google.co.uk/intl/en/help/basics.html
Last accessed 09/04/07.
[Advice for using the Google search engine.]

Anon. *Using the Internet*. Available: http://www.sofweb.vic.edu.au/internet/research.htm
Last accessed 09/04/07.
[Provided by the State of Victoria, Department of Education and Training.]

Brandt, D.S. *Why We Need to Evaluate What We Find on the Internet*. Available: http://www.lib.purdue.edu/research/techman/eval.html
Last accessed 09/04/07.

Dussart, G. (2002) *Biosciences on the Internet*. Wiley, Chichester.

Grassian, E. *Thinking Critically about World Wide Web Resources*. Available:
http://www.library.ucla.edu/libraries/college/help/critical/index.htm
Last accessed 09/04/07.

Isaacs, T. and Isaacs, M. (2000) *Internet Users Guide to Network Resource Tools*, 2000 edn. Addison-Wesley, Harlow.

Winship, I. and McNab, A. (2000) *Students' Guide to the Internet 2000–2001*. Library Association, London.

Study exercises

10.1 Explore the resources of the WWW using a search engine. Pick a search engine (p. 62) and use it to find the answers to the following questions.
  (a) Who is Norman Borlaug, and what should he be widely famous for? What prize did he win and when?
  (b) What is the common name for the species *Conolophus subcristatus*?
  (c) What is the postal address of the Information Office of the Smithsonian Institution?

10.2 Compare results from a variety of search engines. First, think of an appropriate biological keyword or phrase (e.g. a species name) and enter this into several search engines. Make sure that you include meta-search engines such as Dogpile. Compare the outcomes to reveal the strengths and weaknesses of the individual search engines. Work with a colleague to compare different searches on a quantitative (i.e. number of hits) and qualitative (quality of hits) basis.

10.3 Organise your bookmarks. Enter the *'Organize favorites'* menu for your preferred browser and create folders with appropriate headings. Move existing bookmarks to these folders and save any new ones appropriately. Doing this will help you find bookmarks more easily, rather than searching through long lists.

11 Using spreadsheets

The spreadsheet is one of the most powerful and flexible computer applications. It can be described as the electronic equivalent of a paper-based longhand calculation, where the sums are carried out automatically. Spreadsheets provide a dynamic method of storing, manipulating and analysing data sets. Advantages of spreadsheets include:

- Ease and convenience – especially when complex calculations are repeated on different sets of data.
- Accuracy – providing the entry data and cell formulae are correct, the result will be free of calculation errors.
- Improved presentation – data can be produced in graphical or tabular form to a very high quality.
- Integration with other programs – graphs and tables can be exported to other compatible programs, such as a word processor in the same office suite.
- Useful tools – advanced features include hypothesis-testing statistics, database features and macros.

Spreadsheets can be used to:

- manipulate raw data by removing the drudgery of repeated calculations, allowing easy transformation of data and calculation of statistics;
- graph out your data rapidly to get an instant evaluation of results. Printouts can be used in practical and project reports;
- carry out statistical analysis by built-in procedures or by allowing construction of formulae for specific tasks;
- model 'what if' situations where the consequences of changes in data can be seen and evaluated;

The spreadsheet (Fig. 11.1) is divided into rows (identified by numbers) and columns (identified by alphabetic characters). Each individual

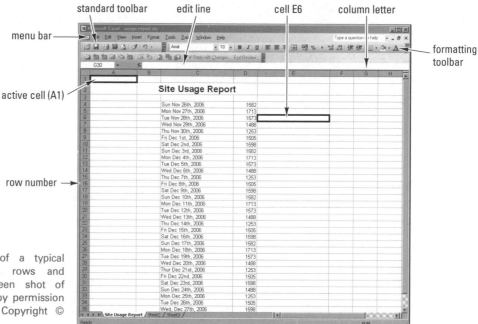

Fig. 11.1 The appearance of a typical spreadsheet, showing cells, rows and columns, toolbars, etc. Screen shot of Microsoft Excel XP reprinted by permission from Microsoft Corporation. Copyright © 2007 Microsoft Corporation.

combination of column and row forms a cell that can contain either a data item, a formula, or a piece of text. Formulae can include scientific and/or statistical functions and/or a reference to other cells or groups of cells (often called a range). Complex systems of data input and analysis can be constructed. The analysis, in part or complete, can be printed out. New data can be added at any time and the sheet will recalculate automatically. The power a spreadsheet offers is directly related to your ability to create arrays of formulae (models) that are accurate and templates that are easy to use.

Data entry

Spreadsheets have built-in commands that allow you to control the layout of data in the cells (see Fig. 11.2). These include number format, the number of decimal places to be shown (the spreadsheet always calculates using eight or more places), the cell width and the location of the entry within the cell (left, right or centre). An auto-entry facility assists greatly in entering large amounts of data by moving the entry cursor either vertically or horizontally as data is entered. Recalculation default is usually automatic so that when a new data value is entered the entire sheet is recalculated immediately.

The parts of a spreadsheet

Labels

These should be used to identify parts of the spreadsheet – for example, stating what data are contained in a particular column or indicating that a cell's contents represent the end point of a calculation. It may be useful to use the *Format > Cells > Border* and *Format > Cells > Patterns* functions to delimit numerical sections of your spreadsheet. Note that spreadsheet programs have been designed to make assumptions about the nature of data entry being made. If the first character is a number, then the entry is

Data output from analytical instruments – *many devices provide output in spreadsheet-compatible form (e.g. a 'comma delimited' file). Once you have uploaded the information into a spreadsheet, you can manipulate, analyse and present it according to your needs. Consult instrument manuals and the spreadsheet help function for details.*

Using hidden (or zero-width) columns – *these are useful for storing intermediate calculations that you do not wish to be displayed on the screen or printout.*

(a)

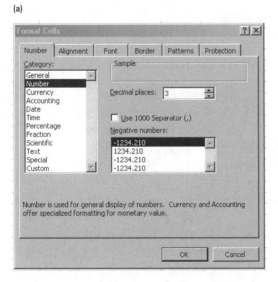

(b)

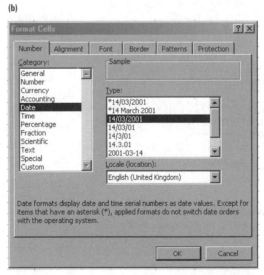

Fig 11.2 Example of cell-formatting options within the Microsoft Excel spreadsheet. These menus are accessed *via* the *Format > Cell* option and would apply to all of a range of selected cells. (a) Use of the number-formatting option to specify that data will be presented to three decimal places (the underlying data will be held to greater accuracy). (b) Use of the date-formatting option to specify that dates will be presented in day/month/year format (spreadsheet dates are stored numerically and converted to appropriate formats. This allows a period between two dates to be calculated more easily.)

treated as numerical data; if it is a letter, then it is treated as a text entry; and if it is a specific symbol ('=' in Microsoft Excel), then what follows is a formula. If you wish to enter text that starts with a number, then you must type a designated character to show this (a single quote mark in Microsoft Excel).

Numbers

You can also enter numbers (values) in cells for use in calculations. Many programs let you enter numbers in more than one way and you must decide which method you prefer. The way you enter the number does not affect the way it is displayed on the screen as this is controlled by the cell format at the point of entry. There are usually special ways to enter data for percentages, currency and scientific notation for very large and small numbers.

Formulae

These are the 'power tools' of the spreadsheet because they do the calculations. A cell can be referred to by its alphanumeric code, e.g. A5 (column A, row 5) and the value contained in that cell manipulated within a formula, e.g. = (A5 + 10) or (A5 + B22) in another cell. Formulae can include pre-programmed functions that can refer to a cell, so that if the value of that cell is changed, so is the result of the formula calculation. They may also include branching options through the use of logical operators (e.g. IF, TRUE, FALSE, OR, etc.).

Functions

A variety of functions is usually offered, but only mathematical and statistical functions will be considered here.

Mathematical functions

Spreadsheets offer a wide range of functions, including trigonometrical functions, angle functions, logarithms and random-number functions. Functions are invaluable for transforming sets of data rapidly and can be used in formulae required for more complex analyses. Spreadsheets work with an order of preference of the operators in much the same way as a standard calculator and this must always be taken into account when operators are used in formulae. They also require a very precise syntax – the program should warn you if you break this.

Statistical functions

Modern spreadsheets incorporate many sophisticated statistical functions, and if these are not appropriate, the spreadsheet can be used to facilitate the calculations required for most of the statistical tests found in textbooks. The descriptive statistics normally available include:

- the sum of all data present in a column, row or block;
- the minimum and maximum of a defined range of cells;
- counts of cells – a useful operation if you have an unknown or variable number of data values;
- averages and other statistics describing location;
- standard deviations and other statistics describing dispersion.

Operators and brackets in spreadsheets – the standard mathematical operators ÷ and × are usually replaced by / and * respectively, while ^ signifies 'to the power'. In complex formulae, brackets should be used to separate the elements, otherwise the results may not be what you expect. For example, Excel will calculate =A1*B1/C1–D1 differently from (A1*B1)/(C1–D1).

Definition

Function – a pre-programmed code for the transformation of values (mathematical or statistical functions) or selection of text characters (string functions).

Example = sin(A5) is an example of a function in Excel. If you write this in a cell, the spreadsheet will calculate the sine of the number in cell A5 (assuming it to be an angle in radians) and write it in the cell. Different programs may use a slightly different syntax.

Empty cells – note that these may be given the value 0 by the spreadsheet for certain functions. This may cause errors e.g. by rendering a minimum value inappropriate. Also, an 'error return' may result for certain functions if the cell content is zero.

Statistical calculations – make sure you understand whether any functions you employ are for populations or samples (see p. 420).

Using text functions – these allow you to manipulate text within your spreadsheet and include functions such as 'search and replace' and alphabetical or numerical 'sort'.

A useful function where you have large numbers of data allows you to create frequency distributions using predefined class intervals.

The hypothesis-testing statistical functions are usually reasonably powerful (e.g. *t*-test, ANOVA, regressions) and they often return the *probability* (P) of obtaining the test statistic when the null hypothesis (p. 426) is true (where $0 < P < 1$), so there may be no need to refer to statistical tables. Again, check on the effects of including empty cells.

Database functions

Many spreadsheets can be used as simple databases and offer a range of functions to support this, including filtering and sorting options. The rows and columns of the spreadsheet are used as the fields and records of the database (see Chapter 12). For many biological purposes, this form of database is perfectly adequate and should be seriously considered before using a full-feature database product.

Copying

All programs provide a means of copying (replicating) formulae or cell contents when required, and this is a very useful feature. This is usually accomplished by 'dragging' a cell's contents to a new range using the mouse. When copying, references to cells may be either relative, changing with the row/column as they are copied, or absolute, remaining a fixed cell reference and not changing as the formulae are copied (Fig. 11.3).

(a)

	Cell	Formula	
Original → cell	A1	=B1+C1	← Original formula
Copied cells	A2	=B2+C2	Copied formulae (relative)
	A3	=B3+C3	
↓	A4	=B4+C4	

(b)

	Cell	Formula	
Original → cell	A1	=B1/$C$1	← Original formula
Copied cells	A2	=B2/$C$1	Copied formulae (mixed relative and absolute)
	A3	=B3/$C$1	
↓	A4	=B4/$C$1	

Fig. 11.3 Illustration of relative (a) and absolute (b) copying. In Excel, the $ sign before and after the column letter makes the cell reference absolute, as shown in (b).

 KEY POINT *The distinction between relative and absolute cell references is very important and must be understood; it provides one of the most common forms of error when copying formulae.*

In Excel, copying is normally *relative* and if you wish a cell reference to be *absolute* when copied, this is done by putting a dollar ($) sign before and after the column reference letter, e.g. $C$56.

Naming blocks

When a group of cells (a block) is carrying out a particular function, it is often easier to give the block a name that can then be used in all formulae referring to that block. This powerful feature also allows the spreadsheet to be more readable.

Spreadsheet templates

A template is a preconstructed spreadsheet containing the formulae required for repeated data analysis. Data are added when they become available, and results are available as soon as the last item is entered. To create a template, the sequence of operations is:

1. Determine what information/statistics you want to produce.
2. Identify the variables you will need to use, both for original data that will be entered and for any intermediate calculations that might be required.

Templates – these should contain:
- a data input section
- data transformation and/or calculation sections
- a results section, which can include graphics
- text in the form of headings and annotations
- a summary section.

3. Set up areas of the spreadsheet for data entry, calculation of intermediate values (statistical values such as sums of squares, etc.), calculation of final parameters/statistics and, if necessary, a summary area.

4. Establish the format of the numeric data if this is different from the default values. This can be done globally (affecting the entire spreadsheet) or locally (affecting only a specified part of the spreadsheet).

5. Establish the column widths required for the various activities.

6. Add text (labels) to identify input, intermediate formulae and output cells. This is valuable in error-tracking and when carrying out further development work. Text can be entered in designated cells, or cells can be annotated using the 'comments' feature (*Insert > Comment*).

7. Enter a test set of values to use during formula entry: use a fully worked example to check that formulae are working correctly.

8. Enter the formulae required to make all the calculations, both intermediate and final. Check that results are correct using the test data.

The spreadsheet is then ready for use. Delete all of the test-data values and you have created your template. Save the template to a disk and it is then available for repeated operations.

Constructing a spreadsheet – start with a simple model and extend it gradually, checking for correct operation as you go.

Graphics display

Most spreadsheets now offer a wide range of graphics facilities that are easy to use, and this represents an ideal way to examine your data sets rapidly and comprehensively (Chapter 61). The quality of the final graphics output (to a printer) is variable but is usually perfectly sufficient for data exploration and analysis. Many of the options are business graphics styles but there are usually histogram, bar charts, $X - Y$ plotting, line and area graphics options available. Note that some spreadsheet graphics may not come up to the standards expected for the formal presentation of scientific data, unless you manipulate the initial output appropriately (see Box 62.2, p. 391).

Printing spreadsheets

This is usually a straightforward menu-controlled procedure, made difficult only by the fact that your spreadsheet may be too big to fit on one piece of paper. Try to develop an area of the sheet that contains only the data that you will be printing, e.g. perhaps a summary area. Remember that columns can usually be hidden for printing purposes and you can control whether the printout is in portrait or landscape mode, and for continuous paper or single sheets (depending on printer capabilities). Use a screen-preview option, if available, to check your layout before printing. A 'print to fit' option is also available in some programs, making the output fit the page dimensions.

Source for further study

Harvey, G. (2003) *Excel 2003 for Dummies*. Wiley, New York.

Hart-Davies, G. (2003) *How to Do Everything with Microsoft Office Excel 2003*. McGraw-Hill Education, Berkeley, California.

(And similar texts for other release versions.)

Study exercises

The instructions and tips for these problems assume that you have Excel (97 or later) available. If not, they should be readily modified for most advanced spreadsheet programs. If you have problems with any of the tasks, consult Box 62.2 or try using the program's *Help* facility.

11.1 Create a spreadsheet and graph (introductory).
(i) Copy the information in the table below into a spreadsheet. Name and save the spreadsheet file appropriately.
(ii) From the copied information, create a pie chart using the *Chart Wizard* function.
(iii) Adjust the colours selected so the chart will print out in black and white. Save the final version of your spreadsheet. Print the chart out directly from Excel.

Visits of birds to feeding area (% daily visits)

Species	% of visits
Blackbird	23
Song thrush	9
Sparrow	17
Starling	45
Other	6
Total	100

11.2 Create a spreadsheet and graph (advanced).
(i) Copy the data in the table above right into a spreadsheet. Name and save the file appropriately.
(ii) Use the spreadsheet and chart-making facilities to explore by eye which of the

Time course of cell population growth

Time (hours)	Cell count (per unit volume)
1	2.00
2	6.00
6	50.0
12	6.00×10^3
18	2.20×10^5
24	1.65×10^7

following transformations (see pp. 384–5) would result in the best linear fit for these data: reciprocal, square root, cube root or log.
(iii) Add a linear trend-line to the chart for the most appropriate transformation.
(iv) Copy the graph to a file in Word and print out.

11.3 Use a spreadsheet as a simple database. Copy the data in the table below into cells within a spreadsheet. Modify the column widths so you can see all of the text on a single screen. Now sort the data in the following ways:
(a) by subject, in alphabetical order;
(b) by date and then by time of day;
(c) by topic, in reverse alphabetical order.

Use the *column hide* function (*Format* menu in Microsoft Excel) so that the information in columns 4 and 6 is not displayed. Find out how to undo this operation.

My exam timetable

Subject	Date	Time	Paper	Location	Question style
Biochemistry	3 Jun	Morning	1	Great Hall	Multiple-choice
Biochemistry	17 Jun	Morning	2	Exam Hall 5	Essay paper
Biochemistry	2 Jun	Afternoon	3	Main Laboratory	Information processing
Ecology	3 Jun	Afternoon	A	Small Hall	Short-answer questions
Ecology	14 Jun	Afternoon	B	Exam Hall 5	Essay paper
Biology A	4 Jun	Morning	1	Small Hall	Short-answer questions
Biology B	1 Jul	Afternoon	1	Exam Hall 3	Short-answer questions
Biology A	13 Jun	Afternoon	2	Exam Hall 5	Essay paper
Biology B	2 Jun	Morning	2	Main Laboratory	Essay paper

Word processors

Word processing is a transferable skill valuable beyond the immediate requirements of your biology course. The word processor has facilitated writing because of the ease of revising text. Using a word processor should improve your writing skills and speed because you can create, check and change your text on the screen before printing it as 'hard copy' on paper. Once entered and saved, multiple uses can be made of a piece of text with little effort.

When using a word processor you can:

- refine material many times before submission;
- insert material easily, allowing writing to take place in any sequence;
- use a spellchecker to check your text;
- use a thesaurus when composing your text;
- carry out ongoing checks of the word count;
- produce high-quality final copies;
- reuse part or all of the text in other documents.

The potential disadvantages of using a word processor include:

- lack of ready access to a computer, software and/or a printer;
- time taken to learn the operational details of the program;
- the temptation to make 'trivial' revisions;
- loss of files due to computer breakdown, or disk loss or failure.

Word processors come as 'packages' comprising the program and a manual, often with a tutorial program. Examples are WordPerfect and Microsoft Word. Most word processors have similar general features but differ in operational detail; it is best to pick one and stick to it as far as possible so that you become familiar with it. Learning to use the package is like learning to drive a car – you need only to know how to drive the computer and its program, not to understand how the engine (program) and transmission (data transfer) work, although a little background knowledge is often helpful and will allow you to get the most from the program.

In most word processors, the appearance of the screen realistically represents what the printout on paper will look like (WYSIWYG – what you see is what you get). Because of variation in operational details, only general and strategic information is provided in this chapter: you must learn the details of your word processor through use of the appropriate manual and 'help' facilities.

Before starting you will need:

- the program (usually installed on a hard disk or available via a network);
- a medium for storage, retrieval and back-up of your own files when created;
- a draft page-layout design: in particular you should have decided on page size, page margins, typeface (font) and size, type of text justification, and format of page numbering;
- an outline of the text content;

The computerised office – many word processors are sold as part of an integrated suite, e.g. Corel WordPerfect Office and Microsoft Office, with the advantage that they share a common interface in the different components (word processor, spreadsheet, database, etc.) and allow ready exchange of information (e.g. text, graphics) between component programs.

Using textbooks, manuals and tutorials – most programs no longer come with paper-based manuals, and support information is usually provided in one or more of the following ways: as a help facility within the program; as a help file on the program CD; or as an online help support site. It is still often worthwhile investing in one of the commercial textbooks that support specific programs.

- access to a suitable printer: this need not be attached to the computer you are using since your file can be taken or sent to an office where a printer is available, providing that it has the same word processing program.

Laying out (formatting) your document

Although you can format your text at any time, it is good practice to enter the basic commands at the start of your document: entering them later can lead to considerable problems due to reorganization of the text layout. If you use a particular set of layout criteria regularly, e.g. an A4 page with space for a letterhead, make a template containing the appropriate codes that can be called up whenever you start a new document. Note that various printers may respond differently to particular codes, resulting in a different spacing and layout.

Typing the text

If new to word processing, think of the screen as a piece of typing paper. The cursor marks the position where your text/data will be entered and can be moved around the screen by use of the cursor-control keys. When you type, don't worry about running out of space on the line because the text will wrap around to the next line automatically. Do not use a carriage return (usually the ENTER or ↵ key) unless you wish to force a new line, e.g. when a new paragraph is wanted. If you make a mistake when typing, correction is easy. You can usually delete characters or words or lines and the space is closed automatically. You can also insert new text in the middle of a line or word. You can insert special codes to carry out a variety of tasks, including changing text appearance such as underlining, **emboldening** and *italics*. Paragraph indentations can be automated using TAB or ⇥ as on a typewriter, but you can also indent or bullet whole blocks of text using special menu options. The function keys are usually pre-programmed to assist in many of these operations.

Editing features

Word processors usually have an array of features designed to make editing documents easy. In addition to the simple editing procedures described above, the program usually offers facilities to allow blocks of text to be moved ('cut and paste'), copied or deleted.

An extremely valuable editing facility is the find or search procedure: this can rapidly scan through a document looking for a specified word, phrase or punctuation. This is particularly valuable when combined with a replace facility so that, for example, you could replace the word 'test' with 'trial' throughout your document simply and rapidly.

Most WYSIWYG word processors have a command that reveals the normally hidden codes controlling the layout and appearance of the printed text. When editing, this can be a very important feature, since some changes to your text will cause difficulties if these hidden codes are not taken into account; in particular, make sure that the cursor is at the correct point before making changes to text containing hidden code, otherwise your text will sometimes change in apparently mystifying ways.

Using a word processor – take full advantage of the differences between word processing and 'normal' writing (which necessarily follows a linear sequence and requires more planning):

- Simply jot down your initial ideas for a plan, preferably at paragraph topic level. The order can be altered easily and if a paragraph grows too much it can easily be split.
- Start writing wherever you wish and fill in the rest later.
- Just put down your ideas as you think, confident in the knowledge that it is the concepts that are important to note; their order and the way you express them can be adjusted later.
- Don't worry about spelling and use of synonyms – these can (and should) be checked during a separate revision run through your text, using the spell-checker first to correct obvious mistakes, then the thesaurus to change words for style or to find the mot juste.
- Don't forget that a draft printout may be required to check (a) for pace and spacing – difficult to correct for on-screen; and (b) to ensure that words checked for spelling fit the required sense.

Deleting and restoring text – because deletion can sometimes be made in error, there is usually an 'undelete' or 'restore' feature that allows the last deletion to be recovered.

Fonts and line spacing

Most word processors offer a variety of fonts depending upon the printer being used. Fonts come in a wide variety of types and sizes, but they are defined in particular ways as follows:

- Typeface: the term for a family of characters of a particular design, each of which is given a particular name. The most commonly used for normal text is Times Roman (as used here for the main text) but many others are widely available, particularly for the better-quality printers. They fall into three broad groups: serif fonts with curves and flourishes at the ends of the characters (e.g. Times Roman); sans serif fonts without such flourishes, providing a clean, modern appearance (e.g. Helvetica, also known as Swiss); and decorative fonts used for special purposes only, such as the production of newsletters and notices.
- Size: measured in points. A point is the smallest typographical unit of measurement, there being 72 points to the inch (about 28 points per cm). The standard sizes for text are 10, 11 and 12 point, but typefaces are often available up to 72 point or more.
- Appearance: many typefaces are available in a variety of styles and weights. Many of these are not designed for use in scientific literature but for desktop publishing.
- Spacing: can be either fixed, where every character is the same width, or proportional, where the width of every character, including spaces, is varied. Typewriter fonts such as Elite and Prestige use fixed spacing and are useful for filling in forms or tables, but proportional fonts make the overall appearance of text more pleasing and readable.
- Pitch: specifies the number of characters per horizontal inch of text. Typewriter fonts are usually 10 or 12 pitch, but proportional fonts are never given a pitch value since it is inherently variable.
- Justification is the term describing the way in which text is aligned vertically. Left justification is normal, but for formal documents, both left and right justification may be used (as here).

You should also consider the vertical spacing of lines in your document. Drafts and manuscripts are frequently double-spaced. If your document has unusual font sizes, this may well affect line spacing, although most word processors will cope with this automatically.

Table construction

Tables can be produced by a variety of methods:

- Using the tab key ⇥ as on a typewriter: this moves the cursor to predetermined positions on the page, equivalent to the start of each tabular column. You can define the positions of these tabs as required at the start of each table.
- Using special table-constructing procedures (see Box 63.2). Here the table construction is largely done for you and it is much easier than using tabs, providing you enter the correct information when you set up the table.
- Using a spreadsheet to construct the table and then copying it to the word processor (see Box 63.2). This procedure requires considerably more manipulation than using the word processor directly and is best reserved for special circumstances, such as the presentation of a very

---

Presenting your documents – it is good practice not to mix typefaces too much in a formal document; also the font size should not differ greatly for different headings, sub-headings and the text.

Preparing draft documents – use double spacing to allow room for your editing comments on the printed page.

Preparing final documents – for most work, use a 12-point proportional serif typeface, with spacing dependent upon the specifications for the work.

large or complex table of data, especially if the data are already stored as a spreadsheet.

Graphics and special characters

Many word processors can incorporate graphics from other programs into the text of a document. Files must be compatible (see your manual) but if this is so, it is a relatively straightforward procedure. For professional documents this is a valuable facility, but for most undergraduate work it is probably better to produce and use graphics as a separate operation, e.g. using a spreadsheet (see Box 62.2).

You can draw lines and other graphical features directly within most word processors, and special characters (e.g. Greek characters) may be available dependent upon your printer's capabilities.

Inserting special characters – Greek letters and other characters are available using the 'Insert' and 'Symbols' features in Word.

Tools

Many word processors also offer you special tools, the most important of which are:

- Macros: special sets of files you can create when you have a frequently repeated set of keystrokes to make. You can record these keystrokes as a 'macro' so that it can provide a short cut for repeated operations.
- Thesaurus: used to look up alternative words of similar or opposite meaning while composing text at the keyboard.
- Spellcheck: a very useful facility that will check your spellings against a dictionary provided by the program. This dictionary is often expandable to include specialist words that you use in your work. The danger lies in becoming too dependent upon this facility, as they all have limitations: in particular, they will not pick up incorrect words that happen to be correct in a different context (i.e. 'was' typed as 'saw' or 'see' rather than 'sea'). Be aware of American spellings in programs from the USA, e.g. 'color' instead of 'colour'. The rule, therefore, is to use the spellcheck first and then carefully read the text for errors that have slipped through.
- Word count: useful when you are writing to a prescribed limit.

Using a spellcheck facility – do not rely on this to spot all errors. Remember that spellcheck programs do not correct grammatical errors.

Printing from your program

If more than one printer is attached to your PC or network, you will need to specify which one to use from the word processor's print menu. Most printers offer choices as to text and graphics quality, so choose draft (low) quality for all but your final copy since this will save both time and materials.

Use a print preview option to show the page layout if it is available. Assuming that you have entered appropriate layout and font commands, printing is a straightforward operation carried out by the word processor at your command. Problems usually arise because of some incompatibility between the criteria you have entered and the printer's own capabilities. Make sure that you know what your printer offers before starting to type: although settings are modifiable at any time, changing the page size, margin size, font size, etc., all cause your text to be rearranged, and this can be frustrating if you have spent hours carefully laying out the pages.

Using the print preview mode – this can reveal errors of several kinds, e.g. spacing between pages, that can prevent you wasting paper and printer ink unnecessarily.

KEY POINT *It is vital to save your work frequently to a memory stick, hard drive or network drive. This should be done every 10 minutes or so. If you do not save regularly, you may lose hours or days of work. Many programs can be set to 'autosave' every few minutes.*

Databases

A database is an electronic filing system whose structure is similar to a manual record-card collection. Its collection of records is termed a file. The individual items of information on each record are termed fields. Once the database is constructed, search criteria can be used to view files through various filters according to your requirements. The computerised catalogues in your library are just such a system; you enter the filter requirements in the form of author or subject keywords.

You can use a database to catalogue, search, sort, and relate collections of information. The benefits of a computerised database over a manual card-file system are:

- The information content is easily amended/updated.
- Printout of relevant items can be obtained.
- It is quick and easy to organise through sorting and searching/ selection criteria, to produce subgroups of relevant records.
- Record displays can easily be redesigned, allowing flexible methods of presenting records according to interest.
- Relational databases can be combined, giving the whole system immense flexibility. The older 'flat-file' databases store information in files that can be searched and sorted, but cannot be linked to other databases.

Choosing between a database and a spreadsheet – use a database only after careful consideration. Can the task be done better within a spreadsheet? A database program can be complex to set up and usually needs to be updated regularly.

Relatively simple database files can be constructed within spreadsheets using the columns and rows as fields and records respectively. These are capable of reasonably advanced sorting and searching operations and are probably sufficient for the types of databases you are likely to require as an undergraduate. You may also make use of a bibliographic database specially constructed for that purpose.

Statistical analysis packages

Statistical packages vary from small programs designed to carry out very specific statistical tasks to large sophisticated packages (SYSTAT, SigmaStat, Minitab, etc.) intended to provide statistical assistance, from experimental design to the analysis of results. Consider the following features when selecting a package:

Using spreadsheet statistics functions – before using a specific statistics package, check whether your spreadsheet is capable of carrying out the form of analysis you require (see Boxes 65.3 and 66.3), as this can often be the simpler option.

- The data entry and editing section should be user-friendly, with options for transforming data.
- Data exploration options should include descriptive statistics and exploratory data analysis techniques.
- Hypothesis-testing techniques should include ANOVA, regression analysis, multivariate techniques and parametric and non-parametric statistics.
- The program should provide assistance with experimental design and sampling methods.
- Output facilities should be suitable for graphical and tabular formats.

Presentation using computer packages – though many computer programs enhance presentational aspects of your work, there are occasions when they can make your presentation worse. Take care to avoid the following common pitfalls:

● Default or 'chart wizard' settings for graphs may result in output that is non-standard for the sciences (see Box 62.2).

● Fonts in labels and legends may not be consistent with other parts of your presentation.

● Some programs cannot produce Greek symbols (e.g. μ); do not use 'u' as a substitute. The same applies to scientific notation and superscripts: do not use 14C for ^{14}C, and replace, e.g., $1.4E+09$ with 1.4×10^9. First try cutting and pasting symbols from Word or, if this fails, draw correct symbols by hand.

Some programs have very complex data entry systems, limiting the ease of using data in different tests. The data entry and storage system should be based on a spreadsheet system, so that subsequent editing and transformation operations are straightforward.

 KEY POINT *Make sure that you understand the statistical basis for your test and the computational techniques involved before using a particular program.*

Graphics/presentation packages

Microsoft Office programs can be used to achieve most coursework tasks, e.g. PowerPoint is useful for creating posters (Box 13.1) and for oral presentations (Box 14.1). Should you need more advanced features, additional software may be available on your network: for example:

● SigmaPlot can produce graphs with floating axes;
● Macromedia Freehand is useful for designing complex graphics;
● DreamWeaver enables you to produce high-quality Web pages;
● MindGenius can be used to produce Mind Maps.

Important points regarding the use of such packages are:

● the learning time required for some of the more complex operations can be considerable;
● the quality of your printer will limit the quality of your output;
● not all files will readily import into a word processor such as Microsoft Word – you may need to save your work in a particular format. The different types of file are distinguished by the three-character filename extension, e.g. .jpg. and .bmp.

 KEY POINT *Computer graphics are not always satisfactory for scientific presentation. You should not accept the default versions produced – make appropriate changes to suit scientific standards and style. Box 62.1 gives a checklist for graph drawing and Box 62.2 provides guidelines for adapting Microsoft Excel output.*

Image storage and manipulation

With the widespread use of digital images (see Chapter 48), programs that facilitate the storage and manipulation of electronic image files have become increasingly important. These programs create a library of your stored images and provide a variety of methods for organising and selecting images. The industry-standard program, Adobe Photoshop, is one of many programs for image manipulation that vary widely in capability, cost and associated learning time. Many are highly sophisticated programs intended for graphic artists. For most scientific purposes, however, relatively limited functions are required.

Sources for further study

Gookin, D. (2003) *Word 2003 for Dummies.* Wiley, New York.

Kaufield, J. (2003) *Access 2003 for Dummies.* Wiley, New York.

Wang. W. (2003) *Office 2003 for Dummies.* Wiley, New York.

(And similar texts for other packages and release versions.)

Word processors, databases and other packages

Study exercises

12.1 Investigate intermediate/advanced Word features. The tasks in the following list are likely to useful in preparing assignments and report writing within the life sciences. Can you carry out all of the tasks? If not, use either a manual or the online *Help* feature to find out how to accomplish them. Tips are given in the answer section.

(a) Sort information in a list into alphabetical order.

(b) Replace a text string word or phrase with a new text string throughout your document.

(c) Replace a text string in normal font with the same text string in italics throughout your document.

(d) Add a 'header' and 'footer' to your document, the former showing the document's title and the latter containing page numbers in the bottom centre of the page.

(e) Adjust the margins of the page to give a 5 cm margin on the left and a 2 cm margin on the right.

(f) Change the type of bullets used in a list from standard (● or ■) to a different form (e.g. −, ◆ or ☑).

(g) Use the 'thesaurus' option to find a different or more suitable word to express your meaning. Try, for example, to find alternatives to the word 'alternative'.

(h) Carry out a spellcheck on your document.

(i) Carry out a word count on your document and on a selected part of it.

(j) Open two documents and switch between them.

12.2 Make precise copies of tables. Copy the following tables using a word processor such as Word

Test organism	Results of analysis (units)		
	August	September	October
X			
Y			
Z			

Test organism	Results of analysis (units)		
	August	September	October
X			
Y			
Z			

12.3 Investigate what programs and packages are available to you as a student. Test each program with appropriate data, images, etc.

Communicating
information

A scientific poster is a visual display of the results of an investigation, usually mounted on a rectangular board. Posters are used in undergraduate courses, to display project results or assignment work, and at scientific meetings to communicate research findings.

In a written report you can include a reasonable amount of specific detail and the reader can go back and reread difficult passages. However, if a poster is long-winded or contains too much detail, your reader is likely to lose interest.

 KEY POINT *A poster session is like a competition – you are competing for the attention of people in a room. Because you need to attract and hold the attention of your audience, make your poster as interesting as possible. Think of it as an advertisement for your work and you will not go far wrong.*

Preliminaries

Before considering the content of your poster, you should find out:

- the linear dimensions of your poster area, typically up to 1.5 m wide by 1 m high;
- the composition of the poster board and the method of attachment, whether drawing pins, Velcro tape, or some other form of adhesive; and whether these will be provided – in any case, it is safer to bring your own;
- the time(s) when the poster should be set up and when you should attend;
- the room where the poster session will be held.

Design

Plan your poster with your audience in mind, as this will dictate the appropriate level for your presentation. Aim to make your poster accessible to a broad audience. Since a poster is a *visual* display, you must pay particular attention to the presentation of information: work that may have taken hours to prepare can be ruined in a few minutes by the ill-considered arrangement of items (Fig. 13.1). Begin by making a draft sketch of the major elements of your poster. It is worth discussing your intended design with someone else, as constructive advice at the draft stage will save a lot of time and effort when you prepare the final version (or consult Simmonds and Reynolds, 1994).

Layout

One approach is to divide the poster into several smaller areas, perhaps six or eight in all, and prepare each as a separate item on a piece of card. Alternatively, you can produce a single large poster on one sheet of paper or card and store it inside a protective cardboard tube. However, a single large poster may bend and crease, making it difficult to flatten out. In addition, photographs and text attached to the backing sheet may work loose; a large poster with embedded images is an alternative approach.

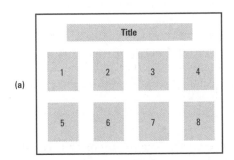
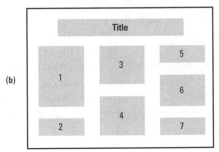
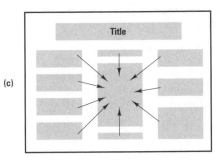

Fig. 13.1 Poster design. (a) An uninspiring design: sub-units of equal area, reading left to right, are not recommended. (b) This design is more interesting and the text will be easier to read (column format). (c) An alternative approach, with a central focus and arrows to guide the reader.

Subdividing your poster means that each smaller area can be prepared on a separate piece of paper or card, of A4 size or slightly larger, making transport and storage easier. It also breaks the reading matter up into smaller pieces, looking less formidable to a potential reader. By using pieces of card of different colours you can provide emphasis for key aspects, or link text with figures or photographs.

You will need to guide your reader through the poster and headings/ sub-headings will help with this aspect. It may be appropriate to use either a numbering system, with large, clear numbers at the top of each piece of card, or a system of arrows (or thin tapes), to link sections within the poster (see Fig. 13.1). Make sure that the relationship is clear and that the arrows or tapes do not cross.

Title

Your chosen title should be concise (no more than eight words), specific and interesting, to encourage people to read the poster. Make the title large and bold – it should run across the top of your poster, in letters at least 4 cm high, so that it can be read from the other side of the room. Coloured spirit-based marker and block capitals drawn with a ruler work well, as long as your writing is readable and neat (the colour can be used to add emphasis). Alternatively, you can print out each word in large type, using a word processor. Details of authors, together with their addresses (if appropriate), should be given, usually across the top of the poster in somewhat smaller lettering than the title.

Text

Write in short sentences and avoid verbosity. Keep your poster as visual as possible and make effective use of the spaces between the blocks of text. Your final text should be double-spaced and should have a minimum capital-letter height of 8 mm (minimum type size 36 point), preferably greater, so that the poster can be read at a distance of 1 m. One method of obtaining text of the required size is to photo-enlarge standard typescript (using a good-quality photocopier), or use a high-quality (laser) printer. It is best to avoid continuous use of text in capitals, since it slows reading and makes the text less interesting to the reader. Also avoid italic, 'balloon' or decorative styles of lettering.

 KEY POINT Keep text to a minimum – aim to have a maximum of 500 words in your poster.

Subtitles and headings

These should have a capital-letter height of 12–20 mm, and should be restricted to two or three words. They can be produced by word processor, photo-enlargement, by stencilling, or by hand, using pencilled guidelines (but make sure that no pencil marks are visible on your finished poster).

Colour

Consider the overall visual effect of your chosen display, including the relationship between your text, diagrams and the backing board. Colour can be used to highlight key aspects of your poster. However, it is very easy to ruin a poster by the inappropriate choice and application of colour. Careful use of two, or at most three, complementary colours and

Presenting a poster at a formal conference – it can be useful to include your photograph for identification purposes, e.g. in the top right-hand corner of the poster.

Making up your poster – text and graphics printed on good-quality paper can be glued directly onto a contrasting mounting card: use photographic spray mountant or Pritt rather than liquid glue. Trim carefully using a guillotine to give equal margins, parallel with the paper. Photographs should be placed in a window mount to avoid the tendency for their corners to curl. Another approach is to trim pages or photographs to their correct size, then encapsulate in plastic film: this gives a highly professional finish and is easy to transport.

Producing composite material for posters – PowerPoint is generally more useful than Word when you wish to include text, graphics and/or images on the same page. It is possible to use PowerPoint to produce a complete poster (Box 13.1), although it can be expensive to have this printed out commercially to A1 or A0 size.

shades will be easier on the eye and should aid comprehension. Colour can be used to link the text with the visual images (e.g. by picking out a colour in a photograph and using the same colour on the mounting board for the accompanying text). For PowerPoint posters, careful choice of colours for the various elements will enhance the final product (Box 13.1). Use coloured inks or water-based paints to provide colour in diagrams and figures, as felt pens rarely give satisfactory results.

Content

The typical format is that of a scientific report (see Box 17.1), i.e. with the same headings, but with a considerably reduced content. Keep references within the text to a minimum – interested parties can always ask you for further information. Also note that most posters have a summary/conclusions section at the end, rather than an abstract.

Introduction
This should give the reader background information on the broad field of study and the aims of your own work. It is vital that this section is as interesting as possible, to capture the interest of your audience. It is often worth listing your objectives as a series of numbered points.

Materials and Methods
Keep this short, and describe only the principal techniques used. You might mention any special techniques, or problems of general interest.

Results
Don't present your raw data: use data reduction wherever possible, i.e. figures and simple statistical comparisons. Graphs, diagrams, histograms and pie charts give clear visual images of trends and relationships and should be used in place of data tables (see p. 390). Final copies of all figures should be produced so that the numbers can be read from a distance of 1 m. Each should have a concise title and legend, so that it is self-contained: if appropriate, a series of numbered points can be used to link a diagram with the accompanying text. Where symbols are used, provide a key on each graph (symbol size should be at least 5 mm). Avoid using graphs straight from a written version, e.g. a project report, textbook, or a paper, without considering whether they need modification to meet your requirements.

Conclusions
This is where many readers will begin, and they may go no further unless you make this section sufficiently interesting. This part needs to be the strongest part of your poster, summarising the main points. Refer to your figures here to draw the reader into the main part of your poster. A slightly larger or bolder typeface may add emphasis, though too many different typefaces can look messy. For references, smaller type can be used.

The poster session

A poster session may be organised as part of the assessment of your coursework, and this usually mirrors those held at most scientific conferences and meetings. Staff and fellow students (delegates at conferences) will mill around, looking at the posters and chatting to their authors, who are usually expected to be in attendance. If you stand at the side of your poster throughout, you are likely to discourage some readers, who may not wish to become involved in a detailed conversation

Presenting at a scientific meeting – never be tempted to spend the minimum amount of time converting a piece of scientific writing into poster format – the least interesting posters are those where the author simply displays pages from a written communication (e.g. a journal article) on a poster board.

Designing the materials and methods section – photographs or diagrams of apparatus can help to break up the text of this section and provide visual interest. It is sometimes worth preparing this section in a smaller typeface.

Keeping graphs and diagrams simple – avoid composite graphs with different scales for the same axis, or with several trend lines (use a maximum of three trend lines per graph).

Listing your conclusions – a series of numbered points is a useful approach, if your findings fit this pattern.

Consider providing a handout – this is a useful way to summarise the main points of your poster, so that your readers have a permanent record of the information you have presented.

Box 13.1   How to create a poster using PowerPoint

Software such as PowerPoint can be used to produce a high-quality poster, providing you have access to a good colour printer. However, you should avoid the standard templates available on the Web as they encourage unnecessary uniformity and stifle creativity, leading to a less satisfying end result. The following steps give practical advice on creating a poster as a single PowerPoint slide:

1. **Sketch out your plans.** Decide on the main poster elements (images, graphs, tables and text sections) and their relationship with each other and draw out a one-page 'storyboard' (see Fig 13.1). Think about colours for background, text and graphics (use two or three complementary colours): dark text on a light background is clearer (high contrast), and uses less ink when printing. Also consider how you will link the elements in sequence, to guide readers through your 'story'.

2. **Get your material ready.** Collect together individual files for pictures, figures and tables. Make any required adjustments to images, graphs or tables before you import them into your poster.

3. **Create a new/blank slide.** Open PowerPoint and select *Blank presentation.* From *File > New*, then use the *File > Page setup* menu to select either *Landscape or Portrait* orientation and to set the correct page size (use *Width* and *Height* commands, or select a standard size, such as A4, A3, A2, etc.). Right-click on the slide and select *Ruler* and *Guides* (to help position elements within the slide – the horizontal and vertical guidelines can be dragged to different positions at later stages, as required) and also select an appropriate *Background* colour. In general, avoid setting a picture as your background as this tends to detract from the content of the poster. Before going further, save your work. Repeat this frequently and in more than one location (e.g. hard drive and USB memory stick).

4. **Add graphics.** For images, use the drop-down *Insert* menu, select *Picture, From File* and browse to *Insert* the correct file. The *Insert, Object* command performs a similar function for Excel charts (graphs). Alternatively, use the copy-and-paste functions of complementary software. Once inserted, resize using the *sizing handles* in one of the corners (for photographs, take care not to alter one dimension relative to the other, or the image will be distorted). To reposition, put the mouse pointer over the image, left-click and hold, then drag to new location). While the *Drawing* toolbar offers standard shapes and other useful features, you should avoid clipart (jaded and over-used) and poor-quality images from the Web (always use the highest resolution possible) – if you do not have your final images, use blank text boxes to show their position within the draft poster.

5. **Add text.** Use either the *Drawing* toolbar to select a *Text box* and place this on your slide, then either type in your text (use the *Enter* key to provide line spacing within the box) or copy-and-paste text from a word-processed file. You will need to consider the type size for the printed poster (e.g. for an A0 poster (size 1189 × 841 mm), a printed type size of 24 point is appropriate for the main text, with larger sizes for headings and titles. If you find things difficult to read on-screen, use the *Zoom* function (either select a larger percentage in the *Zoom box* on the standard toolbar, or hold down the [Ctrl] key and use the mouse wheel to scroll up (*zoom*) or down (*reduce*). Use a separate text box for each element of your poster and don't be tempted to type too much text into each box – write in succinct phrases, using bullet points and numbered lists to keep text concise (aim for no more than 50 words per text box). Select appropriate font styles and colours using the *Format > Font* menu. For a background colour or surrounding line, right-click and use the *Format Text Box* command (line thickness and colour can then be altered using the *Drawing toolbar*). Present supplementary text elements in a smaller type: for example, details of methodology, references cited.

6. **Add boxes, lines and/or arrows** to link elements of the poster and guide the reader (see Fig. 13.1(c)). These features are available from the *Drawing* toolbar. Note that new inserts are overlaid over older inserts – if this proves to be a problem, select the relevant item and use the *Draw > Order* functions to change its relative position.

7. **Review your poster.** Get feedback from another student or your tutor, e.g. on a small printed version, or use a projector to view your poster without printing (adjust the distance between projector and screen to give the correct size).

8. **Revise and edit your poster.** Revisit your work and remove as much unnecessary text as possible. Delete any component that is not essential to the message of the poster. Keep graphs simple and clear (p. 387 gives further advice). 'White space' is important in providing structure.

9. **Print the final version.** Use a high-resolution colour printer (this may be costly, so you should wait until you are sure that no further changes are needed).

Coping with questions in assessed poster sessions – you should expect to be asked questions about your poster, and to explain details of figures, methods, etc.

about the poster. Stand nearby. Find something to do – talk to someone else, or browse among the other posters, but remain aware of people reading your poster and be ready to answer any queries they may raise. Do not be too discouraged if you aren't asked lots of questions: remember, the poster is meant to be a self-contained, visual story, without need for further explanation.

A poster display will never feel like an oral presentation, where the nervousness beforehand is replaced by a combination of satisfaction and relief as you unwind after the event. However, it can be a very satisfying means of communication, particularly if you follow these guidelines.

Text reference

Simmonds, D. and Reynolds, L. (1994) *Data Presentation and Visual Literacy in Medicine and Science*. Butterworth-Heineman, London.

Sources for further study

Alley, M. (2003) *The Craft of Scientific Presentations: Critical Steps to Succeed and Critical Errors to Avoid.* Springer-Verlag, New York.

Briscoe, M.H. (2000) *Preparing Scientific Illustrations: a Guide to Better Posters, Presentations and Publications,* 2nd edn. Springer-Verlag, New York.

Davis, M.F. (2005) *Scientific Papers and Presentations,* 2nd edn. Academic Press, New York.

Gosling, P.J. (1999) *Scientist's Guide to Poster Presentations.* Kluwer, New York.

Hess, G. and Liegel, L. *Creating Effective Poster Presentations.* Available: http://www.ncsu.edu/project/posters/ Last accessed: 09/04/07.

Study exercises

13.1 Design a poster. Working with one or more partners from your year group, decide on a suitable topic (perhaps something linked to your current teaching programme). Working individually, make an outline plan of the major elements of the poster, with appropriate sub-headings and a brief indication of the content and relative size of each element (including figures, diagrams and images). Exchange draft plans with your partners and arrange a session where you can discuss their merits and disadvantages.

13.2 Prepare a checklist for assessing the quality of a poster presentation. After reading through this chapter, prepare a 10-point checklist of assessment criteria under the heading 'What makes a good poster presentation?'. Compare your list with the one that we have provided (p. 442) – do you agree with our criteria, or do you prefer your own list (can you justify your preferences)?

13.3 Evaluate the posters in your university. Most universities have a wide range of academic posters on display. Some may cover general topics (e.g. course structures), while others may deal with specific research topics (e.g. poster presentations from past conferences). Consider their good and bad features (if you wish to make this a group exercise, you might compare your evaluation with that of other students in a group discussion session).

Most students feel very nervous about giving talks. This is natural, since very few people are sufficiently confident and outgoing that they look forward to speaking in public. Additionally, the technical nature of the subject matter may give you cause for concern, especially if you feel that some members of the audience have a greater knowledge than you have. However, this is a fundamental method of scientific communication and an important transferable skill, therefore it forms an important component of many courses.

The comments in this chapter apply equally to informal talks, e.g. those based on assignments and project work, and to more formal conference presentations. It is hoped that the advice and guidance given below will encourage you to make the most of your opportunities for public speaking, but there is no substitute for practice. Do not expect to find all of the answers from this, or any other, book. Rehearse, and learn from your own experience.

 KEY POINT *The three 'Rs' of successful public speaking are: reflect – give sufficient thought to all aspects of your presentation, particularly at the planning stage; rehearse – to improve your delivery; revise – modify the content and style of your material in response to your own ideas and to the comments of others.*

Preparation

Preliminary information

Begin by marshalling the details needed to plan your presentation, including:

- the duration of the talk;
- whether time for questions is included;
- the size and location of the room;
- the projection/lighting facilities provided, and whether pointers or similar aids are available.

It is especially important to find out whether the room has the necessary equipment for digital projection (e.g. PC, projector and screen, black-out curtains or blinds, appropriate lighting) or overhead projection before you prepare your audio-visual aids. If you concentrate only on the spoken part of your presentation at this stage, you are inviting trouble later on. Have a look around the room and try out the equipment at the earliest opportunity, so that you are able to use the lights, projector, etc., with confidence. For digital projection systems, check that you can load/present your material. Box 14.1 gives advice on using PowerPoint.

Audio-visual aids

If you plan to use overhead transparencies, find out whether your department has facilities for their preparation, whether these facilities are available for your use, and the cost of materials. Adopt the following guidelines:

- Keep text to a minimum: present only the key points, with up to 20 words per slide/transparency.

Box 14.1 Tips on preparing and using PowerPoint slides in a spoken presentation

Microsoft PowerPoint can be used to produce high-quality visual aids, assuming a computer and digital projector are available in the room where you intend to speak. The presentation is produced as a series of electronic 'slides' on to which you can insert images, diagrams and text. When creating your slides, bear the following points in mind:

- **Plan the structure of your presentation.** Decide on the main topic areas and sketch out your ideas on paper. Think about what material you will need (e.g. pictures, graphs) and what colours to use for background and text.

- **Choose slide layouts according to purpose.** Once PowerPoint is running, from the *Insert* menu select *New Slide > Choose an Autolayout*. You can then add material to each new slide to suit your requirements.

- **Select your background with care.** Many of the preset background templates available within the *Format* menu (*Apply design template* option) are best avoided, since they are overused and fussy, diverting attention from the content of the slides. Conversely, flat, dull backgrounds may seem uninteresting, while brightly coloured backgrounds can be garish and distracting. Choose whether to present your text as a light-coloured type on a dark background (more restful but perhaps less engaging if the room is dark) or a dark-coloured type on a light background (more lively).

- **Use visual images throughout.** Remember the Chinese proverb 'a picture is worth ten thousand words'. A presentation composed entirely of text-based slides will be uninteresting: adding images and diagrams will brighten up your talk considerably (use the *Insert* menu, *Picture* option). Images can be taken with a digital camera (see Chapter 48), scanned in from a printed version or copied and pasted from the Web, but you should take care not to break copyright regulations. 'Clip art' is copyright-free, but should be used sparingly, as most people will have seen the images before and they are rarely wholly relevant. Diagrams can be made from components created using the *Drawing* toolbar, and graphs and tables can be imported from other programs, e.g. Excel (Box 13.1 gives further specific practical advice on adding graphics, saving files, etc.).

- **Keep text to a minimum.** Aim for no more than 20 words on a single slide (e.g. four/five lines containing a few words per line). Use headings and sub-headings to structure your talk: write only key words or phrases as 'prompts' to remind you to cover a particular point during your talk – never be tempted to type whole sentences as you will then be reduced to reading these from the screen during your presentation, which is boring.

- **Use a large, clear font.** Use the *Slide Master* option within the *View* menu to set the default font to a non-serif style such as Arial, or Comic Sans MS. Default fonts for headings and bullet points are intentionally large, for clarity. Do not reduce these to anything less than 28-point type size (preferably larger), to cram in more words: if you have too much material, create a new slide and divide up the information.

- **Animate your material.** The *Slide Show* menu provides a *Custom Animation* function that enables you to introduce the various elements within a slide, e.g. text can be made to *Appear* one line at a time, to prevent the audience from reading ahead and to help maintain their attention.

- **Don't overdo the special effects.** PowerPoint has a wide range of features that allow complex slide transitions and animations, additional sounds, etc., but these quickly become irritating to an audience unless they have a specific purpose within your presentation.

- **Always edit your slides before use.** Check through your slides and cut out any unnecessary words, adjust the layout and animation. Remember the maxim 'less is more' – avoid too much text; too many bullet points; too many distracting visual effects or sounds.

When presenting your talk:

- **Work out the basic procedures beforehand.** Practise, to make sure that you know how to move forwards and backwards, turn the screen on and off, hide the mouse pointer, etc.

- **Don't forget to engage your audience.** Despite the technical gadgetry, *you* need to play an active role in their presentation, as explained elsewhere in this chapter.

- **Don't go too fast.** Sometimes, new users tend to deliver their material too quickly: try to speak at a normal pace, and practise beforehand.

- **Consider whether to provide a handout.** PowerPoint has several options, including some that provide space, for notes (e.g. Fig. 4.3). However, a handout should not be your default option, as there is a cost involved.

Using audio-visual aids – *don't let equipment and computer gadgetry distract you from the essential rules of good speaking (pp. 88, 92). Remember that you are the presenter.*

- Make sure the text is readable: try out your material beforehand.
- Use several simpler figures rather than a single complex graph.
- Avoid too much colour on overhead transparencies: blue and black are easier to read than red or green.
- Don't mix slides and transparencies as this is often distracting.
- Use spirit-based pens for transparencies: use alcohol for corrections.
- Transparencies can be produced from typewritten or printed text using a photocopier, often giving a better product than pens. Note that you must use special heat-resistant acetate sheets for photocopying.

Audience

You should consider your audience at the earliest stage, since they will determine the appropriate level for your presentation. If you are talking to fellow students you may be able to assume a common level of background knowledge. In contrast, a research lecture given to your department, or a paper at a meeting of a scientific society, will be presented to an audience from a broader range of backgrounds. An oral presentation is not the place for a complex discussion of specialized information: build up your talk from a low level. The speed at which this can be done will vary according to your audience. As long as you are not boring or patronising, you can cover basic information without losing the attention of the more knowledgeable members in your audience.

Pitching your talk at the right level – *the general rule should be: 'do not overestimate the background knowledge of your audience'. This sometimes happens in student presentations, where fears about the presence of 'experts' can encourage the speaker to include too much detail, overloading the audience with facts.*

Content

Although the specific details in your talk will be for you to decide, most spoken presentations share some common features of structure, as described below.

Introductory remarks

It is vital to capture the attention of your audience at the outset. Consequently, you must make sure your opening comments are strong, otherwise your audience will lose interest before you reach the main message. Remember it takes a sentence or two for an audience to establish a relationship with a new speaker. Your opening sentence should be some form of preamble and should not contain any key information. For a formal lecture, you might begin with 'Thank you for that introduction. My talk today is about …' then restate the title and acknowledge other contributors, etc. You might show a transparency or slide with the title printed on it, or an introductory photograph, if appropriate. This should provide the necessary settling-in period.

Getting the introduction right – *a good idea is to have an initial slide giving your details and the title of your talk, and a second slide telling the audience how your presentation will be structured. Make eye contact with all sections of the audience during the introduction.*

After these preliminaries, you should introduce your topic. Begin your story on a strong note – avoid timid or apologetic phrases.

Opening remarks are unlikely to occupy more than 10 per cent of the talk. However, because of their significance, you might reasonably spend up to 25 per cent of your preparation time on them.

What to cover in your introductory remarks – *you should:*

- *explain the structure of your talk;*
- *set out your aims and objectives;*
- *explain your approach to the topic.*

 KEY POINT *Make sure you have practised your opening remarks so that you can deliver the material in a flowing style, with fewer chances of mistakes.*

Allowing time for slides – as a rough guide you should allow at least two minutes per illustration, although some diagrams may need longer, depending on content. Make a note of the half-way point to help you check timing/pace.

The main message

This section should include the bulk of your experimental results or literature findings, depending on the type of presentation. Keep details of methods to the minimum needed to explain your data. This is *not* the place for a detailed description of equipment and experimental protocol (unless it is a talk about methodology). Results should be presented in an easily digested format.

 KEY POINT *Do not expect your audience to cope with large amounts of data; use a maximum of six numbers per slide. Remember that graphs and diagrams are usually better than tables of raw data, since the audience will be able to see the visual trends and relationships in your data (p. 390).*

Present summary statistics (Chapter 65) rather than individual results. Show the final results of any analyses in terms of the statistics calculated, and their significance (p. 427), rather than dwelling on details of the procedures used. Figures should not be crowded with unnecessary detail. Every diagram should have a concise title and the symbols and trend lines should be clearly labelled, with an explanatory key where necessary. When presenting graphical data (Chapter 62) always 'introduce' each graph by stating the units for each axis and describing the relationship for each trend line or data set.

 KEY POINT *Use summary slides at regular intervals, to maintain the flow of the presentation and to emphasise the main points.*

Take the audience through your story step by step at a reasonable pace. Try not to rush the delivery of your main message due to nervousness. Avoid complex, convoluted storylines – one of the most distracting things you can do is to fumble backwards through PowerPoint slides or overhead transparencies. If you need to use the same diagram or graph more than once then you should make two (or more) copies. In a presentation of experimental results, you should discuss each point as it is raised, in contrast to written text, where the results and discussion may be in separate sections. The main message typically occupies approximately 80 per cent of the time allocated to an oral presentation (Fig. 14.1).

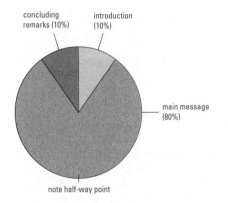

Fig. 14.1 Pie chart showing time allocation for a typical presentation

Concluding remarks

Having captured the interest of your audience in the introduction and given them the details of your story in the middle section, you must now bring your talk to a conclusion. Do not end weakly, e.g. by running out of steam on the last slide. Provide your audience with a clear 'take-home message' by returning to the key points in your presentation. It is often appropriate to prepare a slide or overhead transparency listing your main conclusions as a numbered series.

Signal the end of your talk by saying 'finally ...', 'in conclusion ...', or a similar comment and then finish speaking after that sentence. Your audience will lose interest if you extend your closing remarks beyond this point. You may add a simple end phrase (for example, 'thank you') as you put your notes into your folder, but do not say 'that's all folks!', or make any similar offhand remark. Finish as strongly and as clearly as you started. Box 14.2 gives further advice.

Final remarks – make sure you give the audience sufficient time to assimilate your final slide: some of them may wish to write down the key points. Alternatively, you might provide a handout, with a brief outline of the aims of your study and the major conclusions.

Box 14.2 Hints on spoken presentations

In planning the delivery of your talk, bear the following aspects in mind:

- **Using notes**. Many accomplished speakers use abbreviated notes for guidance, rather than reading word for word from a prepared script. When writing your talk:

  (i) **Consider preparing your first draft as a full script**: write in spoken English and keep the text simple, to avoid a formal, impersonal style. Your aim should be to *talk* to your audience, not to *read* to them.

  (ii) **If necessary, use notecards with key words and phrases**: it is best to avoid using a full script in the final presentation. As you rehearse and your confidence improves, a set of notecards may be an appropriate format. Mark the position of slides/key points, etc.: each notecard should contain details of structure as well as content. Your notes should be written/printed in text large enough to be read easily during the presentation (also check that the lecture room has a lectern light or you may have problems reading your notes if the lights are dimmed). Each note-card or sheet should be clearly numbered, so that you do not lose your place.

  (iii) **Decide on the layout of your talk**: give each subdivision a heading in your notes, so that your audience is made aware of the structure.

  (iv) **Memorise your introductory/closing remarks**: you may prefer to rely on a full written version for these sections, in case your memory fails, or if you suffer 'stage fright'.

  (v) **Using PowerPoint** (Box 14.1): here, you can either use the 'notes' option (*View > Notes Page*), or you may even prefer to dispense with notes entirely, since the slides will help structure your talk, acting as an aide-memoire for your material.

- **Work on your timing**. It is essential that your talk is the right length and the correct pace:

  (i) **Rehearse your presentation**: ask a friend to listen and to comment constructively on those parts that were difficult to follow, to improve your performance.

  (ii) **Use 'split times' to pace yourself**: following an initial run-through, add the times at which you should arrive at the key points of your talk to your notes. These timing marks will help you keep to time during the final presentation.

  (iii) **Avoid looking at your wristwatch when speaking**; this sends a negative signal to the audience. Use a wall clock (where available), or take off your watch and put it beside your notes so that you can glance at it without distracting the audience.

- **Consider your image**. Make sure that the image you project is appropriate for the occasion:

  (i) **Think about what to wear**: aim to be respectable without 'dressing up', otherwise your message may be diminished.

  (ii) **Develop a good posture**: it will help your voice projection if you stand upright, rather than slouching, or leaning over a lectern.

  (iii) **Deliver your material with expression**: project your voice towards the audience at the back of the room and make sure you look round to make eye contact with all sections of the audience. Arm movements and subdued body language will help maintain the interest of your audience. However, you should avoid extreme gestures (it may work for some TV personalities but it is not recommended for the beginner).

  (iv) **Try to identify and control any repetitive mannerisms**: repeated 'empty' words/phrases, fidgeting with pens, keys, etc., will distract your audience. Note-cards held in your hand give you something to focus on, whereas laser pointers will show up any nervous hand tremors. Practising in front of a mirror may help.

- **Think about questions**. Once again, the best approach is to prepare beforehand:

  (i) **Consider what questions are likely to come up, and prepare brief answers**. However, do not be afraid to say 'I don't know': your audience will appreciate honesty, rather than vacillation, if you don't have an answer for a particular question.

  (ii) **If no questions are asked, you might pose a question yourself** and then ask for opinions from the audience: if you use this approach, you should be prepared to comment briefly if your audience has no suggestions, to avoid the presentation ending in an embarrassing silence.

Sources for further study

Alley, M. (2003) *The Craft of Scientific Presentations: Critical Steps to Succeed and Critical Errors to Avoid.* Springer-Verlag, New York.

Capp, C.C. and Capp, G.R. (1989) *Basic Oral Communication*, 5th edn. Prentice Hall, Harlow.

Matthews, C. and Marino, J. (1999) *Professional Interactions: Oral Communication Skills of Science, Technology and Medicine.* Pearson, Harlow.

Radel, J. *Oral Presentations.* Available: http://www.biology.eku.edu/RITCHISO/oralpres.html Last accessed 09/04/07.

Study exercises

14.1 Prepare a checklist for assessing the quality of an oral presentation. After reading through this chapter, prepare a 10-point checklist of assessment criteria under the heading 'What makes a good oral presentation?'. Compare your list with the one that we have provided (pp. 442–3) – do you agree with our criteria, or do you prefer your checklist? Can you justify your preferences?

14.2 Evaluate the presentation styles of other speakers. There are many opportunities to assess the strengths and weaknesses of academic 'public speakers', including your university lecturers, seminar speakers, presenters of TV documentaries, etc. Decide in advance how you are going to tackle the evaluation (e.g. with a quantitative marking scheme, or a less formal procedure).

14.3 Rehearse a talk and get feedback on your performance. There are a number of approaches you might take, including: (i) recording and reviewing your presentation using a digital camera; or (ii) giving your talk to a small group of fellow students and asking them to provide constructive feedback.

Monday:	morning	Lectures (University)
	afternoon	Practical (University)
	evening	**Initial analysis and**
		brainstorming (Home)
Tuesday:	morning	Lectures (University)
	afternoon	**Locate sources (Library)**
	evening	**Background reading (Library)**
Wednesday:	morning	**Background reading (Library)**
	afternoon	Squash (Sports hall)
	evening	**Planning (Home)**
Thursday:	morning	Lectures (University)
	afternoon	**Additional reading (Library)**
	evening	**Prepare outline (Library)**
Friday:	morning	Lab class (University)
	afternoon	**Write first draft (Home)**
	evening	**Write first draft (Home)**
Saturday:	morning	Shopping (Town)
	afternoon	**Review first draft (Home)**
	evening	**Revise first draft (Home)**
Sunday:	morning	Free
	afternoon	**Produce final copy (Home)**
	evening	**Proofread and print**
		essay (Home)
Monday:	morning	**Final read-through and check**
		Submit essay
		(deadline midday)

Fig. 15.1 Example timetable for writing a short essay

> **Creating an outline** – *an informal outline can be made simply by indicating the order of sections on a spider diagram (as in Fig. 15.2).*

> **Talking about your work** – *Discussing your topic with a friend or colleague might bring out ideas or reveal deficiencies in your knowledge.*

Written communication is an essential component of all sciences. Most courses include writing exercises in which you will learn to describe ideas and results accurately, succinctly and in an appropriate style and format. The following features are common to all forms of scientific writing.

Organising your time

Making a timetable at the outset helps ensure that you give each stage adequate attention and complete the work on time (e.g. Fig. 15.1). To create and use a timetable:

1. Break down the task into stages.
2. Decide on the proportion of the total time each stage should take.
3. Set realistic deadlines for completing each stage, allowing some time for slippage.
4. Refer to your timetable frequently as you work: if you fail to meet one of your deadlines, make a serious effort to catch up as soon as possible.

 KEY POINT *The appropriate allocation of your time to reading, planning, writing and revising will differ according to the task in hand (see Chapters 16–18).*

Organising your information and ideas

Before you write, you need to gather and/or think about relevant material (Chapters 8 and 9). You must then decide:

- what needs to be included and what doesn't;
- in what order it should appear.

Start by jotting down headings for everything of potential relevance to the topic (this is sometimes called 'brainstorming'). A spider diagram (Fig. 15.2) or a Mind Map (Fig. 4.2) will help you organise these ideas. The next stage is to create an outline of your text (Fig. 15.3). Outlines are valuable because they:

- force you to think about and plan the structure;
- provide a checklist so nothing is missed out;
- ensure the material is balanced in content and length;
- help you organise figures and tables by showing where they will be used.

 KEY POINT *A suitable structure is essential to the narrative of your writing, and should be carefully considered at the outset.*

In an essay or review, the structure of your writing should help the reader to assimilate and understand your main points. Subdivisions of the topic could simply be related to the physical nature of the subject matter (e.g. zones of an ecosystem) and should proceed logically (e.g. low-water mark to high-water mark). A chronological approach is good for

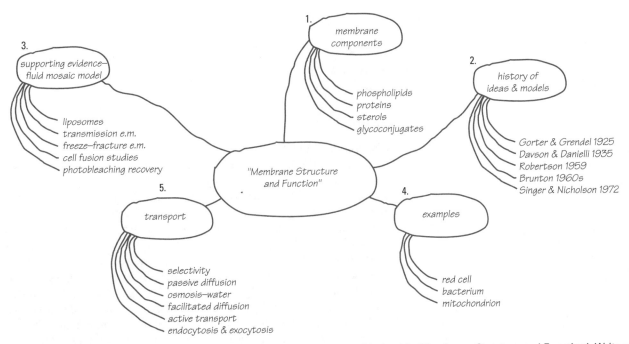

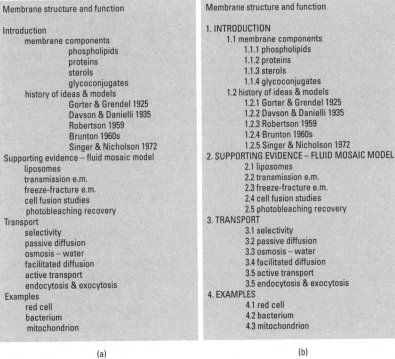

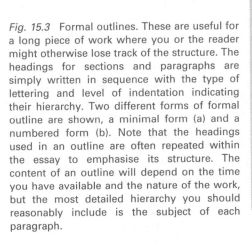

Fig. 15.2 Spider diagram showing how you might 'brainstorm' an essay with the title 'Membrane Structure and Function'. Write out the essay title in full to form the spider's body, and as you think of possible content, place headings around this to form its legs. Decide which headings are relevant and which are not and use arrows to note connections between subjects. This may influence your choice of order and may help to make your writing flow because the links between paragraphs will be natural. You can make an informal outline directly on a spider diagram by adding numbers indicating a sequence of paragraphs (as shown). This method is best when you must work quickly, as with an essay written under exam conditions.

Fig. 15.3 Formal outlines. These are useful for a long piece of work where you or the reader might otherwise lose track of the structure. The headings for sections and paragraphs are simply written in sequence with the type of lettering and level of indentation indicating their hierarchy. Two different forms of formal outline are shown, a minimal form (a) and a numbered form (b). Note that the headings used in an outline are often repeated within the essay to emphasise its structure. The content of an outline will depend on the time you have available and the nature of the work, but the most detailed hierarchy you should reasonably include is the subject of each paragraph.

Membrane structure and function

Introduction
    membrane components
        phospholipids
        proteins
        sterols
        glycoconjugates
    history of ideas & models
        Gorter & Grendel 1925
        Davson & Danielli 1935
        Robertson 1959
        Brunton 1960s
        Singer & Nicholson 1972
Supporting evidence – fluid mosaic model
    liposomes
    transmission e.m.
    freeze-fracture e.m.
    cell fusion studies
    photobleaching recovery
Transport
    selectivity
    passive diffusion
    osmosis – water
    facilitated diffusion
    active transport
    endocytosis & exocytosis
Examples
    red cell
    bacterium
    mitochondrion

(a)

Membrane structure and function

1. INTRODUCTION
    1.1 membrane components
        1.1.1 phospholipids
        1.1.2 proteins
        1.1.3 sterols
        1.1.4 glycoconjugates
    1.2 history of ideas & models
        1.2.1 Gorter & Grendel 1925
        1.2.2 Davson & Danielli 1935
        1.2.3 Robertson 1959
        1.2.4 Brunton 1960s
        1.2.5 Singer & Nicholson 1972
2. SUPPORTING EVIDENCE – FLUID MOSAIC MODEL
    2.1 liposomes
    2.2 transmission e.m.
    2.3 freeze-fracture e.m.
    2.4 cell fusion studies
    2.5 photobleaching recovery
3. TRANSPORT
    3.1 selectivity
    3.2 passive diffusion
    3.3 osmosis – water
    3.4 facilitated diffusion
    3.5 active transport
    3.5 endocytosis & exocytosis
4. EXAMPLES
    4.1 red cell
    4.2 bacterium
    4.3 mitochondrion

(b)

evaluation of past work (e.g. the development of the concept of evolution), whereas a step-by-step comparison might be best for certain exam questions (e.g. 'Discuss the differences between prokaryotes and eukaryotes'). There is little choice about structure for practical and project reports (see p. 105).

Writing

Adopting a scientific style

Your main aim in developing a scientific style should be to get your message across directly and unambiguously. Although you can try to achieve this through a set of 'rules' (see Box 15.1), you may find other requirements driving your writing in a contradictory direction. For instance, the need to be accurate and complete may result in text littered with technical terms, and the flow may be continually interrupted by references to the literature. The need to be succinct also affects style and readability through the use of, for example, stacked noun-adjectives (e.g. 'restriction fragment length polymorphism') and acronyms (e.g. 'RFLP'). Finally, style is very much a matter of taste and each tutor, examiner, supervisor or editor will have pet loves and hates that you may have to accommodate. Different assignments will need different styles; Box 15.2 gives further details.

Developing technique

Writing is a skill that can be improved, but not instantly. You should analyse your deficiencies with the help of feedback from your tutors, be prepared to change work habits (e.g. start planning your work more carefully), and willing to learn from some of the excellent texts that are available on scientific writing (p. 100).

Improving your writing skills – You need to take a long-term view if you wish to improve this aspect of your work. An essential preliminary is to invest in and make full use of a personal reference library (see Box 15.3).

Getting started

A common problem is 'writer's block' – inactivity or stalling brought on by a variety of causes. If blocked, ask yourself these questions:

- Are you comfortable with your surroundings? Make sure you are seated comfortably at a reasonably clear desk and have minimised the possibility of interruptions and distractions.
- Are you trying to write too soon? Have you clarified your thoughts on the subject? Have you done enough preliminary reading?
- Are you happy with the underlying structure of your work? If you haven't made an outline, try this. If you are unhappy because you can't think of a particular detail at the planning stage, just start writing – it is more likely to come to you while you are thinking of something else.
- Are you trying to be too clever? Your first sentence doesn't have to be earth-shattering in content or particularly smart in style. A short statement of fact or a definition is fine. If there will be time for revision, first get your ideas down on paper and then revise grammar, content and order later.
- Do you really need to start writing at the beginning? Try writing the opening remarks after a more straightforward part. For example, with reports of practical work, the Materials and Methods section may be the easiest place to start.
- Are you too tired to work? Don't try to 'sweat it out' by writing for long periods at a stretch: stop frequently for a rest.

Writing with a word processor – use the dynamic/interactive features of the word processor (Chapter 12) to help you get started: first make notes on structure and content, then expand these to form a first draft and finally revise/improve the text.

Box 15.1 How to achieve a clear, readable style

Words and phrases

- Choose short, clear words and phrases rather than long ones: e.g. use 'build' rather than 'fabricate'; 'now' rather than 'at the present time'. At certain times, technical terms must be used for precision, but don't use jargon if you don't have to.
- Don't worry too much about repeating words, especially when to introduce an alternative might subtly alter your meaning.
- Where appropriate, use the first person to describe your actions ('We decided to'; 'I conclude that'), but not if this is specifically discouraged by your supervisor.
- Favour active forms of writing ('the observer completed the survey in ten minutes') rather than a passive style ('the survey was completed by the observer in ten minutes').
- Use tenses consistently. Past tense is always used for Materials and Methods ('samples were taken from...') and for reviewing past work ('Smith (1990) concluded that...'). The present tense is used when describing data ('Fig. 1 shows...'), for generalisations ('Most authorities agree that...') and conclusions ('To conclude, ...').
- Use statements in parentheses sparingly – they disrupt the reader's attention to your central theme.
- Avoid clichés and colloquialisms – they are usually inappropriate in a scientific context.

Punctuation

- Try to use a variety of types of punctuation, to make the text more interesting to read.
- Decide whether you wish to use 'closed' punctuation (frequent commas at the end of clauses) or 'open' punctuation (less frequent punctuation) – be consistent.
- Don't link two sentences with a comma. Use a full stop, this is an example of what *not* to do.
- Pay special attention to apostrophes, using the following rules:

  - To indicate possession, use an apostrophe before an 's' for a singular word (e.g. the rat's temperature was...') and after the s for a plural word ending in s (e.g. the rats' temperatures were = the temperatures of the rats were). If the word has a special plural (e.g. woman → women) then use the apostrophe before the s (the women's temperatures were...).
  - When contracting words, use an apostrophe (e.g. do not = don't; it is = it's), but remember that contractions are generally *not* used in formal scientific writing.
  - Do *not* use an apostrophe for 'its' as the possessive form of 'it' (e.g. 'the university and its surroundings') Note that 'it's' is reserved for 'it is'. This is an exception to the general rule and a very common mistake.
  - Never use an apostrophe to indicate plurals of any kind, including abbreviations.

Sentences

- Don't make them overlong or complicated.
- Introduce variety in structure and length.
- If unhappy with the structure of a sentence, try chopping it into a series of shorter sentences.

Paragraphs

- Get the paragraph length right – five sentences or so. Do *not* submit an essay that consists of a single paragraph, nor one that contains many single sentence paragraphs.
- Make sure each paragraph is logical, dealing with a single topic or theme.
- Take care with the first sentence in a paragraph (the 'topic' sentence); this introduces the theme of the paragraph. Further sentences should then develop this theme, e.g. by providing supporting information, examples, or contrasting cases.
- Use 'linking' words or phrases to maintain the flow of the text within a paragraph (e.g. 'for example'; 'in contrast'; 'however'; 'on the other hand').
- Make your text more readable by adopting modern layout style. The first paragraph in any section of text is usually *not* indented, but following paragraphs may be (by the equivalent of three character spaces). In addition, the space between paragraphs should be slightly larger than the space between lines. Follow departmental guidelines if these specify a format.
- Group paragraphs in sections under appropriate headings and sub-headings to reinforce the structure underlying your writing.
- Think carefully about the first and last paragraphs in any piece of writing: these are often the most important as they respectively set the aims and report the conclusions.

Note: If you're not sure what is meant by any of the terms used here, consult a guide on writing (see p. 100).

Box 15.2 Using appropriate writing styles for different purposes (with examples)

Note that courses tend to move from assignments that are predominantly descriptive in the early years to a more analytical approach towards the final year (see Chapter 5). Also, different styles may be required in different sections of a write-up, e.g. descriptive for introductory historical aspects, becoming more analytical in later sections.

Descriptive writing

This is the most straightforward style, providing factual information on a particular subject, and is most appropriate:

- in essays where you are asked to 'describe' or 'explain' (p. 102);
- when describing the results of a practical exercise, e.g.: 'The experiment shown in Figure 1 confirmed that enzyme activity was strongly influenced by temperature, as the rate observed at 37°C was more than double that seen at 20°C.'

However, in literature reviews and essays where you are asked to 'discuss' (p. 102) a particular topic, the descriptive approach is mostly inappropriate, as in the following example, where a large amount of specific information from a single scientific paper has been used without any attempt to highlight the most important points:

> In a study carried out between July and October 2002, a total of 225 sputum samples from patients attending 25 different clinics in England and Wales were screened. Bacteria were isolated from 67.6% of these samples, with 47.42% of the samples giving *Pseudomonas aeruginosa*, 11.76% *Burkholderia cepacia* and 8.59% *Stentrophomonas maltophilia* (Grey and Gray, 2003).

In the most extreme examples, whole paragraphs or pages of essays may be based on descriptive factual detail from a single source, often with a single citation at the end of the material, as above. Such essays often score low marks in essays where evidence of deeper thinking is required (Chapter 5).

Comparative writing

This technique is an important component of academic writing, and it will be important to develop your comparative writing skills as you progress through your course. Its applications include:

- answering essay questions and assignments of the 'compare and contrast' type (p. 102);
- comparing your results with previously published work in the Discussion section of a practical report.

To use this style, first decide on those aspects you wish to compare and then consider the material (e.g. different literature sources) from these aspects – in what ways do they agree or disagree with each other?

One approach is to compare/contrast a different aspect in each paragraph. At a practical level, you can use 'linking' words and phrases to help orientate your reader as you move between aspects where there is agreement and disagreement. These include, for agreement: 'in both cases'; 'in agreement with'; 'is also shown by the study of'; 'similarly'; 'in the same way'; and for disagreement: 'however'; 'although'; 'in contrast to'; 'on the other hand'; 'which differs from'. The comparative style is fairly straightforward, once you have decided on the aspects to be compared. The following brief example compares two different studies using this style:

> While Grey and Gray (2003) reported that *Pseudo-monas aeruginosa* was present in 47.4% of 225 UK sputum samples, Black and White (2006) showed that 89.1% of sputum samples from 2592 patients were positive for this bacterium.

Comparative text typically makes use of two or more references per paragraph.

Analytical writing

Typically, this is the most appropriate form of writing for:

- a review of scientific literature on a particular topic;
- an essay where you are asked to 'discuss' (p. 102) different aspects of a particular topic;
- evaluating a number of different published sources within the Discussion section of a final-year project dissertation.

By considering the significance of the information provided in the various sources you have read, you will be able to take a more critical approach. Your writing should evaluate the importance of the material in the context of your topic (see also Chapter 9). In analytical writing, you need to demonstrate critical thinking (p. 55) and personal input about the topic in a well-structured text that provides clear messages, presented in a logical order and demonstrating synthesis from a number of sources by appropriate use of citations (p. 47). Detailed information and relevant examples are used only to explain or develop a particular aspect, and not simply as 'padding' to bulk up the essay, as in the following example:

> *Pseudomonas aeruginosa* is often isolated from sputum samples of cystic fibrosis patients: a short-term UK study with a relatively small size sample (225 patients) isolated this bacterium from around half of all samples (Grey and Gray, 2003), while a longer-term study with a far larger sample size (2592 patients) gave an isolation rate of almost 90% (Black and White, 2006).

Analytical writing is based on a broad range of sources, typically with several citations per paragraph.

Box 15.3   Improve your writing ability by consulting a personal reference library

Using dictionaries

We all know that a dictionary helps with spelling and definitions, but how many of us use one effectively? You should:

- Keep a dictionary beside you when writing and always use it if in any doubt about spelling or definitions.
- Use it to prepare a list of words that you have difficulty in spelling: apart from speeding up the checking process, the act of writing out the words helps commit them to memory.
- Use it to write out a personal glossary of terms. This can help you memorise definitions. From time to time, test yourself.

Not all dictionaries are the same! Ask your tutor or supervisor whether he/she has a preference and why. Try out the *Oxford Advanced Learner's Dictionary*, which is particularly useful because it gives examples of use of all words and helps with grammar, e.g. by indicating which prepositions to use with verbs. Dictionaries of biology tend to be variable in quality, possibly because the subject is so wide and new terms are continually being coined. *Henderson's Dictionary of Biological Terms* (Addison Wesley Longman) is a useful example.

Using a thesaurus

A thesaurus contains lists of words of similar meaning grouped thematically; words of opposite meaning always appear nearby.

- Use a thesaurus to find a more precise and appropriate word to fit your meaning, but check definitions of unfamiliar words with a dictionary.
- Use it to find a word or phrase 'on the tip of your tongue' by looking up a word of similar meaning.
- Use it to increase your vocabulary.

Roget's Thesaurus is the standard. Collins publish a combined dictionary and thesaurus.

Using guides for written English

These provide help with the use of words.

- Use guides to solve grammatical problems such as when to use 'shall' or 'will', 'which' or 'that', 'effect' or 'affect', 'can' or 'may', etc.
- Use them for help with the paragraph concept and the correct use of punctuation.
- Use them to learn how to structure writing for different tasks.

Revising your text

Wholesale revision of your first draft is strongly advised for all writing, apart from in exams. Using a word processor, this can be a simple process. Where possible, schedule your writing so you can leave the first draft to 'settle' for at least a couple of days. When you return to it fresh, you will see more easily where improvements can be made. Try the following structured revision process, each stage being covered in a separate scan of your text:

Learning from others – ask a colleague to read through your draft and comment on its content and overall structure.

Revising your text – to improve clarity and shorten your text, 'distil' each sentence by taking away unnecessary words and 'condense' words or phrases by choosing a shorter alternative.

1. Examine content. Have you included everything you need to? Is all the material relevant?
2. Check the grammar and spelling. Can you spot any 'howlers'?
3. Focus on clarity. Is the text clear and unambiguous? Does each sentence really say what you want it to say?
4. Be succinct. What could be missed out without spoiling the essence of your work? It might help to imagine an editor has set you the target of reducing the text by 15 per cent.
5. Improve style. Could the text read better? Consider the sentence and paragraph structure and the way your text develops to its conclusion.

Common errors

These include (with examples):

- Problems over singular and plural words ('a bacteria is'; 'the results shows').
- Verbose text ('One definition that can be employed in this situation is given in the following sentence').
- Misconstructed sentences ('Health and safety regulations should be made aware of ').
- Misuse of punctuation, especially commas and apostrophes (for examples, see Box 15.1).
- Poorly constructed paragraphs (for advice/examples, see Box 15.1).

Sources for further study

Burchfield, R.W. (ed.) (2004) *Fowler's Modern English Usage*, revised 3rd edn. Oxford University Press, Oxford.

Clark, R. *The English Style Book. A Guide to the Writing of Scholarly English*. Available: http://www.litency.com/stylebook/stylebook.php Last accessed: 09/04/07.

Kane, T.S. (1994) *The New Oxford Guide to Writing*. Oxford University Press, New York.
[This is excellent for the basics of English – it covers grammar, usage and the construction of sentences and paragraphs.]

Lindsay, D. (1995) *A Guide to Scientific Writing*, 2nd edn. Longman, Harlow.

McMillan, K.M. and Weyers, J.D.B. (2006) *The Smarter Student: Study Skills and Strategies for Success at University*. Pearson, Harlow.

Partridge, E. (1978) *You Have a Point There*. Routledge, London.
[This covers punctuation in a very readable manner.]

Pechnick, J.A. and Lamb, B.C. (1994) *How to Write about Biology*. HarperCollins, London.

Tichy, H.J. (1988) *Effective Writing for Engineers, Managers and Scientists*, 2nd edn. Wiley, New York.
[This is strong on scientific style and clarity in writing.]

Study exercises

15.1 'Brainstorm' an essay title. Pair up with a partner in your class. Together, pick a suitable essay title from a past exam paper. Using the spider diagram or another technique, individually 'brainstorm' the title. Meet afterwards, compare your ideas, and discuss their relative merits and disadvantages.

15.2 Improve your writing technique. From the following checklist, identify the *three* weakest aspects of your writing, either in your own opinion or from essay/assignment feedback:

- grammar;
- paragraph organisation;
- presentation of work;
- punctuation;
- scientific style;
- sentence structure/variety;
- spelling;
- structure and flow;
- vocabulary.

Now either borrow a book from a local library or buy a book that deals with your weakest aspects of writing. Read the relevant chapters or sections and for each aspect write down some tips that should help you in future.

15.3 Improve your spelling and vocabulary with two lists. Create a pair of lists and pin these up beside your desk. One should be entitled *Spelling Mistakes* and the other *New Words*. Now, whenever you make a mistake in spelling or have to look up how to spell a word in a dictionary, add the problem word to your spelling list, showing where you made the mistake. Also, whenever you come across a word whose meaning is unclear to you, look it up in a dictionary and write the word and its meaning in the 'new words' vocabulary list.

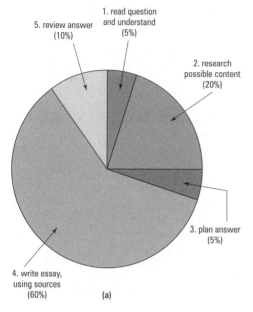

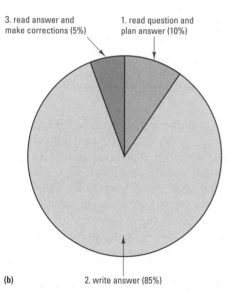

Fig. 16.1 Typical division of time for an essay written as part of in-course assessment (a) or under exam conditions (b).

The function of an essay is to show how much you understand about a topic and how well you can organise and express your knowledge.

Organising your time

The way you should divide your time when producing an essay depends on whether you are writing it for in-course assessment or under exam conditions (Fig. 16.1). Essays written over a long period with access to books and other resources will probably involve a research element, not only before the planning phase but also when writing (Fig. 16.1a). For exams, it is assumed that you have revised appropriately (Chapter 5) and essentially have all the information at your fingertips. To keep things uncomplicated, the time allocated for each essay should be divided into three components – planning, writing and reviewing (Fig. 16.1b), and you should adopt time-saving techniques whenever possible (Box 6.2).

Making a plan for your essay

Dissect the meaning of the essay question or title
Read the title very carefully and think about the topic before starting to write. Consider the definitions of each of the important nouns (this can help in approaching the introductory section). Also think about the meaning of the verb(s) used and try to follow each instruction precisely (see Table 16.1). Don't get sidetracked because you know something about one word or phrase in the title: consider the whole title and all its ramifications. If there are two or more parts to the question, make sure you give adequate attention to each part.

Consider possible content and examples
Research content using the methods described in Chapters 8 and 9. If you have time to read several sources, consider their content in relation to the essay title. Can you spot different approaches to the same subject? Which do you prefer as a means of treating the topic in relation to your title? Which examples are most relevant to your case, and why?

 KEY POINT *Most marks for essays are lost because the written material is badly organised or is irrelevant. An essay plan, by definition, creates order and, if thought about carefully, should ensure relevance.*

Construct an outline
Every essay should have a structure related to its title. Your plan should be written down (but scored through later if written in an exam book). Think about an essay's content in three parts:

1. The introductory section, in which you should include definitions and some background information on the context of the topic being considered. You should also tell your reader how you plan to approach the subject.

Considering essay content – it is rarely enough simply to lay down facts for the reader – you must analyse them and comment on their significance (see p. 98).

Ten Golden Rules for essay writing – *these are framed for in-course assessments (p. 29), though many are also relevant to exams (see also Box 6.2).*

1. *Read the question carefully, and decide exactly what the assessor wants you to achieve in your answer.*
2. *Make sure you understand the question by considering all aspects – discuss your approach with colleagues or a tutor.*
3. *Carry out the necessary research (using books, journals, WWW), taking appropriate notes. Gain an overview of the topic before getting involved with the details.*
4. *Always plan your work in outline before you start writing. Check that your plan covers the main points and that it flows logically.*
5. *Introduce your essay by showing that you understand the topic and stating how you intend to approach it.*
6. *As you write the main content, ensure it is relevant by continually looking back at the question.*
7. *Use headings and sub-headings to organise and structure your essay.*
8. *Support your statements with relevant examples, diagrams and references where appropriate.*
9. *Conclude by summarising the key points of the topic, indicating the present state of knowledge, what we still need to find out and how this might be achieved.*
10. *Always reread the essay before submitting it. Check grammar and spelling and confirm that you have answered all aspects of the question.*

Using diagrams – *give a title and legend to each diagram so that it makes sense in isolation and point out in the text when the reader should consult it (e.g. 'as shown in Fig. 1 ...' or 'as can be seen in the accompanying diagram, ...').*

Table 16.1 Instructions often used in essay questions and their meanings. When more than one instruction is given (e.g. compare and contrast; describe and explain), make sure you carry out *both* or you may lose a large proportion of the available marks (see also Table 5.1)

Account for:	give the reasons for
Analyse:	examine in depth and describe the main characteristics of
Assess:	weigh up the elements of and arrive at a conclusion about
Comment:	give an opinion on and provide evidence for your views
Compare:	bring out the similarities between
Contrast:	bring out dissimilarities between
Criticise:	judge the worth of (give both positive and negative aspects)
Define:	explain the exact meaning of
Describe:	use words and diagrams to illustrate
Discuss:	provide evidence or opinions about, arriving at a balanced conclusion
Enumerate:	list in outline form
Evaluate:	weigh up or appraise; find a numerical value for
Explain:	make the meaning of something clear
Illustrate:	use diagrams or examples to make clear
Interpret:	express in simple terms, providing a judgement
Justify:	show that an idea or statement is correct
List:	provide an itemised series of statements about
Outline:	describe the essential parts only, stressing the classification
Prove:	establish the truth of
Relate:	show the connection between
Review:	examine critically, perhaps concentrating on the stages in the development of an idea or method
State:	express clearly
Summarise:	without illustrations, provide a brief account of
Trace:	describe a sequence of events from a defined point of origin

2. The middle of the essay, where you develop your answer and provide relevant examples. Decide whether a broad analytical approach is appropriate or whether the essay should contain more factual detail.
3. The conclusion, which you can make quite short. You should use this part to summarise and draw together the components of the essay, without merely repeating previous phrases. You might mention such things as: the broader significance of the topic; its future; its relevance to other important areas of biology. Always try to mention both sides of any debate you have touched on, but beware of 'sitting on the fence'.

 KEY POINT *Use paragraphs to make the essay's structure obvious. Emphasise them with headings and sub-headings unless the material beneath the headings would be too short or trivial.*

Start writing

- Never lose track of the importance of content and its relevance. Repeatedly ask yourself: 'Am I really answering this question?' Never waffle just to increase the length of an essay. Quality, rather than quantity, is important.
- Illustrate your answer appropriately. Use examples to make your points clear, but remember that too many similar examples can stifle the flow of an essay. Use diagrams where a written description would be difficult or take too long. Use tables to condense information.

- Take care with your handwriting. You can't get marks if your writing is illegible. Try to cultivate an open form of handwriting, making the individual letters large and distinct. If there is time, make out a rough draft from which a tidy version can be copied.

Reviewing your answer

Learning from lecturers' and tutors' comments – ask for further explanations if you don't understand a comment or why an essay was less successful than you thought it should have been.

Make sure that you leave enough time to:
- reread the question to check that you have answered all points;
- reread your essay to check for errors in punctuation, spelling and content. Make any corrections obvious. In an exam don't panic if you suddenly realise you've missed a large chunk out as the reader can be redirected to a supplementary paragraph if necessary.

Sources for further study

Anon. (2004) *Essay and Report Writing Skills*. Open University, Milton Keynes.

Anon. Yahoo! Directory: Writing > Essays and Research Papers. Available: http://dir.yahoo.com/Social_Science/Communications/Writing/Essays_and_Research_Papers/
Last accessed 09/04/07.
[An extensive directory of Web resources.]

Good, S. and Jensen, B. (1995) *The Student's Only Survival Guide to Essay Writing*. Orca Book Publishers, Victoria, USA.

Study exercises

16.1 Practise dissecting essay titles. Use past exam papers, or make up questions based on learning objectives for your course and your lecture notes. Take each essay title and carefully 'dissect' the wording, working out exactly what you think the assessor expects you to do (see e.g. Table 16.1).

16.2 Write essay plans under self-imposed time limits. Continuing from Table 16.1, outline plans for essays from a past exam paper. Allow yourself a maximum of 5 minutes per outline. Within this time your main goal is to create an essay plan. To do this, you may need to 'brainstorm' the topic. Alternatively, if you allocate 10 minutes per essay, you may be able to provide more details, e.g. list the examples you could describe.

16.3 Practise reviewing your work carefully. For the next assignment you write, review it fully as part of the writing process. This will require you to finish the first draft about one week before the hand-in date, e.g. by setting yourself an earlier deadline than the submission date. This exercise is best done with a word-processed essay. Don't worry if it is a little over the word limit at this stage.

(a) Print out a copy of the essay. Do not look at it for at least two days after finishing this version.

(b) Review 1: spelling, grammar and sense. Read through the draft critically (try to imagine it had been written by someone else) and correct any obvious errors that strike you. Does the text make sense? Do sentences/paragraphs flow smoothly?

(c) Review 2: structure and relevance. Consider again the structure of the essay, asking yourself whether you have really answered the question that was set (see p. 32). Are all the parts in the right order? Is anything missed out? Have you followed precisely the instruction(s) in the title? Are the different parts of the essay linked together well?

(d) Review 3: shorten and improve style. Check the word count. Shorten the essay if required. Look critically at phrasing and, even if the essay is within the word limit, ask yourself whether any of the words are unnecessary or whether the text could be made more concise, more precise or more apt.

Practical reports, project reports, theses and scientific papers differ greatly in depth, scope and size, but they all have the same basic structure. Some variation is permitted, however (see Box 17.1), and you should always follow the advice or rules provided by your department.

Additional parts may be specified: for project reports, dissertations and theses, a Title Page is often required and a List of Figures and Tables as part of the Contents section. When work is submitted for certain degrees, you may need to include declarations and statements made by the student and supervisor. In scientific papers, a list of Keywords is often added following the Abstract: this information may be combined with words in the title for computer cross-referencing systems.

> *Basic structure of scientific reports* – this usually follows the 'IMRAD' acronym: Introduction, Materials and Methods, Results and Discussion.

 KEY POINT *Department, school or faculty regulations may specify a precise format for producing your report or thesis. Obtain a copy of these rules at an early stage and follow them closely, to avoid losing marks.*

Practical and project reports

These are exercises designed to make you think more deeply about your experiments and to practise and test the skills necessary for writing up research work. Special features are:

- Introductory material is generally short and unless otherwise specified should outline the aims of the experiment(s) with a minimum of background material.
- Materials and methods may be provided by your supervisor for practical reports. If you make changes to this, you should state clearly what you did. With project work, your lab notebook (see p. 192) should provide the basis for writing this section.
- Great attention in assessment will be paid to presentation and analysis of data. Take special care over graphs (see p. 387 for further advice). Make sure your conclusions are justified by the evidence you present.

> *Options for discussing data* – the main optional variants of the general structure include combining Results and Discussion *into a single section and adding a separate* Conclusions *section.*
>
> - *The main advantage of a joint* Results and Discussion *section is that you can link together different experiments, perhaps explaining why a particular result led to a new hypothesis and the next experiment. However, a combined* Results and Discussion *section may contravene your department's regulations, so you should check before using this approach.*
> - *The main advantage of having a separate* Conclusions *section is to draw together and emphasise the chief points arising from your work, when these may have been 'buried' in an extensive* Discussion *section.*

Theses and dissertations

These are submitted as part of the examination for a degree following an extended period of research. They act to place on record full details about your experimental work and will normally only be read by those with a direct interest in it – your examiners or colleagues. Note the following:

- You are allowed scope to expand on your findings and to include detail that might otherwise be omitted in a scientific paper.
- You may have problems with the volume of information that has to be organised. One method of coping with this is to divide your thesis into chapters, each having the standard format (as in Box 17.1). A General Introduction can be given at the start and a General Discussion at the end. Discuss this with your supervisor.

> *Oral assessments* – there may be an oral exam ('viva voce') associated with the submission of a thesis or dissertation. The primary aim of the examiners will be to ensure that you understand what you did and why you did it.

Box 17.1   The structure of reports of experimental work

Undergraduate practical and project reports are generally modelled on this arrangement or a close variant of it, because this is the structure used for nearly all research papers and theses. The more common variations include Results and Discussion combined into a single section and Conclusions appearing separately as a series of points arising from the work. In scientific papers, a list of Keywords (for computer cross-referencing systems) may be included following the Abstract. Acknowledgements may appear after the Contents section, rather than near the end. Department or faculty regulations for producing theses and reports may specify a precise format; they often require a Title Page to be inserted at the start and a List of Figures and Tables as part of the Contents section, and may specify declarations and statements to be made by the student and supervisor.

Part (in order)	Contents/purpose	Checklist for reviewing content
Title	Explains what the project was about	Does it explain what the text is about succinctly?
Authors plus their institutions	Explains who did the work and where; also where they can be contacted now	Are all the details correct?
Abstract/Summary	Synopsis of methods, results and conclusion of work described. Allows the reader to grasp quickly the essence of the work	Does it explain why the work was done? Does it outline the whole of your work and your findings?
Contents	Shows the organisation of the text (not required for short papers)	Are all the sections covered? Are the page numbers correct?
Abbreviations	Lists all the abbreviations used (but not those of SI, chemical elements, or standard biochemical terms)	Have they all been explained? Are they all in the accepted form? Are they in alphabetical order?
Introduction	Orientates the reader, explains why the work has been done and its context in the literature, why the methods used were chosen, why the experimental organisms were chosen. Indicates the central hypothesis behind the experiments	Does it provide enough background information and cite all the relevant references? Is it of the correct depth for the readership? Have all the technical terms been defined? Have you explained why you investigated the problem? Have you outlined your aims and objectives? Have you explained your methodological approach? Have you stated your hypothesis?
Materials and Methods	Explains how the work was done. Should contain sufficient detail to allow another competent worker to repeat the work	Is each experiment covered and have you avoided unnecessary duplication? Is there sufficient detail to allow repetition of the work? Are proper scientific names and authorities given for all organisms? Have you explained where you got them from? Are the correct names, sources and grades given for all chemicals?
Results	Displays and describes the data obtained. Should be presented in a form that is easily assimilated (graphs rather than tables, small tables rather than large ones)	Is the sequence of experiments logical? Are the parts adequately linked? Are the data presented in the clearest possible way? Have SI units been used properly throughout? Has adequate statistical analysis been carried out? Is all the material relevant? Are the figures and tables all numbered in the order of their appearance? Are their titles appropriate? Do the figure and table legends provide all the information necessary to interpret the data without reference to the text? Have you presented the same data more than once?
Discussion/ Conclusions	Discusses the results: their meaning, their importance; compares the results with those of others; suggests what to do next	Have you explained the significance of the results? have you compared your data with other published work? Are your conclusions justified by the data presented?
Acknowledgements	Gives credit to those who helped carry out the work	Have you listed everyone that helped, including any grant-awarding bodies?
Literature Cited (Bibliography)	Lists all references cited in appropriate format: provides enough information to allow the reader to find the reference in a library	Do all the references in the text appear on the list? Do all the listed references appear in the text? Do the years of publications and authors match? Are the journal details complete and in the correct format? Is the list in alphabetical order, or correct numerical order?

Steps in the production of a practical report or thesis

Choose the experiments you wish to describe and decide how best to present them

Try to start this process before your lab work ends, because at the stage of reviewing your experiments, a gap may become apparent (e.g. a missing control) and you might still have time to rectify the deficiency. Irrelevant material should be ruthlessly eliminated, at the same time bearing in mind that negative results can be extremely important (see p. 184). Use as many different forms of data presentation as are appropriate, but avoid presenting the same data in more than one form. Relegate large tables of primary data to an appendix and summarise the important points within the main text (with a cross-reference to the appendix). Make sure that the experiments you describe are representative: always state the number of times they were repeated and how consistent your findings were.

Make up plans or outlines for the component parts

The overall structure of practical and project reports is well defined (see Box 17.1), but individual parts will need to be organised as with any other form of writing (see Chapter 15).

Write

The Materials and Methods section is often the easiest to write once you have decided what to report. Remember to use the past tense and do not allow results or discussion to creep in. The Results section is the next easiest as it should only involve description. At this stage, you may benefit from jotting down ideas for the Discussion – this may be the hardest part to compose as you need an overview both of your own work and of the relevant literature. It is also liable to become wordy, so try hard to make it succinct. The Introduction shouldn't be too difficult if you have fully understood the aims of the experiments. Write the Abstract and complete the list of references at the end. To assist with the latter, it is a good idea as you write to jot down the references you use or to pull out their cards from your index system.

Revise the text

Once your first draft is complete, try to answer all the questions given in Box 17.1. Show your work to your supervisors and learn from their comments. Let a friend or colleague who is unfamiliar with your subject read your text; they may be able to pinpoint obscure wording and show where information or explanation is missing. If writing a thesis, double-check that you are adhering to your institution's thesis regulations.

Prepare the final version

Markers appreciate neatly produced work but a well-presented document will not disguise poor science! If using a word processor, print the final version with the best printer available. Make sure figures are clear and in the correct size and format.

Submit your work

Your department will specify when to submit a thesis or project report, so plan your work carefully to meet this deadline or you may lose marks. Tell your supervisor early of any circumstances that may cause delay and check to see whether any forms must be completed for late submission, or evidence of extenuating circumstances.

Choosing between graphs and tables – graphs are generally easier for the reader to assimilate, whereas tables can be used to condense a lot of data into a small space.

Repeating your experiments – remember, if you do an experiment twice, you have repeated it only once.

Presenting your results – remember that the order of results presented in a report need not correspond with the order in which you carried out the experiments: you are expected to rearrange them to provide a logical sequence of findings.

Using the correct tense – always use the past tense to describe the methodology used in your work, since it is now complete. Use the present tense only for generalisations and conclusions (p. 97).

Producing a scientific paper

Scientific papers are the means by which research findings are communicated to others. Peer-reviewed papers are published in journals; each covers a well-defined subject area and publishes details of the format they expect.

Peer review – *the process of evaluation and review of a colleague's work. In scientific communication, a paper is reviewed by two or more expert reviewers for comments on quality and significance as a key component of the validation procedure.*

 KEY POINT *Peer review is an important component of the process of scientific publication; only those papers whose worth is confirmed by the peer-review process will be published.*

It would be very unusual for an undergraduate to submit a paper on his or her own – this would normally be done in collaboration with your project supervisor and only then if your research has satisfied appropriate criteria. However, it is important to understand the process whereby a paper comes into being (Box 17.2), as this can help you understand and interpret the primary literature.

Box 17.2 Steps in producing a scientific paper

Scientific papers are the lifeblood of any science and it is a major landmark in your scientific career to publish your first paper. The main steps in doing this should include the following:

Assessing potential content
The work must be of an appropriate standard to be published and should be 'new, true and meaningful'. Therefore, before starting, the authors need to review their work critically under these headings. The material included in a scientific paper will generally be a subset of the total work done during a project, so it must be carefully selected for relevance to a clear central hypothesis – if the authors won't prune, the referees and editors of the journal certainly will!

Choosing a journal
There are thousands of journals covering biology and each covers a specific area (which may change through time). The main factors in deciding on an appropriate journal are the range of subjects it covers, the quality of its content and the number and geographical distribution of its readers. The choice of journal always dictates the format of a paper since authors must follow to the letter the journal's 'Instructions to Authors'.

Deciding on authorship
In multi-author papers, a contentious issue is often who should appear as an author and in what order they should be cited. Where authors make an equal contribution, an alphabetical order of names may be used. Otherwise, each author should have made a substantial contribution to the paper and should be prepared to defend it in public. Ideally, the order of appearance will reflect the amount of work done rather than seniority. This may not happen in practice!

Writing
The paper's format will be similar to that shown in Box 17.1, and the process of writing will include outlining, reviewing, etc., as discussed elsewhere in this chapter. Figures must be finished to an appropriate standard and this may involve preparing photographs or digital images of them.

Submitting
When completed, copies of the paper are submitted to the editor of the chosen journal with a simple covering letter. A delay of one to two months usually follows while the manuscript is sent to two or more anonymous referees who will be asked by the editor to check that the paper is novel, scientifically correct and that its length is justified.

Responding to referees' comments
The editor will send on the referees' comments to the authors who will then have a chance to respond. The editor will decide on the basis of the comments and replies to them whether the paper should be published. Sometimes quite heated correspondence can result if the authors and referees disagree!

Checking proofs and waiting for publication
If a paper is accepted, it will be sent off to the typesetters. The next the authors see of it is the proofs (first printed version in style of journal), which have to be corrected carefully for errors and returned. Eventually, the paper will appear in print, but a delay of six months following acceptance is not unusual. Nowadays, papers are often available electronically, via the Web, in pdf format – see p. 49 for advice on how to cite 'online early' papers.

Sources for further study

Berry, R. (2004) *The Research Project: How to Write it*, 5th edn. Routledge, London.

Davis, M. (2005) *Scientific Papers and Presentations*. Academic Press, London.

Day, R.A. and Gastel, B. (2006) *How to Write and Publish a Scientific Paper*, 6th edn. Cambridge University Press, Cambridge.

Lobban, C.S. and Schefter, M. (1992) *Successful Lab Reports: a Manual for Science Students*. Cambridge University Press, Cambridge.

Luck, M. (1999) *Your Student Research Project*. Gower, London.

Luey, B. (2002) *Handbook for Academic Authors*, 4th edn. Cambridge University Press, Cambridge.

Matthews, J.R., Bowen, J.M. and Matthews, R. (2000) *Successful Science Writing: A Step-by-step Guide for the Biological and Medical Sciences,* 2nd edn. Cambridge University Press, Cambridge.

Valiela, I. (2001) *Doing Science: Design, Analysis and Communication of Scientific Research*. Oxford University Press, Oxford.
[Covers scientific communication, graphical presentations and aspects of statistics.]

Study exercises

17.1 Write a formal 'Materials and Methods' section. Adopting the style of a research paper (e.g. past tense, all relevant detail reported such that a competent colleague could repeat your work), write out the Materials and Methods for a practical you have recently carried out. Ask a colleague or tutor to comment on what you have written.

17.2 Describe a set of results in words. Again adopting the style of a research paper, write a paragraph describing the results contained in a particular table or graph. Ask a colleague or tutor to comment on your description, to identify what is missing or unclear.

17.3 Write an abstract for a paper. Pair up with a colleague. Each of you should independently choose a different research paper in a current journal. Copy the paper, but mask over the abstract section, having first counted the words used. Swap papers. Now, working to the same amount of words as in the original, read the paper and provide an abstract of its contents. Then, compare this with the real abstract.

The literature survey or review is a specialised form of essay that summarises and reviews the evidence and concepts concerning a particular area of research.

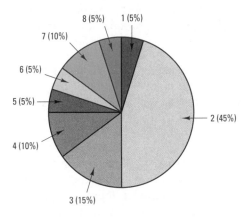

Fig. 18.1 Pie chart showing how you might allocate time for a literature survey:
1. select a topic;
2. scan the literature;
3. plan the review;
4. write first draft;
5. leave to settle;
6. prepare a structured review of text;
7. write final draft;
8. produce top copy.

Creating a glossary – one barrier to developing an understanding of a new topic is the jargon used. To overcome this, create your own glossary. You may wish to cross-reference a range of sources to ensure the definitions are reliable and context-specific. Remember to note your sources in case you wish to use the definition within your review.

Using index cards – these can help you organise large numbers of references. Write key points and author information on each card – this helps when considering where the reference fits into the literature. Arrange the cards in subject piles, eliminating irrelevant ones. Order the cards in the sequence in which you wish to write.

> **KEY POINT** A literature review should **not** *be a simple recitation of facts. The best reviews are those that analyse information rather than simply describe it.*

Making up a timetable

Figure 18.1 illustrates how you might divide up your time for writing a literature survey. There are many subdivisions in this chart because of the size of the task: in general, for lengthy tasks, it is best to divide up the work into manageable chunks. Note also that proportionally less time is allocated to writing itself than with an essay. In a literature survey, make sure that you spend adequate time on research and revision.

Selecting a topic

You may have no choice in the topic to be covered, but if you do, carry out your selection as a three-stage process:

1. Identify a broad subject area that interests you.
2. Find and read relevant literature in that area. Try to gain a broad impression of the field from books and general review articles. Discuss your ideas with your supervisor.
3. Select a relevant and concise title. The wording should be considered very carefully as it will define the content expected by the reader. A narrow subject area will cut down on the amount of literature you will be expected to review, but will also restrict the scope of the conclusions you can make (and vice versa for a wide subject area).

Scanning the literature and organising your references

You will need to carry out a thorough investigation of the literature before you start to write. The key problems are as follows:

- Getting an initial toe-hold in the literature. Seek help from your supervisor, who may be willing to supply a few key papers to get you started. Hints on expanding your collection of references are given on p. 45.
- Assessing the relevance and value of each article. This is the essence of writing a review, but it is difficult unless you already have a good understanding of the field. Try reading earlier reviews in your area and discussing the topic with your supervisor or other academic staff.
- Clarifying your thoughts. Subdividing the main topic and assigning your references to these smaller subject areas may help you gain a better overview of the literature.

Deciding on structure and content

The general structure and content of a literature survey are described below. The *Annual Review* series (available in most university libraries) provides good examples of the style expected in reviews of biosciences.

Introduction

Defining terms – the Introduction is a good place to explain the meaning of the key terms used in your survey or review.

The introduction should give the general background to the research area, concentrating on its development and importance. You should also make a statement about the scope of your survey; as well as defining the subject matter to be discussed, you may wish to restrict the period being considered.

Main body of text

The review itself should discuss the published work in the selected field and maybe subdivided into appropriate sections. Within each portion of a review, the approach is usually chronological, with appropriate linking phrases (e.g. 'Following on from this, …'; 'Meanwhile, Bloggs (1980) tackled the problem from a different angle …'). However, a good review is much more than a chronological list of work done. It should:

- allow the reader to obtain an overall view of the current state of the research area, identifying the key areas where knowledge is advancing;
- show how techniques are developing and discuss the benefits and disadvantages of using particular organisms or experimental systems;
- assess the relative worth of different types of evidence – this is the most important aspect (see Chapter 9). Do not be intimidated from taking a critical approach as the conclusions you may read in the primary literature aren't always correct;
- indicate where there is conflict in findings or theories, suggesting if possible which side has the stronger case;
- indicate gaps in current knowledge.

Balancing opposing views – even if you favour one side of a disagreement in the literature, your review should provide a balanced and fair description of all the published views of the topic. Having done this, if you do wish to state a preference, give reasons for your opinion.

You do not need to wait until you have real all the sources available to you before starting to write the main body. Word processors allow you to modify and move pieces of text at any point and it will be useful to write paragraphs about key sources, or groups of related papers, as you read them. Try to create a general plan for your review as soon as possible. Place your draft sections of text under an appropriate set of sub-headings that reflects your plan, but be prepared to rearrange these and retitle or reorder sections as you proceed. Not only will working in this way help to clarify your thoughts, but it may help you avoid a last-minute rush of writing near to the submission date.

Conclusions

The conclusions should draw together the threads of the preceding parts and point the way forward, perhaps listing areas of ignorance or where the application of new techniques may lead to advances.

References, etc.

Making citations – a review of literature poses stylistic problems because of the need to cite large numbers of papers; in the Annual Review series this is overcome by using numbered references (see p. 48).

The References or Literature Cited section should provide full details of all papers referred to in the text (see p. 48). The regulations for your department may also specify a format and position for the title page, contents page, acknowledgements, etc.

Source for further study

Rudner, L.M. and Schafer, W.D. (1999) How to write a scholarly research report, *Practical Assessment, Research & Evaluation*, 6(13). Available: http://pareonline.net/getvn.asp?v = 6&n = 13.
Last accessed: 09/04/07.

Study exercises

18.1 Summarise the main differences between a review and a scientific paper. From the many subject areas in the *Annual Review* series (find via your library's periodical indexing system), pick one that matches your subject interests, and within this find a review that seems relevant or interesting. Read the review and write down *five* ways in which the writing style and content differs from that seen in scientific papers.

18.2 Gather a collection of primary sources for a topic. From the journal section of the library, select an interesting scientific paper published about 5–10 years ago. First, work *back* from the references cited by that paper: can you identify from the text or the article titles which are the most important and relevant to the topic? List five of these, using the proper conventions for citing articles in a reference list (see Chapter 8). Note that each of these papers will also cite other articles, always going back in time. Now, using the Institute of Scientific Information (ISI) Science Citation Index or a similar system (see p. 46), work *forward* and find out who has cited your selected article in the time since its publication. Again list the five most important articles found.

18.3 Write a synopsis of a review. Again using one of the *Annual Review* series as a source, allow yourself just *five* single-sentence bullet points to summarise the key points reported in a particular review.

Fundamental laboratory techniques

All knowledge and theory in science has originated from practical observation and experimentation: this is equally true for disciplines as diverse as microscopy and molecular genetics. Practical work is an important part of most courses and often accounts for a significant proportion of the assessment marks. The abilities developed in practical classes will continue to be useful throughout your course and beyond, some within science and others in any career you choose (see Chapter 1).

Developing practical skills – these will include:

- *designing experiments*
- *observing and measuring*
- *recording data*
- *analysing and interpreting data*
- *reporting/presenting.*

Being prepared

 KEY POINT *You will get the most out of practicals if you prepare well in advance. Do not go into a practical session assuming that everything will be provided, without any input on your part.*

The main points to remember are:

- Read any handouts in advance: make sure you understand the purpose of the practical and the particular skills involved. Does the practical relate to, or expand upon, a current topic in your lectures? Is there any additional preparatory reading that will help?
- Take along appropriate textbooks, to explain aspects in the practical.
- Consider what safety hazards might be involved, and any precautions you might need to take, before you begin (p. 120).
- Listen carefully to any introductory guidance and note any important points: adjust your schedule/handout as necessary.
- During the practical session, organise your bench space – make sure your lab book is adjacent to, but not within, your working area. You will often find it easiest to keep clean items of glassware etc. on one side of your working space, with used equipment on the other side.
- Write up your work as soon as possible, and submit it on time, or you may lose marks.
- Catch up on any work you have missed as soon as possible – preferably, before the next practical session.

Using textbooks in the lab – take this book along to the relevant classes, so that you can make full use of the information during the practical sessions.

Ethical and legal aspects

You will need to consider the ethical and legal implications of biological work at several points during your studies:

- Safe working means following a code of safe practice, supported by legislation, alongside a moral obligation to avoid harm to yourself and others, as discussed in Chapter 20.
- Any laboratory work that involves working with animals must be carefully considered; Chapter 40 covers this in relation to dissection.
- Fieldwork can have a legal aspect, e.g. in relation to specimen collection (Chapter 34), since legislation may protect particular habitats and/or species. Ethical and human aspects might include: possible damage to fauna and flora; safety of others within your group; implications of your work to others, e.g. landowners or other users (Chapter 29). These must all be considered before any work is carried out.

 SAFETY NOTE *Mobile phones should never be used in a lab class, as there is a risk of contamination from hazardous substances. Always switch off your mobile phone before entering a laboratory. Conversely, they are an extremely useful accessory for fieldwork.*

Getting to grips with bioethics – in addition to any moral implications of your lab practicals and field work, you may have the opportunity to address broader issues within your course (see Box 19.1). Professional scientists should always consider the consequences of their work, and it is therefore important that you develop your appreciation of these issues alongside your academic studies.

Box 19.1 Bioethics

Contemporary bioscience degree programmes place increasing emphasis on the ethical and social impacts of scientific advances, and on the need for scientists engaged in potentially controversial work to communicate their ideas to the general public. Bioscience research raises many moral and legal dilemmas, requiring difficult choices to be made (e.g. animal testing of medical products), and students are likely to be asked to reflect on bioethical topics, e.g. in group discussions and debates on current issues such as:

- environmental ethics (e.g. use of genetically modified plants, hunting of endangered species);
- animal ethics (e.g. factory farming, transgenic animals, xenotransplantation);
- medical and social ethics (e.g. human cloning, embryo research, genetic testing).

In discussing such topics (sometimes referred to as ethical, legal and social issues, ELSI), you will find that there is rarely a 'right' or 'wrong' answer. However, it is important to be able to consider these issues in a logical manner, and to provide a reasoned argument in support of a particular viewpoint. Gaining experience in such debates should also help you understand some of the issues linked to the public understanding of science, and how these can be addressed (see: http://www.copus.org.uk/pubs_guides.html). Although a full exposition is beyond the scope of this book, the following provides a framework of principles for considering particular topics:

- Beneficence – the obligation to do good (for example, if it is possible to prevent suffering by a particular course of action, then it should be carried out).
- Non-maleficence – the duty to cause no harm (contained within the Hippocratic Oath of medical practitioners).
- Justice – the obligation to treat all people fairly and impartially (for example, lack of discrimination between people on the grounds of race or sex).
- Autonomy – the duty to allow an individual to make their own choices, without constraints (this principle underlies the notion of informed consent in medical research).

- Respect – the need to show due regard for others (for example, by taking into account the rights and beliefs of all people equally).
- Rationality – the notion that a particular action or choice should be based on reason and logic (many scientists would argue that the scientific method is an example of rationality).
- Precautionary principle – the notion that it is better not to carry out an action if there is any risk of harm (for example, in deciding that the risks of building nuclear power stations outweigh their potential benefits).

Understanding the various theories of ethics may also help you formulate your ideas. These include:

- Utilitarianism – the notion that it is ethical to choose the action that produces the greatest good for the greatest number.
- Deontology – a theory that states than a particular action is either intrinsically good (right) or bad (wrong). According to deontological theory, decisions should be based on the actions themselves, rather than on their consequences.
- Virtue theory – the notion that making decisions according to established virtues (e.g. honesty, wisdom, justice) will lead to ethically valid choices.
- Objectivism – the theory that what is right and wrong is intrinsic and applies equally to all people, places and times (the alternative is that morality is subjective, being dependent on the views of each individual)

The principles and issues of bioethics are considered on a number of websites, including:

- introductory bioethics (e.g. http://www.access excellence.org/RC/AB/IE/);
- bioethics resources (e.g. http://bioethics.od.nih.gov/, http://www.ethicsweb.ca/resources/ and http://www.beep.ac.uk/content/index.php/);
- ethics discussion groups (e.g. http://www.beep.ac.uk/content/491.0.html/ and http://www-hsc.usc.edu/~mbernste/).

The following introductory texts give further information and guidance: Bryant et al. (2005), Gert et al. (2006) and Mepham (2005).

Basic requirements

Recording practical results

An A4 loose-leaf ring binder offers flexibility, since you can insert laboratory handouts or lined and graph paper at appropriate points. The danger of losing one or more pages from a loose-leaf system is the main drawback.

Presenting results – though you don't need to be a graphic designer to produce work of a satisfactory standard, presentation and layout are important and you will lose marks for poorly presented work. Chapter 62 gives further practical advice.

Using inexpensive calculators – many unsophisticated calculators have a restricted display for exponential numbers and do not show the 'power of 10', e.g. displaying 2.4×10^{-5} as 2.4^{-05}, or even $2.4E{-}05$, or $2.4 - 05$.

Using calculators for numerical problems – Chapter 64 gives further advice.

Presenting graphs and diagrams – ensure these are large enough to be easily read: a common error is to present graphs or diagrams that are too small, with poorly chosen scales (see p. 388).

Bound books avoid this problem, although those containing alternating lined/graph or lined/blank pages tend to be wasteful – it is often better to paste sheets of graph paper into a bound book, as required.

A good quality HB pencil or propelling pencil is recommended for recording your raw data, making diagrams, etc., as mistakes are easily corrected. Buy a black, spirit-based (permanent) marker for labelling experimental glassware, Petri plates, etc. Fibre-tipped fine-line drawing/lettering pens are useful for preparing final versions of graphs and diagrams for assessment purposes. Use a see-through ruler (with an undamaged edge) for graph drawing, so that you can see data points and information below the ruler as you draw.

Calculators

These range from basic machines with no pre-programmed functions and only one memory, to sophisticated programmable portable computers with many memories. The following may be helpful when using a calculator:

- Power sources. Choose a battery-powered machine, rather than a mains-operated or solar-powered type. You will need one with basic mathematical/scientific operations, including powers, logarithms (p. 411), roots and parentheses (brackets), together with statistical functions such as sample means and standard deviations (Chapter 65).
- Mode of operation. The older operating system used by e.g. Hewlett-Packard calculators is known as the reverse Polish notation: to calculate the sum of two numbers, the sequence is 2 [enter] 4 + and the answer 6 is displayed. The more usual method of calculating this equation is as $2 + 4 =$, which is the system used by the majority of modern calculators. Most newcomers find the latter approach to be more straightforward. Spend some time finding out how a calculator operates, e.g. does it have true algebraic logic ($\sqrt{}$ then number, rather than number then $\sqrt{}$)? How does it deal with (and display) scientific notation and logarithms (pp. 410–11)?
- Display. Some calculators will display an entire mathematical operation (e.g. '$2 + 4 = 6$'), while others simply display the last number/operation. The former type may offer advantages in tracing errors.
- Complexity. In the early stages, it is usually better to avoid the more complex machines, full of impressive-looking, but often unused pre-programmed functions – go for more memory, parentheses, or statistical functions rather than engineering or mathematical constants. Programmable calculators may be worth considering for more advanced studies. However, it is important to note that such calculators are often unacceptable for exams.

Presenting more advanced practical work

In some practical reports and in project work, you may need to use more sophisticated presentation equipment. Computer-based graphics packages can be useful – choose easily read fonts such as Arial or Helvetica for posters and consider the layout and content carefully (p. 76). Alternatively, you could use fine-line drawing pens and dry-transfer lettering/symbols, although this can be more time-consuming than computer-based systems, e.g. using Microsoft Excel (p. 72 and 391–3).

> **Printing on acetates** – standard overhead transparencies are not suitable for use in laser printers or photocopiers: you need to make sure that you use the correct type.

To prepare overhead transparencies for spoken presentations, you can use spirit-based markers and acetate sheets. An alternative approach is to print directly from a computer-based package, using a laser printer and special acetates, or directly on to 35 mm slides. You can also photocopy on to special acetates. Further advice on content and presentation is given in Chapter 14.

Text references

Bryant, J., Baggott le Velle, L. and Searle, J. (2005) *Introduction to Bioethics*. Wiley, Chichester.

Gert, B., Culver, C.M. and Clouser, K.D. (2006) *Bioethics: A Systematic Approach,* 2nd edn. Oxford University Press, Oxford.

Mepham, B. (2005) *Bioethics: An Introduction for the Biosciences*. Oxford University Press, Oxford.

Sources for further study

Barnard, C.J., Gilbert, F.S. and MacGregor, P.K. (2001) *Asking Questions in Biology: Key Skills for Practical Assessments and Project Work*, 2nd edn. Prentice Hall, Harlow.

Howell, J.H., Sale, W.F. and Callahan, D. (2000) *Life Choices: A Hastings Center Introduction to Bioethics*, 2nd edn. Georgetown University Press, Washington DC.

Mier-Jedrzejowicz, W.A.C. (1999) *A Guide to HP Handheld Calculators and Computers*. Wilson-Barnett, Tustin, California.
[Provides further guidance on the use of Hewlett-Packard calculators (reverse Polish notation).]

Singer, P. and Kuhse, H. (2006) *Bioethics: An Anthology*. Blackwell, Oxford.

Study exercises

19.1 Consider the value of practical work. Spend a few minutes thinking about the purpose of practical work within a specific part of your course (e.g. a particular first-year module) and then write a list of the six most important points. Compare your list with the generic list that we have provided on p. 443, which is based on our experience as lecturers – does it differ much from your list, which is drawn up from a student perspective?

19.2 Make a list of items required for a particular practical exercise. This exercise is likely to be most useful if you can relate it to an appropriate practical session on your course. However, we have given a model list for a snail dissection, to help you get started. (Note – see also Chapter 29, for a similar study exercise related to fieldwork.)

19.3 Check your calculator skills. Carry out the following mathematical operations, using either a hand-held calculator or a PC with appropriate 'calculator' software.
(a) $5 \times (2 + 6)$
(b) $(8.3 \div [6.4 - 1.9]) \times 24$ (to four significant figures)
(c) $(1 \div 32) \times (5 \div 8)$ (to three significant figures)
(d) $1.2 \times 10^5 + 4.0 \times 10^4$ in scientific notation (see p. 408)
(e) $3.4 \times 10^{-2} - 2.7 \times 10^{-3}$ in 'normal' notation (i.e. conventional notation, not scientific format) and to three decimal places.
(See also the numerical exercises in Chapter 64.)

Health and safety legislation – *In the UK, the* Health & Safety at Work, etc. Act 1974 *provides the main legal framework for health and safety. The* Control of Substances Hazardous to Health (COSHH) Regulations 2002 *impose specific legal requirements for risk assessment wherever hazardous chemicals or biological agents are used, with Approved Codes of Practice for the control of hazardous substances, carcinogens and biological agents, including pathogenic microbes.*

Health and safety law requires institutions to provide a working environment that is safe and without risk to health. Where appropriate, training and information on safe working practices must be provided. Students and staff must take reasonable care to ensure the health and safety of themselves and of others, and must not misuse any safety equipment.

> **KEY POINT** *All practical work must be carried out with safety in mind, to minimise the risk of harm to yourself and to others – safety is everyone's responsibility.*

Risk assessment

The most widespread approach to safe working practice involves the use of risk assessment, which aims to establish:

1. The intrinsic chemical, biological and physical hazards, together with any maximum exposure limits (MELs) or occupational exposure standards (OESs), where appropriate. Chemical manufacturers provide data sheets listing the hazards associated with particular chemical compounds, while pathogenic (disease-causing) microbes are categorised according to their ability to cause illness (pp. 267–9).
2. The risks involved, by taking into account the amount of substance to be used, the way in which it will be used and the possible routes of entry into the body (Fig. 20.1).

> **KEY POINT** *It is important to distinguish between the intrinsic* **hazards** *of a particular substance and the* **risks** *involved in its use in a particular exercise.*

Definitions

Hazard – the ability of a substance or biological agent to cause harm.

Risk – the likelihood that a substance or biological agent might be harmful under specific circumstances.

Distinguishing between hazard and risk – *one of the hazards associated with water is drowning. However, the risk of drowning in a few drops of water is negligible.*

3. The persons at risk, and the ways in which they might be exposed to hazardous substances, including accidental exposure (spillage).
4. The steps required to prevent or control exposure. Ideally, a non-hazardous or less hazardous alternative should be used. If this is not feasible, adequate control measures must be used, e.g. a fume cupboard or other containment system. Personal protective equipment (e.g. lab coats, safety glasses) must be used in addition to such containment measures. A safe means of disposal will be required.

The outcome of the risk-assessment process must be recorded and appropriate safety information must be passed on to those at risk. For most practical classes, risk assessments will have been carried out in advance by the person in charge: the information necessary to minimise the risks to students may be given in the practical schedule. Make sure you know how your department provides such information and that you have read the appropriate material before you begin your practical work. You should also pay close attention to the person in charge at the beginning of the practical session, as they may emphasise the major hazards and risks. In project work, you will need to be involved in the risk assessment process along with your supervisor, before you carry out any practical work.

In addition to specific risk assessments, most institutions will have a safety handbook, giving general details of safe working practices, together with the

inhalation

ingestion

inoculation or absorption

absorption from spillage

Fig. 20.1 Major routes of entry of harmful substances into the body

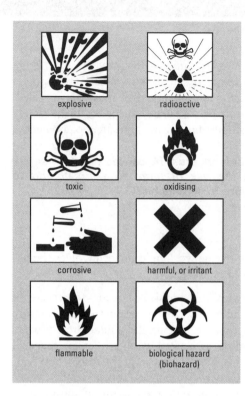

Fig. 20.2 Warning labels for specific chemical hazards. These appear on suppliers' containers and on tape used to label working vessels.

Table 20.1 International distress signals

On land

1. *Six* long whistle blasts, torch flashes, arm waves or shouts for help in succession. *Pause* for about 1 minute, then repeat, as necessary

On water

1. Using a whistle or torch

   The Morse code signal 'SOS' should be signalled, as follows:

   3 *short* blasts/flashes

   then 3 *long* blasts/flashes

   then 3 *short* blasts/flashes.

   Pause, then repeat, as necessary

2. Raise and lower arms slowly and repeatedly

3. Wave an oar with a brightly coloured cloth tied to it, slowly from side to side

4. Fire *red* flares or *orange* smoke

names and telephone numbers of safety personnel, first aiders, hospitals, etc. Make sure you read this and follow any instructions given.

Basic rules for laboratory work

- Make sure you know what to do in case of fire, including exit routes, how to raise the alarm, and where to gather on leaving the building. Remember that the most important consideration at all times is human safety: do not attempt to fight a fire unless it is safe to do so.
- All laboratories display notices telling you where to find the first aid kit and who to contact in case of accident/emergency. Report all accidents, even those appearing insignificant – your department will have a formal recording procedure to comply with safety legislation.
- Wear appropriate protective clothing at all times – a clean lab coat (buttoned up), plus safety glasses if there is any risk to the eyes.
- Never smoke, eat or drink in any laboratory, because of the risks of contamination by inhalation or ingestion (Fig. 20.1).
- Never mouth-pipette any liquid. Use a pipette filler (see p. 122) or, if appropriate, a pipettor (p. 123).
- Take care when handling glassware – see p. 126 for details.
- Know the chemical hazards warning symbols (Fig. 20.2).
- Use a fume cupboard for hazardous chemicals. Make sure that it is working and then open the front only as far as is necessary: many fume cupboards are marked with a maximum opening.
- Always use the minimum quantity of any hazardous materials.
- Work in a logical, tidy manner and minimise risks by thinking ahead.
- Always clear up at the end of each session. This is an important aspect of safety, encouraging responsible laboratory work.
- Dispose of waste in appropriate containers. Most labs will have bins for sharps, glassware, hazardous solutions and radioactive waste.

Basic rules for fieldwork safety

- Make sure you understand your objectives, the potential hazards and appropriate responses to them, before you set out.
- Your work must be designed carefully, to allow for the experience of the participants and the locations visited. Don't overestimate what can be achieved – fieldwork is often physically demanding.
- Any physical disabilities must be brought to the attention of the organiser, so that appropriate precautions can be taken.
- A comprehensive first aid kit must be carried: at least two participants should have training in first aid.
- Never work alone without permission from your organiser.
- Make sure you can read a map and use a compass: carry both.
- Your clothing and equipment must be suitable for all of the weather conditions likely to be encountered during the work.
- Check the weather forecast before departure: look out for changes in the weather at all times and do not hesitate to turn back if necessary.
- Leave full details of your intended working locations, routes and times. Never change these arrangements without informing someone.
- Make sure that you know the international distress signals (Table 20.1). Chapter 29 covers general aspects of fieldwork (p. 175).

Sources for further study

American Chemical Society (2003) *Safety in Academic Chemistry Laboratories,* 7th edn. Available: http://membership.acs.org/c/ccs/pubs/SACL_Students.pdf Last accessed: 09/04/07.

Anon. *BUBL Link – Biochemical Safety.* Available: http://bubl.ac.uk/link/b/biochemicalsafety.htm Last accessed 09/04/07.
[Provides a single-step link to a number of databases giving information on chemical and microbiological hazards.]

Anon. *Howard Hughes Medical Institute Online Safety Course.* Available: http://www.hhmi.org/about/labsafe/safescience.html/ Last accessed 09/04/07.

[This site, designed for laboratory workers, enables you to test your knowledge and understanding of laboratory safety.]

Furr, A.K. (2000) *CRC Handbook of Laboratory Safety.* CRC Press, Boca Raton, Florida.

Health and Safety Executive (2005) *Control of Substances Hazardous to Health Regulations.* HSE, London.

Study exercises

20.1 Test your knowledge of safe working procedures. After reading the appropriate sections of this book, can you remember the following:

(a) The four main steps involved in the process of risk assessment?

(b) The major routes of entry of harmful substances into the body?

(c) The warning labels for the major chemical hazard symbols (either describe them or draw them from memory)?

(d) The international symbol for a biohazard?

(e) The international symbol for radioactivity?

20.2 Locate relevant health and safety features in a laboratory. Find each of the following in one of the laboratories used as part of your course (draw a simple location map, if this seems appropriate):

(a) fire exit(s);

(b) fire-fighting equipment;

(c) first aid kit;

(d) 'sharps' container;

(e) container for disposal of broken glassware;

(f) eye wash station (where appropriate).

20.3 Investigate the health and safety procedures in operation at your university. Can you find out the following?

(a) your university's procedure in case of fire;

(b) the colour coding for fire extinguishers available in your department and their recommendations for use;

(c) the accident-reporting procedure used in your department;

(d) your department's Code of Safe Practice relating to a specific aspect of bioscience, e.g. working with micro-organisms.

20.4 Carry out risk assessments for specific chemical hazards. Look up the hazards associated with the use of the following chemicals and list the appropriate protective measures required to minimise risk during use in a lab class:

(a) formaldehyde solution, used as a preservative for animals collected during a marine biology field trip (Chapter 29);

(b) acetone, used as a solvent for the quantitative analysis of plant pigments (Chapter 59);

(c) sodium hydroxide, used in solid form to prepare a dilute solution to be used for pH adjustment (Chapter 24).

Reading any volumetric scale – *make sure your eye is level with the bottom of the liquid's meniscus and take the reading from this point.*

Measuring and dispensing liquids

The equipment you should choose to measure out liquids depends on the volumes being dispensed, the accuracy required and the number of times the job must be done (Table 21.1).

Table 21.1 Criteria for choosing a method for measuring out a liquid

Method	Best volume range	Accuracy	Usefulness for repetitive measurement
Pasteur pipette	0.03–2 ml	Low	Convenient
Conical flask/beaker	25–5000 ml	Very low	Convenient
Measuring cylinder	5–2000 ml	Medium	Convenient
Volumetric flask	5–2000 ml	High	Convenient
Burette	1–100 ml	High	Convenient
Glass pipette	1–100 ml	High	Convenient
Mechanical pipettor	5–1000 µl	High*	Convenient
Syringe	0.5–20 µl	Medium**	Convenient
Microsyringe	0.5–50 µl	High	Convenient
Weighing	Any (depends on accuracy of balance)	Very high	Inconvenient

* If correctly calibrated and used properly (see p. 124).
** Accuracy depends on width of barrel: large volumes are less accurate.

Certain liquids may cause problems:

- High-viscosity liquids are difficult to dispense: allow time for all the liquid to transfer.
- Organic solvents may evaporate rapidly, making measurements inaccurate: work quickly; seal containers without delay.
- Solutions prone to frothing (e.g. protein and detergent solutions) are difficult to measure and dispense: avoid forming bubbles due to overagitation; do not transfer quickly.
- Suspensions (e.g. cell cultures) may sediment: thoroughly mix them before dispensing.

Pasteur pipettes

Hold correctly during use (Fig. 21.1) – keep the pipette vertical, with the middle fingers gripping the barrel while the thumb and index finger provide controlled pressure on the bulb. Squeeze gently to dispense individual drops. To avoid the risk of cross-contamination, take care not to draw up solution into the bulb or to lie the pipette on its side. Alternatively, use a plastic disposable 'Pastette.'

Measuring cylinders and volumetric flasks

These must be used on a level surface so that the scale is horizontal; you should first fill with solution until just below the desired mark; then fill slowly (e.g. using a Pasteur pipette) until the meniscus is level with the mark. Allow time for the solution to run down the walls of the vessel.

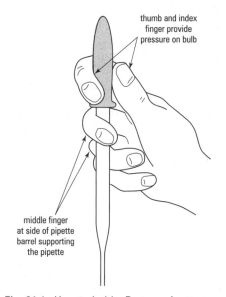

thumb and index finger provide pressure on bulb

middle finger at side of pipette barrel supporting the pipette

Fig. 21.1 How to hold a Pasteur pipette

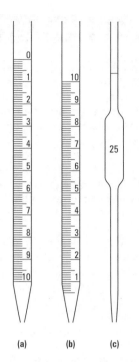

Fig. 21.2 Glass pipettes – graduated pipette, reading from zero to shoulder (a); graduated pipette, reading from maximum to tip, by gravity (b); bulb (volumetric) pipette, showing volume (calibration mark to tip, by gravity) above the bulb (c).

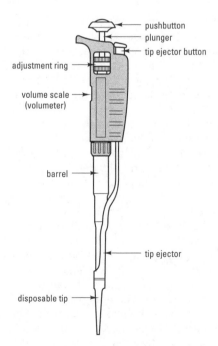

Fig. 21.3 A pipettor – the Gilson Pipetman

Burettes

Burettes should be mounted vertically on a clamp stand – don't overtighten the clamp. First ensure the tap is closed and fill the body with solution using a funnel. Open the tap and allow some liquid to fill the tubing below the tap before first use. Take a meniscus reading, noting the value in your notebook. Dispense the solution via the tap and measure the new meniscus reading. The volume dispensed is the difference between the two readings. Titrations are usually performed on a magnetic stirrer (pp. 129, 133).

Pipettes

These come in various designs, including graduated and bulb (volumetric) pipettes (Fig. 21.2). Take care to check the volume scale before use: some empty from full volume to zero, others from zero to full volume; some scales refer to the shoulder of the tip, others to the tip either by gravity or after blowing out.

KEY POINT *For safety reasons, never mouth-pipette – various aids are available such as the Pi-pump.*

Pipettors (autopipettors)

These come in two basic types:

- Air displacement pipettors. For routine work with dilute aqueous solutions. One of the most widely used examples is the Gilson Pipetman (Fig. 21.3). Box 21.1 gives details on its use.
- Positive displacement pipettors. For non-standard applications, including dispensing viscous, dense or volatile liquids, or certain procedures in molecular genetics, e.g. the PCR (p. 335), where an air displacement pipettor might create aerosols, leading to errors.

Air displacement and positive displacement pipettors may be:

- Fixed volume: capable of delivering a single factory-set volume.
- Adjustable: where the volume is determined by the operator across a particular range of values.
- Preset: movable between a limited number of values.
- Multi-channel: able to deliver several replicate volumes at the same time.

Whichever type you use, you must ensure that you understand the operating principles of the volume scale and the method for changing the volume delivered – some pipettors are easily misread.

A pipettor must be fitted with the correct disposable tip before use: each manufacturer produces different tips to fit particular models. Specialised tips are available for particular applications, e.g. PCR (p. 335).

Syringes

Syringes should be used by placing the tip of the needle in the solution and drawing the plunger up slowly to the required point on the scale. Check the barrel to make sure no air bubbles have been drawn up. Expel slowly and touch the syringe on the edge of the vessel to remove any liquid adhering to the end of the needle. Microsyringes should

Box 21.1 Using a pipettor to deliver accurate, reproducible volumes of liquid

A pipettor can be used to dispense volumes with accuracy and precision, by following this step-wise procedure:

1. **Select a pipettor that operates over the appropriate range.** Most adjustable pipettors are accurate only over a particular working range and should not be used to deliver volumes below the manufacturer's specifications (minimum volume is usually 10–20 per cent of maximum value). Do not attempt to set the volume above the maximum limit, or the pipettor may be damaged.

2. **Set the volume to be delivered.** In some pipettors, you 'dial up' the required volume. Types like the Gilson Pipetman have a system where the scale (or 'volumeter') consists of three numbers, read from top to bottom of the barrel, and adjusted using the black knurled adjustment ring (Fig. 21.3). This number gives the first three digits of the volume scale and thus can only be understood by establishing the maximum volume of the Pipetman, as shown on the push-button on the end of the plunger (Fig. 21.3). The following examples illustrate the principle for two common sizes of Pipetman:

P1000 Pipetman
(maximum volume 1000 μl)
if you dial up

P20 Pipetman
(maximum volume 20 μl)
if you dial up

| 1 |
| 0 |
| 0 |

| 1 |
| 0 |
| 0 |

the volume is set at 1000 μl   the volume is set at 10.0 μl

Note: The Pipetman scale is **not** a percentage one.

3. **Fit a new disposable tip to the end of the barrel.** Make sure that it is the appropriate type for your pipettor and that it is correctly fitted. Press the tip on firmly using a slight twisting motion – if not, you will take up less than the set volume and liquid will drip from the tip during use. Tips are often supplied in boxes, for ease of use: if sterility is important, make sure you use appropriate sterile technique at all times (p. 266). *Never, ever, try to use a pipettor without its disposable tip.*

4. **Check your delivery.** Confirm that the pipettor delivers the correct volume by dispensing volumes of distilled water and weighing on a balance, assuming $1\,mg = 1\,\mu l = 1\,mm^3$. The value should be within 1 per cent of the selected volume. For small

volumes, measure several 'squirts' together, e.g. 20 'squirts' of $5\,\mu l = 100\,mg$. If the pipettor is inaccurate (p. 157) giving a biased result (e.g. delivering significantly more or less than the volume set), you can make a temporary correction by adjusting the volumeter scale down or up accordingly (the volume *delivered* is more important than the value *displayed* on the volumeter), or have the pipettor recalibrated. If the pipettor is imprecise (p. 157), delivering a variable amount of liquid each time, it may need to be serviced. After calibration, fit a clean (sterile) tip if necessary.

5. **Draw up the appropriate volume.** Holding the pipettor *vertically*, press down on the plunger/push-button until a resistance (spring-loaded stop) is met. Then place the end of the tip in the liquid. Keeping your thumb on the plunger/push-button, release the pressure slowly and evenly: watch the liquid being drawn up into the tip, to confirm that no air bubbles are present. Wait a second or so, to confirm that the liquid has been taken up, then withdraw the end of the tip from the liquid. Inexperienced users often have problems caused by drawing up the liquid too quickly/carelessly. If you accidentally draw liquid into the barrel, seek assistance from your demonstrator or supervisor as the barrel will need to be cleaned before further use.

6. **Make a quick visual check on the liquid in the tip.** Does the volume seem reasonable? (e.g. a 100 μl volume should occupy approximately half the volume of a P200 tip). The liquid will remain in the tip, without dripping, as long as the tip is fitted correctly and the pipettor is not tilted too far from a vertical position.

7. **Deliver the liquid.** Place the end of the tip against the wall of the vessel at a slight angle (10–15° from vertical) and press the plunger/push-button slowly and smoothly to the first (spring-loaded) stop. Wait a second or two, to allow any residual liquid to run down the inside of the tip, then press again to the final stop, dispensing any remaining liquid. Remove from the vessel with the plunger/push-button still depressed.

8. **Eject the tip.** Press the tip ejector button if present (Fig. 21.3). If the tip is contaminated, eject directly into an appropriate container, e.g. a beaker of disinfectant, for microbiological work, or a labelled container for hazardous solutions (p. 120). For repeat delivery, fit a new tip if necessary and begin again at step 5 above. Always make sure that the tip is ejected before putting a pipettor on the bench.

always be cleaned before and after use by repeatedly drawing up and expelling pure solvent. The dead space in the syringe needle can occupy up to 4 per cent of the nominal syringe volume. A way of avoiding such problems is to fill the dead space with an inert substance (e.g. silicone oil) after sampling. Alternatively, use a syringe where the plunger occupies the needle space (small volumes only).

Balances

These can be used to weigh accurately (p. 133) how much liquid you have dispensed. Convert mass to volume using the equation:

$$\text{mass/density} = \text{volume} \qquad [21.1]$$

Densities of common solvents can be found in Lide (2002). You will also need to know the liquid's temperature, as density is temperature-dependent.

Example Using eqn [21.1], 9 g of a liquid with a density of $1.2\,\text{g}\,\text{ml}^{-1}$ = 7.5 ml.

Holding and storing liquids

Test tubes

These are used for colour tests, small-scale reactions, holding cultures, etc. The tube can be sterilised by heating (p. 264) and maintained in this state with a cap or cotton-wool plug.

Beakers

Beakers are used for general purposes, e.g. heating a solvent while the solute dissolves, carrying out a titration, etc.

Working with beakers and flasks – remember that volume graduations, where present, are often inaccurate and should be used only where approximations will suffice.

Conical (Erlenmeyer) flasks

These are used for storage of solutions: their wide base makes them stable, while their small mouth reduces evaporation and is easily sealed.

Bottles and vials

These are used when the solution needs to be sealed for safety, sterility or to prevent evaporation or oxidation. They usually have a screw top or ground-glass stopper to prevent evaporation and contamination. Many types are available, including 'bijou', 'McCartney', 'universal', and 'Winkler'.

You should clearly label all stored solutions (see p. 133), including relevant hazard information, preferably marking with hazard warning tape (p. 120). Seal vessels in an appropriate way, e.g. using a stopper or a sealing film such as Parafilm or Nescofilm to prevent evaporation. To avoid degradation store your solution in a fridge, but allow it to reach room temperature before use.

Storing light-sensitive chemicals – use a coloured vessel or wrap aluminium foil around a clear vessel.

Storing an aqueous solution containing organic constituents – unless this has been sterilised or is toxic, microbes will start growing, so store for short periods in a refrigerator; older solutions may not give reliable results.

Creating specialised apparatus

Glassware systems incorporating ground-glass connections such as Quickfit are useful for setting up combinations of standard glass components, e.g. for chemical reactions. In project work, you may need to adapt standard forms of glassware for a special need. A glassblowing service (often available in chemistry departments) can make special items to order.

Choosing between glass and plastic

Bear in mind the following points:

- Reactivity. Plastic vessels often distort at relatively low temperatures; they may be flammable, may dissolve in certain organic solvents and may be affected by prolonged exposure to ultraviolet (UV) light. Some plasticisers may leach from vessels and have been shown to have biological activity. Glass may adsorb ions and other molecules and then leach them into solutions, especially in alkaline conditions. Pyrex glass is stronger than ordinary soda glass and can withstand temperatures up to 500 °C.

- Rigidity and resilience. Plastic vessels are not recommended where volume is critical as they may distort through time: use class A volumetric glassware for accurate work, e.g. preparing solutions (Chapter 23). Glass vessels are more easily broken than plastic, which is particularly important for centrifugation (see p. 364).

- Opacity. Both glass and plastic absorb light in the UV range of the EMR spectrum (Table 21.2) Quartz should be used where this is important, e.g. in cuvettes for UV spectrophotometry (see p. 368).

- Disposability. Plastic items may be cheap enough to make them disposable, an advantage where there is a risk of chemical or microbial contamination.

Table 21.2 Spectral cut-off values for glass and plastics (λ_{50} = wavelength at which transmission of EMR is reduced to 50 per cent)

Material	λ_{50} (nm)
Routine glassware	340
Pyrex glass	292
Polycarbonate	396
Acrylic	342
Polyester	318
Quartz	220

Box 21.2 Safe working with glass

Many minor accidents in the laboratory are due to lack of care with glassware. You should follow these general precautions:

- **Always wear safety glasses when there is any risk of glass breakage** – e.g. when using low pressures, or heating solutions.
- **Take care when attaching tubing to glass tubes and when putting glass tubes into bungs** – always hold the tubing and glassware close together, as shown in Fig. 21.4, and wear thick gloves when appropriate.
- **Use a 'soft' Bunsen flame when heating glassware** – this avoids creating a hot spot, where cracks may start: always use tongs or special heat-resistant gloves when handling hot glassware (never use a rolled-up paper towel).
- **Do not use chipped or cracked glassware** – it may break under very slight strain and should be disposed of in the broken glassware bin.
- **Never carry large glass bottles/flasks by their necks** – support them with a hand underneath or, better still, carry them in a basket.
- **Do not force bungs too firmly into bottles** – these can be extremely difficult to remove. If you need a tight seal, use a screw-top bottle with a rubber or plastic seal.

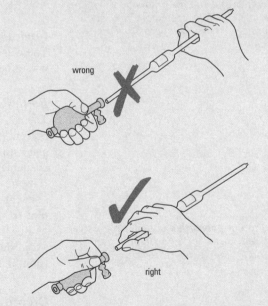

Fig. 21.4 Handling glass pipettes and tubing

- **Dispose of broken glass thoroughly and carefully** – use disposable paper towels and wear thick gloves. Always put pieces of broken glass in the correct bin.

SAFETY NOTE *Special cleaning of glass – for an acid wash use dilute acid, e.g. 100 mmol l⁻¹ (100 mol m⁻³) HCl. Rinse thoroughly at least three times with distilled or deionised water. Glassware that must be exceptionally clean (e.g. for a micro-nutrient study) should be washed in a chromic acid bath, but this involves toxic and corrosive chemicals and should only be used under supervision.*

Cleaning glass and plastic

Take care to avoid the possibility of contamination arising from prior use of chemicals or inadequate rinsing following washing. A thorough rinse with distilled or deionized water immediately before use will remove dust and other deposits and is good practice in quantitative work, but ensure that the rinsing solution is not left in the vessel. 'Strong' basic detergents (e.g. Pyroneg) are good for solubilising acidic deposits. If there is a risk of basic deposits remaining, use an acid wash. If there is a risk of contamination from organic deposits, a rinse with Analar grade ethanol is recommended. Glassware can be disinfected by washing with a sodium hypochlorite bleach such as Chloros, with sodium metabisulphite or a blended commercial product such as Virkon – dilute as recommended before use and rinse thoroughly with sterile water after use. Alternatively, to sterilise, heat glassware to at least 121 °C for 15 minutes, in an autoclave or 160 °C for 3 hours in an oven.

Text references

Lide, D.R. (ed.) (2006) *CRC Handbook of Chemistry and Physics*, 84th edn. CRC Press, Boca Raton, Florida.

Sources for further study

Boyer, R. (2000) *Modern Experimental Biochemistry*, 3rd edn. Benjamin Cummings, San Francisco.

Hendrickson, C., Byrd, L.C., Hunter, N.W. and Henrickson, C. (2005) *A Laboratory Manual for General, Organic and Biochemistry*, 5th edn. McGraw-Hill, New York.

Seidman, L.A. and Moore, C.J. (2000) *Basic Laboratory Methods for Biotechnology: Textbook and Laboratory Reference*. Prentice-Hall, New Jersey.

Study exercises

21.1 Decide on the appropriate methods and equipment for the following procedures:

(a) Preparing one litre of ethanol at approximately 70 per cent v/v in water, for use as a general preservative (Chapter 35).

(b) Adding 10 µl of a sample to the well of an agarose gel during a molecular biology procedure (Chapter 54).

(c) Preparing a calibration standard of 100 ml of DNA, to contain 200.0 µg ml⁻¹, for spectrophotometry (Chapter 59).

(d) Carrying out a titration curve for a buffer solution (Chapter 24).

21.2 Write a protocol for calibrating and using a pipettor. After reading this chapter, prepare a detailed stepwise protocol explaining how to use a pipettor to deliver a specific volume, say of 500 µl (e.g. using a Gilson Pipetman, or an alternative if your department does not use this type). Ask another student to evaluate your protocol and provide you with written feedback – either simply by reading through your protocol, or by trying it out with a pipettor as part of a class exercise (check with a member of staff before you attempt this in a laboratory).

21.3 Determine the accuracy and precision of a pipette. Using the following data for three different models of pipettor, determine which pipettor is most *accurate* and which is most *precise* (check p. 157 if you are unsure of the definitions of these two terms). All three pipettors were set to deliver 1000 µl (1.000 ml), and ten repetitive measurements of the weight of the volume of water in grammes delivered were made using a three-place balance:

Model A pipettor: 0.986; 0.971; 0.993; 0.964; 0.983; 0.996; 0.977; 0.969; 0.982; 0.974

Model B pipettor: 1.013; 1.011; 1.010; 1.009; 1.011; 1.010; 1.011; 1.009; 1.011; 1.012

Model C pipettor: 0.985; 1.022; 1.051; 1.067; 0.973; 0.982; 0.894; 1.045; 1.062; 0.928

In your answer, you should support your conclusions with appropriate numerical (statistical) evidence (see Chapter 65 for appropriate measures of location and dispersion).

Finding out about chemicals – The Merck Index (O'Neil et al., 2006) and the CRC Handbook of Chemistry and Physics (Lide, 2006) are useful sources of information on the physical and biological properties of chemicals, including melting and boiling points, solubility, toxicity, etc. (see Fig. 22.1).

8599. Sodium Chloride. [7647-14-5] Salt; common salt. ClNa; mol wt 58.44. Cl 60.67%, Na 39.34%. NaCl. The article of commerce is also known as *table salt*, *rock salt* or *sea salt*. Occurs in nature as the mineral *halite*. Produced by mining (rock salt), by evaporation of brine from underground salt deposits and from sea water by solar evaporation: *Faith, Keyes & Clark's Industrial Chemicals*, F. A. Lowenheim, M. K. Moran, Eds. (Wiley-Interscience, New York, 4th ed., 1975) pp 722-730. Toxicity studies: E. M. Boyd, M. N. Shanas, *Arch. Int. Pharmacodyn.* **144**, 86 (1963). Comprehensive monograph: D. W. Kaufmann, *Sodium Chloride*, ACS Monograph Series no. **145** (Reinhold, New York, 1960) 743 pp.

Cubic, white crystals, granules, or powder; colorless and transparent or translucent when in large crystals. d 2.17. The salt of commerce usually contains some calcium and magnesium chlorides which absorb moisture and make it cake. mp 804° and begins to volatilize at a little above this temp. One gram dissolves in 2.8 ml water at 25°, in 2.6 ml boiling water, in 10 ml glycerol; very slightly sol in alcohol. Its soly in water is decreased by HCl. Almost insol in concd HCl. Its aq soln is neutral. pH: 6.7-7.3. d of satd aq soln at 25° is 1.202. A 23% aq soln of sodium chloride freezes at −20.5°C (5°F). LD$_{50}$ orally in rats: 3.75 ±0.43 g/kg (Boyd, Shanas).

Note: **Blusalt**, a brand of sodium chloride contg trace amounts of cobalt, iodine, iron, copper, manganese, zinc is used in farm animals.

USE: Natural salt is the source of chlorine and of sodium as well as of all, or practically all, their compds, e.g., hydrochloric acid, chlorates, sodium carbonate, hydroxide, etc.; for preserving foods; manuf soap, to salt out dyes; in freezing mixtures; for dyeing and printing fabrics, glazing pottery, curing hides; metallurgy of tin and other metals.

THERAP CAT: Electrolyte replenisher; emetic; topical anti-inflammatory.

THERAP CAT (VET): Essential nutrient factor. May be given orally as emetic, stomachic, laxative or to stimulate thirst (prevention of calculi). Intravenously as isotonic solution to raise blood volume, to combat dehydration. Locally as wound irrigant, rectal douche.

Fig. 22.1 Example of typical *Merck Index* entry showing type of information given for each chemical. From *The Merck Index: An Encyclopedia of Chemicals, Drugs, and Biologicals,* Fourteenth Edition, Maryadele J. O'Neil, Patricia E. Heckelman, Cherie B. Koch, Kristin J. Roman, Eds. (Merck & Co., Inc., Whitehouse Station, NJ, USA, 2006). Reproduced with permission from *The Merck Index*, Fourteenth Edition. Copyright © 2006 by Merck & Co., Inc., Whitehouse Station, NJ, USA. All rights reserved.

Using chemicals

Safety aspects

In practical classes, the person in charge has a responsibility to inform you of any hazards associated with the use of chemicals. For routine practical procedures, a risk assessment (p. 119) will have been carried out by a member of staff and relevant safety information will be included in the practical schedule: an example is shown in Table 22.1.

In project work, your first duty when using an unfamiliar chemical is to find out about its properties, especially those relating to safety. Your department must provide the relevant information to allow you to do this. If your supervisor has filled out the form, read it carefully before signing. Box 22.1 gives further advice.

KEY POINT *Before you use any chemical you must find out whether safety precautions need to be taken and complete the appropriate forms confirming that you appreciate the risks involved.*

Selection

Chemicals are supplied in various degrees of purity and this is always stated on the manufacturer's containers. Suppliers differ in the names given to the grades and there is no conformity in purity standards. Very pure chemicals cost more, sometimes a lot more, and should only be used if the situation demands. If you need to order a chemical, your department will have a defined procedure for doing this.

Preparing solutions

Solutions are usually prepared with respect to their molar concentrations (e.g. mmol l^{-1}, or mol m^{-3}), or mass concentrations (e.g. g l^{-1}, or kg m^{-3}):

Table 22.1 Representative risk-assessment information for a practical exercise in molecular biology, involving the isolation of DNA

Substance	Hazards	Comments
Sodium dodecyl sulphate (SDS)	Irritant Toxic	Wear gloves
Sodium hydroxide (NaOH)	Highly corrosive Severe irritant	Wear gloves
Isopropanol	Highly flammable Irritant/corrosive Potential carcinogen	No naked flames Wear gloves
Phenol	Highly toxic Causes skin burns Potential carcinogen	Use in fume hood Wear gloves
Chloroform	Volatile and toxic Irritant/corrosive Potential carcinogen	Use in fume hood Wear gloves

Box 22.1   Safe working with chemicals

You should always treat chemicals as potentially dangerous, following these general precautions:

- **Do not use any chemical until you have considered the risks involved** – for lab classes, you should carefully read all hazard and risk information provided *before* you start work. In project work, you may need to be involved in the risk-assessment process with your supervisor.
- **Wear a laboratory coat at all times** – the coat should be fully fastened and cleaned appropriately, should any chemical compound be spilled on it. Closed-toe footwear will protect your feet should any spillages occur.
- **Make sure you know where the safety apparatus is kept before you begin working** – this apparatus includes eye bath, fire extinguishers and blanket, first aid kit.
- **Wear safety glasses and gloves when working with toxic, irritant or corrosive chemicals, and for any substances where the hazards are not yet fully characterised** – make sure you understand the hazard warning signs (p. 120) along with any specific

hazard-coding system used in your department. Carry out procedures with solid material in a fume cupboard.

- **Use aids such as pipette fillers to minimise the risk of contact with hazardous solutions** – these aids are further detailed on pp. 122-3.
- **Never smoke, eat, drink or chew gum in a lab where chemicals are handled** – this will minimise the risk of ingestion.
- **Label all solutions appropriately** – use the appropriate hazard warning information (see pp. 120, 133).
- **Report all spillages of chemicals/solutions** – make sure that spillages are cleaned up properly.
- **Store hazardous chemicals only in the appropriate locations** – for example, a spark-proof fridge is required for flammable liquids; acids and solvents should not be stored together.
- **Dispose of chemicals in the correct manner** – if unsure, ask a member of staff (do not assume that it is safe to use the lab waste bin or the sink for disposal).
- **Wash hands after any direct contact with chemicals or biochemical material** – always wash your hands at the end of a lab session.

Examples   Using eqn [22.1], 25 g of a substance dissolved in 400 ml of water would have a mass concentration (p. 138) of $25 \div 400 = 0.0625$ g ml^{-1} ($\equiv 62.5$ mg ml$^{-1} \equiv 62.5$ g l^{-1})

Using eqn [22.1], 0.4 mol of a substance dissolved in 0.5 litres of water would have a molar concentration of $0.4 \div 0.5 = 0.8$ mol l^{-1} ($\equiv 800$ mmol l^{-1}).

Solving solubility problems – if your chemical does not dissolve after a reasonable time:
- *check the limits of solubility for your compound (see* Merck Index, *O'Neil et al. 2006),*
- *check the pH of the solution – solubility often changes with pH, e.g. you may be able to dissolve the compound in an acidic or basic solution.*

both can be regarded as an amount of *substance* per unit volume of *solution*, in accordance with the relationship:

$$\text{Concentration} = \frac{\text{amount}}{\text{volume}} \qquad [22.1]$$

The most important aspect of eqn [22.1] is to recognise clearly the units involved, and to prepare the solution accordingly: for molar concentrations, you will need the relative molecular mass of the compound, so that you can determine the mass of substance required. Further advice on concentrations and interconversion of units is given on p. 139.

Box 22.2 shows the steps involved in making up a solution. The concentration you require is likely to be defined by a protocol you are following and the grade of chemical and supplier may also be specified. Success may depend on using the same source and quality, e.g. with enzyme work. To avoid waste, think carefully about the volume of solution you require, though it is always a good idea to err on the high side because you may spill some or make a mistake when dispensing it. Try to choose one of the standard volumes for vessels, as this will make measuring-out easier.

Use distilled or deionised water to make up aqueous solutions and stir to make sure all the chemical is dissolved. Magnetic stirrers are the most convenient means of doing this: carefully drop a clean magnetic stirrer bar ('flea') in the beaker, avoiding splashing; place the beaker centrally on the stirrer plate, switch on the stirrer and gradually increase the speed of stirring. When the crystals or powder have completely dissolved, switch off and retrieve the flea with a magnet or another flea. Take care not to contaminate your solution when you do this and rinse the flea with distilled water.

Box 22.2 How to make up an aqueous solution of known concentration from solid material

1. **Find out or decide the concentration of chemical required** and the degree of purity necessary.

2. **Decide on the volume of solution required.**

3. **Find out the relative molecular mass of the chemical (M_r).** This is the sum of the atomic (elemental) masses of the component elements and can be found on the container. If the chemical is hydrated, i.e. has water molecules associated with it, these must be included when calculating the mass required.

4. **Work out the mass of chemical that will give the concentration desired in the volume required.**
   Suppose your procedure requires you to prepare 250 ml of 0.1 mol l^{-1} NaCl.

   (a) Begin by expressing all volumes in the same units, either millilitres or litres (e.g. 250 ml as 0.25 litres).

   (b) Calculate the number of moles required from eqn [22.1]: 0.1 = amount (mol) ÷ 0.25.
   By rearrangement, the required number of moles is thus 0.1 × 0.25 = 0.025 mol.

   (c) Convert from mol to g by multiplying by the relative molecular mass (M_r for NaCl = 58.44)

   (d) Therefore, you need to make up 0.025 × 58.44 = 1.461 g to 250 ml of solution, using distilled water.

   In some instances, it may be easier to work in SI units, though you must be careful when using exponential numbers (p. 410).
   Suppose your protocol states that you need 100 ml of 10 mmol l^{-1} KCl.

   (a) Start by converting this to 100×10^{-6} m^3 of 10 mol m^{-3} KCl.

   (b) The required number of mol is thus $(100 \times 10^{-6}) \times (10) = 10^3$.

   (c) Each mol of KCl weighs 72.56 g (M_r, the relative molecular mass).

   (d) Therefore you need to make up 72.56 $\times 10^{-3}$ g = 72.56 mg KCl to 100×10^{-6} m^3 (100 ml) with distilled water.

   See Box 23.1 for additional information.

5. **Weigh out the required mass of chemical to an appropriate accuracy.** If the mass is too small to weigh to the desired degree of accuracy, consider the following options:

   (a) Make up a greater volume of solution.

   (b) Make up a stock solution that can be diluted at a later stage (see below).

   (c) Weigh the mass first, and calculate what volume to make the solution up to afterwards using eqn [22.1].

6. **Add the chemical to a beaker or conical flask then add a little less water than the final amount required.** If some of the chemical sticks to the paper, foil or weighing boat, use some of the water to wash it off.

7. **Stir and, if necessary, heat the solution to ensure all the chemical dissolves.** You can visually determine when this has happened by observing the disappearance of the crystals or powder.

8. **If required, check and adjust the pH of the solution when cool** (see p. 146).

9. **Make up the solution to the desired volume.** If the concentration needs to be accurate, use a class A volumetric flask; if a high degree of accuracy is not required, use a measuring cylinder (class B).

   (a) Pour the solution from the beaker into the measuring vessel using a funnel to avoid spillage.

   (b) Make up the volume so that the meniscus comes up to the appropriate measurement line (p. 122). For accurate work, rinse out the original vessel and use this liquid to make up the volume.

10. **Transfer the solution to a reagent bottle or a conical flask and label the vessel clearly.**

'Obstinate' solutions may require heating, but do this only if you know that the chemical will not be damaged at the temperature used. Use a stirrer-heater to keep the solution mixed as you heat it. Allow the solution to cool before you measure volume or pH as these are affected by temperature.

Stock solutions

Stock solutions are valuable when making up a range of solutions containing different concentrations of a reagent or if the solutions have some common ingredients. They also save work if the same solution is used over a prolonged period (e.g. a nutrient solution). The stock solution

Table 22.2 Use of stock solutions. Suppose you need a set of solutions 10 ml in volume containing differing concentrations of KCl, with and without reagent Q. You decide to make up a stock of KCl at twice the maximum required concentration ($50\,mmol\,l^{-1} = 50\,mol\,m^{-3}$) and a stock of reagent Q at twice its required concentration. The table shows how you might use these stocks to make up the media you require. Note that the total volumes of stock you require can be calculated from the table (end column).

Stock solutions	Volume of stock required to make required solutions (ml)						Total volume of stock required (ml)
	No KCl plus Q	No KCl minus Q	$15\,mmol\,l^{-1}$ KCl plus Q	$15\,mmol\,l^{-1}$ KCl minus Q	$25\,mmol\,l^{-1}$ KCl plus Q	$25\,mmol\,l^{-1}$ KCl minus Q	
$50\,mmol\,l^{-1}$ KCl	0	0	3	3	5	5	16
[reagent Q] $\times$ 2	5	0	5	0	5	0	15
Water	5	10	2	7	0	5	29
Total	10	10	10	10	10	10	60

is more concentrated than the final requirement and is diluted as appropriate when the final solutions are made up. The principle is best illustrated with an example (Table 22.2).

Preparing dilutions

Making a single dilution

You may need to dilute a stock solution to give a particular mass concentration, or molar concentration. Use the following procedure:

1. Transfer an accurate volume of stock solution to a volumetric flask, using appropriate equipment (Table 21.1).
2. Make up to the calibration mark with solvent – add the last few drops from a pipette or solvent bottle, until the meniscus is level with the calibration mark.
3. Mix thoroughly, either by repeated inversion (holding the stopper firmly) or by prolonged stirring, using a magnetic stirrer. Make sure you add the magnetic flea *after* the volume adjustment step.

For routine work using dilute aqueous solutions where the highest degree of accuracy is not required, it may be acceptable to substitute test tubes or conical flasks for volumetric flasks. In such cases, you would calculate the volumes of stock solution and diluent required, with the assumption that the final volume is determined by the sum of the individual volumes of stock and diluent used (e.g. Table 22.2). Thus, a two-fold dilution would be prepared using 1 volume of stock solution and 1 volume of diluent. The dilution factor is obtained from the ratio of the initial concentration of the stock solution and the final concentration of the diluted solution. The dilution factor can be used to determine the volumes of stock and diluent required in a particular instance. For example, suppose you wanted to prepare 100 ml of a solution of NaCl at $0.2\,mol\,l^{-1}$. Using a stock solution containing $4.0\,mol\,l^{-1}$ NaCl, the dilution factor is $0.2 \div 4.0 = 0.05 = 1/20$ (a twenty-fold dilution). Therefore, the amount of stock solution required is 1/20th of 100 ml = 5 ml and the amount of diluent needed is 19/20th of 100 ml = 95 ml.

Preparing a dilution series

Dilution series are used in a wide range of procedures, including the preparation of standard curves for calibration of analytical instruments (p. 303), and in microbiology and immunoassay, where a range of

Making a dilution – use the relationship $[C_1]V_1 = [C_2]V_2$ to determine volume or concentration (see p. 139).

Removing a magnetic flea from a volumetric flask – use a strong magnet to bring the flea to the top of the flask, to avoid contamination during removal.

Using the correct volumes for dilutions – it is important to distinguish between the volumes of the various liquids: a one-in-ten dilution is obtained using 1 volume of stock solution plus 9 volumes of diluent (1 + 9 = 10). Note that when this is shown as a ratio, it may represent the initial and final volumes (e.g. 1:10) or, sometimes, the volumes of stock solution and diluent (e.g. 1:9).

Using diluents – various liquids are used, including distilled or deionised water, salt solutions, buffers, Ringer's solution (p. 279), etc., according to the specific requirements of the procedure.

dilutions of a particular sample is often required (pp. 276, 312). A variety of different approaches can be used:

Linear dilution series Here, the concentrations are separated by an equal amount, e.g. a series containing protein at 0, 0.2, 0.4, 0.6, 0.8, 1.0 µg ml^{-1}. Such a dilution series might be used to prepare a calibration curve for spectrophotometric assay of protein concentration (Box 52.1), or an enzyme assay (p. 319). Use $[C_1]V_1 = [C_2]V_2$ (p. 139) to determine the amount of stock solution required for each member of the series, with the volume of diluent being determined by subtraction.

Logarithmic dilution series Here, the concentrations are separated by a constant proportion, often referred to as the step interval. This type of serial dilution is useful when a broad range of concentrations is required, e.g. for titration of biologically active substances (p. 295), making a plate count of a suspension of microbes (p. 276), or when a process is logarithmically related to concentration (p. 312).

The most common examples are:

- Doubling dilutions – where each concentration is half that of the previous one (two-fold step interval, $\log_2$ dilution series). First, make up the most concentrated solution at twice the volume required. Measure out half of this volume into a vessel containing the same volume of diluent, mix thoroughly and repeat, for as many doubling dilutions as are required. The concentrations obtained will be 1/2, 1/4, 1/8, 1/16, etc., times the original (i.e. the dilutions will be two, four, eight and sixteen-fold, etc.).
- Decimal dilutions – where each concentration is one-tenth that of the previous one (ten-fold step interval, $\log_{10}$ dilution series). First, make up the most concentrated solution required, with at least a 10 per cent excess. Measure out one-tenth of the volume required into a vessel containing nine times as much diluent, mix thoroughly and repeat. The concentrations obtained will be 1/10, 1/100, 1/1000, etc., times the original (i.e. dilutions of 10^{-1}, 10^{-2}, 10^{-3}, etc.). To calculate the actual concentration of solute, multiply by the appropriate dilution factor.

When preparing serial doubling or decimal dilutions, it is often easiest to add the appropriate amount of diluent to several vessels beforehand, as shown in the worked example in Fig. 22.2. When preparing a dilution series, it is essential that all volumes are dispensed accurately, e.g. using calibrated pipettors (p. 123), otherwise any inaccuracies will be compounded, leading to gross errors in the most dilute solutions.

Harmonic dilution series Here, the concentrations in the series take the values of the reciprocals of successive whole numbers, e.g. 1, 1/2, 1/3, 1/4, 1/5, etc. The individual dilutions are simply achieved by a stepwise increase in the volume of diluent in successive vessels, e.g. by adding 0, 1, 2, 3, 4 and 5 times the volume of diluent to a set of test tubes, then adding a constant unit volume of stock solution to each vessel. Although there is no dilution transfer error between individual dilutions, the main disadvantage is that the series is non-linear, with a step interval that becomes progressively smaller as the series is extended.

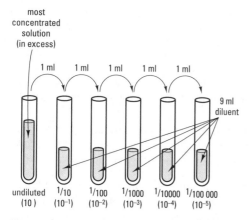

Fig. 22.2 Preparation of a dilution series. The example shown is a decimal dilution series, down to 1/100 000 (10^{-5}) of the solution in the first (left-hand) tube. Note that all solutions must be mixed thoroughly before transferring the volume to the next in the series. In microbiology and cell culture, sterile solutions and appropriate aseptic technique will be required (p. 264).

Preparing a dilution series using pipettes or pipettors – use a fresh pipette or disposable tip for each dilution, to prevent carry-over of solutions.

Solutions must be thoroughly mixed before measuring out volumes for the next dilution. Use a fresh measuring vessel for each dilution to avoid contamination, or wash your vessel thoroughly between dilutions. Clearly label the vessel containing each dilution when it is made: it is easy to get confused! When deciding on the volumes required, allow for the aliquot removed when making up the next member in the series. Remember to discard any excess from the last in the series if volumes are critical.

Mixing solutions and suspensions

Various devices may be used, including:

- Magnetic stirrers and fleas. Magnetic fleas come in a range of shapes and sizes, and some stirrers have integral heaters. During use, stirrer speed may increase as the instrument warms up.
- Vortex mixers. For vigorous mixing of small volumes of solution, e.g. when preparing a dilution series in test tubes. Take care when adjusting the mixing speed – if the setting is too low, the test tube will vibrate rather than creating a vortex, giving inadequate mixing. If the setting is too high, the test tube may slip from your hand.
- Orbital shakers and shaking water baths. These are used to provide controlled mixing at a particular temperature, e.g. for long-term incubation and cell-growth studies (p. 272).
- Bottle rollers. For cell-culture work, ensuring gentle, continuous mixing.

 SAFETY NOTE *When using a vortex mixer with open and capped test tubes – do not vortex too vigorously or liquid will spill from the top of the tube, creating a contamination risk.*

 SAFETY NOTE *You must always clean up any spillages of chemicals, as you are the only person who knows the risks from the spilled material.*

Storing chemicals and solutions

Labile chemicals may be stored in a fridge or freezer. Take special care when using chemicals that have been stored at low temperature: the container and its contents must be warmed up to room temperature before use, otherwise water vapour will condense on the chemical. This may render any weighing you do meaningless and it could ruin the chemical. Other chemicals may need to be kept in a desiccator, especially if they are deliquescent.

 KEY POINT *Label all stored chemicals clearly with the following information: the chemical name (if a solution, state solute(s), concentration(s) and pH if measured), plus any relevant hazard-warning information, the date made up, and your name.*

Using balances

Electronic balances with digital readouts are now favoured over mechanical types: they are easy to read and their self-taring feature means the mass of the weighing boat or container can be subtracted automatically before weighing an object. The most common type offers accuracy down to 1 mg over the range 1 mg to 160 g, which is suitable for most biological applications.

To operate a standard self-taring balance:

1. Check that it is level, using the adjustable feet to centre the bubble in the spirit level (usually at the back of the machine). For accurate work, make sure a draught shield is on the balance.
2. Place an empty vessel in the middle of the balance pan and allow the reading to stabilise. *If the object is larger than the pan, take care that*

Deciding on which balance to use – select a balance that weighs to an appropriate number of decimal places (p. 125). For example, you should use a top-loading balance weighing to one decimal place for less accurate work. Note that a weight of 6.4 g on such a balance may represent a true value of between 6.350 g and 6.449 g (to three decimal places).

Box 22.3   How to use Vernier calipers

Note that numbers on the scale refer to centimetres. Vernier scales consist of two numerical scales running side by side, the moving one being shorter with ten divisions compressed into the length for nine on the longer, static one. Use Vernier calipers to measure objects to the nearest 0.1 mm:

1. **Clamp the stops lightly over the object** as in Fig. 22.3(a), taking care not to deform it.

2. **Read off the number of whole millimetres** by taking the value on the fixed scale lying to the left of the first line on the moving (short) scale, i.e. 8 mm in Fig. 22.3(b).

3. **Read off 0.1 mm value** by finding which line in the moving scale corresponds most closely with a line on the fixed scale, i.e. 0.5 mm in Fig. 22.3(b). If the zero of the short scale corresponded to a whole number on the static scale, then record 0.0 mm as this shows fully the precision of the measurement.

4. **Add these numbers to give the final reading**, i.e. 8.5 mm in Fig. 22.3(b).

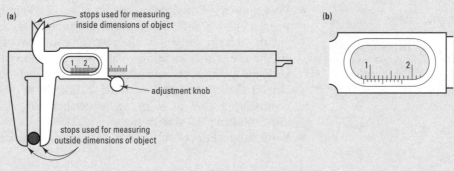

Fig. 22.3   (a) Vernier caliper   (b) Vernier measurement scale

---

Weighing – *never weigh anything directly on a balance's pan: you may contaminate it for other users. Use a weighing boat or a slip of aluminium foil. Otherwise, choose a suitable vessel such as a beaker, conical flask or aluminium tray.*

Table 22.3 Suitability of devices for measuring linear dimensions

Measurement device	Suitable lengths	Degree of precision
Eyepiece graticule (light microscopy)	1 μm to 10 mm	0.5 μm
Vernier calipers	1–100 mm	0.1 mm
Ruler	10 mm to 1 m	1.0 mm
Tape measure	10 mm to 30 m	1.0 mm
Optical surveying devices	1 m to 100 m	0.1 m

See Box 22.3 for method of using Vernier calipers

no part rests on the body of the balance or the draught shield as this will invalidate the reading. Press the tare bar to bring the reading to zero.

3. Place the chemical or object carefully in the vessel (powdered chemicals should be dispensed with a suitably sized clean spatula). Take care to avoid spillages.

4. Allow the reading to stabilise and make a note of the value.

5. If you add excess chemical, take great care when removing it. Switch off if you need to clean any deposit accidentally left on or around the balance.

Larger masses should be weighed on a top-loading balance to an appropriate degree of accuracy. Take care to note the limits for the balance: while most have devices to protect against overloading, you may damage the mechanism. In the field, spring or battery-operated balances may be preferred. Try to find a place out of the wind to use them. For extremely small masses, there are electrical balances that can weigh down to 1 μg, but these are very delicate and must be used under supervision.

Measuring length and area

When measuring linear dimensions, the device you need depends on the size of object you are measuring and the precision demanded (Table 22.3).

For many regularly shaped objects, area can be estimated from linear dimensions (see p. 409). The areas of irregular shapes can be measured with an optical measuring device or a planimeter. These have the benefits of speed and ease of use; instructions are machine-specific. A simple 'low tech' method is to trace objects on to good-quality paper or to photocopy them. If the outline is then cut round, the area can be estimated by weighing the cutout and comparing to the mass of a piece of the same paper of known area. Avoid getting moisture from the specimen on to the paper as this will affect the reading.

Measuring and controlling temperature

Heating specimens

Care is required when heating specimens – there is a danger of fire whenever organic material is heated and a danger of scalding from heated liquids. Safety glasses should always be worn. Use a thermostatically controlled electric stirrer-heater if possible. If using a Bunsen burner, keep the flame well away from yourself and your clothing (tie back long hair). Use a non-flammable mat beneath a Bunsen to protect the bench. Switch off when no longer required. To light a Bunsen, close the air hole first, then apply a lit match or lighter. Open the air hole if you need a hotter, more concentrated flame: the hottest part of the flame is just above the apex of the blue cone in its centre.

Ovens and drying cabinets may be used to dry specimens or glassware. They are normally thermostatically controlled. If drying organic material for dry weight measurement, do so at about 80 °C to avoid caramelising the specimen. Always state the actual temperature used as this affects results. Check that all water has been driven off by weighing until a constant mass is reached.

Cooling specimens

Fridges and freezers are used for storing stock solutions and chemicals that would either break down or become contaminated at room temperature. Normal fridge and freezer temperatures are about 4 °C and −15 °C respectively. Ice baths can be used when reactants must be kept close to 0 °C. Most bioscience departments will have a machine that provides flaked ice for use in these baths. If common salt is mixed with ice, temperatures below 0 °C can be achieved. A mixture of ethanol and solid CO_2 will provide a temperature of −72 °C if required. To freeze a specimen quickly, immerse in liquid N_2 (−196 °C) using tongs and wearing an apron and thick gloves, as splashes will damage your skin. Always work in a well-ventilated room.

Maintaining cultures or specimens at constant temperature

Thermostatically controlled temperature rooms and incubators can be used to maintain temperature at a desired level. Always check with a thermometer or thermograph that the thermostat is accurate enough for your study. To achieve a controlled temperature on a smaller scale, e.g. for an oxygen electrode (p. 355), use a water bath. These usually incorporate heating elements, a circulating mechanism and a thermostat. Baths for sub-ambient temperatures have a cooling element.

Controlling atmospheric conditions

Gas composition

The atmosphere may be 'scrubbed' of certain gases by passing through a U-tube or Dreschel bottle containing an appropriate chemical or solution.

For accurate control of gas concentrations, use cylinders of pure gas; the contents can be mixed to give specified concentrations by controlling individual flow rates. The cylinder-head regulator (Fig. 22.4) allows you to

SAFETY NOTE *Heating/cooling glass vessels – take care if heating or cooling glass vessels rapidly as they may break when heat-stressed. Freezing aqueous solutions in thin-walled glass vessels is risky because ice expansion may break the glass.*

Using thermometers – *some are calibrated for use in air, others require partial immersion in liquid and others total immersion – check before use.*

SAFETY NOTE *If a mercury thermometer is broken, report the spillage, as mercury is a poison.*

Example Water vapour can be removed by passing gas over dehydrated $CaCO_3$ and CO_2 may be removed by bubbling through KOH solution.

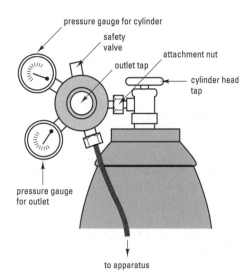

pressure gauge for cylinder

safety valve

outlet tap

attachment nut

cylinder head tap

pressure gauge for outlet

to apparatus

Fig. 22.4 Parts of a cylinder-head regulator. The regulator is normally attached by tightening the attachment nut clockwise; the exception is with cylinders of hydrogen, where the special regulator is tightened *anticlockwise* to avoid the chance of this potentially explosive gas being incorrectly used.

Using a timer – *always set the alarm before the critical time, so that you have adequate time to react.*

control the pressure (and hence flow rate) of gas; adjust using the controls on the regulator or with spanners of appropriate size. Before use, ensure the regulator outlet tap is off (turn anticlockwise), then switch on at the cylinder (turn clockwise) – the cylinder dial will give you the pressure reading for the cylinder contents. Now switch on at the regulator outlet (turn clockwise) and adjust to desired pressure/flow setting. To switch off, carry out the above directions in reverse order.

Pressure

Many forms of pump are used to pressurise or provide a partial vacuum, usually to force gas or liquid movement. Each has specific instructions for use. Many laboratories are supplied with 'vacuum' (suction) and pressurised air lines that are useful for procedures such as vacuum-assisted filtration. Make sure you switch off the taps after use. Take special care with glass items kept at very low or high pressures. These should be contained within a metal cage to minimise the risk of injury.

Measuring time

Many experiments and observations need to be carefully timed. Large-faced stopclocks allow you to set and follow 'experimental time' and remove the potential difficulties in calculating this from 'real time' on a watch or clock. Some timers incorporate an alarm that you can set to warn when readings or operations must be carried out; 24-hour timers are available for controlling light and temperature regimes.

Miscellaneous methods for treating samples

Homogenising

This involves breaking up and mixing specimens to give a uniform preparation. Blenders are used to homogenise animal and plant material and work best when an adequate volume of liquid is present: buffer solution may be added to specimens for this purpose. Use in short bursts to avoid overheating the motor and the sample. A pestle and mortar is used for grinding up specimens. Acid-washed sand grains can be added to help break up the tissues. For quantitative work with brittle samples, care must be taken not to lose material when the sample breaks into fragments.

Separation of components of mixtures and solutions

Particulate solids (e.g. soils) can be separated on the basis of size using sieves. These are available in stacking forms that fit on automatic shakers. Sieves with the largest pores are placed at the top and the assembly is shaken for a fixed time until the sample separates. Suspensions of solids in liquids may be separated out by centrifugation (see p. 360) or filtration. Various forms of filter paper are available having different porosities and purities. Vacuum-assisted filtration speeds up the process and is best carried out with a filter funnel attached to a filter flask. Filtration through pre-sterilised membranes with very small pores (e.g. the Millipore type) is an excellent method of sterilising small volumes of solution. Solvents can be removed from solutes by heating, using rotary film evaporation under low pressure and, for water, by freeze drying. The last two are especially useful for heat-labile solutes – refer to the manufacturers' specific instructions for use.

Text references

Lide, D.R. (ed.) (2006) *CRC Handbook of Chemistry and Physics*, 84th edn. CRC Press, Boca Raton, Florida.

O'Neil, M.J., Smith, A., and Henckelman, P.E. (eds.) (2006) *The Merck Index: An Encyclopedia of Chemicals, Drugs and Biologicals*, 14th edn. Merck & Co., Inc., Whitehouse Station, New Jersey.

Sources for further study

Jack, C.R. (1995) *Basic Biochemical Laboratory Procedures and Computing*. Oxford University Press, Oxford.

Seidman, L.A. and Moore, C.J. (2000) *Basic Laboratory Methods for Biotechnology: Textbook and Laboratory Reference*. Prentice-Hall, New Jersey.

Study exercises

22.1 Practise the calculations involved in preparing specific volumes of aqueous solutions (see also Study exercise 23.1). What mass of substance would be required to prepare each of the following (answer in each case to three significant figures):

(a) 100 ml of NaCl at 50 mmol l^{-1}? (M_r of NaCl = 58.44.)

(b) 250 ml of mannitol at 0.10 mol l^{-1}? (M_r of mannitol = 182.17.)

(c) 200 ml of a bovine serum albumin solution at 800 µg ml^{-1}?

(d) 0.5 litres of $MgCl_2$ prepared from the hexahydrate salt at 22.5 mmol l^{-1}? (M_r of $MgCl_2.6H_2O$ = 203.30.)

(e) 400 ml of DNA at 20 ng µl^{-1}?

22.2 Practise the calculations involved in preparing dilutions (answer in each case to three significant figures).

(a) If you added 1.0 ml of an aqueous solution of NaCl at 0.4 mol l^{-1} to 9.0 ml of water, what would be the final concentration of NaCl in mmol l^{-1}?

(b) If you added 25 ml of an aqueous solution of DNA at 10 µg ml^{-1} to a 500 ml volumetric flask and made it up to the specified volume with water, what would be the final concentration of DNA, in ng ml^{-1}?

(c) If you added 10 µl of an aqueous solution of sucrose at 200 mmol l^{-1} to a 250 ml volumetric flask and made it up to the specified volume with water, what would be the final concentration of sucrose, in nmol ml^{-1}?

(d) How would you prepare 250 ml of KCl at a final concentration of 20.0 mmol l^{-1} from a solution containing KCl at 0.2 mol l^{-1}?

(e) How would you prepare 1×10^{-3} m^3 of glucose at a final concentration of 50 µmol m^{-3} from a stock solution containing glucose at 20.0 g m^{-3} (M_r of glucose = 180.16)?

22.3 Practise the calculations involved in using stock solutions. Suppose you had the following stock solutions: NaCl 100.0 mmol l^{-1}; KCl 200.0 mmol l^{-1}; $CaCl_2$ 160.0 mmol l^{-1}; glucose 5.0 mmol l^{-1}. Calculate the volumes of each stock solution and the volume of water required to prepare each of the following (answer in each case to three significant figures):

(a) 1.0 ml of a solution containing only KCl at 10.0 mmol l^{-1}

(b) 50 ml of a solution containing NaCl at 2.5 mmol l^{-1} and glucose at 0.5 mmol l^{-1}

(c) 100 ml of a solution containing NaCl at 5.0 mmol l^{-1}, KCl at 2.5 mmol l^{-1}, $CaCl_2$ at 40.0 mmol l^{-1} and glucose at 0.25 mmol l^{-1}

(d) 10 ml of a solution containing NaCl, $CaCl_2$ and KCl, all at 20.0 µmol ml^{-1}

(e) 25 ml of a solution containing $CaCl_2$ at 8.0 mmol l^{-1}, KCl at 50.0 mmol l^{-1} and glucose at 20.0 nmol ml^{-1}.

Definitions

Electrolyte – a substance that dissociates, either fully or partially, in water to give two or more ions.

Relative atomic mass (A_r) – the mass of an atom relative to $^{12}C = 12$.

Relative molecular mass (M_r) – the mass of a compound's formula unit relative to $^{12}C = 12$.

Mole (of a substance) – the equivalent in mass to relative molecular mass in grams.

Expressing solute concentrations – you should use SI units wherever possible. However, you are likely to meet non-SI concentrations and you must be able to deal with these units too.

Example A 1.0 molar solution of NaCl would contain 58.44 g NaCl (the relative molecular mass) per litre of solution.

A solution is a homogeneous liquid, formed by the addition of solutes to a solvent (usually water in biological systems). The behaviour of solutions is determined by the types of solutes involved and by their proportions, relative to the solvent. Many laboratory exercises involve calculation of concentrations, e.g. when preparing an experimental solution at a particular concentration, or when expressing data in terms of solute concentration (see pp. 139–141; p. 145). Make sure that you understand the basic principles set out in this chapter before you tackle such exercises.

Solutes can affect the properties of solutions in several ways, including:

Electrolytic dissociation

This occurs where individual molecules of an electrolyte dissociate to give charged particles (ions). For a strong electrolyte, e.g. NaCl, dissociation is essentially complete. In contrast, a weak electrolyte, e.g. acetic acid, will be only partly dissociated, depending upon the pH and temperature of the solution (p. 146).

Osmotic effects

These are the result of solute particles lowering the effective concentration of the solvent (water). These effects are particularly relevant to biological systems since membranes are far more permeable to water than to most solutes. Water moves across biological membranes from the solution with the higher effective water concentration to that with the lower effective water concentration (osmosis).

Ideal/non-ideal behaviour

This occurs because solutions of real substances do not necessarily conform to the theoretical relationships predicted for dilute solutions of so-called ideal solutes. It is often necessary to take account of the non-ideal behaviour of real solutions, especially at high solute concentrations (see Lide, 2006 and Robinson and Stokes, 2002, for appropriate data).

Concentration

In SI units (p. 162), the concentration of a solute in a solution is expressed in $mol\,m^{-3}$, which is convenient for most biological purposes. The concentration of a solute is usually symbolised by square brackets, e.g. [NaCl]. Details of how to prepare a solution using SI and non-SI units are given on p. 130.

A number of alternative ways of expressing the relative amounts of solute and solvent are in general use, and you may come across these terms in your practical work, or in the literature:

Molarity

This is the term used to denote molar concentration, $[C]$, expressed as moles of solute per litre volume of solution ($mol\,l^{-1}$). This non-SI term continues to find widespread usage, in part because of the familiarity of working scientists with the term, but also because laboratory glassware is calibrated in millilitres and litres, making the preparation of molar and

Box 23.1 Useful procedures for calculations involving molar concentrations

1. **Preparing a solution of defined molarity.** For a solute of known relative molecular mass, M_r, the following relationship can be applied:

$$[C] = \frac{\text{mass of solute/relative molecular mass}}{\text{volume of solution}} \quad [23.1]$$

So, if you wanted to make up 200 ml (0.2 l) of an aqueous solution of NaCl (M_r 58.44) at a concentration of 500 mmol l^{-1} (0.5 mol l^{-1}), you could calculate the amount of NaCl required by inserting these values into eqn [23.1]:

$$0.5 = \frac{\text{mass of solute/58.44}}{0.2}$$

which can be rearranged to

$$\text{mass of solute} = 0.5 \times 0.2 \times 58.44 = 5.844\,\text{g}$$

The same relationship can be used to calculate the concentration of a solution containing a known amount of a solute, e.g. if 21.1 g of NaCl were made up to a volume of 100 ml (0.1 l), this would give

$$[\text{NaCl}] = \frac{21.1/58.44}{0.1} = 3.61\,\text{mol l}^{-1}$$

2. **Dilutions and concentrations.** The following relationship is very useful if you are diluting (or concentrating) a solution:

$$[C_1]V_1 = [C_2]V_2 \quad [23.2]$$

where $[C_1]$ and $[C_2]$ are the initial and final concentrations, while V_1 and V_2 are their respective volumes: each pair must be expressed in the same units. Thus, if you wanted to dilute 200 ml of 0.5 mol l^{-1} NaCl to give a final molarity of 0.1 mol l^{-1}, then, by substitution into eqn [23.2]:

$$0.5 \times 200 = 0.1 \times V_2$$

Thus $V_2 = 1\,000$ ml (in other words, you would have to add water to 200 ml of 0.5 mol l^{-1} NaCl to give a final volume of 1 000 ml to obtain a 0.1 mol l^{-1} solution).

3. **Interconversion.** A simple way of interconverting amounts and volumes of any particular solution is to divide the amount and volume by a factor of 10^3: thus a molar solution of a substance contains 1 mol l^{-1}, which is equivalent to 1 mmol ml^{-1}, or 1 µmol µl^{-1}, or 1 nmol nl^{-1}, etc. You may find this technique useful when calculating the amount of substance present in a small volume of solution of known concentration, e.g. to calculate the amount of NaCl present in 50 µl of a solution with a concentration (molarity) of 0.5 mol l^{-1} NaCl:

   (a) this is equivalent to 0.5 µmol µl^{-1};

   (b) therefore 50 µl will contain $50 \times 0.5\,\text{µmol} = 25\,\text{µmol}$.

Alternatively, you may prefer to convert to primary SI units, for ease of calculation (see Box 26.1).

The unitary method' (p. 412) is an alternative approach to these calculations.

---

millimolar solutions relatively straightforward. However, the symbols in common use for molar (M) and millimolar (mM) solutions are at odds with the SI system and many people now prefer to use mol l^{-1} and mmol l^{-1} respectively, to avoid confusion. Box 23.1 gives details of some useful approaches to calculations involving molarities.

Molality

This is used to express the concentration of solute relative to the *mass* of solvent, i.e. mol kg^{-1}. Molality is a temperature-independent means of expressing solute concentration, rarely used except when the osmotic properties of a solution are of interest (p. 141).

Example A 0.5 molal solution of NaCl would contain $58.44 \times 0.5 = 29.22$ g NaCl per kg of water.

Per cent composition (% w/w)

This is the solute mass (in g) per 100 g solution. The advantage of this expression is the ease with which a solution can be prepared, since it simply requires each component to be pre-weighed (for water, a volumetric measurement may be used, e.g. using a measuring cylinder) and then mixed together. Similar terms are parts per thousand (‰), i.e. mg g^{-1}, and parts per million (ppm), i.e. µg g^{-1}.

Example A 5% w/w sucrose solution contains 5 g sucrose and 95 g water (= 95 ml water, assuming a density of 1 g ml^{-1}) to give 100 g of solution.

Per cent concentration (% w/v and % v/v)

For solutes added in solid form, this is the number of grams of solute per 100 ml solution. This is more commonly used than per cent composition, since solutions can be accurately prepared by weighing out the required amount of solute and then making this up to a known volume using a volumetric flask. The equivalent expression for liquid solutes is % v/v.

The principal use of mass/mass or mass/volume terms (including $g l^{-1}$) is for solutes whose relative molecular mass is unknown (e.g. cellular proteins), or for mixtures of certain classes of substance (e.g. total salt in sea water). You should *never* use the per cent term without specifying how the solution was prepared, i.e. by using the qualifier w/w, w/v or v/v. For mass concentrations, it is simpler to use mass per unit volume, e.g. $mg l^{-1}$, $\mu g \mu l^{-1}$, etc.

Parts per million concentration (ppm)

This is a non-SI weight per volume (w/v) concentration term commonly used in quantitative analysis such as flame photometry, atomic absorption spectroscopy and gas chromatography, where low concentrations of solutes are analysed. The term ppm is equivalent to the expression of concentration as $\mu g ml^{-1}$ (10^{-6} $g ml^{-1}$) and a 1.0 ppm solution of a substance will have a concentration of 1.0 $\mu g ml^{-1}$ (1.0×10^{-6} $g ml^{-1}$).

Parts per billion (ppb) is an extension of this concentration term as $ng ml^{-1}$ (10^{-9} $g ml^{-1}$) and is commonly used to express concentrations of very dilute solutions. For example, the allowable concentration of arsenic in water is 0.05 ppm, but it is more conveniently expressed as 50 ppb.

Activity (a)

This is a term used to describe the *effective* concentration of a solute. In dilute solutions, solutes can be considered to behave according to ideal (thermodynamic) principles, i.e. they will have an effective concentration equivalent to the actual concentration. However, in concentrated solutions ($\geqslant 500 mol m^{-3}$), the behaviour of solutes is often non-ideal, and their effective concentration (activity) will be less than the actual concentration [C]. The ratio between the effective concentration and the actual concentration is called the activity coefficient (γ) where

$$\gamma = \frac{a}{[C]} \qquad [23.3]$$

Equation [23.3] can be used for SI units ($mol m^{-3}$), molarity ($mol l^{-1}$) or molality ($mol kg^{-1}$). In all cases, γ is a dimensionless term, since a and [C] are expressed in the same units. The activity coefficient of a solute is effectively unity in dilute solution, decreasing as the solute concentration increases (Table 23.1). At high concentrations of certain ionic solutes, γ may increase to become greater than unity.

 KEY POINT *Activity is often the correct expression for theoretical relationships involving solute concentration (e.g. where a property of the solution is dependent on concentration). However, for most practical purposes, it is possible to use the actual concentration of a solute rather than the activity, since the difference between the two terms can be ignored for dilute solutions.*

The particular use of the term 'water activity' is considered below, since it is based on the mole fraction of solvent, rather than the effective concentration of solute.

Example A 5% w/v sucrose solution contains 5 g sucrose in 100 ml of solution. A 5% v/v glycerol solution would contain 5 ml glycerol in 100 ml of solution.

Note that when water is the solvent this is often not specified in the expression, e.g. a 20% v/v ethanol solution contains 20% ethanol made up to 100 ml of solution using water.

Example The concentration of a NaCl solution is stated as 3 ppm. This is equivalent to $3 \mu g ml^{-1}$ ($3 mg l^{-1}$). The relative molecular mass of NaCl is 58.44 g mol^{-1}, so the solution has a concentration of $3 \times 10^{-6} \div 58.44$ mol $ml^{-1} = 5.13 \times 10^{-8}$ mol $ml^{-1} = 0.0513 \mu mol ml^{-1} = 51.3 \mu mol l^{-1}$.

Table 23.1 Activity coefficient of NaCl solutions as a function of molality. Data from Robinson and Stokes (2002).

Molality	Activity coefficient at 25 °C
0.1	0.778
0.5	0.681
1.0	0.657
2.0	0.668
4.0	0.783
6.0	0.986

Example A solution of NaCl with a molality of 0.5 mol kg^{-1} has an activity coefficient of 0.681 at 25 °C and a molal activity of $0.5 \times 0.681 = 0.340$ mol kg^{-1}.

Equivalent mass (equivalent weight)

Equivalence and normality are outdated terms, although you may come across them in older texts. They apply to certain solutes whose reactions involve the transfer of charged ions, e.g. acids and alkalis (which may be involved in H^+ or OH^- transfer), and electrolytes (which form cations and anions that may take part in further reactions). These two terms take into account the valency of the charged solutes. Thus the equivalent mass of an ion is its relative molecular mass divided by its valency (ignoring the sign), expressed in grams per equivalent (eq) according to the relationship:

$$\text{equivalent mass} = \frac{\text{relative molecular mass}}{\text{valency}} \qquad [23.4]$$

Examples For carbonate ions (CO_3^{2-}), with a relative molecular mass of 60.00 and a valency of 2, the equivalent mass is $60.00/2 = 30.00 \, g \, eq^{-1}$.

For sulphuric acid (H_2SO_4, relative molecular mass 98.08), where 2 hydrogen ions are available, the equivalent mass is $98.08/2 = 49.04 \, g \, eq^{-1}$.

For acids and alkalis, the equivalent mass is the mass of substance that will provide 1 mol of either H^+ or OH^- ions in a reaction, obtained by dividing the molecular mass by the number of available ions (n), using n instead of valency as the denominator in eqn [23.4].

Normality

A 1 normal solution (1 N) is one that contains one equivalent mass of a substance per litre of solution. The general formula is:

$$\text{normality} = \frac{\text{mass of substance per litre}}{\text{equivalent mass}} \qquad [23.5]$$

Example A 0.5 N solution of sulphuric acid would contain $0.5 \times 49.04 = 24.52 \, g \, l^{-1}$.

Osmolarity

This non-SI expression is used to describe the number of moles of osmotically active solute particles per litre of solution ($osmol \, l^{-1}$). The need for such a term arises because some molecules dissociate to give more than one osmotically active particle in aqueous solution.

Example Under ideal conditions, 1 mol of NaCl dissolved in water would give 1 mol of Na^+ ions and 1 mol of Cl^- ions, equivalent to a theoretical osmolarity of $2 \, osmol \, l^{-1}$.

Osmolality

This term describes the number of moles of osmotically active solute particles per unit mass of solvent ($osmol \, kg^{-1}$). For an ideal solute, the osmolality can be determined by multiplying the molality by n, the number of solute particles produced in solution (e.g. for NaCl, $n = 2$). However, for real (i.e. non-ideal) solutes, a correction factor (the osmotic coefficient, ϕ) is used:

$$\text{osmolality} = \text{molality} \times n \times \phi \qquad [23.6]$$

Example A $1.0 \, mol \, kg^{-1}$ solution of NaCl has an osmotic coefficient of 0.936 at 25 °C and an osmolality of $1.0 \times 2 \times 0.936 = 1.872 \, osmol \, kg^{-1}$.

If necessary, the osmotic coefficients of a particular solute can be obtained from tables (e.g. Table 23.2): non-ideal behaviour means that ϕ may have values >1 at high concentrations. Alternatively, the osmolality of a solution can be measured using an osmometer.

Colligative properties and their use in osmometry

Several properties vary in direct proportion to the effective number of osmotically active solute particles per unit mass of solvent and can be used to determine the osmolality of a solution. These colligative properties include freezing point, boiling point and vapour pressure.

An osmometer is an instrument that measures the osmolality of a solution, usually by determining the freezing point depression of the solution in relation to pure water, a technique known as cryoscopic

Table 23.2 Osmotic coefficients of NaCl solutions as a function of molality. Data from Robinson and Stokes (2002).

Molality	Osmotic coefficient at 25 °C
0.1	0.932
0.5	0.921
1.0	0.936
2.0	0.983
4.0	1.116
6.0	1.271

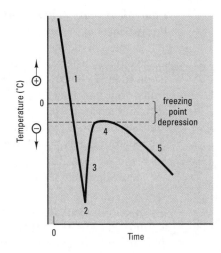

Fig. 23.1 Temperature responses of a cryoscopic osmometer. The response can be subdivided into:
1. initial supercooling
2. initiation of crystallisation
3. crystallisation/freezing
4. plateau, at the freezing point
5. slow temperature decrease.

Using an osmometer – *it is vital that the sample holder and probe are clean, otherwise small droplets of the previous sample may be carried over, leading to inaccurate measurement.*

osmometry. A small amount of sample is cooled rapidly and then brought to the freezing point (Fig. 23.1), which is measured by a temperature-sensitive thermistor probe calibrated in mosmol kg^{-1}. An alternative method is used in vapour pressure osmometry, which measures the relative decrease in the vapour pressure produced in the gas phase when a small sample of the solution is equilibrated within a chamber.

Osmotic properties of solutions

Several interrelated terms can be used to describe the osmotic status of a solution. In addition to osmolality, you may come across the following:

Osmotic pressure
This is based on the concept of a membrane permeable to water, but not to solute molecules. For example, if a sucrose solution is placed on one side and pure water on the other, then a passive driving force will be created and water will diffuse across the membrane into the sucrose solution, since the effective water concentration in the sucrose solution will be lower (see Fig. 23.2). The tendency for water to diffuse into the sucrose solution could be counteracted by applying a hydrostatic pressure equivalent to the passive driving force. Thus, the osmotic pressure of a solution is the excess hydrostatic pressure required to prevent the net flow of water into a vessel containing the solution. The SI unit of osmotic pressure is the pascal, Pa $(= kg\,m^{-1}\,s^{-2})$. Older sources may use atmospheres, or bars, and conversion factors are given in Box 26.1 (p. 161). Osmotic pressure and osmolality can be interconverted using the expression 1 osmol kg^{-1} = 2.479 MPa at 25 °C.

The use of osmotic pressure has been criticised as misleading, since a solution does not exhibit an 'osmotic pressure' unless it is placed on the other side of a selectively permeable membrane from pure water.

Example A 1.0 mol kg^{-1} solution of NaCl at 25 °C has an osmolalilty of 1.872 osmol kg^{-1} and an osmotic pressure of 1.872 × 2.479 = 4.641 MPa.

Water activity (a_w)
This is a term often used to describe the osmotic behaviour of microbial cells. It is a measure of the relative proportion of water in a solution, expressed in terms of its mole fraction, i.e. the ratio of the number of moles of water (n_w) to the total number of moles of all substances (i.e. water and

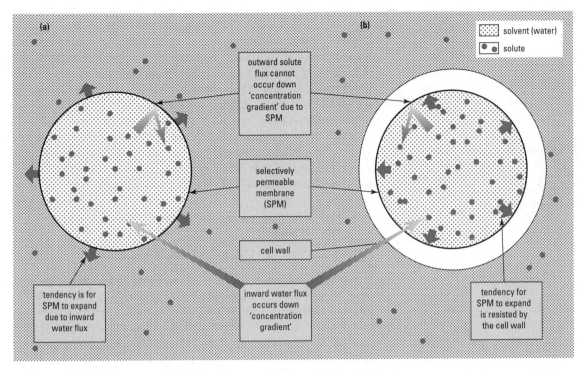

Fig. 23.2 Illustration of forces driving solvent (water) and solute movement across a selectively permeable membrane (SPM). Energetically, both solutes and solvents tend to move down their respective 'concentration gradient' (strictly, down their chemical potential gradient). However, solute molecules cannot leave the model cells illustrated because they cannot pass through the SPM. In the situation illustrated in (a), water will tend to move from outside the cell to within because the solute molecules have effectively 'diluted' the water in within the cell (illustrated by the density of point shading), creating a gradient in 'concentration' and because this molecule is able to pass through the SPM. The result will be an expansion of this model cell (short arrows). The osmotic pressure is the (theoretical) pressure that would need to be applied to prevent this. If the model cell were surrounded by a cell wall, as in (b), this would resist expansion, leading to internal pressurisation (turgor pressure, p. 144).

Table 23.3 Water activity (a_w) of NaCl solutions as a function of molality. Data from Robinson and Stokes (2002).

Molality	a_w
0.1	0.997
0.5	0.984
1.0	0.967
2.0	0.932
4.0	0.852
6.0	0.760

solutes) in solution (n_t), taking into account the molal activity coefficient of the solvent, water (i.e. γ_w):

$$a_w = \gamma_w \frac{n_w}{n_t}$$

[23.7]

The water activity of pure water is unity, decreasing as solutes are added. One disadvantage of a_w is the limited change that occurs in response to a change in solute concentration: a $1.0\,mol\,kg^{-1}$ solution of NaCl has a water activity of 0.967 (Table 23.3).

Osmolality, osmotic pressure and water activity are measurements based solely on the osmotic properties of a solution, with no regard for any other driving forces, e.g. hydrostatic and gravitational forces. In circumstances where such other forces are important, you will need to measure a variable that takes into account these aspects of water status, namely water potential.

Water potential (hydraulic potential) and its applications

Water potential, Ψ_w, is the most appropriate measure of osmotic status in many areas of bioscience. It is a term derived from the chemical potential of water. It expresses the difference between the chemical potential of water in the test system and that of pure water under standard conditions and has units of pressure (i.e. Pa). It is a more appropriate term than osmotic pressure because it is based on sound theoretical principles and because it can be used

Examples A $1.0\,mol\,kg^{-1}$ solution of NaCl has a (negative) water potential of $-4.641\,MPa$.

Pure water at 0.2 MPa pressure (about 0.1 MPa above atmospheric pressure) has a (positive) water potential of 0.1 MPa.

to predict the direction of passive movement of water, since water will flow down a gradient of chemical potential (i.e. osmosis occurs from a solution with a higher water potential to one with a lower water potential). A solution of pure water at 20 °C and at 0.1 MPa pressure (i.e. ≈ atmospheric) has a water potential of zero. The addition of solutes will lower the water potential (i.e. make it negative), while the application of pressure, e.g. from hydrostatic or gravitational forces, will raise it (i.e. make it positive).

Often, the two principal components of water potential are referred to as the solute potential, or osmotic potential (Ψ_s, sometimes symbolised as Ψ_π or π) and the hydrostatic pressure potential (Ψ_p) respectively. For a solution at atmospheric pressure, the water potential is due solely to the presence of osmotically active solute molecules (osmotic potential) and may be calculated from the measured osmolality (osmol kg^{-1}) at 25 °C, using the relationship:

$$\Psi_w (\text{MPa}) = \Psi_s (\text{MPa}) = -2.479 \times \text{osmolality} \qquad [23.8]$$

For aquatic microbial cells, e.g. algae, fungi and bacteria, equilibrated in their growth medium at atmospheric pressure, the water potential of the external medium will be equal to the cellular water potential ('isotonic') and the latter can be derived from the measured osmolality of the medium (eqn [23.8]) by osmometry (pp. 141–2). The water potential of such cells can be subdivided into two major parts, the cell solute potential (Ψ_s) and the cell turgor pressure (Ψ_p) as follows:

$$\Psi_w = \Psi_s + \Psi_p \qquad [23.9]$$

To calculate the relative contribution of the osmotic and pressure terms in eqn [23.9], an estimate of the internal osmolality is required, e.g. by measuring the freezing point depression of expressed intracellular fluid. Once you have values for Ψ_w and Ψ_s, the turgor pressure can be calculated by substitution into eqn [23.9].

For terrestrial plant cells, the water potential may be determined directly using a vapour pressure osmometer, by placing a sample of the material within the osmometer chamber and allowing it to equilibrate. If Ψ_s of expressed sap is then measured, Ψ_p can be determined from eqn [23.9].

The van't Hoff relationship can be used to estimate Ψ_s, by summation of the osmotic potentials due to the major solutes, determined from their concentrations, as:

$$\Psi_s = -RTn\phi[C] \qquad [23.10]$$

where RT is the product of the universal gas constant and absolute temperature (2 479 J mol^{-1} at 25 °C), n and ϕ are as previously defined and $[C]$ is expressed in SI terms as mol m^{-3}.

Measuring water potential – eqn [23.9] ignores the effects of gravitational forces – for systems where gravitational effects are important an additional term is required (Nobel, 2005).

Text references

Lide, D.R. (ed.) (2006) *CRC Handbook of Chemistry and Physics*, 87th edn. CRC Press, Boca Raton, Florida.

Nobel, P.S. (2005) *Physicochemical and Environmental Plant Physiology*, 3rd edn. Academic Press, New York.

Robinson, R.A. and Stokes, R.H. (2002) *Electrolyte Solutions*, 2nd edn. Dover Publications, New York.

Sources for further study

Burtis, C.A. and Ashwood, E.R. (2001) *Fundamentals of Clinical Chemistry*, 5th edn. Saunders, Philadelphia.

Chapman, C. (1998) *Basic Chemistry for biology*. McGraw-Hill, New York.

Seidman, L.A. and Moore, C.J. (2000) *Basic Laboratory Methods for Biotechnology: Textbook and Laboratory Reference*. Prentice-Hall, New Jersey.

Study exercises

23.1 Practise calculations involving molar concentrations (see also Study exercises 22.1 and 22.2). What mass of substance would be required to prepare each of the following aqueous solutions (answer in grams, to three decimal places in each case):
- (a) 1 litre of NaCl at a concentration of 1 molar? (M_r of NaCl = 58.44.)
- (b) 250 ml of $CaCl_2$ at 100 mmol l^{-1}? (M_r of $CaCl_2$ = 110.99.)
- (c) 2.5 l of mannitol at 10 nmol μl^{-1}? (M_r of mannitol = 182.17.)
- (d) 400 ml of KCl at 5% w/v?
- (e) 250 ml of glucose at 2.50 mol m^{-3}? (M_r of glucose = 180.16.)

23.2 Practise expressing concentrations in different ways. Express all answers to three significant figures:
- (a) What is 5 g l^{-1} sucrose, expressed in terms of molarity? (M_r of sucrose = 342.3.)
- (b) What is 1.0 mol m^{-3} NaCl, expressed in g l^{-1}? (M_r of NaCl = 58.44.)
- (c) What is 5% v/v ethanol, expressed in terms of molarity? (M_r of ethanol = 46.06 and density of ethanol at 25°C = 0.789 gml^{-1}.)
- (d) What is 150 mmol l^{-1} glucose, expressed in terms of per cent concentration (% w/v)? (M_r of glucose = 180.16.)
- (e) What is a 1.0 molal solution of KCl, expressed as per cent composition (% w/w)? (M_r of KCl = 74.55.)

23.3 Calculate osmolality and osmotic potentials.
- (a) Assuming NaCl, KCl and $CaCl_2$ behave according to ideal thermodynamic principles, what would be the predicted osmolality of a solution containing:
  - (i) NaCl alone, at 50 mmol kg^{-1}?
  - (ii) KCl at 200 mmol kg^{-1} and $CaCl_2$ at 40 mmol kg^{-1}?
  - (iii) NaCl at 100 mmol kg^{-1}, KCl at 60 mmol kg^{-1} and $CaCl_2$ at 75 mmol kg^{-1}?
- (b) What is the predicted osmotic pressure and osmotic potential of each of the solutions in (a) at 25°C? (answer to three significant figures in all cases).

Definitions

Acid – a compound that acts as a proton donor in aqueous solution.

Base – a compound that acts as a proton acceptor in aqueous solution.

Conjugate pair – an acid together with its corresponding base.

Alkali – a compound that liberates hydroxyl ions when it dissociates. Since hydroxyl ions are strongly basic, this will reduce the proton concentration.

Ampholyte – a compound that can act as both an acid and a base. Water is an ampholyte since it may dissociate to give a proton and a hydroxyl ion (amphoteric behaviour).

 SAFETY NOTE *Safe working with strong acids or alkalis – these can be highly corrosive; rinse with plenty of water, if spilled.*

Table 24.1 Effects of temperature on the ion product of water (K_w), H^+ ion concentration and pH at neutrality. Values calculated from Lide (2006).

Temp. (°C)	K_w (mol^2 l^{-2})	[H$^+$] at neutrality (nmol l^{-1})	pH at neutrality
0	0.11×10^{-14}	33.9	7.47
4	0.17×10^{-14}	40.7	7.39
10	0.29×10^{-14}	53.7	7.27
20	0.68×10^{-14}	83.2	7.08
25	1.01×10^{-14}	100.4	7.00
30	1.47×10^{-14}	120.2	6.92
37	2.39×10^{-14}	154.9	6.81
45	4.02×10^{-14}	199.5	6.70

Example *Human blood plasma has a typical H^+ concentration of approximately 0.4×10^{-7} mol l^{-1} ($= 10^{-7.4}$ mol l^{-1}), giving a pH of 7.4.*

pH is a measure of the amount of hydrogen ions (H^+) in a solution: this affects the solubility of many substances and the activity of most biological systems, from individual molecules to whole organisms. It is usual to think of aqueous solutions as containing H^+ ions (protons), though protons actually exist in their hydrated form, as hydronium ions (H_3O^+). The proton concentration of an aqueous solution [H^+] is affected by several factors:

● Ionisation (dissociation) of water, which liberates protons and hydroxyl ions in equal quantities, according to the reversible relationship:

$$H_2O \rightleftharpoons H^+ + OH^- \tag{24.1}$$

● Dissociation of acids, according to the equation:

$$H–A \rightleftharpoons H^+ + A^- \tag{24.2}$$

where H–A represents the acid and A^- is the corresponding conjugate base. The dissociation of an acid in water will increase the amount of protons, reducing the amount of hydroxyl ions as water molecules are formed (eqn [24.1]). The addition of a base (usually, as its salt) to water will decrease the amount of H^+, due to the formation of the conjugate acid (eqn [24.2]).

● Dissociation of alkalis, according to the relationship:

$$X–OH \rightleftharpoons X^+ + OH^- \tag{24.3}$$

where X–OH represents the undissociated alkali. Since the dissociation of water is reversible (eqn [24.1]), in an aqueous solution the production of hydroxyl ions will effectively act to 'mop up' protons, lowering the proton concentration.

Many compounds act as acids, bases or alkalis: those that are almost completely ionised in solution are usually called strong acids or bases, while weak acids or bases are only slightly ionised in solution (p. 148).

In an aqueous solution, most of the water molecules are not ionised. In fact, the extent of ionisation of pure water is constant at any given temperature and is usually expressed in terms of the ion product (or ionisation constant) of water, K_w:

$$K_w = [H^+][OH^-] \tag{24.4}$$

where [H^+] and [OH^-] represent the molar concentration (strictly, the activity) of protons and hydroxyl ions in solution, expressed as mol l^{-1}. At 25 °C, the ion product of pure water (Table 24.1) is 10^{-14} mol^2 l^{-2} (i.e. 10^{-8} mol^2 m^{-6}). This means that the concentration of protons in solution will be 10^{-7} mol l^{-1} (10^{-4} mol m^{-3}), with an equivalent concentration of hydroxyl ions (eqn [24.1]). Since these values are very low and involve negative powers of 10, it is customary to use the pH scale, where:

$$pH = -\log_{10}[H^+] \tag{24.5}$$

and [H^+] is the proton activity in mol l^{-1} (see p. 140).

KEY POINT *While pH is strictly the negative logarithm (to the base 10) of H^+ activity, in practice H^+ concentration in $mol\,l^{-1}$ (equivalent to $kmol\,m^{-3}$ in SI terminology) is most often used in place of activity, since the two are virtually the same, given the limited dissociation of H_2O. The pH scale is not SI: nevertheless, it continues to be used widely in biological science.*

The value where an equal amount of H^+ and OH^- ions are present is termed neutrality: at 25 °C the pH of pure water at neutrality is 7.0. At this temperature, pH values below 7.0 are acidic while values above 7.0 are alkaline.

Always remember that the pH scale is a logarithmic one, not a linear one: a solution with a pH of 3.0 is not twice as acidic as a solution of pH 6.0, but one thousand times as acidic (i.e. contains 1000 times the amount of H^+ ions). Therefore, you may need to convert pH values into proton concentrations before you carry out mathematical manipulations (see Box 65.2, p. 422). For similar reasons, it is important that pH change is expressed in terms of the original and final pH values, rather than simply quoting the difference between the values: a pH change of 0.1 has little meaning unless the initial or final pH is known.

Measuring pH

pH electrodes

Accurate pH measurements can be made using a pH electrode, coupled to a pH meter. The pH electrode is usually a combination electrode, comprising two separate systems: an H^+-sensitive glass electrode and a reference electrode which is unaffected by H^+ ion concentration (Fig. 24.1). When this is immersed in a solution, a pH-dependent voltage between the two electrodes can be measured using a potentiometer. In most cases, the pH electrode assembly (containing the glass and reference electrodes) is connected to a separate pH meter by a cable, although some hand-held instruments (pH probes) have the electrodes and meter within the same assembly, often using an H^+-sensitive field effect transistor in place of a glass electrode, to improve durability and portability.

Box 24.1 gives details of the steps involved in making a pH measurement with a glass pH electrode and meter.

pH indicator dyes

These compounds (usually weak acids) change colour in a pH-dependent manner. They may be added in small amounts to a solution, or they can be used in paper strip form. Each indicator dye usually changes colour over a restricted pH range, typically 1–2 pH units (Table 24.2): universal indicator dyes/papers make use of a combination of individual dyes to measure a wider pH range. Dyes are not suitable for accurate pH measurement as they are affected by other components of the solution including oxidizing and reducing agents and salts. However, they are useful for:

- estimating the approximate pH of a solution;
- determining a change in pH, for example at the end-point of a titration or the production of acids during bacterial metabolism (pp. 140–1);
- establishing the approximate pH of intracellular compartments, for example the use of neutral red as a 'vital' stain (p. 248).

Measuring pH – the pH of a neutral solution changes with temperature (Table 24.1), due to the enhanced dissociation of water with increasing temperature. This must be taken into account when measuring the pH of any solution and when interpreting your results.

SAFETY NOTE *Preparing a dilute acid solution using concentrated acid – always slowly add the concentrated acid to water, not the reverse, since the strongly exothermic process can trigger a violent reaction with water.*

SAFETY NOTE *Preparing an alkali solution – typically, the alkali will be in solid form (e.g. NaOH) and addition to water will rapidly raise the temperature of the solution: use only heat-resistant glassware, cooled with water if necessary.*

Table 24.2 Properties of some pH indicator dyes

Dye	Acid-base colour change	Useful pH range
Thymol blue (acid)	red–yellow	1.2–6.8
Bromophenol blue	yellow–blue	1.2–6.8
Congo red	blue–red	3.0–5.2
Bromocresol green	yellow–blue	3.8–5.4
Resazurin	orange–violet	3.8–6.5
Methyl red	red–yellow	4.3–6.1
Litmus	red–blue	4.5–8.3
Bromocresol purple	yellow–purple	5.8–6.8
Bromothymol blue	yellow–blue	6.0–7.6
Neutral red	red–yellow	6.8–8.0
Phenol red	yellow–red	6.8–8.2
Thymol blue (alkaline)	yellow–blue	8.0–9.6
Phenol-phthalein	none–red	8.3–10.0

Box 24.1   Using a glass pH electrode and meter to measure the pH of a solution

The following procedure should be used whenever you make a pH measurement: consult the manufacturer's handbook for specific information, where necessary. Do not be tempted to miss out any of the steps detailed below, particularly those relating to the effects of temperature, or your measurements are likely to be inaccurate.

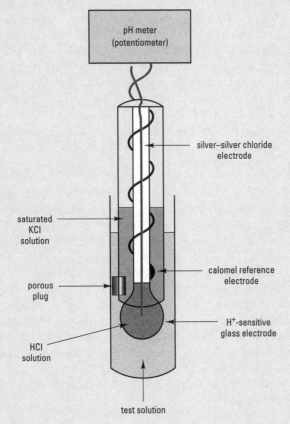

pH meter
(potentiometer)

silver–silver chloride
electrode

saturated
KCl
solution

calomel reference
electrode

porous
plug

HCl
solution

H⁺-sensitive
glass electrode

test solution

Fig. 24.1 Measurement of pH using a combination pH electrode and meter. The electrical potential difference recorded by the potentiometer is directly proportional to the pH of the test solution.

1. **Stir the test solution thoroughly before you make any measurement:** it is often best to use a magnetic stirrer. Leave the solution for sufficient time to allow equilibration at lab temperature.

2. **Record the temperature of every solution you use,** including all calibration standards and samples, since this will affect K_w, neutrality and pH.

3. **Set the temperature compensator on the meter to the appropriate value.** This control makes an allowance for the effect of temperature on the electrical potential difference recorded by the meter: it does *not* allow for the other temperature-dependent effects mentioned elsewhere. Basic instruments have no temperature compensator, and should only be used at a specified temperature, either 20 °C or 25 °C, otherwise they will not give an accurate measurement. More sophisticated systems have automatic temperature compensation.

4. **Rinse the electrode assembly with distilled water** and gently dab off the excess water on to a clean tissue: check for visible damage or contamination of the glass electrode (consult a member of staff if the glass is broken or dirty). Also check that the solution within the glass assembly is covering the metal electrode.

5. **Calibrate the instrument:** set the meter to 'pH' mode, if appropriate, and then place the electrode assembly in a standard solution of known pH, usually pH 7.00. This solution may be supplied as a liquid, or may be prepared by dissolving a measured amount of a calibration standard in water: calibration standards are often provided in tablet form, to be dissolved in water to give a particular volume of solution. Adjust the calibration control to give the correct reading. Remember that your calibration standards will only give the specified pH at a particular temperature, usually either 20 °C or 25 °C. If you are working at a different temperature, you must establish the actual pH of your calibration standards, either from the supplier, or from literature information.

6. **Remove the electrode assembly from the calibration solution and rinse again with distilled water:** dab off the excess water. Basic instruments have no further calibration steps (single-point calibration), while the more refined pH meters have additional calibration procedures.

   If you are using a basic instrument, you should check that your apparatus is accurate over the appropriate pH range by measuring the pH of another standard whose pH is close to that expected for the test solution. If the standard does not give the expected reading, the instrument is not functioning correctly: consult a member of staff.

   If you are using an instrument with a slope control function, this will allow you to correct for any deviation in electrical potential from that predicted by the theoretical relationship (at 25 °C,

   (continued)

Box 24.1   (continued)

a change in pH of 1.00 unit should result in a change in electrical potential of 59.16 mV) by performing a two-point calibration. Having calibrated the instrument at pH 7.00, immerse in a second standard at the same temperature as that of the first standard, usually buffered to either pH 4.00 or pH 9.00, depending upon the expected pH of your samples. Adjust the slope control until the exact value of the second standard is achieved (Fig. 24.2). A pH electrode and meter calibrated using the two-point method will give accurate readings over the pH range from 3 to 11: laboratory pH electrodes are not accurate outside this range, since the theoretical relationship between electrical potential and pH is valid.

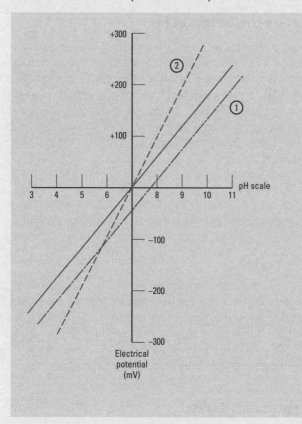

Fig. 24.2 The relationship between electrical potential and pH. The solid line shows the response of a calibrated electrode while the other plots are for instruments requiring calibration: 1 has the correct slope but incorrect isopotential point (calibration control adjustment is needed); 2 has the correct isopotential point but incorrect slope (slope control adjustment is needed).

7. **Once the instrument is calibrated, measure the pH of your solution(s)**, making sure that the electrode assembly is rinsed thoroughly between measurements. You should be particularly aware of this requirement if your solutions contain organic biological material, e.g. soil, tissue fluids, protein solutions, etc., since these may adhere to the glass electrode and affect the calibration of your instrument. If your electrode becomes contaminated during use, check with a member of staff before cleaning: avoid touching the surface of the glass electrode with abrasive material. Allow sufficient time for the pH reading to stabilise in each solution before taking a measurement: for unbuffered solutions, this may take several minutes, so do not take inaccurate pH readings due to impatience!

8. **After use, the electrode assembly must not be allowed to dry out.** Most pH electrodes should be stored in a neutral solution of KCl, either by suspending the assembly in a small beaker, or by using an electrode cap filled with the appropriate solution (typically 1.0 mol l^{-1} KCl buffered at pH 7.0). However, many labs simply use distilled water as a storage solution, leading to loss of ions from the interior of the electrode assembly. In practice, this means that pH electrodes stored in distilled water will take far longer to give a stable reading than those stored in KCl.

9. **Switch the meter to zero (where appropriate), but do not turn off the power:** pH meters give more stable readings if they are left on during normal working hours.

   Problems (and solutions) include: inaccurate and/or unstable pH readings caused by cross-contamination (rinse electrode assembly with distilled water and blot dry between measurements); development of a protein film on the surface of the electrode (soak in 1% w/v pepsin in 0.1 mol l^{-1} HCl for at least an hour); deposition of organic or inorganic contaminants on the glass bulb (use an organic solvent, such as acetone, or a solution of 0.1 mol l^{-1} disodium ethylenediamine-tetraacetic acid, respectively); drying out of the internal reference solutions (drain, flush and refill with fresh solution, then allow to equilibrate in 0.1 mol l^{-1} HCl for at least an hour); cracks or chips to the surface of the glass bulb (use a replacement electrode).

Buffers

Rather than simply measuring the pH of a solution, you may wish to *control* the pH, e.g. in metabolic experiments, or in a growth medium for cell culture (p. 279). In fact, you should consider whether you need to control pH in any experiment involving a biological system, whether whole organisms, isolated cells, subcellular components or biomolecules. One of the most effective ways to control pH is to use a buffer solution.

A buffer solution is usually a mixture of a weak acid and its conjugate base. Added protons will be neutralised by the anionic base while a reduction in protons, e.g. due to the addition of hydroxyl ions, will be counterbalanced by dissociation of the acid (eqn [24.2]); thus the conjugate pair acts as a 'buffer' to pH change. The innate resistance of most biological fluids to pH change is due to the presence of cellular constituents that act as buffers, e.g. proteins, which have a large number of weakly acidic and basic groups in their amino acid side chains.

Buffer capacity and the effects of pH

The extent of resistance to pH change is called the buffer capacity of a solution. The buffer capacity is measured experimentally at a particular pH by titration against a strong acid or alkali: the resultant curve will be strongly sigmoidal, with a plateau where the buffer capacity is greatest (Fig. 24.3). The mid-point of the plateau represents the pH where equal quantities of acid and conjugate base are present, and is given the symbol pK_a, which refers to the negative logarithm (to the base 10) of the acid dissociation constant, K_a, where

$$K_a = \frac{[H^+][A^-]}{[HA]} \qquad [24.6]$$

By rearranging eqn [24.6] and taking negative logarithms, we obtain:

$$pH = pK_a + \log_{10}\frac{[A^-]}{[HA]} \qquad [24.7]$$

This relationship is known as the Henderson–Hasselbalch equation and it shows that the pH will be equal to the pK_a when the ratio of conjugate base to acid is unity, since the final term in eqn [24.7] will be zero. Consequently, pK_a is an important factor in determining buffer capacity at a particular pH. In practical terms, a buffer solution will work most effectively at pH values about one unit either side of the pK_a.

Selecting an appropriate buffer

When selecting a buffer, you should be aware of certain limitations to their use. Citric acid and phosphate buffers readily form insoluble complexes with divalent cations, while phosphate can also act as a substrate, activator or inhibitor of certain enzymes. Both of these buffers contain biologically significant quantities of cations, e.g. Na^+ or K^+. TRIS (Table 24.3) is often toxic to biological systems: due to its high lipid solubility it can penetrate membranes, uncoupling electron transport reactions in whole cells and isolated organelles. In addition, it is markedly affected by temperature, with a tenfold increase in H^+ concentration from $4\,°C$ to $37\,°C$. A number of zwitterionic molecules (having both positive and negative groups) have

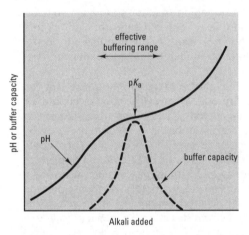

Fig. 24.3 Theoretical pH titration curve for a buffer solution. pH change is lowest and buffer capacity is greatest at the pK_a of the buffer solution.

An ideal buffer for biological purposes would possess the following characteristics:

- *impermeability to biological membranes;*
- *biological stability and lack of interference with metabolic and biological processes;*
- *lack of significant absorption of ultraviolet or visible light;*
- *lack of formation of insoluble complexes with cations;*
- *minimal effect of ionic composition or salt concentration;*
- *limited pH change in response to temperature.*

Table 24.3 pK_a values at 25°C and M_r of some acids and bases (upper section) and some large organic zwitterions (lower section) commonly used in buffer solutions. For polyprotic acids, where more than one proton may dissociate, the pK_a values are given for each ionisation step. Only the trivial acronyms of the larger molecules are provided: their full names can be obtained from the catalogues of most chemical suppliers.

Acid or base	pK_a value(s)	M_r
Acetic acid	4.8	60.1
Boric acid	9.2	61.8
Citric acid	3.1, 4.8, 5.4	191.2
Glycylglycine	3.1, 8.2	132.1
Phosphoric acid	2.1, 7.1, 12.3	98.0
Phthalic acid	2.9, 5.5	166.1
Succinic acid	4.2, 5.6	118.1
TRIS(base)*	8.3	121.1
CAPS (free acid)	10.4	221.3
CHES (free acid)	9.3	207.3
HEPES (free acid)	7.5	238.3
MES (free acid)	6.1	213.2
MOPS (free acid)	7.2	209.3
PIPES (free acid)	6.8	302.4
TAPS (free acid)	8.4	243.3
TRICINE (free acid)	8.1	179.2

*Note that this compound is hygroscopic and should be stored in a desiccator; also see text regarding its potential toxicity (p. 150).

Table 24.4 Preparation of sodium phosphate buffer solutions for use at 25°C. Prepare separate stock solutions of (a) disodium hydrogen phosphate and (b) sodium dihydrogen phosphate, both at 200 mol m^{-3}. Buffer solutions (at 100 mol m^{-3}) are then prepared at the required pH by mixing together the volume of each stock solution shown in the table, then diluting to a final volume of 100 ml using distilled or deionised water.

Required pH (at 25°C)	Volume of stock (a) Na$_2$HPO$_4$ (ml)	Volume of stock (b) NaH$_2$PO$_4$ (ml)
6.0	6.2	43.8
6.2	9.3	40.7
6.4	13.3	36.7
6.6	18.8	31.2
6.8	24.5	25.5
7.0	30.5	19.5
7.2	36.0	14.0
7.4	40.5	9.5
7.6	43.5	6.5
7.8	45.8	4.2
8.0	47.4	2.6

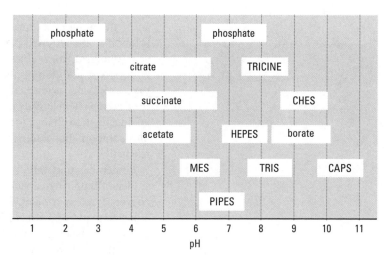

Fig. 24.4 Useful pH ranges of some commonly used buffers

been introduced to overcome some of the disadvantages of traditional buffers. These newer compounds are often referred to as 'Good buffers', to acknowledge the work of Dr N.E. Good: HEPES is one of the most useful zwitterionic buffers, with a pK_a of 7.5 at 25°C.

These zwitterionic substances are usually added to water as the free acid: the solution must then be adjusted to the correct pH with a strong alkali, usually NaOH or KOH. Alternatively, they may be used as their sodium or potassium salts, adjusted to the correct pH with a strong acid, e.g. HCl. Consequently, you may need to consider what effects such changes in ion concentration may have in a solution where zwitterions are used as buffers. In addition, zwitterionic buffers can interfere with protein determinations (e.g. Lowry method, p. 320).

Figure 24.4 shows a number of traditional and zwitterionic buffers and their effective pH ranges. When selecting one of these buffers, aim for a pK_a that is in the direction of the expected pH change (Table 24.3). For example, HEPES buffer would be a better choice of buffer than PIPES for use at pH 7.2 for experimental systems where a pH increase is anticipated, while PIPES would be a better choice where acidification is expected.

Preparation of buffer solutions

Having selected an appropriate buffer, you will need to make up your solution to give the desired pH. You will need to consider two factors:

- The ratio of acid and conjugate base required to give the correct pH.
- The amount of buffering required; buffer capacity depends upon the absolute quantities of acid and base, as well as their relative proportions.

In most instances, buffer solutions are prepared to contain between 10 mmol l^{-1} and 200 mmol l^{-1} of the conjugate pair. Although it is possible to calculate the quantities required from first principles using the Henderson–Hasselbalch equation, there are sources that tabulate the amount of substance required to give a particular volume of solution with a specific pH value for a range of buffers (e.g. Anon., 2006a, b). For traditional buffers, it is customary to mix stock solutions of acidic and basic components in the correct proportions to give the required pH (Table 24.4). For zwitterionic acids, the usual procedure is to add the compound to water, then bring the solution to the required pH by adding a specific amount of strong alkali or acid (obtained from tables). Alternatively, the required pH can be obtained by dropwise addition of alkali or acid, using a meter to check the pH, until the correct value is reached. When preparing solutions of zwitterionic buffers, the acid may be relatively insoluble. Do not wait for it to dissolve fully before adding alkali to change the pH – the addition of alkali will help bring the acid into solution (but make sure it has all dissolved before the desired pH is reached).

Remember that buffer solutions will only work effectively if they have sufficient buffering capacity to resist the change in pH expected during the course of the experiment. Thus a weak solution of HEPES (e.g. 10 mmol l^{-1}, adjusted to pH 7.0 with NaOH) will not be able to buffer the growth medium of a dense suspension of cells for more than a few minutes.

Finally, when preparing a buffer solution based on tabulated information, always confirm the pH with a pH meter before use.

Text references

Anon. (2006a) *BioBasics Technical Library: Phosphate Buffer Table*.
Available: http://www.sigmaaldrich.com/Area_of_ Interest/ Research_Essentials/Biochemicals/Key_Resources/ Technical_Library.html
Last accessed 09/04/07.

Anon. (2006b) *pH Theory and Practice: A Radiometer Analytical Guide*.

Available: http://www.radiometer-analytical.com/news/ en_ph_theory.asp?s = ovuk&OVRAW = pH% 20measurement&OVKEY = ph%20measurement& OVMTC = standard
Last accessed 09/04/07.

Lide, D.R. (ed.) (2006) *CRC Handbook of Chemistry and Physics*, 84th edn. CRC Press, Boca Raton, Florida.

Study exercises

24.1 Practise interconverting pH values and proton concentrations. Express all answers to three significant figures.

(a) What is pH 7.4 expressed as [H^+] in mol l^{-1}?

(b) What is pH 4.1, expressed as [H^+] in mol m^{-3}?

(c) What is the pH of a solution containing H^+ at 2×10^{-5} mol l^{-1}?

(d) What is the pH of a solution containing H^+ at $10^{-12.5}$ mol l^{-1}?

(e) What is the pH of a solution containing H^+ at 2.8×10^{-5} mol m^{-3}?

24.2 Decide on a suitable buffer to use. In the following instances, choose a buffer that would be suitable:

(a) Maintaining the pH at 8.5 during an enzyme assay of a cell-free extract at 25°C.

(b) Keeping a stable pH of 6.5 in an experiment to measure the uptake of radiolabelled glucose by a dense suspension of *E. coli*.

(c) Carrying out an assay of photosynthetic activity at pH 7.2 at temperatures of 10°C, 20°C and 30°C.

(d) Stabilising pH at 5.5 during enzyme extraction, in a solution where you intend to measure total protein concentration at a later stage.

24.3 Practise using the Henderson–Hasselbalch equation. What are the relative proportions of deprotonated (A^-) and protonated (HA) forms of each substance at the following pH values:

(a) acetic acid ($pK_a = 4.8$) for use in an experiment at pH 3.8?

(b) boric acid ($pK_a = 9.2$) for use in an experiment at pH 9.5?

(c) HEPES ($pK_a = 7.5$) for use in an experiment at pH 8.1?

The investigative approach

The term data (singular = datum, or data value) refers to items of information, and you will use different types of data from a wide range of sources during your practical work. Consequently, it is important to appreciate the underlying features of data collection and measurement.

Variables

Biological variables (Fig. 25.1) can be classified as follows:

Quantitative variables

These are characteristics whose differing states can be described by means of a number. They are of two basic types:

- Continuous variables, such as length; these are usually measured against a numerical scale. Theoretically, they can take any value on the measurement scale. In practice, the number of significant figures of a measurement is directly related to the precision of your measuring system; for example, dimensions measured with Vernier calipers will provide readings of greater precision than a millimetre ruler (p. 134).
- Discontinuous (discrete) variables, such as the number of eggs in a nest; these are always obtained by counting and therefore the data values must be whole numbers (integers). There are no intermediate values – for example, you never find 1.25 eggs in a nest.

Working with discontinuous variables – note that while the original data values must be integers, derived data and statistical values do not have to be whole numbers. Thus, it is perfectly acceptable to express the mean number of children per family as 2.4.

Ranked variables

These provide data that can be listed in order of magnitude (i.e. ranked). A familiar example is the abundance of an organism in a sample, which is often expressed as a series of ranks, e.g. rare = 1, occasional = 2, frequent = 3, common = 4, and abundant = 5. When such data are given numerical ranks, rather than descriptive terms, they are sometimes called 'semi-quantitative data'. Note that the difference in magnitude between ranks need not be consistent. For example, regardless of whether there was a one-year or a five-year gap between offspring in a family, their rank in order of birth would be the same.

Qualitative variables (attributes)

These are non-numerical and descriptive; they have no order of preference, and therefore are not measured on a numerical scale nor ranked in order of magnitude, but are described in terms of categories. Examples include viability (i.e. dead or alive) and shape (e.g. round, flat, elongated, etc.).

Variables may be independent or dependent. Usually, the variable under the control of the experimenter (e.g. time) is the independent variable, while the variable being measured is the dependent variable (p. 185). Sometimes it is not appropriate to describe variables in this way, and they are then referred to as interdependent variables (e.g. the length and breadth of an organism).

The majority of data values are recorded as direct measurements, readings or counts, but there is an important group, called derived (or computed), that results from calculations based on two or more data values, e.g. ratios, percentages, indices and rates.

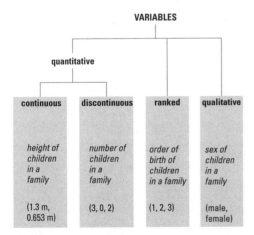

Fig. 25.1 Examples of the different types of variables as used to describe some characteristics of families

Measurement scales

Variables may be measured on different types of scale:

- Nominal scale: this classifies objects into categories based on a descriptive characteristic. It is the only scale suitable for qualitative data.
- Ordinal scale: this classifies by rank. There is a logical order in any number scale used.
- Interval scale: this is used for quantitative variables. Numbers on an equal-unit scale are related to an arbitrary zero point.
- Ratio scale: this is similar to the interval scale, except that the zero point now represents an absence of that character (i.e. it is an absolute zero). In contrast to the interval scale, the ratio of two values is meaningful (e.g. a temperature of 200K is twice that of 100K).

The measurement scale is important in determining the mathematical and statistical methods used to analyse your data. Table 25.1 presents a summary of the important properties of these scales. Note that you may be able to measure a characteristic in more than one way, or you may be able to convert data collected in one form to a different form. For instance, you might measure light in terms of the photon flux density (p. 343) between particular wavelengths of the EMR spectrum (ratio scale), or simply as 'blue' or 'red' (nominal scale); you could find out the dates of birth of individuals (interval scale) but then use this information to rank them in order of birth (ordinal scale). Where there are no other constraints, you should use a ratio scale to

Table 25.1 Some important features of scales of measurement

	Measurement scale			
	Nominal	Ordinal	Interval	Ratio
Type of variable	Qualitative (Ranked)* (Quantitative)*	Ranked (Quantitative)*	Quantitative	Quantitative
Examples	Species Sex Colour	Abundance scales Reproductive condition Optical assessment of colour development	Fahrenheit temperature scale Date (BC/AD)	Kelvin temperature scale Weight Length Response time Most physical measurements
Mathematical properties	Identity	Identity Magnitude	Identity Magnitude Equal intervals	Identity Magnitude Equal intervals True zero point
Mathematical operations possible on data	None	Rank	Rank Addition Subtraction	Rank Addition Subtraction Multiplication Division
Typical statistics used	Only those based on frequency of counts made: contingency tables, frequency distributions, etc. Chi-square test	Non-parametric methods, sign tests. Mann–Whitney U-test	Almost all types of test, t-test, analysis of variance (ANOVA), etc. (check distribution before using, Chapter 66)	Almost all types of test, t-test, ANOVA, etc. (check distribution before using, Chapter 66)

*In some instances (see text for examples).

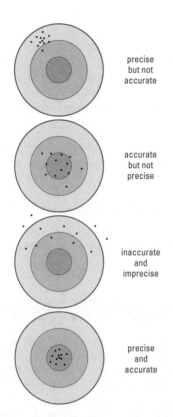

precise
but not
accurate

accurate
but not
precise

inaccurate
and
imprecise

precise
and
accurate

Fig. 25.2 'Target' diagrams illustrating precision and accuracy

measure a quantitative variable, since this will allow you to use the broadest range of mathematical and statistical procedures (Table 25.1).

Accuracy and precision

Accuracy is the closeness of a measured or derived data value to its true value, while precision is the closeness of repeated measurements to each other (Fig. 25.2). A balance with a fault in it (i.e. a bias, see below) could give precise (i.e. very repeatable) but inaccurate (i.e. untrue) results. Unless there is bias in a measuring system, precision will lead to accuracy and it is precision that is generally the most important practical consideration, if there is no reason to suspect bias. You can investigate the precision of any measuring system by repeated measurements of individual samples.

Absolute accuracy and precision are impossible to achieve, due to both the limitations of measuring systems for continuous quantitative data and the fact that you are usually working with incomplete data sets (samples, p. 179). It is particularly important to avoid spurious accuracy in the presentation of results; include only those digits that the accuracy of the measuring system implies (p. 408). This type of error is common when changing units (e.g. inches to metres) and in derived data, especially when calculators give results to a large number of decimal places.

Bias (systematic error) and consistency

Bias is a systematic or non-random distortion and is one of the most troublesome difficulties in using numerical data. Biases may be associated with incorrectly calibrated instruments, e.g. a faulty pipettor, or with experimental manipulations, e.g. shrinkage during the preservation of a specimen. Bias in measurement can also be subjective, or personal, e.g. an experimenter's preconceived ideas about an 'expected' result.

Bias can be minimised by using a carefully standardised procedure, with fully calibrated instruments. You can investigate bias in 'trial runs' by measuring a single variable in several different ways, to see whether the same result is obtained.

If a personal bias is possible, 'blind' measurements should be made where the identity of individual samples is not known to the operator, e.g. using a coding system.

Measurement error

All measurements are subject to error, but the dangers of misinterpretation are reduced by recognising and understanding the likely sources of error and by adopting appropriate protocols and calculation procedures.

A common source of measurement error is carelessness, e.g. reading a scale in the wrong direction or parallax errors. This can be reduced greatly by careful recording and may be detected by repeating the measurement. Other errors arise from faulty or inaccurate equipment, but even a perfectly functioning machine has distinct limits to the accuracy and precision of its measurements. These limits are often quoted in manufacturers' specifications and are applicable when an instrument is new; however, you should allow for some deterioration with age. Further errors are introduced when the subject being studied is open to influences outside your control. Resolving such problems requires appropriate experimental design (Chapter 31) and sampling procedures (Chapter 30).

Minimising errors – determine early in your study what the dominant errors are likely to be and concentrate your time and effort on reducing these.

Working with derived data – special effort should be made to reduce measurement errors because their effects can be magnified when differences, ratios, indices or rates are calculated.

One major influence virtually impossible to eliminate is the effect of the investigation itself: even putting a thermometer in a liquid may change the temperature of the liquid. The very act of measurement may give rise to a confounding variable (p. 186) as discussed in Chapter 31.

Sources for further study

Anon. *Measurement*. Available: http://wikipedia.org/wiki/Measurement
Last accessed: 09/04/07.

Erikson, B.H. and Nosanchuk, T.A. (1992) *Understanding Data*, 2nd edn. Open University Press, Milton Keynes. [A text aimed at social science students but with clear explanations of issues that are generic, including information on analysis of data.]

Friedrich, G.W. *Basic Principles of Measurement. Methods of Inquiry*. Available: http://www.scils.rutgers.edu/~gusf/measurement.html
Last accessed: 09/04/07.
[Course notes covering diverse aspects of enquiry.]

National Instruments *Measurement Encyclopedia*. Available: http://zone.ni.com/devzone/nidzgloss.nsf/glossary/
Last accessed: 09/04/07.

Study exercises

25.1 Classify variables. Decide on the type of variables used for the following measures, indicating whether they are quantitative or qualitative, continuous or discontinuous, and the type of scale that would be used.

(a) Number of organisms in a population.
(b) Length of individuals in a population.
(c) Colour of flowers.
(d) Species present in a sample.
(e) Date of a sample.
(f) Reproductive condition of an animal.

25.2 Assess errors and accuracy of a set of measurements. Assume that you are asked to measure the length, breadth and height of each of a sample of 20 limpet shells (see figure) using (a) a ruler and (b) a pair of Vernier calipers (p. 134). Identify the sources of error likely to be present in your measurements using each tool and the precision to which you would be able to quote your data. Devise a protocol for taking these measurements that would help you to minimise the errors and maximise the accuracy and precision.

25.3 Investigate types of errors. A student weighed a set of standard masses on two electronic balances and obtained the readings shown in the table below. Explain these results in terms of the type of error involved in each case.

Lateral (a) and ventral (b) views of a limpet to show measurement variables length (L), breadth (B) and height (H). Note also the uneven edge of the shell.

Comparison of weights of masses on two balances

	Standard mass (g)				
	10	25	50	100	250
Reading (balance A)	10.050	25.049	50.051	100.048	250.052
Reading (balance B)	10.004	25.011	50.021	100.039	250.102

When describing a measurement, you normally state both a number and a unit (e.g. 'the length is 1.85 metres'). The number expresses the ratio of the measured quantity to a fixed standard, while the unit identifies that standard measure or dimension. Clearly, a single unified system of units is essential for efficient communication of such data within the scientific community. The Système International d'Unités (SI) is the internationally ratified form of the metre-kilogram-second system of measurement and represents the accepted scientific convention for measurements of physical quantities.

Another important reason for adopting consistent units is to simplify complex calculations where you may be dealing with several measured quantities (see pp. 405 and 407). Although the rules of the SI are complex and the scale of the base units is sometimes inconvenient, to gain the full benefits of the system you should observe its conventions strictly.

The description of measurements in SI involves:

- seven base units and two supplementary units, each having a specified abbreviation or symbol (Table 26.1);
- derived units, obtained from combinations of base and supplementary units, which may also be given special symbols (Table 26.2);
- a set of prefixes to denote multiplication factors of 10^3, used for convenience to express multiples or fractions of units (Table 26.3).

Table 26.1 The base and supplementary SI units

Measured quantity	Name of SI unit	Symbol
Base units		
Length	metre	m
Mass	kilogram	kg
Amount of substance	mole	mol
Time	second	s
Electric current	ampere	A
Temperature	kelvin	K
Luminous intensity	candela	cd
Supplementary units		
Plane angle	radian	rad
Solid angle	steradian	sr

Table 26.2 Some important derived SI units

Measured quantity	Name of unit	Symbol	Definition in base units	Alternative in derived units
Energy	joule	J	$m^2\,kg\,s^{-2}$	N m
Force	newton	N	$m\,kg\,s^{-2}$	$J\,m^{-1}$
Pressure	pascal	Pa	$kg\,m^{-1}\,s^{-2}$	$N\,m^{-2}$
Power	watt	W	$m^2\,kg\,s^{-3}$	$J\,s^{-1}$
Electric charge	coulomb	C	A s	$J\,V^{-1}$
Electric potential difference	volt	V	$m^2\,kg\,A^{-1}\,s^{-3}$	$J\,C^{-1}$
Electric resistance	ohm	Ω	$m^2\,kg\,A^{-2}\,s^{-3}$	$V\,A^{-1}$
Electric conductance	siemens	S	$s^3\,A^2\,kg^{-1}\,m^{-2}$	$A\,V^{-1}$ or Ω^{-1}
Electric capacitance	farad	F	$s^4\,A^2\,kg^{-1}\,m^{-2}$	$C\,V^{-1}$
Luminous flux	lumen	lm	cd sr	
Illumination	lux	lx	$cd\,sr\,m^{-2}$	$lm\,m^{-2}$
Frequency	hertz	Hz	s^{-1}	
Radioactivity	becquerel	Bq	s^{-1}	
Enzyme activity	katal	kat	mol substrate s^{-1}	

Table 26.3 Prefixes used in the SI

Multiple	Prefix	Symbol	Multiple	Prefix	Symbol
10^{-3}	milli	m	10^{3}	kilo	k
10^{-6}	micro	μ	10^{6}	mega	M
10^{-9}	nano	n	10^{9}	giga	G
10^{-12}	pico	p	10^{12}	tera	T
10^{-15}	femto	f	10^{15}	peta	P
10^{-18}	atto	a	10^{18}	exa	E
10^{-21}	zepto	z	10^{21}	zetta	Z
10^{-24}	yocto	y	10^{24}	yotta	Y

Recommendations for describing measurements in SI units

Basic format

- Express each measurement as a number separated from its units by a space. If a prefix is required, no space is left between the prefix and the unit it refers to. Symbols for units are only written in their singular

> **Example** n stands for nano and N for newtons.

> **Example** 1 982 963.192 309 kg (perhaps better expressed as 1.982 963 192 309 Gg).

form and do not require full stops to show that they are abbreviated or that they are being multiplied together.

- Give symbols and prefixes appropriate upper- or lower-case initial letters as this may define their meaning. Upper-case symbols are named after persons but when written out in full they are not given initial capital letters.
- Show the decimal sign as a full point on the line. Some metric countries continue to use the comma for this purpose and you may come across this in the literature: commas should not therefore be used to separate groups of thousands. In numbers that contain many significant figures, you should separate multiples of 10^3 by spaces rather than commas.

Compound expressions for derived units

- Take care to separate symbols in compound expressions by a space to avoid the potential for confusion with prefixes. Note, for example, that 200 m s (metre-seconds) is different from 200 ms (milliseconds).
- Express compound units by using negative powers rather than a solidus (/): for example, write $mol\, m^{-3}$ rather than mol/m^3. The solidus is reserved for separating a descriptive label from its units (see p. 387).
- Use parentheses to enclose expressions being raised to a power if this avoids confusion: for example, a photosynthetic rate might be given in $mol\, CO_2\, (mol\, photons)^{-1}\, s^{-1}$.
- Where there is a choice, select relevant (natural) combinations of derived and base units: e.g. you might choose units of $Pa\, m^{-1}$ to describe a hydrostatic pressure gradient rather than $kg\, m^{-2}\, s^{-2}$, even though these units are equivalent and the measurements are numerically the same.

Use of prefixes

- Use prefixes to denote multiples of 10^3 (Table 26.3) so that numbers are kept between 0.1 and 1000.
- Treat a combination of a prefix and a symbol as a single symbol. Thus, when a modified unit is raised to a power, this refers to the whole unit including the prefix.
- Avoid the prefixes deci (d) for 10^{-1}, centi (c) for 10^{-2}, deca (da) for 10 and hecto (h) for 100 as they are not strictly SI.
- Express very large or small numbers as a number between 1 and 10 multiplied by a power of 10 if they are outside the range of prefixes shown in Table 26.3.
- Do not use prefixes in the middle of derived units: they should be attached only to a unit in the numerator (the exception is in the unit for mass, kg).

> **Examples**
>
> 10 μm is preferred to 0.000 01 m or 0.010 mm.
>
> $1 mm^2 = 10^{-6} m^2$ (not one-thousandth of a square metre).
>
> $1 dm^3$ (1 litre) is more properly expressed as $1 \times 10^{-3} m^3$.
>
> The mass of a neutrino is 10^{-36} kg.
>
> State as $MW\, m^{-2}$ rather than $W\, mm^{-2}$.

 KEY POINT *For the foreseeable future, you will need to make conversions from other units to SI units, as much of the literature quotes data using imperial, c.g.s. or other systems. You will need to recognise these units and find the conversion factors required. Examples relevant to biology are given in Box 26.1. Table 26.4 provides values of some important physical constants in SI units.*

Box 26.1　Conversion factors between some redundant units and the SI

Quantity	SI unit/symbol	Old unit/symbol	Multiply number in old unit by this factor for equivalent in SI unit*	Multiply number in SI unit by this factor for equivalent in old unit*
Area	square metre/m^2	acre	4.04686×10^3	0.247105×10^{-3}
		hectare/ha	10×10^3	0.1×10^{-3}
		square foot/ft^2	0.092903	10.7639
		square inch/in^2	645.16×10^{-9}	1.55000×10^6
		square yard/yd^2	0.836127	1.19599
Angle	radian/rad	degree/°	17.4532×10^{-3}	57.2958
Energy	joule/J	erg	0.1×10^{-6}	10×10^6
		kilowatt hour/kWh	3.6×10^6	0.277778×10^{-6}
		calorie/cal	4.1868	0.2388
Length	metre/m	Ångstrom/Å	0.1×10^{-9}	10×10^9
		foot/ft	0.3048	3.28084
		inch/in	25.4×10^{-3}	39.3701
		mile	1.60934×10^3	0.621373×10^{-3}
		yard/yd	0.9144	1.09361
Mass	kilogram/kg	ounce/oz	28.3495×10^{-3}	35.2740
		pound/lb	0.453592	2.20462
		stone	6.35029	0.157473
		hundredweight/cwt	50.8024	19.6841×10^{-3}
		ton (UK)	1.01605×10^3	0.984203×10^{-3}
Pressure	pascal/Pa	atmosphere/atm	101325	9.86923×10^{-6}
		bar/b	100000	10×10^{-6}
		millimetre of mercury/mmHg	133.322	7.50064×10^{-3}
		torr/Torr	133.322	7.50064×10^{-3}
Radioactivity	becquerel/Bq	curie/Ci	37×10^9	27.0270×10^{-12}
Temperature	kelvin/K	centigrade (Celsius) degree/°C	°C $+ 273.15$	K $- 273.15$
		Fahrenheit degree/°F	(°F $+ 459.67) \times 5/9$	(K $\times 9/5) - 459.67$
Volume	cubic metre/m^3	cubic foot/ft^3	0.0283168	35.3147
		cubic inch/in^3	16.3871×10^{-6}	61.0236×10^3
		cubic yard/yd^3	0.764555	1.30795
		UK pint/pt	0.568261×10^{-3}	1759.75
		US pint/liq pt	0.473176×10^{-3}	2113.38
		UK gallon/gal	4.54609×10^{-3}	219.969
		US gallon/gal	3.78541×10^{-3}	264.172

*In the case of temperature measurements, use formulae shown.

Table 26.4　Some physical constants in SI terms

Physical constant	Symbol	Value and units
Avogadro's constant	N_A	$6.022174 \times 10^{23} \, \text{mol}^{-1}$
Boltzmann's constant	k	1.380626×10^{-23}
Charge of electron	e	$1.602192 \times 10^{-19} \, \text{C}$
Gas constant	R	$8.31443 \, \text{J K}^{-1} \, \text{mol}^{-1}$
Faraday's constant	F	$9.648675 \times 10^4 \, \text{C mol}^{-1}$
Molar volume of ideal gas at STP	V_0	$0.022414 \, \text{m}^3 \, \text{mol}^{-1}$
Speed of light in vacuo	c	$2.997924 \times 10^8 \, \text{m s}^{-1}$
Planck's constant	h	$6.626205 \times 10^{-34} \, \text{J s}$

Some implications of SI in biology

Volume

The SI unit of volume is the cubic metre, m^3, which is rather large for practical purposes. The litre (l) and the millilitre (ml) are technically obsolete, but are widely used and glassware is still calibrated using them. Note also that the US spelling is liter. You may find litre given the symbol L, rather than l to avoid confusion with 1 and I.

Mass

The SI unit for mass is the kilogram (kg) rather than the gram (g): this is unusual because the base unit has a prefix applied.

Amount of substance

You should use the mole (mol, i.e. Avogadro's constant, see Table 26.4) to express very large numbers. The mole gives the number of atoms in the atomic mass, a convenient constant. Always specify the elementary unit referred to in other situations (e.g. mol photons $m^{-2} s^{-1}$).

Concentration

The SI unit of concentration, $mol\,m^{-3}$, is quite convenient for biological systems. It is equivalent to the non-SI term 'millimolar' ($mM \equiv mmol\,l^{-1}$), while 'molar' ($M \equiv mol\,l^{-1}$) becomes $kmol\,m^{-3}$. Note that the symbol M in the SI is reserved for mega and hence should not be used for concentrations. If the solvent is not specified, then it is assumed to be water (see Chapter 23).

Time

In general, use the second (s) when reporting physical quantities having a time element (e.g. give photosynthetic rates in mol $CO_2\,m^{-2}\,s^{-1}$). Hours (h), days (d) and years should be used if seconds are clearly absurd (e.g., samples were taken over a 5-year period). Note, however, that you may have to convert these units to seconds when doing calculations.

Temperature

The SI unit is the kelvin, K. The degree Celsius scale has units of the same magnitude, °C, but starts at 273.15 K, the melting point of ice at STP. Temperature is similar to time in that the Celsius scale is in widespread use, but note that conversions to K may be required for calculations. Note also that you must not use the degree sign (°) with K and that this symbol must be in upper case to avoid confusion with k for kilo; however, you *should* retain the degree sign with °C to avoid confusion with the coulomb, C.

Light

While the first six base units in Table 26.1 have standards of high precision, the SI base unit for luminous intensity, the candela (cd) and the derived units lm and lx (Table 26.2), are defined in 'human' terms. They are, in fact, based on the spectral responses of the eyes of 52 American GIs measured in 1923! Clearly, few organisms 'see' light in the same way as this sample of humans. Also, light sources differ in their spectral quality. For these reasons, it is better to use expressions based on energy or photon content (e.g. $W\,m^{-2}$ or mol photons $m^{-2}\,s^{-1}$) in studies other than those on human vision. Ideally you should specify the photon wavelength spectrum involved (see Chapter 55).

Sources for further study

Institute of Biology (1997) *Biological Nomenclature: Recommendations on Terms, Units and Symbols*, 2nd edn. Institute of Biology, London.

The NIST Reference on Constants, Units and Uncertainty: International System of Units (SI). Available: http://physics.nist.gov/cuu/Units/
Last accessed 09/04/07.

Pennycuick, C.J. (1988) *Conversion Factors: SI Units and Many Others: Over 2100 Conversion Factors for Biologists and Mechanical Engineers Arranged in 21*

Quick-Reference Tables. University of Chicago Press, Chicago.

Rowlett, R. *How Many? A Dictionary of Units of Measurement*. Available: http://www.unc.edu/~rowlett/units/
Last accessed 09/04/07.

Tapson, F. *A Dictionary of Units*. Available: http://www.ex.ac.uk/cimt/dictunit/dictunit.htm
Last accessed 09/04/07.

Study exercises

26.1 Practise converting between units. Using Box 26.1 as a source, convert the following amounts into the units shown. Give your answers to three significant figures.
(a) 101 000 Pa into atmospheres.
(b) One square yard into square millimetres.
(c) One UK pint into millilitres.
(d) 37 °C into Kelvin.
(e) 11 stone 6 pounds into kilograms.

26.2 Practise using prefixes appropriately. Simplify the following number/unit combinations using an appropriate prefix so that the number component lies between 0.1 and 1000.
(a) 10 000 mm
(b) 0.015 ml
(c) 5×10^9 J
(d) 65 000 m s^{-1}
(e) 0.000 000 0001 g

26.3 Check units in an equation. The Hagan–Poiseuille equation describes water flux in smooth cylindrical pipes assuming laminar flow. This equation can be expressed as:

$$J_v = \frac{r^2 \times \delta P}{8\eta \times \delta x}$$

where:

r = the radius of the cylindrical pipe (m);
η ('eta') = the viscosity of the liquid (Pa s);
δP = the pressure difference across the ends of the pipe (Pa);
δx = the length of the pipe (m).

Verify, by putting relevant units in place of the variables in this equation and simplifying the resulting relationship, that appropriate units for J_v are m s^{-1} (mean flow rate).

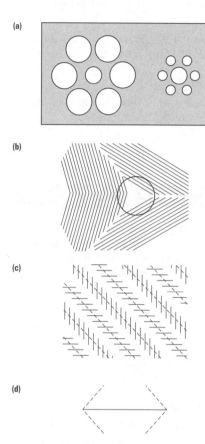

Observations provide the basic information leading to the formulation of hypotheses, an essential step in the scientific method (see Fig. 31.1). Observations are obtained either directly by our senses or indirectly through the use of instruments that extend our senses and may be either:

- Qualitative: described by words or terms rather than by numbers and including subjective descriptions in terms of variables such as colour, shape and smell; often recorded using photographs and drawings.
- Quantitative: numerical values derived from counts or measurements of a variable (see Chapter 25), frequently requiring use of some kind of instrument.

KEY POINT *Although qualitative and quantitative observations are useful in biology, you should try to make numerical counts or measurements wherever possible, as this allows you to define your observations more rigorously and make objective comparisons using statistical tools.*

Factors influencing the quality of observations

Perception

Observation is highly dependent upon the perception of the observer (Fig. 27.1). Perception involves both visual and intuitive processes, so your interpretation of what you see is very dependent upon what you already know or have seen before. Thus, two persons observing the same event or object may 'see' it differently, a good example of bias. This is frequently true in microscopy where experience is an important factor in interpretation.

When you start biology, your knowledge base will be limited and your experience restricted. Practical training in observation provides the opportunity to develop both aspects of your skills in a process that is effectively a positive feedback loop – the more you know/see as a result of practice, the better will your observations become.

Precision and error

Obviously very important for interpretive accuracy, with both human and non-human components. These are dealt with in Chapter 25.

Artefacts

These are artificial features introduced usually during some treatment process such as chemical fixation prior to microscopic examination. They may be included in the interpretive process if their presence is not recognised – again, prior experience and knowledge are important factors in spotting artefacts (see Chapter 44 and especially Fig. 44.1 in relation to microscopy).

Developing observational skills

You must develop your knowledge and observational skills to benefit properly from your practical work. The only way to acquire these skills is through extensive practice.

Fig. 27.1 Examples of 'optical illusions' caused by problems of perception. Image (a) shows how the sizes of adjacent objects can distort a simple comparison of size: the central circles in either hexagonal pattern are the same size. Image (b) reveals how shapes can be distorted by adjacent linear objects: the inner shape is a perfect circle. Image (c) illustrates how directional cues can lead to confusion: the dashed lines are parallel. Image (d) shows how adjacent shapes can make comparison of simple linear dimensions difficult: the two solid lines are the same length. In all cases, the correct perception can only be confirmed with a measurement aid such as a ruler or compass.

Observer effects – remember that your presence, or the act of observation itself, may influence the event you are observing. Take appropriate precautions, like using a hide when observing animal behaviour.

Make sure your observations are:

- relevant, i.e. directed towards a clearly defined objective;
- accurate, i.e. related to a scale whenever possible;
- repeatable, i.e. as error free (precise) as possible.

Much biological work attempts to relate structure to function, often through careful analysis of structure at different levels. One of the best ways to develop observational skills is by making accurate drawings or diagrams, forcing you to look more carefully than is usual (see Chapter 28). An important observational skill to develop is the interpretation of two-dimensional images, such as sections through plant/animal material and photographs, in terms of the three-dimensional forms from which they are derived. This requires a clear understanding of the nature of the image in terms of both scale and orientation (see Chapter 44).

Counting

Counting is an observational skill that requires practice to become both accurate and efficient. It is easy to make errors or lose count when working with large numbers of objects. Use a counting aid whenever human error might be significant. There are many such aids such as tally counters, tally charts and specialised counting devices like colony counters. It is important to avoid counting items twice. For example, when counting microbial colonies on Petri plates, each colony can be marked off as it is counted on the base of the plate using a spirit-based marker pen.

Another valuable technique is to use a grid system to organise the counting procedure. This has become formalised in equipment such as the haemocytometer used for counting blood cells (see Box 46.1, p. 275). Remember that you must decide on a protocol for sampling, particularly with regard to the direction of counting within the grid and for dealing with boundary overlaps to prevent double counting at the edges of the grid squares (see Fig. 30.4, p. 181).

Observation during examinations

Making appropriate observations during practical examinations often causes difficulty, particularly when qualitative observations are needed, e.g. when asked to classify a specimen, giving reasons. Answering such questions clearly requires biological knowledge but also requires a strategy to provide the relevant observations. Thus for the above example, observations relevant to determining each taxonomic level should be made and recorded: set out your observations in a logical sequence so that you show the examiner how you arrived at your conclusion.

A similar approach is needed when asked to make observations related to structure or function. For example, if comments on a method of locomotion are required, do not make observations on irrelevant structures. You will obtain maximum marks only if your answers are concise and relevant (Chapter 6).

Preparation – thorough theoretical groundwork before a practical class or examination is vitally important for improving the quality of your observations (see Fig. 27.1).

Perceptual illusions – beware deceptions such as holes appearing as bumps in photographs, misinterpretations due to lack of information on scale and other simple but well-known tricks of vision (see Fig. 27.1).

Making counts – acetate sheets can be used as overlays for photographs, drawings and other images, then as each object is counted it can be marked off using a water-based marker pen. Remember to mark identification points in case the sheet slips!

Text references and sources for further study

Akins, K. (ed.) (1996) *Perception*. Vancouver Studies in Cognitive Science, Vol. 5. Oxford University Press, Oxford.
[A detailed examination of the problems of perception from a psychological perspective.]

Edwards, B. (2000). *The New Drawing on the Right Side of your Brain*. Souvenir Press, London.

Raulin, M.L. *Naturalistic Observation and Case-Study Research*. Available: http://www.abacon.com/graziano/ch06/index.htm
Last accessed 09/04/07.

Study exercises

27.1 Assess the difficulties of making counts of large numbers. The dots in the area shown in the figure below represent bacterial colonies on a Petri dish. To appreciate the difficulties of counting numbers accurately, make three rapid independent attempts at counting the dots by eye (without using any aids). Now subdivide the area, such that the number of colonies per subdivision is more easily counted (i.e. is a number between 1 and 20). Again carry out three further estimates using these subdivisions, summing the counts obtained each time. Compare the results using both methods – which is the more precise and which the more accurate?

27.2 Extrapolate from 2-D to 3-D. Each of the diagrams on the next page represents a series of nine transverse sections taken at regular intervals through an organism or part of an organism. Use these diagrams to construct a 3-D representation of the structure in each case.

27.3 Trick your brain into letting you do better drawings. One of the problems with making drawings is that the brain thinks it 'knows' how things should look. This may cause distortion in diagrams. One way to produce realistic drawings of relatively complex objects is to turn the object upside down, then try to draw it. Because the brain no longer recognises the object, it focuses on the relationships between spaces and lines and allows you to make a reasonable copy (Edwards, 2000). Try this out by taking a photograph of an object that you would normally find difficult to draw, e.g. a person's face or an animal, and using this technique to copy it.

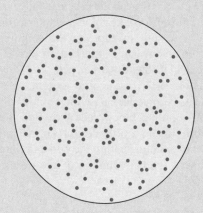

Bacterial colonies on a Petri dish

(continued)

Study exercises (continued)

(a)

(b)

(c)

Sections taken at regular intervals through organisms or parts of organisms

Drawing has an important place in biological teaching because of its role in developing observation skills. You need to look at a specimen very carefully to be able to draw it accurately, while labelling a diagram forces you to think about the component structures and their positions. If your observation of a specimen is poor, so too will be your diagram.

Strictly, a *drawing* is a detailed and accurate representation of a specimen, requiring no previous biological knowledge. This level of artwork is never required for normal practical work. A *diagram*, on the other hand, needs to be accurate in its general proportions, but is otherwise very stylised, showing only the most important features. Biological knowledge is required to select items for inclusion and to decide what detail to ignore. Diagrams are often called figures in formal scientific writing, but may sometimes be referred to loosely as drawings, since the above distinction is frequently ignored.

> **Figure numbering in reports** – *if producing several diagrams, graphs, etc., number them consecutively as Figure 1, Figure 2, etc., and use this system to refer to them in the text.*

 KEY POINT *You may not feel confident about being able to produce quality artwork in practicals, especially when the time allowed is limited. However, the requirements of biological drawing are not as demanding as you might think and the skills required can be learned. By following the guidelines and techniques explained below, most students should be able to produce good diagrams.*

The main types of figure

Cell diagrams

In a cell diagram, your aim is to show accurately the details of the individual cells in a tissue. You would normally draw a cell diagram from a specimen viewed by light microscopy at a magnification of 200× or more. Your diagram should be detailed, but need not comprise more than a few cells, especially if they are all similar (Fig. 28.1(a)). You may be asked to draw only one complete cell, plus *part* of any neighbouring cells, so that the interrelationship between cells can be seen. Use labels to note any structures that are stained, and what this means in terms of cell chemistry. For cells with thin walls, you may need to exaggerate the wall thickness to show it on your diagram. If this is necessary, add an explanatory note to your diagram (see Fig. 28.1(a)).

Tissue diagrams or maps

The purpose of a tissue diagram is to show how the tissues are placed in a section through a whole organism or one of its organs (compare Figs 28.1(b) and 28.1(c)). You should not include any cellular detail unless specifically instructed, and shading or hatching should be kept to a minimum. The main difficulty you will encounter is deciding where to draw the boundary line between tissues: cell differentiation is rarely discrete, so cells at the boundary may show characteristics of both tissues. A certain amount of background knowledge and interpretive skill is required for this.

> **Definitions**
>
> When drawing cell diagrams, tissue maps and body plans, it might help to bear in mind the following definitions:
>
> **Cells** – the 'building blocks' of life. Always separated from their environment and other cells by a membrane ± a cell wall. Typical dimensions: 10 to 100 µm. Examples: leaf mesophyll cell; kidney nephron cell; yeast cell.
>
> **Tissues** – collections of cells having similar structure and function. From one cell thick to many cell layers. Examples: palisade mesophyll in a leaf; renal cortex in the kidney.
>
> **Organs** – structures consisting of several tissues working as a group to perform a specific function. Examples: leaf; kidney.

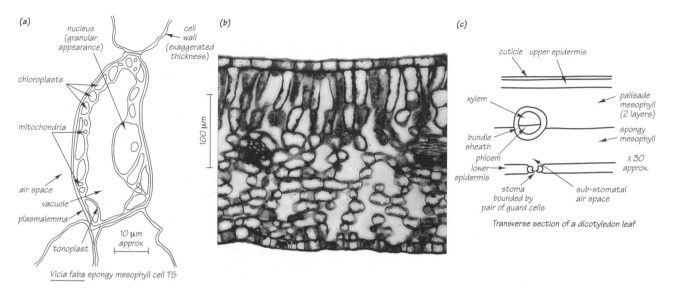

Fig. 28.1 Cell and tissue diagrams compared with a photomicrograph. (a) Cell diagram showing representative spongy mesophyll cells from an electron microscope image of a leaf section; (b) a photomicrograph of a transverse section through a leaf (light microscopy). Different scales are used, but in both cases they are indicated by a bar (see p. 259). Note how the cell-wall structure is clarified in (a) by slightly exaggerating its thickness. In a light microscope section less detail of organelles would be visible. (c) Tissue diagram of part of a TS leaf section. Note that this diagram is uncluttered by detail and the individual tissues are not shaded – the scale is indicated by a magnification factor relevant to the actual size of the diagram (reduced here) (see p. 258).

Choosing what to include in a morphological diagram – since the essence of this type of drawing is to show only the important features, it is worth making a list of the items to be included, or to highlight these where mentioned in your practical schedule. Don't forget, however, to add enough anatomical detail, e.g. of a body plan, to place outline features in the appropriate context. Always add a scale and effective labelling. Note that if you wish to show fine detail, consider doing this on a separate higher-scale diagram.

Morphological diagrams and body plans

In a morphological diagram, the objective is to provide a lifelike representation indicating the main surface features (Fig. 28.2). For a body plan, your aim is to show the relationships between segments, organs or other body parts, often following a dissection (Fig. 28.3). In both cases, shading should be avoided, unless it serves to highlight a particular feature. You should fully label both types of diagram and add notes where relevant. The main problem you are likely to encounter is keeping the different parts in proportion to each other – this can be solved by using construction lines and frames (see below).

> **KEY POINT** Remember that in drawing some 'artistic licence' is possible, allowing you to merge features seen on different specimens, or show, for example, the inside or underside of one of the parts.

Continuous tone drawings are more refined versions of the above where shading is used to provide a pictorial representation of the item of interest, providing fine detail and realism (Fig. 28.4). These drawings are generally only used for presentation work and are not normally expected from students. It can be argued that a properly taken photograph (Chapter 48) is often more appropriate and less time-consuming.

Apparatus diagrams

Here, your aim is to portray the components of some experimental set-up as a diagram (Fig. 28.5). Note that these figures are normally drawn as a section rather than a perspective drawing. Also, you may be more concerned with the relationship between parts than with showing them

Drawing and diagrams

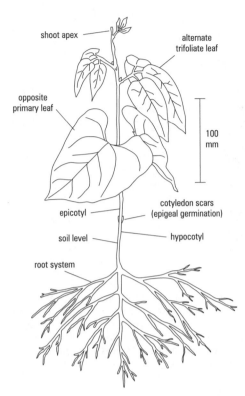

Fig. 28.2 Morphological diagram of a French bean seedling, *Phaseolus vulgaris* L. Note the lack of shading compared with Fig. 28.4.

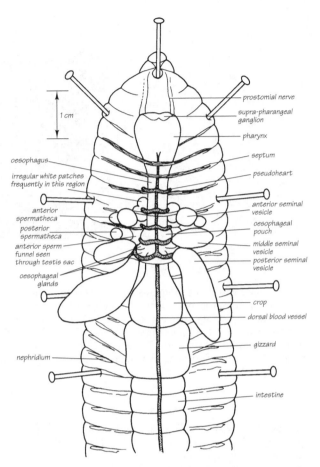

Fig. 28.3 Example of a body plan. This is a diagram of a general dissection of an earthworm, *Lumbricus terrestris* L., dorsal view. Note the lack of shading and the use of labels and annotations to identify clearly what has been drawn.

Fig. 28.4 Continuous tone drawing of a French bean seedling, *Phaseolus vulgaris* L. Compare with Fig. 28.2.

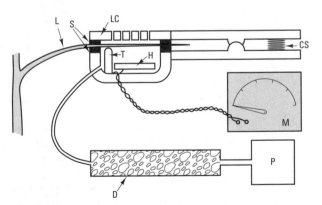

Fig. 28.5 Two-dimensional apparatus diagram of a diffusion porometer (not drawn to uniform scale). Note the use of letters to simplify labelling – these should be explained in the figure legend (e.g. L = leaf, P = pump).

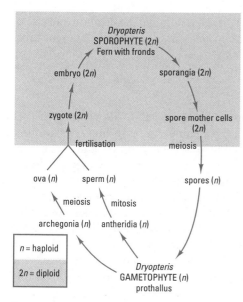

Fig. 28.6 Example of a chart: diagrammatic life cycle of the fern *Dryopteris filix-mas*. Note the use of shading and chromosome number to denote haploid and diploid phases of the life cycle, and how a key is used to explain this.

to a uniform scale. For example, in Fig. 28.5, the leaf clamp (LC) is exaggerated in size compared with other components so that its internal detail can be shown.

Charts

The main purpose of a chart is to organise information (e.g. Fig. 28.6). You can use charts to communicate complex ideas, procedures or lists of facts by simplifying, grouping and appropriate layout. In biology, they are particularly useful for illustrating life cycles, metabolic pathways and organisational hierarchies. Flowcharts (e.g. Fig. 33.1) are a specialised form. To be effective, charts must be logically organised. A good chart should clarify the parts and their relationships and its presentation should be simple, clear and visually pleasing. Make several rough sketches with different arrangements before deciding on the final version. You should use appropriate words or symbols to denote the components and link them with lines or arrows to show sequences or interrelationships. Computer software packages can be used to enhance the quality of presentation.

Graphs and histograms

These are used to display numerical information (data) in a form that is easily assimilated. Chapter 62 covers the main types of graphs and how they should be constructed.

Steps towards drawing a good diagram

To produce good figures, both planning and careful execution are needed (Box 28.1).

Planning

The first stage of any drawing is to decide exactly what to draw – this may seem obvious, but until you have focused your thoughts, you will not be able to decide on the answers to the following questions:

- What is the purpose of the drawing?
- What type of drawing is required?
- What should go into it?
- What magnification or reduction is required?

Box 28.1 Checklist for making a good diagram

1. **Decide exactly what you are going to draw and why.**

2. **Decide how large the diagram should be.**

3. **Decide where you are going to place the diagram on the page.**

4. **Start drawing:**
   - **(a)** Draw what you see, not what you expect to see.
   - **(b)** Use carefully measured construction lines to provide the correct proportions.
   - **(c)** Avoid shading.
   - **(d)** Avoid excess detail, especially in tissue diagrams.
   - **(e)** Use conventions where appropriate e.g. cut edges represented by a double line.

5. **Label the drawing/diagram carefully and comprehensively.**

6. **Give the diagram a title, scale and legend:** include organism, classification, part drawn, orientation, stain(s), magnification, etc.

Once these decisions are made, you can determine the position and size of your diagram. Your diagram should be as large as possible, but remember to leave space for legends and labels.

Materials

Most diagrams for practicals are drawn in pencil, to allow corrections to be made. Propelling pencils are valuable for ensuring constant line thickness but they do not allow you the flexibility to vary line thickness as you can by changing the angle of an ordinary pencil. If you prefer to use an ordinary pencil, sharpen it frequently. Invest in a good-quality eraser – those of poor quality tend to smudge badly – and frequently clean its working surface on a spare piece of paper. Always use plain paper for drawing, and if you are asked to supply your own, make sure it is of good quality. Use pen and ink to create line drawings for illustration purposes in posters, project reports, etc. Such diagrams should be in black and white only. Computer drawing programs can also provide good-quality output suitable for these tasks.

Constructing a diagram

1. Draw a faint rectangle in pencil to show the figure boundaries.
2. Draw very faint 'construction lines' using a 2H pencil with a sharp point to get the basic proportions and outlines correct before progressing. These should be erased once the basic drawing is complete. To lay in construction lines, use a ruler or pair of dividers to determine the actual proportions of the object to be drawn and then, using these dimensions, construct a scaled frame to allow further important reference points to be located (Fig. 28.7).
3. Draw the main outlines faintly with your 2H pencil. When satisfied, go over the lines with a sharp HB or 2B pencil. Draw firm, continuous lines, not hesitant, scratchy ones and make sure that junctions between lines are properly drawn. If you need to distinguish between different regions within your drawing, use hatching or stippling (but not shading), and avoid drawing unnecessary detail such as large numbers of cells (see Fig. 28.8). Always draw what you see, rather than what you think you should see, and seek advice at an early stage if you can't see a particular structure, or if you think your specimen is atypical. When drawing a specimen that is symmetrical or that contains repeated forms, it will save time if you draw its outline, and only provide detail in one of the replicated elements.
4. Complete your drawing by adding labels. This requires you to interpret your observations and helps you to remember what the structures look like. Careful and accurate labelling is as important as the drawing itself. It should be done clearly and neatly using either radiating or horizontal lines ending in arrowheads or large dots to indicate exact label references. The lines should not cross. Labels should be written clearly in one orientation, so that they can be read without needing to turn the paper. Annotations (short explanatory notes in brackets below the labels) are strongly recommended – regard them as notes to yourself about what you have seen, or what has been pointed out by tutors. For practical work, use a pencil for labelling in case your demonstrator or tutor corrects your work.

Producing labels for diagrams in formal reports – *create the text using a word processor and high-quality printer. This can then be stuck on your diagram using e.g. Pritt Stick, taking care to keep all the text parallel. If the diagram is now photocopied or photographed with high contrast, the use of different pieces of paper cannot be detected. Alternatively, scan the diagram and add labels using a suitable program such as PowerPoint or Word.*

Using construction lines – *these are vital for producing well-proportioned drawings.*

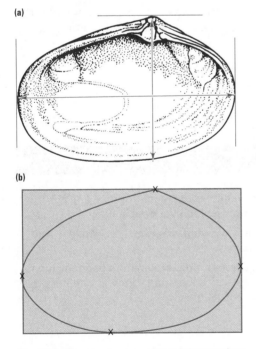

Fig. 28.7 How to draw an object in proportion: (a) determine linear dimensions; (b) construct a frame and outline, using reference points (x) determined from scaled measurements of the original specimen.

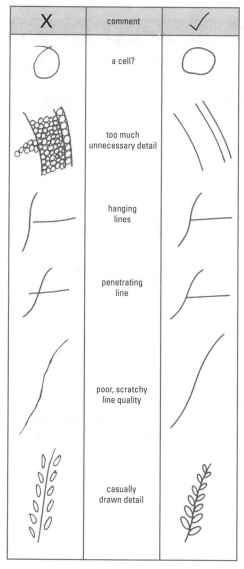

X	comment	✓
	a cell?	
	too much unnecessary detail	
	hanging lines	
	penetrating line	
	poor, scratchy line quality	
	casually drawn detail	

Fig. 28.8 Examples of common errors in biological drawings. Most of these mistakes are due to lack of care or attention to detail – easily solved!

5. Add a title, a scale or magnification factor (see Chapter 44), and a legend. The legend should provide all relevant information, including:

   (a) the binomial Latin name for the organism and a taxonomic classification, if appropriate;

   (b) details of the preparation of the subject, e.g. TS, whole mount, ventral dissection, etc., and any stains used.

Drawing from the microscope

Begin by positioning the paper beside your drawing hand and use the 'opposite' eye for examining the specimen; thus for a right-handed person, the paper is placed on the right of the microscope and you use your left eye. With a binocular microscope, use only one of the eyepieces; if you keep both eyes open, it is possible with practice to learn to draw and see the page with one eye while observing the specimen with the other. For specimens that need to be drawn very accurately in project work, projection devices such as the *camera lucida* may be required.

Avoiding mistakes

There are four main categories of error in student diagrams:

- Incorrect positioning and proportions – solve these problems by following steps 1–3 above.
- Forgetting to add a title, scale or a full set of labels – use a checklist like that provided in Box 28.1 to ensure your diagram is complete.
- Untidiness in presentation – avoid this type of error by using the correct materials as discussed above and by taking care – untidiness is frequently due to lack of attention to detail, as illustrated in Fig. 28.8.
- Biological inaccuracies – this kind of mistake is the most important, and will lose you most marks. Avoiding these errors requires preparation before the practical, so that you know more about what you are drawing and, for instance, have a good idea which parts are which *before* you start. Try to focus clearly on the objectives of the practical, and listen carefully to any tips given by your tutors, which may relate to the particular specimen(s) available on that day rather than to those in your notes or texts.

Finally, it is important to realise that you cannot expect biological drawing skills to develop overnight. This skill, like any other, requires much practice. Try to learn from any feedback your tutor may provide, and if your marks are consistently low without explanation, seek advice.

Sources for further study

Jepson, M. (1942) *Biological Drawings with Notes, Parts 1 and 2*, 5th edn. John Murray, London.
[Although this book is old, you may find it in your library. Provides examples of labelled biological drawings in an exemplary style.]

Leslie, C.W. (1984) *The Art of Field Sketching: A Naturalist's Sketchbook*. Prentice-Hall, Englewood Cliffs, NJ.

Sodt, J. *Botanical Illustration: A Selected Bibliography*. Available: http://www.library.wwu.edu/ref/subjguides/botill.htm

Last accessed 09/04/07.
[A wide-ranging source of information about drawing plant life. Although the focus of this site is on botanical illustration, it contains many useful references.]

Zweifel, F.W. (1988) *A Handbook of Biological Illustration*, 2nd edn. University of Chicago Press, Chicago.

Drawing and diagrams

28.1 Practise making well-proportioned drawings. Choose a picture of an animal or plant from a biology textbook or a printout from the Web. Using the techniques described on pp. 172–3 and Fig 28.7, make your own drawing based on this image.

28.2 Devise a flowchart. Present the instructions given in one of the Boxes in this text (for example, Boxes 39.1, 46.1 and 52.1) in the form of a flowchart.

28.3 Compare and contrast the main types of biological diagrams. Complete the following table.

Comparison of the main types of biological diagrams.

| Feature | Type of diagram | | |
	Cell diagram	Tissue diagram	Morphological diagram
Should the diagram carry a scale?			
What might be an appropriate scale in m?			
Approximate number of cells visible			
Are organelles visible?			
Should shading be used?			
Should labelling be used?			

Fieldwork is often one of the most enjoyable and rewarding aspects of undergraduate courses in biology, although it can also be arduous, especially in bad weather. The aims of fieldwork include:

- introducing a broad range of living organisms, with consideration of their roles in natural communities;
- developing skills in field techniques, with particular reference to sampling strategies and methods, and to the measurement of environmental variables;
- providing experience in the use of identification keys and fostering diagnostic skills;
- developing skills in data handling, analysis, interpretation and presentation, together with the writing of scientific reports;
- experiencing teamwork over an extended period;
- broadening environmental awareness.

To make the most of your time and effort while carrying out field investigations, you will need clear objectives and adequate preparation. You can minimise problems by carefully thinking your activities through beforehand.

 KEY POINT *Field studies are often time-restricted, and access to equipment and apparatus may be limited, so anticipation is the key to successful and safe fieldwork.*

Field excursions

You are likely to make short visits to field sites during the early part of your course. These excursions provide an opportunity to gain an overview of a particular environment in a single visit, e.g. an afternoon excursion to a marine rocky shore. Note the following points:

- Carry out the recommended background reading – you will get much more out of the visit if you read any specific handouts or notes in advance. If you plan to take these into the field, you can put them in separate polythene pockets – if necessary, you can make these fully waterproof by sealing the top opening with adhesive tape. It might also be useful to carry out some general background reading on the area to be visited. You should pay particular attention to any risk assessment information and safety guidance notes.
- Make sure you are well-prepared for the weather and the terrain – listen to the weather forecast and take appropriate clothing to keep you warm and dry. Just because your visit is short, you should not dismiss this aspect of the trip. It is better to take extra clothing that you do not use than to leave it at home. Match your footwear with the environment, e.g. wear waterproof boots, not trainers, on a visit to a rocky shore. Brightly coloured clothing is a good idea, since this will be easier to spot if you become lost, or separated from the main group. Take account of the exposure of the site – in bright sunshine,

---

Typical coursework assessments for field excursions – these include:

- *For specimens you have seen, recalling the names, diagnostic features, habitat, growth form, behaviour and physiology, etc.*
- *Making a diagrammatic representation of the site visited, e.g. a labelled map or transect.*
- *Interpreting key features of a photograph of the area visited.*
- *Interpreting data representing aspects of the location, based on the knowledge you have gained about the organisms and their environment.*

Preventing rain damage to printed information in the field – *use laminated sheets for keys, maps and any other items you wish to protect.*

you may need a hat and sun protection cream, but in chilly and windy conditions a hat, gloves and a warm coat would be required. A large pair of domestic rubber gloves can be worn over woollen gloves for work in cold and wet conditions, e.g. sampling a lake in winter. Students without appropriate clothing may be unable to participate in all or some of the field activities.

● During the excursion, keep up with the field leader – you will need to be close enough to hear any commentary. Take notes as you go, to get maximum value from the visit: your course assessment may test recall and understanding and this may be your only opportunity to visit this location with a knowledgeable 'guide'. Remember to bring a suitable notebook and pencil for making notes and recording data – for further details, see Chapter 32. Rewrite your field notes soon after returning from the excursion, adding additional detail, as required.

● Act responsibly at all times – consider your own safety and that of others in the group, and minimise the impact of your visit on the environment, e.g. keep to paths, where appropriate. Follow at all times any specific safety instructions you are given during the visit.

 SAFETY NOTE *Anticipate possible hazards – for example, there may be a significant risk of slipping on seaweed-covered rocks, on wet grass or on loose scree slopes. Also consider the risks involved in handling plants and animals (poisons, allergies, bites, stings, etc.) – if in doubt, ask your instructor before picking up any unknown specimen.*

Planning and preparation for field-based project work

In an individual or group project involving a fieldwork component, you are likely to carry out a more involved and detailed study, in comparison to a field excursion: Box 29.1 gives details of the key questions to consider.

 KEY POINT *Always construct a checklist (e.g. Table 29.1) of the required fieldwork equipment, paying attention to the smallest detail.*

Table 29.1 A checklist of generic equipment for fieldwork

● Notebook and pencil
● Datasheets
● Indelible marker pens
● Camera
● Food and drink (plus emergency rations)
● First aid equipment
● Safety (protective) gear
● Whistle
● Two-way radio or mobile phone
● Watch or clock
● Torch
● Multi-purpose knife (e.g. Swiss Army knife)
● Maps and compass or GPS (position locating device)
● Hand lens
● Specialised measuring equipment
● Sampling equipment (including spares)
● Specimen storage materials and labels
● Rucksack or other carrying devices
● Laminated guides and plans
● Specific items such as field guides, biological keys, etc.
● Survival bag (e.g. for longer trips, or in remote locations)
● Buoyancy aid and line (e.g. for trips near deep water)

Making a preliminary inspection of a field site

It is unlikely that you will be able to answer the questions in Box 29.1 if you or your instructor have not visited the field area in advance. The location, for example, may be remote, rugged or dangerous; it may be intertidal or subject to problems of access; the area may present social hazards (e.g. in some urban areas) or be culturally sensitive (e.g. nature reserves or archaeologically important sites). All fieldwork programmes, therefore, should be preceded by a 'scoping' or orientation study involving a preliminary inspection of the area. Make sure that the organisms that you wish to study are present and accessible. Any problems relating to access problems should be resolved at this stage and permissions obtained, where appropriate. Always make sure that landowners know who you are, where you are from, what you intend to do, and when you propose to be on their land. Be open and honest with them and involve them in your project.

 KEY POINT *Never enter private land without permission: this is potentially trespassing. After your fieldwork, always thank the landowner for their help and support.*

Box 29.1 Questions you should consider before carrying out a field project

- **What are the aims and objectives of the study?** Your fieldwork must be realistic in its scope and purpose; most people overestimate what can be done in a given time period, especially if the weather is poor. Recognise your own limitations and those of others who might be involved and do not overstretch yourself.

- **How long and how many periods of fieldwork are required?** If your work requires more than one visit, what time interval will be needed between visits? This may affect the logistics and cost of the work.

- ⚠ **What are the safety implications of the work?** It is always essential to take full account of safety issues, and you should read and take note of the basic rules in Chapter 20. Make sure that a risk assessment has been carried out before any field-based project work is carried out: the major hazards should be considered, and any steps required to minimise risk must be identified in advance (see p. 119).

- **Am I likely to encounter any difficult conditions or environments?** Plans should always be based upon a preliminary site inspection and you should always discuss your intentions with more experienced fieldworkers.

- **What samples are needed and how will they be collected?** The number, frequency, nature and spatial or temporal distribution of samples must be consistent with the purpose of the fieldwork and may require considerable planning in advance. Too many samples can be as much of a problem as too few, so determine the minimum required sample size with statistical evaluation in mind. Chapter 30 considers aspects of sampling strategy. The sampling or environmental measurement and recording protocol that you choose will influence your equipment requirements.

- **How should samples be stored for transport to the laboratory?** If the work involves collection of samples or specimens, make sure you include appropriate storage vessels and a means of labelling them (e.g. a spirit-based marker). You may also need to identify temporary storage locations in the field – note that samples may be ruined by inappropriate field storage (e.g. inactivation of bacteria by brief exposure to bright sunlight) and this damage may not be obvious on return to the laboratory. Water samples may require refrigeration to minimise changes in biological composition. Remember too that large numbers of water, sediment or soil samples can be very heavy; make sure that you have considered the best way to transport them. Chapters 34, 35 and 36 give further advice on collecting and preserving biological specimens.

- **What equipment will be required?** You should compile a detailed list well in advance, and then check that everything is available and in working order.

- **What transport will be needed for personnel, equipment and samples?** You may need to arrange for a vehicle or a boat of appropriate size. On land, choose a vehicle appropriate for the terrain. In shallow water, use only an inflatable dinghy or shallow-draught boat. In deep water, make sure that the boat can withstand any possible sea conditions. Take account of the capacity of your means of transport – to overload a vehicle or boat is a risk to safety.

No sampling or survey work is carried out at the inspection stage. However, the field area should be walked over and examined. Appropriate notes should be taken, and, where applicable, the site should be photographed, so that the strategy for subsequent sampling can be determined.

Working safely and responsibly

Always wear appropriate clothing and safety equipment and let someone else know where you will be working and when you expect to return. Do not then alter this plan without informing that person. Remember, too,

SAFETY NOTE Remember your responsibilities during fieldwork – although your activities will normally be organised in consultation with a member of staff, this does not free you from a personal responsibility for your own and the group's safety.

SAFETY NOTE *Safe and responsible fieldwork also means:*
- *Do not drink alcohol or take drugs (other than prescribed medication) in the field.*
- *Do not annoy or antagonise the local population.*
- *Leave gates as you find them.*
- *Do not drop litter.*
- *Avoid fire risks, especially when working in wooded areas, sand dunes or heathland.*
- *Do not frighten or disturb livestock or domestic animals.*
- *Do not attempt to carry too much. Keep rucksack loads below 14 kg. Make several trips rather than risk exhaustion or injury by trying to carry too great a load.*
- *You should also observe the general safety points discussed in Chapter 20.*

that what may be a comfortable environment for one person may not be so for another, depending on their outdoor experience and interests. Do not attempt to work from a boat or climb a rock face without prior training. If your work requires specialist support you must seek assistance from an appropriately qualified individual (e.g. a qualified rock climber or a scuba diver).

Sources for further study

Anon. (1995) *Field Work Code of Practice.* CVCP, London.

Jones, A.M., Duck, R., Reed, R. and Weyers, J.D.B. (2000). *Practical Skills in Environmental Science.* Prentice Hall, Harlow.

Nichols, D. (1999) *Safety in Biological Fieldwork. Guidance Notes for Codes of Practice*, 4th edn. Institute of Biology, London.

Watts, S. and Halliwell, L. (eds) (1996) *Essential Environmental Science: Methods and Techniques.* Routledge, London.

Study exercises

29.1 Check your knowledge of generic fieldwork requirements. What items are missing from the following list of generic fieldwork items? Notebook and pencil; datasheets; camera; food and drink (plus emergency rations); safety (protective) gear; whistle, watch or clock; torch; multi-purpose knife; map(s) and compass/GPS device; hand lens; specialised measuring equipment; specimen storage materials and labels; rucksack or other carrying devices; laminated guides and plans; field guides.

29.2 List specific items that would be needed to carry out an ecological investigation. This could be, for example, of the fauna and flora of a marine rocky shore or of plants in a meadow. Make a note of why each item is required and any problems that might be associated with its use.

29.3 Consider the hazards of a particular field study. List the potential hazards of carrying out a survey of woodlice in an urban estate environment.

Definitions

Biological population – all those individuals within a specified time or space about which inferences are to be made, specified according to some biological definition (perhaps related to life history, growth stage, or sex), and normally investigated at a particular location and time.

Parameter – a numerical constant or mathematical function used to describe a particular population (e.g. the mean height of 18-year-old females).

Statistic – an estimate of a parameter obtained from a sample (e.g. the height of 18-year-old females based on those in your class).

Choosing relevant population factors to specify – these should include:

- *geographical location;*
- *type of habitat;*
- *date and time of sampling;*
- *age, sex, physiological condition and health of sampled organisms;*
- *other details relevant to your work, e.g. an index of pollution.*

This information might apply to all members of the population or from a matrix of data associated with each replicate.

Selecting specimens – do not include data for which the appropriate population specification is unavailable.

When carrying out research, it is unlikely that you can observe or measure every individual in the population of organisms in which you are interested. In practice, statistics obtained from a subset (or sample) are used to estimate relevant parameters for the biological population. Samples consist of data values for a particular variable (e.g. length), each recorded from an individual sampling unit (e.g. a limpet) in a sample of n units (e.g. $n = 50$ limpets) taken from the population under investigation (e.g. those limpets on a particular rocky shore). The term 'replicate' can be applied either to the measurement (i.e. repeated measurement or observation taken at the same time or location) or the actual sampling unit.

Symbols are used to represent each type of sample statistic: these are given Roman-character symbols, e.g. $\overline{Y}$ for the sample mean, while the equivalent population parameter is given a Greek symbol, e.g. μ for the population mean. When estimating population parameters from sample statistics, the sample size is important, larger sample sizes allowing greater statistical confidence. However, the optimum sample size is normally a balance between statistical and practical considerations.

This chapter deals mainly with sampling in fieldwork where natural populations are to be observed under undisturbed conditions; however, the same principles may also apply in a laboratory context (see Chapter 31).

Specifying the population being sampled

At the outset, it is important to provide a complete description of the biological population being sampled. Failure to do this will make your results difficult to interpret or to compare with other observations, including your own. When populations are to be compared, ideally only the variable under consideration should differ if maximum significance is to be placed upon the analysis.

Deciding on a sampling strategy

Selecting a sample involves the formulation of rules and methods (the sampling protocol) by which some members of the population are included in the sample. The chosen sample is then measured using defined procedures to obtain relevant data. Finally, the information so obtained is processed to calculate appropriate statistics. Good sampling design takes into account all of these procedures, and it should relate to the objectives of your investigation by providing valid estimates of features of the population(s) of interest.

Truly representative samples should be:

- taken at random, or in a manner that ensures that every member of the population has an equal chance of being selected;
- large enough to provide sufficient precision in estimation of population characteristics;
- unbiased by the sampling procedure or equipment.

Samples and sampling

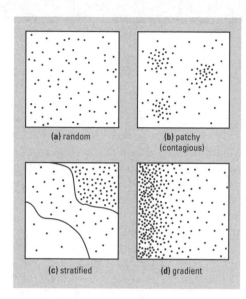

Fig. 30.1 Types of distributions

Definitions

Homogeneous – evenly distributed.

Patchy – showing clustered (contagious) distribution (e.g. numbers of parasites within hosts).

Gradient – a distribution that varies smoothly over the sampling area.

Stratified – showing a distribution with discrete levels or strata (e.g. algae on a rocky shore)

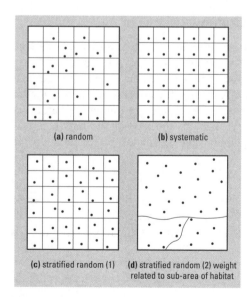

Fig. 30.2 Basic methods of sampling

These requirements may well conflict and there is rarely any unique best answer to a sampling problem. You should, however, take great care to minimise bias, or population parameters inferred from your samples will be unrealistic and this may invalidate your work and its conclusions.

 KEY POINT *Sampling involves choices on the part of the investigator. A 'sampling strategy' should allow you to obtain reliable and useful information about your particular population(s), while using your resources efficiently.*

The sampling protocol

You should decide on a sampling protocol before any investigation proceeds. The main aspects to be determined are: the position of samples; the size and shape of the sampling area; and the number of sampling units in each sample. Before this can be done, however, information is required about the likely distribution of organisms. This can be even (homogeneous), patchy (contagious), stratified (homogeneous within sub-areas) or present as a gradient (Fig. 30.1). You might decide which type applies from a pilot study, published research or by analogy with other systems.

Locating your samples

The positions where sampling takes place can be determined either randomly, systematically or in some stratified manner (Fig. 30.2).

In simple random sampling (Fig. 30.2(a)), a 2-D coordinate grid is superimposed on the area to be investigated. The required number of grid reference data pairs is then obtained using random numbers (p. 188) and samples taken at these points. Every organism in the population thus has an equal chance of selection, but the area may not be covered evenly. This method is best if the distribution of organisms is homogeneous.

Systematic sampling (Fig. 30.2(b)) involves selecting the location of the first sampling position at random and then taking samples at fixed distances from this. This method has the advantage of simplicity and it is often used where the intention is to map data. The disadvantages are firstly, that the results can be biased if the interval between sampling positions coincides with some periodic distribution of the population, and secondly, that there is no reliable method of estimating the standard error of the sample mean.

Stratified random sampling may be preferable if you wish to avoid these disadvantages yet still ensure that each part of the area is represented (e.g. where you suspect there to be stratified micro-habitats, as in a tidal seashore). The area is divided into sub-areas within which random sampling is carried out. These can either be constant in size (Fig. 30.2(c)) or related to known features in the sampling area (Fig. 30.2(d)). If the latter, strata are normally sampled in proportion to their area. 'Weighting' is the general term applied to sampling procedures that allow the calculated statistics to represent the population better by accounting for differences in the distribution of the chosen character. You can analyse data from different strata by a one-way analysis of variance (see Chapter 66).

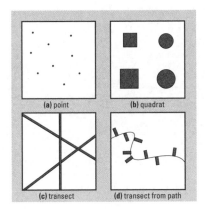

Fig. 30.3 Methods of positioning samples

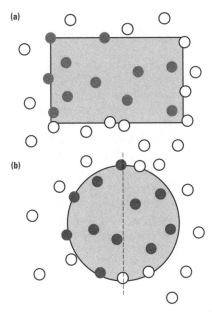

Fig. 30.4 Examples of sampling protocols for reducing edge effects. Filled circles represent objects to be counted or measured, open circles those to be ignored. (a) Rectangular area. All objects touching the top and left-hand sides including the top left-hand corner are included, as well as those clearly within the perimeter. (b) Circular area. All objects clearly within the perimeter are included as well as those touching the perimeter on the left-hand side. In the bottom half, those that would touch both the perimeter and the imaginary plane of symmetry are ignored.

The dimensions of the sampling area

The chief options are:

- point sampling (Fig. 30.3(a));
- quadrat sampling (Fig. 30.3(b));
- transect (traverse) sampling (Fig. 30.3(c)).

Quadrats are usually either circular or square. A circular quadrat has the advantage that its position can be marked as a single (central) point and the area defined by use of a tape measure, whereas a square quadrat may require marking at each corner. Transects are generally used when it is difficult to move through the site to position quadrats. If a defined path is present, transects taken at right angles to the path (Fig. 30.3(d)) will save time in reaching the sites.

The problems involved are best illustrated by reference to the selection of a quadrat for fieldwork sampling. Note first that the maximum number of independent sampling units (quadrats) is equal to the area occupied by the population under study (total theoretical sampling area) divided by the area of the sampling units (quadrat area).

The distribution and size of the organisms must be considered: it is obvious that you would require different-sized quadrats for trees in a forest than for daisies on a lawn. When the distribution is truly random, then all quadrat sizes are equally effective for estimating population parameters (assuming the total number of individuals sampled is equal). If the distribution is patchy, a smaller quadrat size may be more effective than a larger one: too large an area might obscure the true nature of the clumped distribution. Alternatively, you may wish to exclude a patchy distribution from your investigation and should thus choose a relatively large sampling area. If the distribution is stratified or graded, then sampling area is generally less important than sampling position.

Small sample areas have the advantage that more small samples can usually be taken for the same amount of labour. This may result in increased precision and many small areas will cover a wider range of the habitat than few large ones, so the catch can be more representative. However, sampling error at the edge of quadrats is proportionately greater as sample area diminishes, increasing as the scale of the quadrat and the sampled item become closer. To avoid such effects, you need to establish a protocol for dealing with items that overlap the edge of the quadrats (Fig. 30.4). These protocols are also valid when sampling objects in e.g. microscope fields.

Number of sampling units per sample

When small numbers of sampling units are used, this can lead to imprecise estimates of population parameters because the values of the sample statistics will be susceptible to the effects of random variation – this is especially true if the underlying spatial distribution is patchy, as is often the case. You may then be unable to demonstrate statistically that there are differences between the populations. On the other hand, measuring very large numbers of replicates may represent an impractical workload.

To estimate appropriate numbers of sampling units, you can use data from a pilot study to work out the probability of detecting a specified difference in the measured variable at a specified confidence level (see Sokal

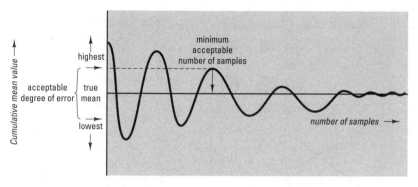

Fig. 30.5 Determination of the number of sampling units required

and Rohlf, 1994). However, the formulae involved are complex. A simpler method, usually performed as a pilot study, is as follows:

1. Take five sampling units at random and calculate the arithmetic mean of the measured variable for this sample.
2. Take five more units and calculate the mean for the ten units you have now collected.
3. Continue sampling in five-unit steps and plot the cumulative mean value against the number of samples. When the mean fluctuates within acceptable limits, say ±5 per cent, a suitable number of sampling units has been reached (Fig. 30.5).

The sequence of events in creating a sampling strategy is shown in Fig. 30.6.

Sampling in time

Sampling in time presents a different set of problems. If examining a phenomenon that fluctuates regularly (e.g. with a period governed by day and night, or high and low tide), then the frequency of sampling has to be determined with that periodicity in mind. In fact, you should consider the possible existence of periodic phenomena, even if they are not immediately obvious. If the phenomenon you are investigating is likely to change through time by some complex function such as the logarithmic decay in pollutant concentration, then your sampling intervals should be spaced according to an appropriate logarithmic series.

Subsampling

If you do not wish to sample the whole of a quadrat, perhaps because the density of sampling units is too high and it would take too long, then you can employ subsampling by studying a defined part of the quadrat – also known as two-stage sampling. If studying the density of organisms, this can simplify counting: assuming the organisms are randomly distributed, and only a small proportion of the population is sampled in each subunit, then the counts should follow a Poisson distribution (see Chapter 66). If this distribution is confirmed, only single subsamples need to be counted to estimate the total numbers in each quadrat. The precision of the estimate then depends on the size of the count.

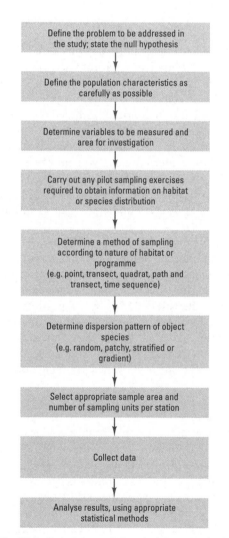

Fig. 30.6 Flowchart outlining decisions required for field sampling studies

Text reference

Sokal, R.R. and Rohlf, F.J. (1994) *Biometry*, 3rd edn. W.H. Freeman, San Francisco.

Sources for further study

Baker, J.M. and Wolff, W.J. (1987) *Biological Surveys of Estuaries and Coasts*. Cambridge University Press, Cambridge.
[Includes chapters on planning biological surveys and safety as well as numerous sections on sampling specific habitats.]

Keith, L.H. (1991) *Environmental Sampling and Analysis: a Practical Guide*. Lewis Publishers Inc., Michigan.
[Advanced text that is particularly good on how to devise protocols.]

New, T.R. (1998) *Invertebrate Surveys for Conservation*. Oxford University Press, Oxford.
[Excellent source for most aspects of sampling.]

Slingsby, D. and Cook, C. (1986) *Practical Ecology*. Macmillan Education, London.
[Contains a general introduction to sampling and other aspects of fieldwork.]

Study exercises

30.1 Calculate the density of subsampled copepods. A sample of plankton was collected using a 5 litre water bottle sampler. It was filtered upon collection through a 200 μm mesh net and the contents retained and preserved. Because the resulting volume of material was too large to count all the individuals present, the material was subsampled by taking ten replicate 1 per cent aliquots. The numbers of copepods collected for each subsample determined using a counting chamber are given in the table below. From these data, calculate the mean density of copepods as (a) density per litre and (b) density per cubic metre of water.

Sub-samples of copepods

1	2	3	4	5	6	7	8	9	10
234	255	342	276	288	295	324	301	299	273

30.2 Devise a protocol for counting limpets on a rocky platform using some form of quadrat. Take into account the size of the organism (up to 50 mm long), their normal density (up to 50 per square metre) and the significance of edge effects at the periphery of your sampling device. Assume that the limpets are randomly distributed.

30.3 Calculate the number of samples required for a given level of accuracy. The table below presents data on the effect of number of samples on the estimate of the mean for a population of limpets. Calculate the minimum number of samples that would be necessary to provide a measure of the mean (sample mean) that is within 10 per cent of the true (population) mean (which is 50).

Estimated mean after taking different numbers of samples

Number of samples	1	3	5	10	15	20	25	30	35	40
Estimated mean (limpets m^{-2})	90	62	42.8	56.1	43.3	54.8	52.7	48.9	50.6	50

Definitions

There are many interpretations of the following terms. For the purposes of this chapter, the following will be used.

Paradigm – theoretical framework so successful and well confirmed that most research is carried out within its context and doesn't challenge it – even significant difficulties can be 'shelved' in favour of its retention.

Theory – a collection of hypotheses that covers a range of natural phenomena – a 'larger-scale' idea than a hypothesis. Note that a theory may be 'hypothetical', in the sense that it is a tentative explanation.

Hypothesis – an explanation tested in a specific experiment or by a set of observations. Tends to involve a 'small-scale' idea.

(Scientific) Law – this concept can be summarised as an equation (law) that provides a succinct encapsulation of a system, often in the form of a mathematical relationship. The term is often used in the physical sciences (e.g. 'Beer's Law', p. 366).

Biology is a body of knowledge based on observation and experiment. Biologists attempt to explain life in terms of theories and hypotheses. They make predictions from these hypotheses and test them by experiment or further observations. The philosophy and sociology that underlie this process are complex topics (see e.g. Chalmers, 1999). Any brief description must involve simplifications.

Figure 31.1 models the scientific process you are most likely to be involved in – testing 'small-scale' hypotheses. These represent the sorts of explanations that can give rise to predictions that can be tested by an experiment or a series of observations. For example, you might put forward the hypothesis that the rate of K^+ efflux from a particular cell type is dependent on the intracellular concentration of calcium ions. This might then lead to a prediction that the application of a substance known to decrease the intracellular concentration of calcium ions would reduce K^+ efflux from the cells. An experiment could be set up to test this hypothesis and the results would either confirm or falsify the hypothesis.

If confirmed, a hypothesis is retained with greater confidence. If falsified, it is either rejected outright as false, or modified and retested. Alternatively, it might be decided that the experiment was not a valid test of the hypothesis (perhaps because it was later found that the applied substance could not penetrate the cell membrane to a presumed site of action).

Nearly all scientific research deals with the testing of small-scale hypotheses. These hypotheses operate within a theoretical framework that has proven to be successful (i.e. is confirmed by many experiments and is consistently predictive). This operating model or 'paradigm' is not changed readily, and even if a result appears that seems to challenge the conventional view, would not be overturned immediately. The conflicting result would be 'shelved' until an explanation was found after further investigation. In the example used above, a relevant paradigm could be the notion that life processes are ultimately chemical in nature.

Although changes in paradigms are rare, they are important, and the scientists who recognise them become famous. For example, a 'paradigm shift' can be said to have occurred when Darwin's ideas about Natural Selection replaced Special Creation as an explanation for the origin of species. Generally, however, results from hypothesis-testing tend to support and develop ('articulate') the paradigm, enhancing its relevance and strengthening its status. Thus, research in the area of population genetics has developed and refined Darwinism.

Where do ideas for small-scale hypotheses come from? They arise from one or more thought processes on the part of a scientist:

- analogy with other systems;
- recognition of a pattern;
- recognition of departure from a pattern;
- invention of new analytical methods;
- development of a mathematical model;
- intuition;
- imagination.

Recently, it has been recognised that the process of science is not an entirely objective one. For instance, the choice of analogy that led to a

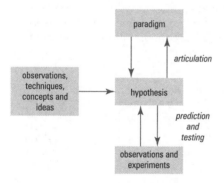

Fig. 31.1 A model of scientific method as used when testing hypotheses on a small scale. Hypotheses can arise as a result of various thought processes on the part of the scientist, and are consistent with the overlying paradigm. Each hypothesis is testable by experiment or observation, leading to its confirmation or rejection. Confirmed hypotheses act to strengthen the status of the paradigm, but rejected ones do not immediately result in the paradigm's replacement.

new hypothesis might well be subjective, depending on past knowledge or understanding. Also, science is a social activity, where researchers put forward and defend viewpoints against those who hold an opposing view; where groups may work together towards a common goal; and where effort may depend on externally dictated financial opportunities and constraints. As with any other human activity, science is bound to involve an element of subjectivity.

No hypothesis can ever be rejected with certainty. Statistics allow us to quantify as vanishingly small the probability of an erroneous conclusion, but we are nevertheless left in the position of never being 100 per cent certain that we have rejected all relevant alternative hypotheses, nor 100 per cent certain that our decision to reject some alternative hypotheses was correct. However, despite these problems, experimental science has yielded and continues to yield many important findings.

 KEY POINT *The fallibility of scientific 'facts' is essential to grasp. No explanation can ever be 100 per cent certain as it is always possible for a new alternative hypothesis to be generated. Our understanding of biology changes all the time as new observations and methods force old hypotheses to be retested.*

Quantitative hypotheses, those involving a mathematical description of the system, have become very important in biology. They can be formulated concisely by mathematical models. Formulating models is often useful because it forces deeper thought about mechanisms and encourages simplification of the system. A mathematical model:

- is inherently testable through experiment;
- identifies areas where information is lacking or uncertain;
- encapsulates many observations;
- allows you to predict the behaviour of the system.

Remember, however, that assumptions and simplifications required to create a model may result in it being unrealistic. Further, the results obtained from any model are only as good as the information put into it.

The terminology of experimentation

In many experiments, the aim is to provide evidence for causality. If *x* causes *y*, we expect, repeatedly, to find that a change in *x* results in a change in *y*. Hence, the ideal experiment of this kind involves measurement of *y*, the dependent (measured) variable, at one or more values of *x*, the independent variable, and subsequent demonstration of some relationship between them. Experiments therefore involve comparisons of the results of treatments – changes in the independent variable as applied to an experimental subject. The change is engineered by the experimenter under controlled conditions.

Subjects given the same treatment are known as replicates (they may be called plots). A block is a grouping of replicates or plots. The blocks are contained in a field, i.e. the whole area (or time) available for the experiment (Fig. 31.2). These terms originated from the statistical analysis of agricultural experiments, but they are now used for all areas of biology.

Deciding whether to accept or reject a hypothesis – this is sometimes clear-cut, as in some areas of genetics, where experiments can be set up to result in a binary outcome (Chapters 53 and 54). In many other cases, the existence of 'biological variation' means that statistical techniques need to be employed (Chapters 65 and 66; Box 53.2).

Definition

Mathematical model – an algebraic summary of the relationship between the variables in a system.

Fig. 31.2 Terminology and physical arrangement of elements in an experiment. Each block should contain the complete range of treatments (treatments may be replicated more than once in each block).

Why you need to control variables in experiments

Interpretation of experiments is seldom clear-cut because uncontrolled variables always change when treatments are given.

Confounding variables

These increase or decrease systematically as the independent variable increases or decreases. Their effects are known as systematic variation. This form of variation can be disentangled from that caused directly by treatments by incorporating appropriate controls in the experiment. A control is really just another treatment where a potentially confounding variable is adjusted so that its effects, if any, can be taken into account. The results from a control may therefore allow an alternative hypothesis to be rejected. There are often many potential controls for any experiment.

The consequence of systematic variation is that you can never be certain that the treatment, and the treatment alone, has caused an observed result. By careful design, you can, however, 'minimise the uncertainty' involved in your conclusion. Methods available include:

- Ensuring, through experimental design, that the independent variable is the only major factor that changes in any treatment.
- Incorporating appropriate controls to show that potential confounding variables have little or no effect.
- Selecting experimental subjects randomly to cancel out systematic variation arising from biased selection.
- Matching or pairing individuals among treatments so that differences in response due to their initial status are eliminated.
- Arranging subjects and treatments randomly so that responses to systematic differences in conditions do not influence the results.
- Ensuring that experimental conditions are uniform so that responses to systematic differences in conditions are minimised. When attempting this, beware 'edge effects' where subjects on the periphery of the layout receive substantially different conditions from those in the centre.

Nuisance variables

These are uncontrolled variables that cause differences in the value of y independently of the value of x, resulting in random variation. Experimental biology is characterised by the high number of nuisance variables that are found and their relatively great influence on results: biological data tend to have large errors. To reduce and assess the consequences of nuisance variables:

- incorporate replicates to allow random variation to be quantified;
- choose subjects that are as similar as possible;
- control random fluctuations in environmental conditions.

Constraints on experimental design

Box 31.1 outlines the important stages in designing an experiment. At an early stage, you should find out how resources may constrain the design. For example, limits may be set by availability of subjects, cost of treatment, availability of a chemical or bench space. Logistics may be a factor (e.g. time taken to record or analyse data).

Example Suppose you wish to investigate the effect of a metal ion on the growth of a culture. If you add the metal as a salt to the culture and then measure the growth, you will immediately introduce at least two confounding variables, compared with a control that has no salt added. Firstly, you will introduce an anion that may also affect growth in its own right, or in combination with the metal ion; secondly, you will alter the osmotic potential of the medium (see p. 144). Both of these effects could be tested using appropriate controls.

Reducing edge effects – one way to do this is to incorporate a 'buffer zone' of untreated subjects around the experiment proper.

Evaluating design constraints – a good way to do this is by processing an individual subject through the experimental procedures – a 'preliminary run' can help to identify potential difficulties.

Box 31.1 Checklist for designing and performing an experiment

1. Preliminaries

(a) **Read background material** and decide on a subject area to investigate.

(b) **Formulate a simple hypothesis to test.** It is preferable to have a clear answer to one question than to be uncertain about several questions.

(c) **Decide which dependent variable you are going to measure and how:** is it relevant to the problem? Can you measure it accurately, precisely and without bias?

(d) **Think about and plan the statistical analysis of your results.** Will this affect your design?

2. Designing

(a) **Find out the limitations on your resources.**

(b) **Choose treatments that alter the minimum of confounding variables.**

(c) **Incorporate as many effective controls as possible.**

(d) **Keep the number of replicates as high as is feasible.**

(e) **Ensure that the same number of replicates is present in each treatment.**

(f) **Use effective randomisation and blocking arrangements.**

3. Planning

(a) **List all the materials you will need.** Order any chemicals and make up solutions; grow, collect or breed the experimental subjects you require; check equipment is available.

(b) **Organise space and/or time** in which to do the experiment.

(c) **Account for the time taken to apply treatments and record results.** Make out a timesheet if things will be hectic.

4. Carrying out the experiment

(a) **Record the results and make careful notes of everything you do** (see Chapter 32). Make additional observations to those planned if interesting things happen.

(b) **Repeat experiment** if time and resources allow.

5. Analysing

(a) **Graph data as soon as possible** (during the experiment if you can). This will allow you to visualise what has happened and make adjustments to the design (e.g. timing of measurements).

(b) **Carry out the planned statistical analysis.**

(c) **Jot down conclusions and new hypotheses** arising from the experiment.

Your equipment or facilities may affect design because you cannot regulate conditions as well as you might desire. For example, you may be unable to ensure that temperature and lighting are equal over an experiment laid out in a glasshouse or you may have to accept a great deal of initial variability if your subjects are collected from the wild. This problem is especially acute for experiments carried out in the field.

Use of replicates

Deciding the number of replicates in each treatment – try to:

- *maximise the number of replicates in each treatment;*
- *make the number of replicates even.*

Replicate results show how variable the response is within treatments. They allow you to compare the differences among treatments in the context of the variability within treatments – you can do this via statistical tests such as analysis of variance (Chapter 66). Larger sample sizes tend to increase the precision of estimates of parameters and increase the chances of showing a significant difference between treatments if one exists. For statistical reasons (weighting, ease of calculation, fitting data to certain tests), it is often best to keep the number of replicates even. Remember that the degree of independence of replicates is important: subsamples cannot act as replicate samples – they tell you about variability in the measurement method but not in the quantity being measured.

If the total number of replicates available for an experiment is limited by resources, you may need to compromise between the number of treatments and the number of replicates per treatment. Statistics can help here, for it is possible to work out the minimum number of replicates you would need to show a certain difference between pairs of means (say 10 per cent) at a specified level of significance (say $P = 0.05$). For this, you need to obtain a prior estimate of variability within treatments (see Sokal and Rohlf, 1994).

Randomisation of treatments

The two aspects of randomisation you must consider are:

- positioning of treatments within experimental blocks;
- allocation of treatments to the experimental subjects.

For relatively simple experiments, you can adopt a completely randomised design; here, the position and treatment assigned to any subject is defined randomly. You can draw lots, use a random number generator on a calculator, or use the random number tables that can be found in most books of statistical tables (see Box 31.2).

Box 31.2   How to use random number tables to assign subjects to positions and treatments

This is one method of many that could be used. It requires two sets of n random numbers – where n is the total number of subjects used.

1. **Number the subjects in any arbitrary order** but in such a way that you know which is which (i.e. mark or tag them).

2. **Decide how treatments will be assigned**, e.g. first five subjects selected, treatment A; second five – treatment B, etc.

3. **Use the first set of random numbers in the sequence obtained to identify subjects and allocate them to treatment groups** in order of selection as decided in (2).

4. **Map the positions for subjects in the block or field. Assign numbers to these positions using the second set of random numbers**, working through the positions in some arbitrary order, e.g. top left to bottom right.

5. **Match the original numbers given to subjects with the position numbers.**

To obtain a sequence of random numbers:

1. **Decide on the range of random numbers you need.**

2. **Decide how you wish to sample the random number tables** (e.g. row by row and top to bottom) and your starting point.

3. **Moving in the selected manner, read the sequence of numbers until you come to a group that fits your needs** (e.g. in the sequence 978186, 18 represents a number between 1 and 20). Write this down and continue sampling until you get a new number. If a number is repeated, ignore it. Small numbers need to have the appropriate number of zeros preceding (e.g. 5 = 05 for a range in the tens, 21 = 021 for a range in the hundreds).

4. **When you come to the last number required, you don't need to sample any more:** simply write it down.

Example: You find the following random number sequence in a table and wish to select numbers between 1 and 10 from it.

9059146823	4862925166	1063260345
1277423810	9948040676	6430247598
8357945137	2490145183	5946242208
6588812379	2325701558	3260726568

Working left to right and top to bottom, the order of numbers found is 5, 10, 3, 9, 4, 6, 2, 1, 8, 7 as indicated by green type. If the table is sampled by working row by row right to left from bottom to top, the order is 6, 10, 7, 2, 9, 3, 4, 8, 1, 5.

A completely randomised layout has the advantage of simplicity but cannot show how confounding variables alter in space or time. This information can be obtained if you use a blocked design in which the degree of randomisation is restricted. Here, the experimental space or time is divided into blocks, each of which accommodates the complete set of treatments (Fig. 31.2). When analysed appropriately, the results for the blocks can be compared to test for differences in the confounding variables and these effects can be separated out from the effects of the treatments. The size and shape (or timing) of the block you choose is important: besides being able to accommodate the number of replicates desired, the suspected confounding variable should be relatively uniform within the block.

A Latin square is a method of placing treatments so that they appear in a balanced fashion within a square block or field. Treatments appear once in each column and row (see Fig. 31.3), so the effects of confounding variables can be 'cancelled out' in two directions at right angles to each other. This is effective if there is a smooth gradient in some confounding variable over the field. It is less useful if the variable has a patchy distribution, where a randomised block design might be better.

Latin square designs are useful in serial experiments where different treatments are given to the same subjects in a sequence (e.g. Fig. 31.4). A disadvantage of Latin squares is the fact that the number of plots is equal to the number of replicates, so increases in the number of replicates can only be made by the use of further Latin squares.

Pairing and matching subjects

The paired comparison is a special case of blocking used to reduce systematic variation when there are two treatments. Examples of its use are:

- 'Before and after' comparison. Here, the pairing removes variability arising from the initial state of the subjects, e.g. weight gain of mice on a diet, where the weight gain may depend on the initial weight.
- Application of a treatment and control to parts of the same subject or to closely related subjects. This allows comparison without complications arising from different origin of subjects, e.g. drug or placebo given to sibling rats, virus-containing or control solution swabbed on left or right halves of a leaf.
- Application of treatment and control under shared conditions. This allows comparison without complications arising from different environments of subjects, e.g. rats in a cage, plants in a pot.

Matched samples represent a restriction on randomisation where you make a balanced selection of subjects for treatments on the basis of some attribute or attributes that may influence results, e.g. age, sex, prior history. The effect of matching should be to 'cancel out' the unwanted source(s) of variation. Disadvantages include the subjective element in choice of character(s) to be balanced, inexact matching of quantitative characteristics, the time matching takes and possible wastage of unmatched subjects.

When analysed statistically, both paired comparisons and matched samples can show up differences between treatments that might otherwise be rejected on the basis of a fully randomised design, but note that the statistical analysis may be different.

Example If you knew that soil type varied in a graded fashion across a field, you might arrange blocks to be long thin rectangles at right angles to the gradient to ensure conditions within the block were as even as possible.

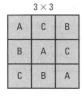

Fig. 31.3 Examples of Latin square arrangements for 3 and 4 treatments. Letters indicate treatments; the number of possible arrangements for each size of square increases greatly as the size increases.

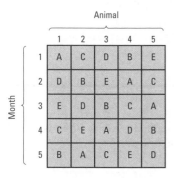

Fig. 31.4 Example of how to use a Latin square design to arrange sequential treatments. The experimenter wishes to test the effect of drugs A–E on weight gain, but only has five animals available. Each animal is fed on control diet for the first 3 weeks of each month, then on control diet plus drug for the last week. Weights are taken at start and finish of each treatment. Each animal receives all treatments.

Definition

Interaction – where the effect of treatments given together is greater or less than the sum of their individual effects.

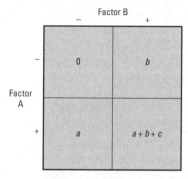

Fig. 31.5 Design of a simple multifactorial experiment. Factors A and B have effects *a* and *b* when applied alone. When both are applied together, the effect is denoted by $a + b + c$.

- If $c = 0$, there is no interaction (e.g. $2 + 2 + c = 4$).
- If c is positive, there is a positive interaction (synergism) between A and B (e.g. $2 + 2 + c = 5$).
- If c is negative, there is a negative interaction (antagonism) between A and B (e.g. $2 + 2 + c = 3$).

Reporting correctly – it is good practice to report how many times your experiments were repeated (in Materials and Methods); in the Results section, you should add a statement saying that the illustrated experiment is representative.

Multifactorial experiments

The simplest experiments are those in which one treatment (factor) is applied at a time to the subjects. This approach is likely to give clear-cut answers, but it could be criticised for lacking realism. In particular, it cannot take account of interactions among two or more conditions that are likely to occur in real life. A multifactorial experiment (Fig. 31.5) is an attempt to do this; the interactions among treatments can be analysed by specialised statistics.

Multifactorial experiments are economical on resources because of 'hidden replication'. This arises when two or more treatments are given to a subject because the result acts statistically as a replicate for each treatment. Choice of relevant treatments to combine is important in multifactorial experiments; for instance, an interaction may be present at certain concentrations of a chemical but not at others (perhaps because the response is saturated). It is also important that the measurement scale for the response is consistent, otherwise spurious interactions may occur. Beware when planning a multifactorial experiment that the numbers of replicates do not get out of hand: you may have to restrict the treatments to 'plus' or 'minus' the factor of interest (as in Fig. 31.5).

Repetition of experiments

Even if your experiment is well designed and analysed, only limited conclusions can be made. Firstly, what you can say is valid for a particular place and time, with a particular investigator, experimental subject and method of applying treatments. Secondly, if your results were significant at the 5 per cent level of probability (p. 427), there is still an approximately one-in-twenty chance that the results did arise by chance. To guard against these possibilities, it is important that experiments are repeated. Ideally, this would be done by an independent scientist with independent materials. However, it makes sense to repeat work yourself so that you can have full confidence in your conclusions. Many scientists recommend that experiments are done three times in total, but this may not be possible in undergraduate work.

Text references

Chalmers, A.F. (1999) *What is this Thing called Science?*, 3rd edn. Open University Press, Buckingham.

Sokal, R.R. and Rohlf, F.J. (1994) *Biometry*, 3rd edn. W.H. Freeman, San Francisco.

Sources for further study

Heath, D. (1995) *An Introduction to Experimental Design and Statistics for Biology.* UCL Press, London.

Quinn, G.P. and Keough, M.J. (2002) *Experimental Design and Data Analysis for Biologists.* Cambridge University Press, Cambridge.

Study exercises

31.1 Create a Latin square design. Treatments W, X, Y and Z are to be applied to potted plants in a glasshouse where the researcher suspects there may be a slight gradation in temperature and light over an oblong bench that can hold a total of 48 pots. You decide to use an experimental design consisting of three 4 × 4 blocks of plants, each arranged in a different Latin square design (see below). Assign treatments to the locations in the diagram below. Explain why this design will help to eliminate the effects of the confounding variables.

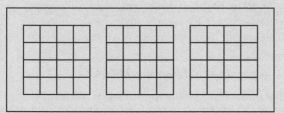

Suspected temperature gradient

Suspected light gradient

Layout for glasshouse experiment

31.2 Generate random numbers. Produce a list of 20 random whole numbers between 1 and 5. This can be carried out using a spreadsheet. If using MS Excel, investigate the RAND() and INT functions. The RAND() function produces a random number between 0 and 1, so you will need to multiply by a constant factor to scale your final output appropriately. Copy your test formula(e) to several cells to test empirically whether it works.

31.3 Investigate possible interactions. Treatments A, B and C involve tests of three different nutrients on proliferation of cells in a culture of carrot cells. Analyse the results in the table below to determine whether any interactions have occurred between the four possible combinations of treatments. No complex statistical analysis is required, simply a calculation of mean experimental effect (treatment minus controls) in each combined treatment and a comparison with the data for the relevant treatments on their own (in other words, assume that observed differences reflect true underlying differences). Classify the results as 'no interaction', 'antagonism' or 'synergism'.

Results of cell proliferation experiment

Treatment	Replicate (growth of callus in g)				
	1	2	3	4	5
Control	4.8	5.2	5.2	4.8	5.0
A	6.5	7.3	7.0	7.1	7.1
B	7.7	8.3	8.5	7.5	8.0
C	10.7	10.0	9.9	9.8	9.6
A+B	6.6	7.2	6.8	7.2	7.2
A+C	11.9	11.7	12.0	12.0	12.4
B+C	17.3	16.8	17.1	16.6	17.2
A+B+C	12.8	12.9	13.1	13.3	12.9

When carrying out advanced lab work or research projects, you will need to master the important skill of managing data and observations and learn how to keep a record of your studies in a lab book. This is important for the following reasons:

- An accurate and neat record helps when using information later, perhaps for exam purposes or when writing a report.
- It allows you to practise important skills such as scientific writing, drawing diagrams, preparing graphs and tables and interpreting results.
- Analysing and writing up your data as you go along prevents a backlog at the end of your study time.
- You can show your work to a future employer to prove you have developed the skills necessary for writing up properly; in industry, this is vital so that others in your team can interpret and develop your work.

Understanding what's expected – especially when taking notes for a lab-based practical, pay special attention to the aims and learning objectives (pp. 22–4) of the session, as these will indicate the sorts of notes you should be taking, including content and diagrams, and the ways in which you should present these for assessment.

 KEY POINT *A good set of lab notes should:*

- *outline the purpose of your experiment or observation;*
- *set down all the information required to describe your materials and methods;*
- *record all relevant information about your results or observations and provide a visual representation of the data;*
- *note your immediate conclusions and suggestions for further experiments.*

Collecting and recording primary data

Individual observations (e.g. laboratory temperature) can be recorded in the text of your notes, but tables are the most convenient way to collect large amounts of information. When preparing a table for data collection, you should:

Recording primary data – never be tempted to jot down data on scraps of paper: you are likely to lose them, or to forget what individual values mean.

1. Use a concise title or a numbered code for cross-referencing.
2. Decide on the number of variables to be measured and their relationship with each other and lay out the table appropriately:
   (a) The first column of your table should show values of the independent (controlled) variable, with subsequent columns for the individual (measured) values for each replicate or sample.
   (b) If several variables are measured for the same organism or sample, each should be given a row.
   (c) In time-course studies, put the replicates as columns grouped according to treatment, with the rows relating to different times.
3. Make sure the arrangement reflects the order in which the values will be collected. Your table should be designed to make the recording process as straightforward as possible, to minimise the possibility of mistakes. For final presentation, a different arrangement may be best (Chapter 63).
4. Consider whether additional columns are required for subsequent calculations. Create a separate column for each mathematical manipulation, so the step-by-step calculations are clearly visible. Use a computer spreadsheet (pp. 68–73) if you are manipulating lots of data.

Designing a table for data collection – use a spreadsheet or the table-creating facility in a word processor to create your table. This will allow you to reorganise it easily if required. Make sure there is sufficient space in each column for the values – if in doubt, err on the generous side.

5. Use a pencil to record data so that mistakes can be easily corrected.
6. Take sufficient time to record quantitative data unambiguously – use large clear numbers, making sure that individual numerals cannot be confused.
7. Record numerical data to an appropriate number of significant figures, reflecting the accuracy and precision of your measurement (p. 408). Do not round off data values, as this might affect the subsequent analysis of your data.
8. Record discrete or grouped data as a tally chart (see p. 383), each row showing the possible values or classes of the variable. Provided that tally marks are of consistent size and spacing, this method has the advantage of providing an 'instant' frequency distribution chart.
9. Prepare duplicated recording tables if your experiments or observations will be repeated.
10. Explain any unusual data values or observations in a footnote. Don't rely on your memory.

Recording details of project work

The recommended system is one where you make a dual record.

Primary record

The primary record is made at the bench or in the field. In this, you must concentrate on the detail of materials, methods and results. Include information that would not be used elsewhere, but that might prove useful in error tracing: for example, if you note how a solution was made up (exact volumes and weights used rather than concentration alone), this could reveal whether a miscalculation had been the cause of a rogue result. Note the origin, type and state of the chemicals and organism(s) used. Make rough diagrams to show the arrangement of replicates, equipment, etc. If you are forced to use loose paper to record data, make sure each sheet is dated and taped to your lab book, collected in a ring binder, or attached together with a treasury tag. The same applies to traces, printouts and graphs.

The basic order of the primary record should mirror that of a research report (see p. 105), including: the title and date; brief introduction; comprehensive materials and methods; the data and short conclusions.

Secondary record

You should make a secondary record concurrently or later in a bound book and it ought to be neater, in both organisation and presentation. This book will be used when discussing results with your supervisor, and when writing up a report or thesis, and may be part of your course assessment. Although these notes should retain the essential features of the primary record, they should be more concise and the emphasis should move towards analysis of the experiment. Outline the aims more carefully at the start and link the experiment to others in a series (e.g. 'Following the results of Expt D24, I decided to test whether...'). You should present data in an easily digested form, e.g. as tables of means or as summary graphs. Use appropriate statistical tests (Chapter 66) to support your analysis of the results. The choice of a bound book ensures that data are not easily lost.

Points to note

The dual method of recording deals with the inevitable untidiness of notes taken at the bench or in the field; these often have to be made rapidly, in awkward positions and in a generally complex environment. Writing a second, neater version forces you to consider again details that might have been overlooked in the primary record and provides a duplicate in case of loss or damage.

If you find it difficult to decide on the amount of detail required in Materials and Methods, the basic ground rule is to record enough information to allow a reasonably competent scientist to repeat your work exactly. You must tread a line between the extremes of pedantic, irrelevant detail and the omission of information essential for a proper interpretation of the data – better perhaps to err on the side of extra detail to begin with. An experienced worker can help you decide which subtle shifts in technique are important (e.g. batch numbers for an important chemical, or when a new stock solution is made up and used). Many important scientific advances have been made because of careful observation and record taking or because coincident data were recorded that did not seem of immediate value.

When creating a primary record, take care not to lose any of the information content of the data: for instance, if you only write down means and not individual values, this may affect your ability to carry out subsequent statistical analyses.

There are numerous ways to reduce the labour of keeping a record. Don't repeat Materials and Methods for a series of similar experiments; use devices such as 'method as for Expt B4'. A photocopy might suffice if the method is derived from a text or article (check with supervisor). To save time, make up and copy a checklist in which details such as chemical batch numbers can be entered.

Special requirements for fieldwork

The main problems you will encounter in the field are the effects of the weather while taking a primary record and the distance you might be from a suitable place to make a neat secondary record. Wind, rain and cold temperatures are not conducive to neat note-taking and you should be prepared for the worst possible conditions at all times. Make sure your clothing allows you to feel comfortable while recording data.

The simplest method of protecting a field notebook is to enclose it in a clear polythene bag large enough for you to take notes inside (Fig. 32.1). Alternatively, you could use a clipboard with a waterproof cover to shield your notes or a special notebook with a waterproof cover. When selecting a field notebook, choose a small size – the dimensions of outside pockets may dictate the upper size limit.

If recording results and observations outdoors:

- Use a pencil as ink pens such as ballpoints smudge in wet conditions, are temperamental in the cold and may not work at awkward angles. Don't forget to take a sharpener.
- Prepare well to enhance the speed and quality of your field note-taking – the date and site details can be written down before setting out and tables can be made out ready for data entry.

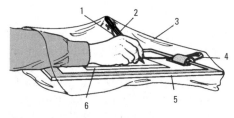

Fig. 32.1 Simple but effective method for keeping notes dry in the field

1. string attaching pencil to clip
2. pencil (not pen) writes on damp paper
3. transparent plastic bag
4. bulldog clip
5. hardboard (at least 31 × 22 cm for A4 paper)
6. record sheet.

- Transcribe field notes to a duplicate primary record at your base each time you return there. There is a very real risk of your losing or damaging a field notebook. Also, poor weather may prevent full note-taking and the necessary extra details should be written up while fresh in your memory.
- Consider using a voice recorder rather than a notebook, in which case voice transcription into written form should also take place while your memory is fresh in case the sound quality is poor.
- Use photographs to set data in context, when appropriate. Develop photographs as soon as possible to check their suitability. Consider using a digital camera when the suitability of a record must be guaranteed at the time of the visit (see Chapter 48).

Field data may be logged automatically, stored temporarily in the instrument's electronic memory, and downloaded to a portable computer ('data logger') when convenient. The information is then transferred to a data bank back at base. If you are using this system, make back up copies of each period's data as soon as possible – if the recording instrument's memory is cleared or overwritten after reading there may be no recourse if the logging machine fails.

Using communal records

If working with a research team, you may need to use their communal databases. These avoid duplication of effort and ensure uniformity in techniques. You will be expected to use the databases carefully and to contribute to them properly. They might include:

- a shared notebook of common techniques (e.g. how to make up media or solutions);
- a set of simplified step-by-step instructions for use of equipment. Manuals are often complex and poorly written and it may help to redraft them, incorporating any differences in procedure adopted by the group;
- an alphabetical list of suppliers of equipment and consumables (perhaps held on a card-index system);
- a list of chemicals required by the group and where they are stored;
- the risk-assessment sheets for dangerous procedures (p. 119);
- the record book detailing the use of radioisotopes and their disposal.

Sources for further study

Anon. *Suggestions for Keeping Laboratory Notebooks.* Available: http://otl.stanford.edu/inventors/resources/labnotebooks.html Last accessed 09/04/07. [A Stanford University website that looks at the laboratory notebook from the patenting perspective.]

Kanare, H.M. (1985) *Writing the Laboratory Notebook.* American Chemical Society, Washington, DC.

Pechenik, J.A.A. (2001) *Short Guide to Writing About Biology*, 4th edn. Longman, Harlow. [Chapter 8 is particularly relevant.]

Slingsby, D. and Cook, C. (1986) *Practical Ecology.* Macmillan Education, London.

Study exercises

32.1 Design a primary data collection sheet for a behavioural study. Imagine you wish to describe the feeding behaviour of birds at their nest. You wish to study the nature of the food, the relative participation of both parents, and the frequency of visits during the daytime with a view to analysis using a spreadsheet.

32.2 Outline the advantages and problems that would be associated with using a tape recorder to record data. Consider the behavioural study described in Study exercise 32.1 as an example.

32.3 Design a secondary record sheet for the collection and analysis of a set of count data for the number of bacterial colonies developed in each of ten replicates for each of five different treatments. Assume that you need to calculate means, variances, etc., and then to compare the results for each treatment.

Research projects are an important component of the final-year syllabus for most degree programmes in the life sciences, while shorter projects may also be carried out during courses in earlier years. Project work can be extremely rewarding, although it does present a number of challenges. The assessment of your project is likely to contribute significantly to your degree grade, so all aspects of this work should be approached in a thorough manner.

Deciding on a topic to study

Assuming you have a choice, this important decision should be researched carefully. Make appointments to visit possible supervisors and ask them for advice on topics that you find interesting. Use library texts and research papers to obtain further background information. Perhaps the most important criterion is whether the topic will sustain your interest over the whole period of the project. Other things to look for include:

- Opportunities to learn new skills. Ideally, you should attempt to gain experience and skills that you might be able to 'sell' to a potential employer.
- Ease of obtaining valid results. An ideal project provides a means to obtain 'guaranteed' data for your report, but also the chance to extend knowledge by doing genuinely novel research.
- Assistance. What help will be available to you during the project? A busy lab with many research students might provide a supportive environment should your potential supervisor be too busy to meet you often; on the other hand, a smaller lab may provide the opportunity for more personal interaction with your supervisor.
- Impact. Your project may result in publishable data: discuss this with your prospective supervisor.

Planning your work

As with any lengthy exercise, planning is required to make the best use of the time allocated (p. 10). This is true on a daily basis as well as over the entire period of the project. It is especially important not to underestimate the time it will take to write and produce your thesis (see below). If you wish to benefit from feedback given by your supervisor, you should aim to have drafts in his/her hands in good time. Since a large proportion of marks will be allocated to the report, you should not rush its production.

If your department requires you to write an interim report, look on this as an opportunity to clarify your thoughts and get some of the time-consuming preparative work out of the way. If not, you should set your own deadlines for producing drafts of the introduction, materials and methods section, etc.

Obtaining ethical approval – if any aspect of your project involves work with human or animal subjects, then you must obtain the necessary ethical clearance before you begin; consult your Department's ethical committee for details.

The Internet as an information source – since many university departments have home pages on the World Wide Web, searches using relevant key words may indicate where research in your area is currently being carried out. Academics usually respond positively to emailed questions about their area of expertise.

Asking around – one of the best sources of information about supervisors, laboratories and projects is past students. Some of the postgraduates in your department may be products of your own system and they could provide an alternative source of advice.

Liaising with your supervisor(s) – this is essential if your work is to proceed efficiently. Specific meetings may be timetabled, e.g. to discuss a term's progress, review your work plan or consider a draft introduction. Most supervisors also have an 'open-door' policy, allowing you to air current problems. Prepare well for all meetings: have a list of questions ready before the meeting; provide results in an easily digestible form (but take your lab notebook along); be clear about your future plans for work.

 KEY POINT *Project work can be very time-consuming at times. Try not to neglect other aspects of your course – make sure your lecture notes are up-to-date and collect relevant supporting information as you go along.*

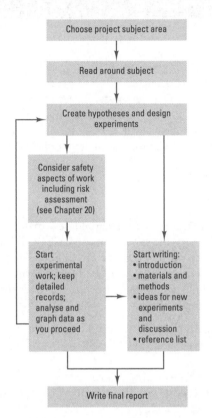

Fig. 33.1 Flowchart showing a recommended sequence of events in carrying out an undergraduate research project

 SAFETY NOTE *It is essential that you follow all the safety rules applying to the laboratory or field site. Make sure you know all relevant procedures – normally there will be prominent warnings about these. If in doubt, ask.*

Getting started

Figure 33.1 is a flowchart illustrating how a project might proceed; at the start, don't spend too long reading the literature and working out a lengthy programme of research. Get stuck in and do an experiment. There's no substitute for 'getting your hands dirty' for stimulating new ideas:

- even a 'failed' experiment will provide some useful information that may allow you to create a new or modified hypothesis;
- pilot experiments may point out deficiencies in experimental technique that will need to be rectified;
- the experience will help you create a realistic plan of work.

Designing experiments or sampling procedures

Design of experiments is covered in Chapter 31, while sampling procedure is dealt with in Chapter 30. Avoid being too ambitious at the start of your work. It is generally best to work with a simple hypothesis and design your experiments or sampling around this. A small pilot experiment or test sample will highlight potential stumbling blocks including resource limitations, whether in materials, or time, or both.

Working in a laboratory environment

During your time as a project student, you are effectively a guest in your supervisor's laboratory.

- Be considerate – keep your 'area' tidy and offer to do your share of lab duties such as calibrating the pH meter, replenishing stock solutions, distilled water, etc., maintaining cultures, tending plants or animals.
- Use instruments carefully – they could be worth more than you'd think. Careless use may invalidate calibration settings and ruin other people's work as well as your own.
- Do your homework on techniques you intend to use – there's less chance of making costly mistakes if you have a good background understanding of the methods you will be using.
- Always seek advice if you are unsure of what you are doing.

Keeping notes and analysing your results

Tidy record keeping is often associated with good research, and you should follow the advice and hints given in Chapter 32. Try to keep copies of all files relating to your project. As you obtain results, you should always calculate, analyse and graph data as soon as you can (see Fig. 33.1). This can reveal aspects that may not be obvious in numerical or readout form. Don't be worried by negative results – these can sometimes be as useful as positive results if they allow you to eliminate hypotheses – and don't be too dispirited if things do not work first time. Thomas Edison's maxim 'Genius is one per cent inspiration and ninety-nine per cent perspiration' certainly applies to research work.

Writing your project report

The structure of scientific reports is dealt with in Chapter 17. The following advice concerns methods of accumulating relevant information.

Brushing up on IT skills – word processors and spreadsheets are extremely useful when producing a thesis. Chapters 11 and 12 detail key features of these programs. You might benefit from attending courses on the relevant programs or studying manuals or texts so that you can use them more efficiently.

Using drawings and photographs – these can provide valuable records of sampling sites or experimental set-ups and could be useful in your report. Plan ahead and do the relevant work at the time of carrying out your research rather than afterwards. Refer to Chapters 28 and 48 for tips on technique.

Introduction   This is a big piece of writing that can be very time-consuming. Therefore, the more work you can do on it early on, the better. You should allocate some time at the start for library work (without neglecting field or bench work), so that you can build up a database of references (Chapter 8). Though photocopying can be expensive, you will find it valuable to have copies of key reviews and references handy when writing away from the library. Discuss proposals for content and structure with your supervisor to make sure your effort is relevant. Leave space at the end for a section on aims and objectives. This is important to orient readers (including assessors), but you may prefer to finalise the content after the results have been analysed.

Materials and methods   You should note as many details as possible *when doing the experiment or making observations*. Don't rely on your memory or hope that the information will still be available when you come to write up. Even if it is, chasing these details can waste valuable time.

Results   Show your supervisor graphed and tabulated versions of your data promptly. These can easily be produced using a spreadsheet (p. 68), but you should seek your supervisor's advice on whether the design and print quality is appropriate to be included in your report. You may wish to access a specialist graphics program to produce publishable-quality graphs and charts: allow some time for learning its idiosyncrasies. If you are producing a poster for assessment (Chapter 13), be sure to mock up the design well in advance. Similarly, think ahead about your needs for any seminar or poster you will present.

Discussion   Because this comes at the end of your report, and some parts can only be written after you have all the results in place, the temptation is to leave the discussion to last. This means that it might be rushed – not a good idea because of the weight attached by assessors to your analysis of data and thoughts about future experiments. It will help greatly if you keep notes of aims, conclusions and ideas for future work as you go along (Fig. 33.1). Another useful tip is to make notes of comparable data and conclusions from the literature as you read papers and reviews.

Acknowledgements   Make a special place in your notebook for noting all those who have helped you carry out the work, for use when writing this section of the report.

References   Because of the complex formats involved (p. 48), these can be tricky to type. To save time, process them in batches as you go along.

 KEY POINT *Make sure you are absolutely certain about the deadline for submitting your report and try to submit a few days before it. If you leave things until the last moment, you may find access to printers, photocopiers and binding machines is difficult.*

Sources for further study

Luck, M. (1999) *Your Student Research Project*. Gower, London.

Marshall, P. (1997) *Research Methods: How to Design and Conduct a Successful Project*. How To Books, Plymouth.

Study exercises

Note: These exercises assume that you have started a research project, or are about to start one, as part of your studies.

33.1 Prepare a project plan. Make a formal plan for your research project, incorporating any milestones dictated by your department, such as interim reports and final submission dates. Discuss your plan with your supervisor and incorporate his or her comments. Refer back to the plan frequently during your project, to see how well you are meeting your deadlines.

33.2 Resolve to write up your work as you go along. Each time you complete an experiment or observation, write up the materials and methods, analyse the data and draw up the graphs as soon as you can. While you may reject some of your work at a later stage, you may wish to modify it; this will spread out the majority of the effort and allow time for critical thinking close to the final submission date.

33.3 Devise a computer database for keeping details of your references. Keeping these records up to date will save you a lot of time when writing up. You will need to decide on an appropriate referencing format, or find out about that followed by your department (see Chapter 8).

Obtaining and identifying specimens

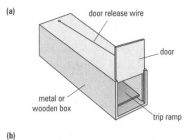

(a)

door release wire

door

metal or wooden box

trip ramp

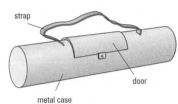

(b)

strap

door

metal case

(c)

net

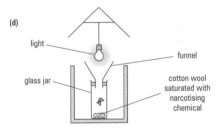

(d)

light

funnel

glass jar

cotton wool saturated with narcotising chemical

(e)

suction applied by mouth

(f)

camoflaged cover

trap

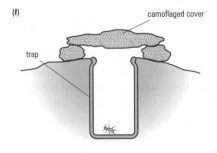

If you are required to collect material you may need to use a formal sampling procedure (see Chapter 30) or simple qualitative collecting. Your choice of equipment will depend upon your objectives. Some of the main reasons for collecting include:

- Obtaining specimens for subsequent laboratory experimentation or observation: this requires collection of living material while causing minimum stress and damage to the organisms.
- Making estimates of population and community parameters: this can be destructive (requiring killing of specimens) or non-destructive, depending upon the objectives of the study and any requirements for subsequent laboratory work, e.g. sorting and identification.
- Gathering specimens for museum-type collections: here, the main objective is to obtain undamaged and representative specimens, usually in a preserved form (see Chapter 35).
- Collecting for subsequent chemical/biochemical analysis: this may require a formal protocol. This objective requires care in both the method of collection and subsequent storage to avoid inducing chemical changes that are artefacts of the collection and storage processes. Deep freezing is usually the preferred storage method when subsequent chemical analysis is likely.

The main 'rules' for collecting are:

- Collect only enough for your purposes.
- Treat animals with respect at all times: do not cause unnecessary stress or suffering. There are formal rules for the handling of many vertebrate species but the same attitude should be taken towards all living organisms.
- Minimise damage and stress during collection and transport: stressed organisms are of little use for realistic experimentation.
- Be aware of the limitations and bias of your collecting equipment: this is particularly important for formal sampling procedures where collecting devices almost always have such problems. This may require specific testing for your particular usage.
- Keep good records of collection details.

> **KEY POINT** *Check any legislation relating to the species or habitats you are intending to use. Obtain any permits required and strictly obey any regulations.*

Equipment for collecting

There is an immense variety of collecting and sampling devices available, see e.g. Fig. 34.1. In general, the collection of remote and/or animal samples presents the greatest problems. The more remote the operator is

Fig. 34.1 Examples of devices used for collecting and sampling. (a) Small vertebrate trap – the animal walks up the ramp towards bait and releases the trap door by a simple trip mechanism. (b) Botanist's vasculum – used in the field for collecting and protecting plant, fungal and animal specimens. (c) Sweep net – used for catching flying insects or aquatic specimens. (d) Light trap – used for catching night-flying insects. (e) 'Pooter' – device for collecting small animals such as insects by suction. (f) Pitfall trap – animal falls into camouflaged trap and cannot escape.

from the point of sampling, the more difficult is the evaluation of the quality of the sample in terms of its representativeness. Animal collection can be difficult because of factors such as mobility and complications introduced when allowing for avoidance behaviour.

Collecting formal samples

Here the objective is to obtain specimens that both qualitatively and quantitatively represent some well-defined habitat(s). Some of the more obvious practical considerations are:

- Are there 'edge effects' associated with the sampling method? Because of such effects you may need to adopt a special protocol, e.g. to determine whether or not a specimen falling on the boundary is included or excluded from the sample (p. 181).
- Will the sample be uniform? For example, remote sampling of marine sediments of different texture using grabs often results in 'bites' of different depth being taken.
- Does the method sample all components of the biota equally well? This may not be important as long as it adequately samples those components in which you are interested.
- How accurately can the location of the sample be defined, especially with regard to other samples intended as replicates? This is particularly important in remote sampling.
- Is the size of the sampling unit adequate for the size and distribution of the object, species or communities being sampled? In remote sampling, this can often be a problem and frequently the method used is a compromise.
- Can interspecies interactions affect the integrity of the sample after collection but before processing? Factors such as predation can be prevented by using an appropriate chemical fixative immediately upon collection (see Chapter 35). Freezing is a poor option here since, upon thawing, many animal and plant tissues tend to disintegrate and the specimens are in poor condition for subsequent identification, etc.

General collecting strategies

For qualitative collecting, where a representative sample of the population or habitat is not the objective, the effectiveness of equipment may be of less significance. Here, suitability for capturing living and undamaged specimens may be the principal criterion in choosing your equipment. For mobile animals this often involves nets and traps of various kinds, combined with narcotising agents such as smoke for insects (see Chapter 35). If specimens can be killed before capture, then spray biocides can be a useful aid. Some useful general points about all collections are:

- Think ahead and be prepared. Your collecting equipment must be appropriate for the task.
- Keep a good record of collection details: this is particularly important if the collection is for museum or herbarium purposes.
- Keep collected plants in a humid atmosphere to ensure good condition. The vasculum originally designed for this purpose has largely been replaced by the polythene bag in most circumstances, except where mechanical damage is likely.

Definition

Edge effect – in this context, any phenomenon associated with the sampling procedure at the edge of a sampling device.

Defining location – *for greatest precision, use the coordinates obtained from a large-scale map; for less precise measurements, coordinates given by satellite-based or radio-based position-fixing may be appropriate.*

Recording information – *the following details should be recorded upon collection: date, time, location (grid ref.), habitat details, collecting technique, preservation technique.*

SAFETY NOTE *Safety issues associated with collecting – risks from the organisms themselves include toxicity of some plant specimens; allergic reactions to stings and bites; venomous bites and stings; and viral and bacterial infections from bites and excreta. Risks associated with the collecting environment and equipment include rough terrain, sharp edges and water hazards. Always follow the safety rules of your department and any more specific advice produced locally. Before you collect in a new environment, make sure you consult an experienced worker who can alert you to potential problems.*

- Keep all living animals in conditions as similar as possible to the environment from which they were collected. Aquatic specimens are usually particularly temperature-sensitive and should be kept in a Thermos flask or 'Coolbox' to prevent rapid temperature changes.
- Rigid containers are better than plastic bags for most purposes since they help to prevent damage to the specimens. Plastic containers should be chosen in preference to glass for most purposes. Make sure that the container seals properly to avoid the loss of specimens and to prevent loss of water.

Sources for further study

Anon. *An Introduction to Collecting Plants.*
Available: http://www.anbg.gov.au/cpbr/herbarium/collecting/index. html
Last accessed: 09/04/07.
[Centre for Plant Biodiversity Research website with links and information relating to collecting plants. Based upon Australian habitats; lots of useful information and suggestions.]

Census of Marine Life *Investigating Marine Life. How do Scientists Collect Organisms?* http://www.coml.org/edu/tech/collect/col1.htm
Last accessed: 09/04/07.
[Well-illustrated guide to marine sampling techniques.]

Lincoln, R.J. and Sheals, J.G. (1979) *Invertebrate Animals: Collection and Preservation.* Cambridge University Press, Cambridge.
[An excellent methodology source for all invertebrate groups. Note that some of the chemicals recommended have been superseded by other, usually less toxic, equivalents.]

Wolberg, D. and Reinard, P. (1997) *Collecting the Natural World: Legal Requirements and Personal Liability for Collecting Plants, Animals, Rocks, Minerals and Fossils.* Geoscience Press, Phoenix.
[Deals with US regulations; gives a good idea of what needs to be considered.]

Study exercises

34.1 Specify collection equipment. What equipment would you use to make a collection of the following organisms and how would you transport them to the laboratory in a living condition?
(a) *Daphnia* (planktonic crustacean)
(b) earthworms
(c) bluebells
(d) ants

34.2 Specify preservation methods. How would you preserve organisms (a)–(d) in Study exercise 34.1 for subsequent examination?

34.3 Specify labelling details. What items would you include on the labels for preserved specimens of organisms (a)–(d) in Study exercise 34.1?

Fixation is a chemical process that stops autolysis and stabilises protein and other major components of tissues so that during subsequent processing, the tissues retain as fully as possible the form they had in life. Preservation allows material to be stored indefinitely by destroying any bacteria, fungi, etc., that could degrade the specimen. Preservation and fixation used to be synonymous in that most of the commonly used preservatives also had a fixative action. This has changed with the introduction of 'phenoxetols'. These are good preservatives but since they do not arrest autolysis, their use must be preceded by treatment with a true fixative: for that reason, they are called post-fixation preservatives.

> *Fixing specimens for histological or cytological work* – the procedures and chemicals used are critical to success. Consult specialist texts such as Kiernan (1999) or materials and methods sections of relevant research papers.

 KEY POINT *Take care when fixing specimens – remember, anything that will fix your specimen is also capable of fixing you. Be very careful in your handling of these materials and obey all safety precautions.*

Your choice of fixative will depend upon the material to be fixed and the purpose for which it is being fixed. Some of the most common fixatives (Table 35.1) may be used alone although more often they are used in mixtures, the object being to combine the virtues of the various ingredients.

Narcotisation

> *Acting humanely* – always narcotise animals before carrying out any procedures that might cause pain.

Narcotisation (relaxation) is usually advisable for animals since many are highly contractile and assume grossly distorted postures if placed straight into fixative; it is also more humane to narcotise before killing. Failure to narcotise may result in contortion, rupture of the body wall or evisceration and reduce the scientific value of the specimen. The need for narcotisation varies with the type of organism and the objectives of the study: some of the most widely used narcotising agents are given in Table 35.2.

Table 35.1 Some of the most widely used fixatives/preservatives and their properties. Note that there are significant safety issues with nearly all these substances (see column 4).

Substance	Fixation	Usage	Notes
Formaldehyde	+	4% v/v	Comes as 40% v/v solution (= formalin): normal dilution 1 + 9. Make up with sea water for marine specimens; buffer for calcareous specimens. Use in a fume cupboard – health hazard
Ethanol	+	70% v/v aqueous	Highly volatile; inflammable; containers must be well sealed to prevent evaporation; causes shrinkage and decolorisation as well as loss of lipids
Acetic acid	+	In mixtures	Pungent vapour
Picric acid	+	In mixtures	Risk of explosion; detonates readily on contact with some metals. Not recommended for routine student use (significant health risk)
Mercuric chloride	+	In mixtures	Extremely poisonous; corrosive to metal implements; tissue will contain mercuric salt deposits
Osmium tetroxide	+	1% v/v in buffer or vapour	Both fluid and vapour highly toxic; use in fume cupboard only. Excellent for cytological detail but is expensive and can only be used for very small specimens due to poor penetration speed. Vapour good for protozoa
Propylene phenoxetol	–	1–2% v/v aqueous	Relatively expensive but innocuous and effective preservative: needs pre-fixation stage
Glutaraldehyde	+	2–4% v/v in buffer	Highly toxic and severe skin irritant. Must be used cold

Table 35.2 Narcotising agents and their characteristics

Agent	Usage	Notes
Cold (chilling)	Cold-blooded animals	Effective form of relaxing many animals such as tropical and sub-tropical invertebrates
Heat (warming)	Slow heat	Works for some animals. Start from ambient but keep time period as short as possible
Magnesium sulphate or	7% w/v in water	Quite effective for many invertebrates but beware of osmotic problems if made up in sea water: keep exposure times fairly short (1–2 h)
Magnesium chloride	20% w/v in water	
Menthol crystals	Float on water	Slow but effective for many aquatic animals
Chloral hydrate	1% w/v in sea water	General narcotising agent
Ethanol	10% v/v dropwise	Slow and rather tedious process for all but very small specimens
MS-222 (Tricaine)	Use as 0.05% w/v aqueous solution	Good for marine and freshwater fish: very rapid effect (15 s–1 min)
Chloretone	0.1–0.5% w/v in water	General narcotising agent
Ethyl acetate	Vapour	Effective for most insects (kills as well); highly volatile
Ether	Vapour	Effective for vertebrates
Chloroform	Vapour	Effective for vertebrates

SAFETY NOTE **Keep safety in mind** – *many fixatives and preservatives are toxic and a risk assessment should be made before they are used. Always use safety equipment when handling these potentially toxic chemicals, including goggles, buttoned lab coat and latex gloves.*

Selecting your fixative/preservative

The main factors you must consider in selecting the preservative/fixative suitable for your materials are:

- Speed of penetration, which determines the size of object that can be fixed. Some fixatives penetrate very slowly (e.g. osmium salts) and only very small pieces of tissue can be fixed. Others such as formaldehyde penetrate very rapidly and allow fixation of relatively large specimens, e.g. whole animals. It may be necessary to inject fixatives into the body cavities of large specimens to ensure adequate fixation.
- Shrinkage: fixatives such as ethanol and mercuric chloride cause tissues to shrink, thus distorting them significantly. The addition of glacial acetic acid is frequently used to reduce this problem.
- Hardening: ethanol is particularly liable to cause tissues to harden on prolonged exposure to concentrations above 70% v/v. For whole specimens, the addition of 3% v/v glycerol to 70% v/v alcohol reduces this, making it a useful medium for specimen storage (not to be used for histological studies). The time in hardening solutions must be carefully controlled for histological material.
- Decolorisation: for whole specimens, the loss of colour may be detrimental and solvents such as ethanol must be used with care since they readily remove many pigments.
- Osmotic problems: distortion of tissues by osmotic movements of water can be rapid and serious. Some fixatives are best made up in sea water if they are to be used for marine specimens to avoid osmotic swelling of tissues. Others such as osmium salts and glutaraldehyde usually need to be made up in a buffer solution isotonic with the tissues.
- Decalcification: acidic fixatives such as unbuffered formalin and those containing picric acid or acetic acid readily dissolve calcareous

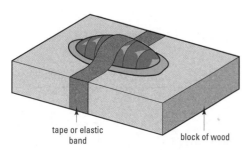

Fig. 35.1 Method for constraining animal specimens liable to curl up or flex during wet preservation

tape or elastic band

block of wood

structures. For histological preparations this may be desirable, but for whole specimens it is highly undesirable. The acidic properties of formaldehyde can be overcome by neutralising with calcium salts.

- Other chemical reactions: fixatives containing mercuric chloride usually result in the deposition of mercuric precipitates in the tissues. For histological preparations, these must be removed by thorough washing and if necessary by post-treating with iodised alcohol. There is no good reason for using mercuric chloride fixatives other than for histological work and the benefits must be weighed against the dangers and difficulties. Note also that you must not use metal implements with such fixatives as they are extremely corrosive – use only wooden or glass implements.

Preservation

Wet preservation

This is used mainly for animals and is usually preceded by narcotisation. The process of fixation is then comparatively straightforward: sometimes it is necessary to arrange the body and appendages of the animal using tapes, elastic bands, etc. before fixation begins (see Fig. 35.1). The solutions most commonly used for wet preservation have been either:

- a 5–10% v/v aqueous solution of formalin (= 40% v/v aqueous formaldehyde), neutralised with calcium carbonate or some other agent to prevent any acidity in the solution resulting in the slow dissolution of calcareous structures (decalcification); or
- 50–70% v/v ethanol.

Formaldehyde is cheap and non-inflammable but tends to stiffen and harden tissues on prolonged exposure: safety considerations make it particularly problematical to use because of its noxious vapour. Ethanol is inflammable, highly volatile and tends to cause shrinkage and decolorisation in soft-bodied animals: pass them slowly through a graded series of concentrations to minimise this problem. Industrial methylated spirit is quite suitable for preservation if ethanol is unavailable.

A 1–2% v/v aqueous solution of propylene phenoxetol has been widely used for vertebrate and invertebrate preservation provided it is preceded by the use of a fixative: it preserves the natural colour well and leaves the material pliable. It is comparatively expensive but is non-flammable and non-volatile.

Wet preservation of plant material is relatively straightforward and has the advantage of preserving the form of the specimen providing it is done carefully. Herbaceous plants should be preserved in a solution with an alcohol base, while for succulents a formaldehyde base is better. Algae, fungi, ferns, lichens and seed plants can be stored wet in formalin acetic acid (FAA) solution: this is made from formalin, acetic acid and ethanol in the ratio 85 : 10 : 5. To retain the green coloration of ferns and seed plants, add a crystal of copper sulphate to the solution and incubate the specimens for 3–10 days, then transfer to ordinary FAA solution. Colours of flowers cannot easily be preserved using wet preservatives.

For marine specimens, make up a stock of 1 part propylene glycol with 1 part formalin. Use this in the ratio of 1 part stock to 8 parts sea water for

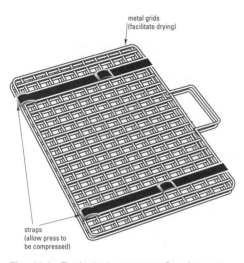

Fig. 35.2 Typical plant press. Specimens are placed between absorbent paper sheets within the metal grids, then the straps are tightened.

Using a plant press – be sure to label each specimen carefully in a way that ensures that the label cannot be lost during paper changes.

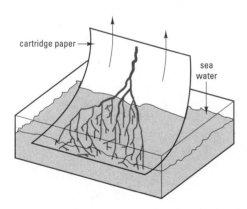

Fig. 35.3 Method for 'floating out' algal specimens. Arrows show direction of lifting

fixation and 1 part stock to 9 parts sea water for subsequent preservation. This markedly reduces the formaldehyde concentration without loss of the fixative/preservative effect and has been shown to be satisfactory even for specimens with calcareous components.

Dry preservation

This is used mainly for vertebrate taxidermy, and for arthropod and plant material, although the development of freeze-drying has made this procedure applicable to almost any type of small or medium-sized organism. Arthropods and vascular plants, with their hard exoskeletons and vascular tissue respectively, are most easily dealt with since the skeletons prevent collapse and loss of form upon drying. Similarly, some types of fungal fruiting bodies such as 'bracket fungi' can be preserved in a dry state.

Flowering plants, or parts of plants, mosses and liverworts, should be dried in a plant press (Fig. 35.2) between sheets of absorbent paper. Interpose a sheet of muslin or greaseproof paper between the paper and the specimen to prevent the specimen adhering to the paper. Great care must be exercised in the cleaning and arrangement of the plant so that essential features are not obscured. Change the drying sheets daily until dry. Drying by heating the press is particularly good for preserving colour: use a low-temperature oven or place on a radiator.

Algae are usually dried in a press after floating them on to a sheet of cartridge paper that is immersed in sea water: this allows careful arrangement of the often delicate algal fronds (see Fig. 35.3). Once suitably arranged, the process is essentially the same as for flowering plants although regular changes of the absorbent paper layers are necessary and the drying procedure usually takes several days. Fungi are best preserved by drying without any attempt at pressing.

Storage

Careful maintenance of specimens after the fixation/preservation processes is very important. Specimens may be stored for long periods provided they are protected from pests such as mites: this may require the use of chemicals such as naphthalene in air-tight containers. Seeds are frequently stored after air-drying only, but they have a finite period of viability.

The storage conditions for all dried materials are critical; these include moderate to low temperatures and very low humidity. Storage of wet material should be in appropriate, vapour-tight vessels: even then the containers will need checking and topping up over time, particularly when volatile preservatives are used. Labelling must be comprehensive and should contain information on the fixation and preservation processes as well as ecological and taxonomic details.

Text reference

Kiernan, J.A. (1999) *Histological and Histochemical Methods: Theory and Practice*, 3rd edn. Scion Publishing Ltd., Bloxham.

Sources for further study

Centre for Plant Biodiversity Research *Plant Collection Procedures and Specimen Preservation*. Available: http://www.anbg.gov.au/cpbr/herbarium/collecting/collection-procedures.html
Last accessed 09/04/07

Lincoln, R.J. and Sheals, J.G. (1979) *Invertebrate Animals: Collection and Preservation*. Cambridge University Press, Cambridge.

Lo Bianco, S. *Classic Resources: The Methods Employed at the Naples Zoological Station for the Preservation of Marine Animals (Trans. E. O. Hovey)*. Available: http://www.mbl.edu/BiologicalBulletin/CLASSICS/

Russell/Russell-pp11-50.html#Preparations
Last accessed 09/04/07.
[Web reprint of a 'classic' resource.]

National Museum of National History, Smithsonian Institution. *Algae Research: Collection and Preservation*. Available: http://www.nmnh.si.edu/botany/projects/algae/collpres.htm
Last accessed 09/04/07.

Smalldon, G. and Lee, E.W. (1979) *A Synopsis of Methods for the Narcotisation of Marine Invertebrates*. Royal Scottish Museum, Edinburgh.

Study exercises

35.1 Write a protocol for collection, preservation and fixation of an animal specimen. This could be for any animal, but an illustrative answer is given for an earthworm specimen that is to be processed for microscopic sectioning.

35.2 Write a protocol for the collection and preservation of a plant specimen. This could be for any plant, but an illustrative answer is given for a marine seaweed to be used as a reference specimen for future identification purposes.

35.3 State the difficulties encountered when fixing materials for microscopic sectioning. Make a list of the safety problems associated with the use of five fixatives used in microscopy.

Microbes can be studied by taking samples for analysis, usually in one of the following ways:

- by direct examination of individual cells of a particular microbe, e.g. using fluorescence microscopy (p. 243);
- by isolating/purifying a particular species or related individuals of a taxonomic group, e.g. the faecal indicator bacterium *Escherichia coli* in sea water;
- by studying microbial processes, rather than individual microbes, either *in situ*, or in the laboratory.

The sampling process

The main factors to be considered when deciding on a suitable sampling procedure are discussed in Chapter 30.

Sampling techniques include the use of swabs, Sellotape, strips and agar contact methods for sampling surfaces, bottles for aquatic habitats, plastic bags and corers for soils and sediments. A wide range of complex apparatus is available for accurately sampling water or soil at particular depths.

 KEY POINT *An important feature of all microbiological sampling protocols is that the sampling apparatus must be sterile; strict aseptic technique must be used throughout the sampling process (see Chapter 45).*

The sampling method must minimise the chance of contamination with microbes from other sources, especially the exterior of the sampling apparatus and the operator. For example, if you are sampling an aquatic habitat, stand downstream of the sampling site. A portable Bunsen burner or spirit lamp can be used to assist sterile technique during field sampling, e.g. while flaming a loop (p. 266). Alternatively, use disposable plastic loops.

Process the sample as quickly as possible, to minimise any changes in microbiological status. As a general guideline, many procedures require samples to be analysed within six hours of collection. Changes in aeration, pH and water content may occur after collection. Some microbes are more susceptible to such effects, e.g. anaerobic bacteria may not survive if the sample is exposed to air. Sunlight can also inactivate bacteria; samples should be shielded from direct sunlight during collection and transport to the laboratory.

Soil and water samples are often kept cool (at 0–5°C) during transport to the laboratory. In contrast, some microbes adapted to grow in association with warm-blooded animals may be damaged by low temperatures. An alternative approach is to keep the sample near the ambient sampling temperature using an insulated vessel (e.g. Thermos flask).

Isolation and purification techniques

Several different approaches may be used to obtain microbes in pure culture. The choice of method will depend upon the microbe to be isolated: some organisms are relatively easy to isolate, while others require more involved procedures.

Alternatives to traditional culture-based methods – microbes can now be studied by molecular methods, including the amplification of specific nucleic acid sequences by PCR (p. 335).

Avoiding contamination during sampling – *always remember that* you are the most important source of contamination of field samples: components of the oral or skin microflora are the most likely contaminants.

Subsampling – to minimise the effects of changes in temperature, aeration and water status during transportation, a primary sample may be returned to the laboratory, where the working sample (subsample) is then taken (e.g. from the centre of a large block of soil).

 SAFETY NOTE *When working with newly isolated microbes, you should always treat them as potentially harmful (p. 264) until they have been identified (Chapter 39).*

Obtaining a pure culture – if a single colony from a primary isolation medium is used to prepare a streak dilution plate and all the colonies on the second plate appear identical, then a pure culture has been established. Otherwise, you cannot assume that your culture is pure and you should repeat the subculture until you have a pure culture.

Using a sonicator – minimise heat damage with short treatment 'bursts' (typically up to 1 min), cooling the sample between bursts, e.g. using ice.

Definitions

Psychrophile – a microbe with an optimum temperature for growth of <20 °C (lit. 'cold-loving').

Psychrotroph – a microbe with an optimum temperature for growth of ⩾ 20 °C, but capable of growing at lower temperature, typically 0–5 °C (lit. 'cold-feeding').

Thermophile – a microbe with an optimum growth temperature of >45 °C (lit. 'heat-loving').

Mesophile – a microbe with an optimum growth temperature of 20–45 °C (lit. 'middle-loving').

Separation methods

Most microbial isolation procedures involve some form of separation to obtain individual microbial cells. The most common approach is to use an agar-based medium for primary isolation, with streak dilution, spread plating or pour plating to produce single colonies, each derived from a single type of microbe (pp. 267–8). It is often necessary to dilute samples before isolation, so that a small number of individual microbial cells is transferred to the growth medium. Strict serial dilution (pp. 131–2) of a known amount of sample is needed for quantitative work.

 KEY POINT *If your aim is to isolate a particular microbe, perhaps for further investigation, you will need to subculture individual colonies from the primary isolation plate to establish a pure culture, also known as an axenic culture.*

Pure cultures of most microbes can be maintained indefinitely, using sterile technique and microbial culture methods (Chapter 45).

Other separation techniques include:

- Dilution to extinction. This involves diluting the sample to such an extent that only one or two microbes are present per millilitre: small volumes of this dilution are then transferred to a liquid growth medium (broth). After incubation, most of the tubes will show no growth, but some tubes will show growth, having been inoculated with a single viable microbe at the outset. This should give a pure culture, though it is wasteful of resources.

- Sonication/homogenisation. This is useful for separating individual microbial cells from each other and from inert particles, prior to isolation. However, some decrease in viability is likely.

- Filtration. This can be useful where the number of microbes is low. Samples can be passed through a sterile cellulose ester filter (pore size 0.2 μm), which is then incubated on the surface of an appropriate solidified medium. Sieving and filtration techniques (p. 136) are often used in soil microbiology to subdivide a sample on the basis of particle size.

- Micromanipulation. It may be possible to separate a microbe from contaminants using a micropipette and dissecting microscope (pp. 255–6). The microbe can then be transferred to an appropriate growth medium, to give a pure culture. However, this is rarely an easy task for the novice.

- Motility. Phototactic microbes (including photosynthetic flagellates and motile cyanobacteria) will move towards a light source; heterotrophic flagellate bacteria will move through a filter of appropriate pore size into a nutrient solution, or away from unfavourable conditions (chemotaxis).

Selective and enrichment methods

Selective and enrichment techniques can be considered together, since they both enhance the growth of a particular microbe when compared with its competitors and they are often combined in specific media (Box 36.1).

The difference between selective and enrichment techniques is that the former use growth conditions unfavourable for competitors while the latter provide improved growth conditions for the chosen microbes.

Laboratory incubation under selective conditions will allow the particular microbes to be isolated in pure culture.

Box 36.1   Differential media for bacterial isolation: an example

MacConkey agar is both a selective and a differential medium, useful for the isolation and identification of intestinal Gram-negative bacteria. Each component in the medium has a particular role:

- **Peptone:** (a meat digest) provides a rich source of complex organic nutrients, to support the growth of non-exacting bacteria.
- **Bile salts:** toxic to most microbes apart from those growing in the intestinal tract (selective agent).
- **Lactose:** present as an additional, specific carbon source (enrichment agent).
- **Neutral red:** a pH indicator dye, to show the decrease in pH that accompanies the breakdown of lactose.
- **Crystal violet:** selectively inhibits the growth of Gram-positive bacteria. (This is only present in certain formations of MacConkey agar.)

Any intestinal Gram-negative bacterium capable of fermenting lactose will grow on MacConkey agar to produce large red–purple colonies, the red coloration being due to the neutral red indicator under acidic conditions while the purple coloration, often accompanied by a metallic sheen, is due to the precipitation of bile salts and crystal violet at low pH.

In contrast, enteric Gram-negative bacteria unable to metabolise lactose will give colonies with no obvious pigmentation. This differential medium has been particularly useful in medical microbiology, since many enteric bacteria are unable to ferment lactose (e.g. *Salmonella*, *Shigella*) while others metabolise this carbohydrate (e.g. *Escherichia coli*, *Klebsiella* spp.). Colonial morphology (Fig. 39.2) on such a medium can give an experienced bacteriologist important clues to the identity of an organism, e.g. capsulate *Klebsiella* spp. characteristically produce large, convex, mucoid colonies with a weak pink coloration, due to the fermentation of lactose, while *E. coli* produces smaller, flattened colonies with a stronger red colouration and a metallic sheen.

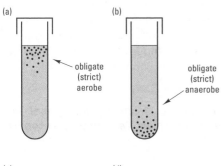

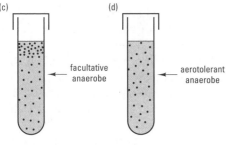
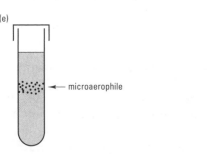

Fig. 36.1 Agar shake tubes. Bacteria are suspended in molten agar at 45–50 °C and allowed to cool. The growth pattern after incubation reflects the atmospheric (oxygen) requirements of the bacterium.

🔑 KEY POINT *Selective methods are based on the use of physico-chemical conditions that will permit the growth of a particular group of microbes while inhibiting others. Enrichment techniques encourage the growth of certain bacteria, usually by providing additional nutrients in the growth medium.*

Methods based on specific physical conditions include:

- Temperature. Psychrophilic and psychrotrophic microbes can be isolated by incubating the growth medium at 4 °C, while thermophilic microbes require temperatures above 45 °C for isolation. Short-term heat treatment of samples can be used to select for endospore-forming bacteria, e.g. 70–80 °C for 5–15 min, prior to isolation.
- Atmosphere. Many eukaryotic microbes are obligate aerobes, requiring an adequate supply of oxygen to grow. Bacteria vary in their responses to oxygen: obligate anaerobes are the most demanding, growing only under anaerobic conditions (e.g. in an anaerobic cabinet or jar). Oxygen requirements can be determined using the agar shake tube method as part of the isolation procedure (Fig. 36.1). Some pathogenic bacteria grow best in an atmosphere with a reduced oxygen status and increased CO_2 concentration: such carboxyphilic bacteria (capnophiles) are grown in an incubator where the gas composition can be adjusted.
- Centrifugation. This can be used to separate buoyant microbes from their non-buoyant counterparts – on centrifugation, such organisms will collect at the surface while the remaining microbes will sediment. Alternatively, density gradient methods may be used (p. 361). Centrifugation can be combined with repeated washing, to separate microbes from contaminants.
- Ultraviolet irradiation. Some microbes are tolerant of UV treatment and can be selected by exposing samples to UV light. However, the survivors may show a greater rate of mutation.
- Illumination. Samples may be enriched for cyanobacteria and microalgae by incubation under a suitable light regime. For dilute

Definitions

Aerotolerant anaerobe – *a microbe that grows by fermentation, but which is insensitive to air/oxygen (in contrast to strict anaerobes, which are typically killed by exposure to air/oxygen).*

Capnophile – *a microbe that thrives in the presence of high levels of atmospheric carbon dioxide.*

Facultative anaerobe – *a microbe that grows by aerobic respiration when oxygen is present, switching to fermentation under anaerobic conditions.*

samples, where the number of photosynthetic microbes is too low to give the sample any visible green coloration, there is a risk of photoinhibition and loss of viability if the irradiance is too high. Such samples need shading during initial growth.

Chemical methods form the mainstay of bacteriological isolation techniques and various media have been developed for the isolation of specific groups of bacteria. The chemicals involved can be subdivided into the following groups:

- Selectively toxic substances: for example, salt-tolerant, Gram-positive cocci can be grown in a medium containing 7.5% w/v NaCl, which prevents the growth of most common heterotrophic bacteria. Several media include dyes as selective agents, particularly against Gram-positive bacteria.
- Antibiotics: for example, the use of antibacterial agents (e.g. penicillin, streptomycin, chloramphenicol) in media designed to isolate fungi, or the use of antifungal agents (e.g. cycloheximide, nystatin) in bacterial media. Some antibacterial agents show a narrow spectrum of toxicity and these can be incorporated into selective isolation media for resistant bacteria, e.g. metronidazole for anaerobic bacteria.
- Nutrients that encourage the growth of certain microbes: including the addition of a particular carbon source, or specific inorganic nutrients.
- Substances that affect the pH of the medium: for example, the use of alkaline peptone water at pH 8.6 for the isolation of *Vibrio* spp.

 KEY POINT *Note that subcultures from a primary isolation medium must be grown in a non-selective medium, to confirm the purity of the isolate.*

Table 36.1 Selective agents in bacteriological media

Substance	Selective for
Azide salts	*Enterococcus* spp.
Bile salts	Intestinal bacteria
Brilliant green	Gram-negative bacteria
Gentian violet	Gram-negative bacteria
Lauryl sulphate	Gram-negative bacteria
Methyl violet	*Vibrio* spp.
Malachite green	*Mycobacterium*
Polymyxin	*Bacillus* spp.
Sodium selenite	*Salmonella* spp.
Sodium chloride	Halotolerant bacteria
	Staphylococcus aureus
Sodium tetrathionate	*Salmonella* spp.
Trypan blue	*Streptococcus* spp.
Tergitol/surfactant	Intestinal bacteria

Many of the selective and enrichment media used in bacteriology are able to distinguish between different types of bacteria: such media are termed differential media or diagnostic media and they are often used in the preliminary stages of an identification procedure. Box 36.1 gives details of the constituents of MacConkey medium, a selective, differential medium used in clinical microbiology (e.g. for the isolation of certain faecal bacteria), while Table 36.1 gives details of selective agents used. Further details on methods can be found in Collins et al. (2004). Note that isolation procedures for a particular microbe often combine several of the techniques described above. For instance, a protocol for isolating food-poisoning bacteria from a foodstuff might involve:

1. homogenisation of a known amount of sample in a suitable diluent;
2. serial decimal dilution;
3. separation procedures using spread or pour plates to quantify the number of bacteria of a particular type present in the foodstuff and provide a viable count (p. 278);
4. selective/enrichment procedures, e.g. specific media/temperatures/atmospheric conditions, depending on the bacteria to be isolated;
5. confirmation of identity: any organism growing on a primary isolation medium would require subculture and further tests, to confirm the preliminary identification (Chapter 39).

Text references

Collins, C.H., Lyne, P.M. and Grange, J.M. (2004) *Microbiological Methods*, 8th edn. Hodder-Arnold, London.

Sources for further study

Atlas, R.M. (2004) *Handbook of Microbiological Media*, 3rd edn. CRC Press, Boca Raton, Florida.
[Gives details of culture media for a broad range of microbes.]

Eaton, A.D., Clesceri, L.S., Rice, E.W. and Greenberg, A.E. (2005) *Standard Methods for Examination of Water and Wastewater*, 21st edn. American Public Health Association, Washington, DC.
[Gives standard US protocols for a range of indicator bacteria.]

Hurst, C.J., Crawford, R.L. and Knudsen, G.R. (2002) *Manual of Environmental Microbiology*, 2nd edn. American Society for Microbiology, Washington, DC.

Levett, P.N. (ed.) (1991) *Anaerobic Microbiology: A Practical Approach*. IRL Press, Oxford.
[Provides details of the methods used to isolate and culture anaerobic micro-organisms.]

Pepper, I.L. and Gerba, C.P. (2005) *Environmental Microbiology: a Laboratory Manual,* 2nd edn. Academic Press, New York.

Rochelle, P.A. (2001) *Environmental Molecular Microbiology: Protocols and Applications*. Horizon Scientific Press, Norwich.
[Covers molecular approaches to the detection and identification of microbes in environmental samples.]

Study exercises

36.1 Plan a collection/sampling strategy for a target group of microbes. What approaches might you take in the following instances?
(a) Collecting representatives of the normal skin microflora.
(b) Sampling psychrotrophic anaerobes from the subsurface mud of an estuary.
(c) Collecting a sample of sea water to enumerate faecal indicator bacteria.
(d) Mapping the microflora on the surface of a leaf.

36.2 Decide on an appropriate isolation procedure for a particular microbe. How might you isolate the following microbes from a sample?
(a) Photosynthetic flagellate algae in a sample of pond water.
(b) Bacteria growing as a biofilm on the surface of sand particles.
(c) Faecal streptococci (enterococci) present at low density (<1 per ml) in a sample of river water.
(d) A distinctively shaped bacterium present at very low numbers in a sample containing a high number of other microbes of different shape but with similar nutritional requirements.

36.3 Investigate the selective basis of microbiological media. Using textbooks on bacteriological methods or the Web, research how each of the following media operates, in terms of their selective and diagnostic (differential) features:
(a) Mannitol salt agar for *Staphylococcus aureus.*
(b) Membrane lauryl sulphate broth for coliforms and *E. coli.*
(c) Slanetz and Bartley medium for faecal streptococci (enterococci).
(d) Mannitol pyruvate egg yolk polymyxin (MPYP) medium for *Bacillus cereus.*

Systematics – the study of the diversity of living organisms and of the evolutionary relationships between them.

Classification (taxonomy) – the study of the theory and methods of organisation of taxa and, therefore, a part of systematics.

Taxon (plural taxa) – an assemblage of organisms sharing some basic features.

Nomenclature – the allocation of names to taxa.

Identification – the placing of organisms into taxa (see Chapters 38 and 39).

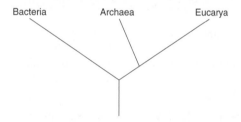

Bacteria   Archaea   Eucarya

Fig. 37.1 Simplified diagram of the three major domains of the Tree of Life (based on the rRNA sequencing work of Carl Woese).

Saprotroph – a heterotrophic organism that feeds on dead organic matter. (Other nutritional terms are defined on p. 271).

The use of scientific names is fundamental to all aspects of biological science since it aims to provide a system of identification that is precise, fixed and of universal application. Without such a system, comparative studies would be impossible.

There are two possible bases for such classification:

- Phenetic taxonomy, which involves grouping on the basis of phenotypic similarity, frequently using complex statistical techniques to obtain objective measures of similarity. The characters used have been largely morphological and anatomical, but biochemical, cytological and other characters are increasingly used, especially for microbes (p. 219) where structural characters are few.
- Phylogenetic (=phyletic) taxonomy, which involves grouping on the basis of presumed evolutionary, and therefore genetic, relationships.

These two systems are often broadly similar in outcome, since closely related organisms are usually fairly similar to each other and because judgements of evolutionary relationships are usually themselves based upon similarities. The situation is made more complex by phenomena such as convergent and divergent evolution; phyletic classifications are also liable to subjective bias.

New classifications have been proposed on the basis of studies of biomolecules, e.g. rRNA sequences, that are robust and objective. One such arrangement divides organisms into three major domains: Bacteria, Archaea and Eucarya (Fig. 37.1).

The hierarchical system

Another approach is to recognise six kingdoms if the viruses are included:

- Virales: the viruses. Very simple in structure, obligate parasites of prokaryotic or eukaryotic cells.
- Prokaryotae (Monera): the bacteria and cyanobacteria. Prokaryotic, non-nucleate and lacking membrane-bound organelles, such as: mitochondria, plastids, endoplasmic reticulum, Golgi apparatus.
- Protista: includes algae, protozoa and slime moulds. Eukaryotic and mainly unicellular. A heterogeneous group.
- Plantae: the plants. Eukaryotic, walled, mainly multicellular and usually photoautotrophs.
- Fungi: eukaryotic, usually syncytial, walled and saprotrophic.
- Animalia: the animals. Eukaryotic, multicellular and heterotrophs.

Six other levels of taxa are generally accepted: phylum, class, order, family, genus and species, although in botany and microbiology, division is used instead of phylum.

 KEY POINT *Whereas the levels of taxa above genus are rather subjective and vary among authorities, the use of genus and species names are governed by strict, internationally agreed conventions called Codes of Nomenclature.*

There are three major Codes of Nomenclature, the Botanical, Zoological and Bacteriological Codes, which operate on similar but not identical principles (Lapage et al., 1992; Wakeham-Dawson et al., 1999; Greuter et al., 2000). Fungi are covered by the International Code of Botanical Nomenclature despite the fact that they are not plants.

The basis of classification

No simple definition of a species is possible, but there are two generally used definitions:

- A group of organisms capable of interbreeding and producing fertile offspring – this, however, excludes all asexual, parthenogenetic and apomictic forms.
- A group of organisms showing a close similarity in phenotypic characteristics – this would include morphological, anatomical, biochemical, ecological and life history characters.

 KEY POINT *The basic unit of classification is the species, which represents a group of recognizably similar individuals, clearly distinct from other such groups.*

When species are compared, groups of species may show a number of features in common; they are then arranged into larger groupings known as genera (singular genus). This process can be repeated at each taxonomic stage to form a hierarchical system of classification whose different levels are known as taxonomic ranks. The number of levels in this system is arbitrary and based upon practical experience – the seven levels normally used have been found to be sufficient to accommodate the majority of the variation observed in nature.

When a generic or specific name is changed as a result of further study, the former name becomes a synonym; you should always try to use the latest name. Where a generic name has been changed recently, the old name is occasionally given in parentheses to allow easy reference to the old name. There are taxonomic reference works available for each discipline or subdiscipline, such as the *Flora Europea* (for plants) and the *Plymouth Marine Fauna* (for British marine animals); these often provide the current versions of a name and often its synonyms.

Nomenclature in practice

The scientific name of an organism is effectively a symbol or cipher that removes the need for repeated use of descriptions. It normally comprises two words and is, therefore, called a binomial term. The name of the genus is followed by a second term that identifies the species, e.g. *Quercus robur*, *Apis mellifera*, *Pseudomonas aeruginosa* or *Saccharomyces cerevisae* (Table 37.1).

DNA-based definition of a species – *in microbiology, members of a bacterial species can be characterised by having a DNA similarity of ≥70 per cent.*

Example When the cockle *Cardium edule* was renamed *Cerastoderma edule*, it was commonly referred to in textbooks as *Cerastoderma (Cardium) edule*: this is strictly not correct practice but can be helpful for non-specialists.

Writing taxonomic names – *always underline or italicise generic and species names to avoid confusion: thus bacillus is a descriptive term for rod-shaped bacteria, while* <u>Bacillus</u> *is a generic name.*

Table 37.1 Example of taxonomies for a plant, an animal and a bacterium

Common name	English oak	Honey-bee	Pseudomonas	Baker's yeast
Kingdom	Plantae	Animalia	Monera	Fungi
Phylum/Division	Anthophyta	Arthropoda	Gracilicutes	Ascomycota
Class	Dicotyledonae	Insecta	Scotobacteria	Incertae sedis[1]
Order	Fagales	Hymenoptera	Pseudomonadales	Saccharomycetales
Family	Fagaceae	Apidae	Pseudomonadaceae	Saccharomycetaceae
Genus	*Quercus*	*Apis*	*Pseudomonas*	*Saccharomyces*
Species	*Q. robur*	*A. mellifera*	*P. aeruginosa*	*S. cerevisiae*

[1] Meaning = 'uncertain taxonomic position'

Box 37.1   Basic rules for the writing of taxonomic names

- **Names of the seven levels of taxa should take lower-case initial letters,** e.g. class Mollusca or kingdom Fungi.
- **The Latin forms of all taxon names except the specific name take initial capital letters,** e.g. 'the Arthropoda ...' but anglicised versions do not, e.g. 'the arthropods ...'.
- **The names of the higher taxa are all plural,** hence 'the Mollusca are ...' while the singular of the anglicised version is used for a single member of that taxon, hence 'a mollusc is ...'.
- **The binomial system gives each species two terms,** the first being the generic name and the second the specific name, which must never be used by itself. The genus and species names are distinguished from the rest of the text either:
  - (a) by being underlined (when handwritten), e.g. Patella vulgata; or
  - (b) by being set in italics (in print or on a word processor), e.g. *Patella vulgata*.
- **The generic name is singular and always takes an initial capital letter.** If you use the generic name this implies that the point being made is a generic characteristic unless the specific name is present. Write the generic name in full when first used in a text, e.g. *Patella vulgata*, but subsequent references can be abbreviated to its initial letter, e.g. *P. vulgata*, unless this will result in confusion with another genus also being considered.

- **The abbreviation 'sp.' should be used in place of the specific name if a single unspecified species of a genus is being referred to,** e.g. *Patella* sp.; it is not underlined or italicised. If more than one unspecified species is meant, then the correct form is 'spp.', e.g. *Patella* spp.
- **Common names should not normally be written with a capital letter,** e.g. limpet.
- **The name of each species should be followed by the authority:** on first usage in formal reports and in titles, the name or names of the person(s) to whom that name is attributed and the date of that description should be quoted. These names may sometimes be abbreviated, e.g. L. for Linnaeus and standard abbreviations must be used. If the species was first described under its current generic name, the authority's name, often in abbreviated form is added, e.g. *Quercus robur* L. If, however, the species was first described under a different genus, the name of the author of the original description is presented in parentheses, e.g. *Escherichia coli* (Migula) Castellani and Chalmers. Note that in zoology, the date the description was published is also included, e.g. *Ischnochiton kermadecensis* Iredale, 1914. The use of authorities should be confined to formal papers, final year project reports, etc.; they would not normally be used in practical reports, short assignments or examinations.

Example   The full, formal name of baker's yeast is *Saccharomyces cerevisiae* Meyen ex Hansen 1883. After first use, this would be abbreviated to *S. cerevisiae.*

Common names are often interesting, but totally unsatisfactory for use in biological nomenclature because of the lack of consistency in their use: the Codes of Nomenclature were established to prevent the ambiguities associated with informal/common names.

The Codes require that all scientific names are either Latin or treated as Latin, written in the Latin alphabet and subject to the rules of Latin grammar. Consequently, you must be very precise in your use of such names. In some cases, the Codes stipulate a standardised ending for the names of all taxa of a given taxonomic rank, e.g. names of all animal families must end in -idae while plant, fungal and bacterial families end in -aceae. When used in a formal scientific context, you should follow the specific name by the authority on which that name is based, i.e. the name of the person describing that species and the date of the description.

Box 37.1 summarises the basic rules for writing taxonomic names.

Taxa below the rank of species

Some use is made of taxa below the rank of species. Within zoology, this is confined to the term subspecies, so the names of subspecies have three components, e.g. *Mus musculus domesticus*, no rank term being necessary.

In bacteriology, the use of subspecies is again acceptable although a word indicating rank is usually inserted, e.g. *Bacillus subtilis* subsp. *niger*. The Bacteriological Code considers ranks from the level of subspecies up to, and including, class: the use of the term variety is discouraged, as the term is synonymous with subspecies. However, other terms are in widespread use for taxa below the species level, especially in medical microbiology and plant pathology, when a particular strain of bacterium has been identified (pp. 231–2). Subspecies identification is often referred to as typing and the following terms apply:

- biovar, or biotype: subdivided according to biochemical characteristics;
- serovar or serotype: subdivided by serological methods, using antibodies (see Chapter 51);
- pathovar: subdivided according to pathogenicity (ability to cause disease);
- phagovar or phage type: subdivided according to susceptibility to particular viruses.

Many micro-organisms are now referred to by their generic and specific names followed by a culture collection reference number, e.g. *Bacillus subtilis* NCTC 10400, where NCTC stands for the National Collection of Type Cultures and 10400 is the reference number of that strain in the collection.

In botany, several categories below the rank of species are recognised and a term of rank is used before the name, e.g. *Salix repens* var. *fusca*: the term var. is short for the Latin word *varietas* and is subordinate to the term subspecies in the Botanical Code. The term cultivar (cv.) is an important modern term frequently used in experimental work and refers to cultivated varieties of plants.

The special case of viruses

The classification and nomenclature of viruses is less advanced than for cellular organisms and the current nomenclature has been arrived at on a piecemeal, ad hoc basis. The International Committee for Virus Taxonomy proposed a unified classification system, dividing viruses into 50 families on the basis of:

- host preference;
- nucleic acid type (i.e. DNA or RNA);
- whether the nucleic acid is single- or double-stranded;
- the presence or absence of a surrounding envelope.

Virus family names end in -viridae and genus names in -virus. (Note that these names are *not* latinised and the genus–species binomial is not now approved). However, this system has not yet been adopted universally and many viruses are still referred to by their trivial names or by code names (sigla), e.g. the bacterial viruses ϕX174, T4, etc. Many of the names used reflect the diseases caused by the virus. Often, a three-letter abbreviation is used, e.g. HIV (for human immunodeficiency virus), TMV (for tobacco mosaic virus).

Text references

Greuter, W., et al. (eds) (2000) *International Code of Botanical Nomenclature (St Louis Code)*. Koeltz Scientific Books, Königstein.

Lapage, et al. (eds) (1992) *International Code of Nomenclature of Bacteria (1990 revision)*. American Society for Microbiology, Washington, DC.

Wakeham-Dawson, A., et al. (eds) (1999) *International Code of Zoological Nomenclature*, 4th edn. The International Trust for Zoological Nomenclature, London.

Sources for further study

Anon. *International Code of Botanical Nomenclature (St Louis Code)*. Available: http://www.bgbm.fu-berlin.de/iapt/nomenclature/code/SaintLouis/0000St.Luistitle.htm Last accessed 09/04/07.
[Web version of Greuter et al. (2000).]

Boone, D.R. *Naming a New Procaryotic Taxon.* Available: http://methanogens.pdx.edu/naming.html Last accessed 09/04/07.
[Succinct indication of procedures for naming a new prokaryote species.]

Buchen-Osmond, C. (2002) *Universal Virus Database of the International Committee on Taxonomy of Viruses.* Available: http://www.ncbi.nlm.nih.gov/ICTVdb/ Last accessed 09/04/07.

[Includes a catalogue of approved virus names, image gallery, descriptions, etc.]

Fauquet, C.M., Mayo, M.A., Maniloff, J., Desselberger, U. and Ball, L.A. (2005) *Virus Taxonomy. Eighth Report of the International Committee on the Taxonomy of Viruses.* Elsevier, London.

Institute of Biology (2000) *Biological Nomenclature: Recommendations on Terms, Units and Symbols.* Institute of Biology, London.
[Includes sections on taxonomy and classification of organisms.]

Jeffrey, C. (1989) *Biological Nomenclature*, 3rd edn. Edward Arnold, London.

Study exercises

37.1 Research full classifications. Provide the full classification of the following species, laid out as in the table in Study exercise 37.2.
  (a) The limpet *Patella vulgata*
  (b) The Great White Shark
  (c) The Giant Redwood
  (d) The earthworm *Lumbricus terrestris*.

37.2 What is wrong with the hierarchical classifications given in the table below?

Examples of hierarchical classifications

Nautilus	Edible mushroom	E. coli bacterium
Animalia	Fungi	Monera (Bacteria)
Mollusca	Basidiomycota	Proteobacteria
Cephalopoda	Basidiomycetes	Enterobacteriales
Nautilidae	Agaricales	Gamma subdivision
Nautilida	Agaricaceae	Enterobacteriaceae
Nautilus pompilius	*Agaricus bisporus*	*Escherichia coli*

37.3 Compare a variety of current textbooks and Internet sites to discover alternative classification schemes at kingdom level. Make your own notes regarding the *evidence* used to support the alternative kingdom classifications.

Fig. 38.1 Research scientists examining a rocky shore habitat at Orkney, Scotland, and making counts of the different species present, using quadrats. This image illustrates the problems facing a field biologist in identifying plants and animals. What are the species present here? How would you start to identify a specimen, say a mollusc or an alga, from this environment, if its identity were unknown to you? In the absence of specialised knowledge, the task is made far easier by using a flora or fauna with a key that will guide you to descriptions of candidate species.

> *Out-of-date guides* – taxonomists frequently change the names of taxa and update their classification. An up-to-date identification requires an up-to-date guide.

> *Observing rare specimens* – take special care not to collect, disturb or destroy rare plants and animals in the course of your observations.

The normal way to identify plants or animals is to use identification guides. These consist of two parts:

1. Written and pictorial descriptions of organisms, which you compare with your unknown specimen to aid in its identification. Good descriptions direct you to the crucial diagnostic features for the relevant taxon, explain the range of variability found and point out biological and ecological characteristics of importance.
2. Keys, which help you find the likely description for your specimen rapidly and simply. Most keys are arranged to present you with a series of choices, usually dichotomous (dividing in two). The paired statements of each 'couplet' are framed to be contrasting and mutually exclusive. Each choice you make narrows down the possibilities for your specimen until you find the appropriate description.

The authors of identification guides assume that you have a live or preserved specimen to hand and the means to observe it closely and measure it (see Fig. 38.1). The terminology in guides is designed to combine precision with brevity. Guides for animal species are called faunas and those for plant species are called floras.

KEY POINT *To use an identification guide properly, you need to know enough of the vocabulary to understand the choices presented to you, but all identification guides provide both a glossary and a list of abbreviations to help with this.*

The best identification guides are those that lead you in the simplest way to a correct identification. If you need to choose one for your area of interest, think about the following questions:

- What degree of prior knowledge is assumed? Some guides are written for novices, whereas others assume an expert's command of terminology. If tempted to go for the former type, consider whether it will always be suitable for your needs.
- What is the scope of the guide? Guides may be restricted in the taxa they consider or in the geographical region that is covered; this will suit you if your interests are similarly narrow. However, if your interests are wide, the relevant guide may be so large as to be unwieldy in the field.
- How well is the guide illustrated? Good-quality illustrations enhance the ease of use of a guide – features can be shown pictorially that might involve an off-putting specialised vocabulary to describe. Accurately coloured illustrations can be helpful, but note that colour can be a variable character: look for good line diagrams that highlight the critical diagnostic points.
- Is the guide divided into parts? Good guides are arranged in short parts dealing with different levels of the taxonomic hierarchy. This speeds up identification by allowing you to skip initial material when you have a fair idea of the specimen's identity.
- Do you like the style of the key? As discussed below, there are several ways in which a key can be presented, one of which may suit you more than the others. If you can't actually test out a key yourself on

real specimens, the next best thing is to ask for the opinion of someone who has used it.

Types of key

Bracketed keys

Here, numbered pairs of adjacent lines in the key present you with a choice and either 'send' you to a new couplet or provide the tentative identification (Box 38.1).

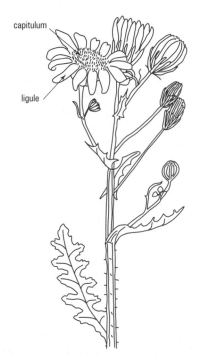

Fig. 38.2 Ragwort flower (*Senecio* spp.). Adapted from figure from *New Flora of the British Isles*, 2nd edition, Cambridge University Press, (Stace, C.A., 1997).

Box 38.1   Example of a bracketed key

Part of key to ragworts (Fig. 38.2), modified after Stace (1997):

1. Ligules <8 mm or 0; capitula cylindrical, about 2× as long as wide 2
   Ligules >8 mm; capitula bell-shaped in flower, about 1.5× as long as wide. 3
2. Ligules usually 0; achenes ⩽ 2.5 mm   *Senecio vulgaris*
   Ligules usually present; achenes <3 mm   *Senecio cambrensis*

If there are no ligules on your specimen or they are less than 8 mm in length, you should proceed to choice 2, but if they are present and greater than 8 mm in length, you should proceed to choice 3.

In this case, the choice at 2 is sufficient to pin down the species; sometimes quite early in a key a distinctive characteristic may allow the specimen to be 'identified' to species level, while for the other options the specimen's identity remains open. Note the use of more than one comparison in each couplet to provide confirmation.

Indented keys

In this method, the pairs of choices are indented and given the same number. They are separated by other choices further down the sequence. Having made a choice, you look at the next couplet below which will be one indent level further in. When a choice is sufficiently distinctive, the tentative identity of your specimen will be given (Box 38.2).

Box 38.2   Example of an indented key

Part of key for common species of true bumblebee (Fig. 38.3), modified after Prys-Jones and Corbet (1987):

1. Thorax with black area(s)
   2. Thorax all black
      3. Pollen baskets with red hairs   *Bombus rudarius*
      3. Pollen baskets with black hairs   *Bombus lapidarius*
   2. Thorax black with yellow or brown patches
      4. Tail white, buff or brown
         5. Scutellum black   *Bombus terrestris*
         5. Scutellum yellow   *Bombus hortorum*
      4. Tail red or orange   *Bombus pratorum*
1. Thorax without any black   *Bombus pascuorum*

If your bumblebee has a black thorax with yellow patches, proceed to choice 4; if its tail is brown, carry on to choice 5, etc.

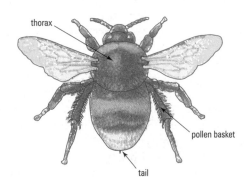

Fig. 38.3 Bumblebee (*Bombus* spp., ♀)

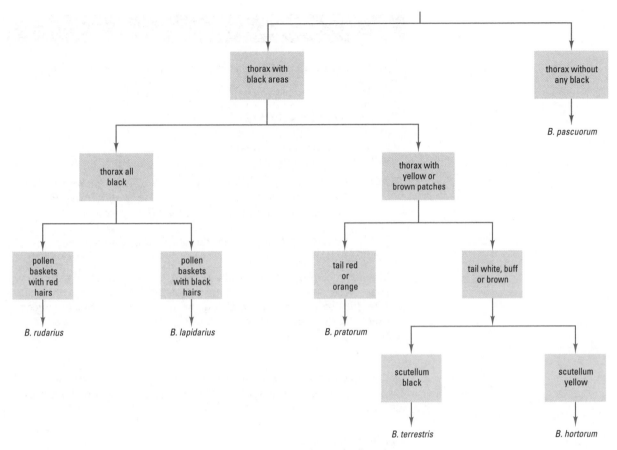

Fig. 38.4 A key for bumblebees (*Bombus* spp.) laid out in the form of a flowchart.

 KEY POINT *Bracketed keys have the advantage that they keep the couplets close together for ready comparison, but indented keys show the relationships between taxa more clearly and allow you to backtrack more easily if an error has been made.*

Flowchart keys

In this form of key, the choices are laid out in the form of a flowchart (Fig. 38.4), which allows easier cross-checking of options but is only feasible where there are a small number of choices. To use this type of key, follow the arrows after making each choice in the sequence; this will lead you on to another choice and eventually to the tentative identification.

Multiaccess keys

These allow you to choose the characters used in the key according to the state of your specimen (Box 38.3). They are useful in situations where:

- important characters are difficult to observe;
- characters are likely to be misinterpreted;
- a single character would be unreliable in isolation;
- a part is missing or seems abnormal.

Another way of presenting a multi-access key is in the form of a table. For instance, the taxa to be distinguished could make up the rows

Box 38.3 Example of a multi-access key

Part of key to species of willowherbs, modified after Stace (1997):

Stigma 4-lobed	A
Stigma club-shaped	B
Seeds minutely uniformly papillose	C
Seeds with longitudinal papillose ridges	D
Stems erect or erect at apex	E
Stems trailing on ground	F

ACE	Petals 10–16 mm, purplish pink	*Epilobium hirsutum*
	Petals 5–9 mm, paler	*Epilobium parviflorum*
BCE	Petioles 4–15 mm, plant perennating by rosettes	
		Epilobium roseum
	Petioles $\leqslant$ 4 mm, plant perennating by stolons ending in tight bud	*Epilobium palustre*

If your specimen had a 4-lobed stigma and erect stems, but no seeds were available to examine, you could 'identify' it as either *E. hirsutum* or *E. parviflorum* and distinguish between these choices on the basis of petal size and colour.

of the table and the relevant characteristics the columns (see Table 39.1). Like the flowchart, this type of key is limited to a small number of choices.

Computerised keys

These simplify the initial stages of identification by providing a series of menu-like choices for the user. They can rapidly provide a list of tentative identifications on even a few positive choices from these menus, ranking these in likelihood of being correct. You can then work down the list, comparing your specimen to a description of the proposed species (see below). At present, computerised keys are more likely to be used in a laboratory or field station than in the field.

Advice for using keys

- Note down the route taken (i.e. the numbering system for the decision tree): this makes it easier to trace back your path through the key.
- At each step, read the full description for both choices before arriving at a decision about which one to take.
- Never guess if you do not know the precise meaning of the terms used – consult the key's glossary and list of abbreviations. Where measurements are required, use a ruler – do not guess sizes.
- If features are very small, use an appropriate lens to inspect them clearly.
- If the key is a multipart one, look carefully at the descriptions for higher levels of taxa before progressing to the species key: this not only acts as a check that you are correct up to this stage, but may also provide definitions of useful terminology.
- If both of a pair of choices seem reasonable, try out each route – one will usually prove to be unsuitable at a later stage.

Problems of identification – *sometimes your best attempts to identify the specimen will be confronted by the existence of sexual dimorphism or polymorphism, juvenile and adult phases (e.g. gametophyte and sporophyte phases for certain plants), local forms, non-native taxa, etc. A good guide will point out these problems where they occur.*

KEY POINT *When you arrive at the end of a key's path,* do not simply accept this as a reliable identification of your specimen. *Compare your specimen with the full description of the species.*

Comparing specimens with descriptions

If the specimen doesn't fit the description properly, follow the instructions outlined below:

- Compare the specimen with neighbouring descriptions: in a well-organised guide, those of similar species will be together.
- Go back along the path of the key and re-examine each decision you have made. Try going down the alternative route for any that might have been questionable.
- Check to see whether you inadvertently went down the wrong pathway even though you made the correct diagnosis.
- Bear in mind the possibility that your specimen is not typical. A good key will use characteristics that are constant, but biological variation will often throw up an oddity to confuse you. Try to obtain another specimen, preferably not genetically related to the original.
- Consider the possibility that it could be outside its normal geographical range or even new to science.

The ultimate check on an identification is a comparison of the specimen with an authentically named specimen in a museum or herbarium. The ultimate comparison would be with the type specimen, the specimen used when the species or subspecies was first described and named. If this is the only specimen collected by the author(s) who named the species, it is called a holotype. Other 'type' specimens include:

- paratypes – those other than the type specimen also used by the author(s) at the time of the original description;
- syntypes – a collection of specimens used for the original collection, but from which no one specimen was defined as the type specimen;
- lectotype – a particular syntype subsequently chosen and designated through publication to act as the type specimen.

Text references

Prys-Jones, O.E. and Corbet, S.A. (1987) *Naturalist's Handbook 6. Bumblebees.* Cambridge University Press, Cambridge.

Stace, C.A. (1997) *New Flora of the British Isles,* 2nd edn. Cambridge University Press, Cambridge.

Sources for further study

Schmidt, D. *Flora and Fauna Field Guides.* Available: http://gateway.library.uiuc.edu/bix/fieldguides/flora.htm Last accessed 09/04/07.
[Listing of floras and faunas across the world; also classified by type of organism, e.g. mammals of Central and South America.]

Thompson Scientific *BiologyBrowser.* Available: http://www.biologybrowser.org/

Last accessed 09/04/07.
[Includes a search engine for information about specific organisms.]

University of Illinois *Flora and Fauna Field Guides.* Available: http://www.library.uiuc.edu/bix/fieldguides/flora.htm
Last accessed 09/04/07.
[A listing of guides by geographical region.]

Study exercises

38.1 Devise a key. The table below gives details of some invertebrate phyla from freshwater habitats, together with illustrations of 'representative' animals. The taxa are presented in no particular order. Use this information to construct a dichotomous key to help with preliminary identification of specimens. Ask a colleague to test your key by correctly assigning the animals to phyla from their pictures alone. Refine the key if problems are encountered during testing.

Members of some invertebrate phyla

Taxonomic division and description	Representative animal
Phylum Mollusca, Class Bivalvia *Soft bodied with hard shell in two parts*	 Zebra mussel, *Dreissena polymorpha*
Phylum Arthropoda Sub-phylum Arachnida *Hard bodied with four pairs of jointed legs*	 Water spider, *Argyroneta aquatica*
Phylum Mollusca, Class Gastropoda *Soft bodied with hard shell in one part*	River snail, *Viviparus viviparus*
Phylum Platyhelminthes *Soft bodied, free moving, flattened and unsegmented, without legs*	Flatworm, *Dugesia lugubris*
Phylum Arthropoda Sub-phylum Insecta *Hard bodied with three pairs of jointed legs*	 Great diving beetle, *Dytiscus marginalis*

(continued)

Study exercises (continued)

Taxonomic division and description	Representative animal
Phylum Annelida **Class Oligochaetae** *Soft bodied, free moving, divided into 15 or more segments (without head capsule or leg-like appendages)*	 Water worms, various sp.
Phylum Nematoda *Soft bodied, free-moving, thread-like and unsegmented, without legs*	 Nematodes, various sp.
Phylum Arthropoda **Sub-phylum Crustacea** *Hard bodied, with four or more pairs of jointed legs*	 Freshwater shrimp, *Crangonyx pseudogracilis*
Phylum Coelenterata **Class Hydrozoa** *Soft bodied, surface-attached, unsegmented and tube-like, without legs*	e.g. *Hydra* sp.

38.2 Use a standard flora or fauna. Visit your university library and take out one of the standard keys for the local flora or fauna. Find a specimen plant or animal whose identity you do not know (take care not to kill or destroy any protected specimens) and use the key and its descriptions to identify it as best you can. Check your identification with a tutor.

38.3 Search for Web-based identification keys and test them. Hypertext (p. 58) lends itself well to dichotomous or multiple choices, taking you to new pages via links based on appropriate options. Search for a Web-based identification key using a search engine and try it out.

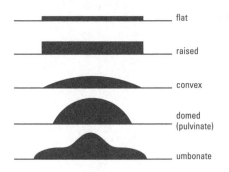

Fig. 39.1 Colony elevations (cross-sectional profile)

flat

raised

convex

domed (pulvinate)

umbonate

Most of the methods described in this chapter were developed for the identification of bacteria, and bacterial examples are used to illustrate the principles involved. While the basic techniques are equally applicable to other types of microbe, the identification systems for some protozoa, fungi and algae rely predominantly on microscopic appearance. Identification of viruses requires electron microscopy or immunological techniques.

 KEY POINT *Identification of bacteria is often based on a combination of a number of different features, including growth characteristics, microscopic examination, physiological or biochemical characterisation, and, where necessary, immunological tests.*

Direct observation

Once a microbe has been isolated (Chapter 36) and cultured in the laboratory (Chapter 46), the visual appearance of individual colonies on the surface of a solidified medium may provide useful information. Bacteria typically produce smooth, glistening colonies, varying in diameter from <1 mm to >1 cm. Actinomycete colonies are often <1 cm, with a shrivelled, powdery surface. Filamentous fungi usually grow as large, spreading mycelia with a matt appearance and are identified by microscopy, using the morphological characteristics of their reproductive structures. Yeasts produce small, glistening colonies; identification usually involves microscopy, combined with physiological and biochemical tests similar to those used for bacteria.

Colony characteristics

When measuring colony size, choose a typical colony, well spaced from any others, as colony size is affected by competition for nutrients. The characteristics of a microbial colony on a particular medium include:

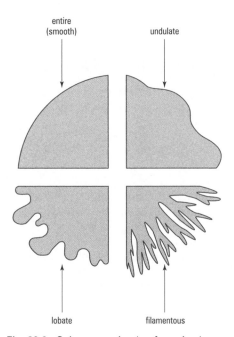

entire (smooth)

undulate

lobate

filamentous

Fig. 39.2 Colony margins (surface view)

- Size: some bacteria produce punctiform colonies, with a diameter of less than 1 mm, while motile bacteria may spread over the entire plate.
- Form: colonies may be circular, irregular, lenticular (spindle-shaped) or filamentous.
- Elevation: colonies may be flat, raised, convex, etc. (see Fig. 39.1).
- Margin: the edge of a colony may be entire (smooth) or more distinctive, e.g. undulate, lobate or filamentous (Fig. 39.2).
- Consistency: colonies may be viscous (or mucoid), butyrous (of similar consistency to butter) or friable (dry and granular), etc.
- Colour: some bacteria produce characteristic pigments. A few pigments are fluorescent under UV light.
- Optical properties: colonies may be translucent or opaque.
- Haemolytic reactions on blood agar: many pathogenic bacteria produce characteristic zones of haemolysis. Alpha haemolysis is a partial breakdown of the haemoglobin from the erythrocytes, producing a green zone around the colony, while beta haemolysis is the complete destruction of haemoglobin, producing a clear zone.

⚠ **SAFETY NOTE** *Never attempt to smell mould cultures, because of the risk of inhaling large numbers of spores.*

- Odour: some actinomycetes and cyanobacteria produce earthy odours, while certain bacteria and yeasts produce fruity or 'off' odours. However, odour is not a reliable characteristic in bacterial identification.

Microscopic examination

Bacteria are usually observed using an oil immersion objective at a total magnification of $\times 1000$ (p. 255).

Motility

Wet mounts can be prepared by placing a small drop of bacterial suspension on a clean, degreased slide, adding a coverslip and examining the film without delay. For aerobes, areas near air bubbles or by the edge of the coverslip give best results, whereas anaerobes show greatest motility in the centre of the preparation, with rapid loss of motility due to oxygen toxicity.

Prepare wet mounts using young cultures in exponential growth in a liquid medium (p. 272). It is best to work with cultures grown at 20 or 25 °C, since those grown at 37 °C may not be actively motile on cooling to room temperature. It is essential to distinguish between the following:

- True motility, due to the presence of flagella: bacteria dart around the field of view, changing direction in zigzag, tumbling movements.
- Brownian motion: non-motile bacteria show a localised, vibratory, random motion, due to bombardment of bacterial cells by molecules in the solution.
- Passive motion, due to currents within the suspension: all cells will be swept in the same direction at a similar rate of movement.
- Gliding motility: a slower, intermittent movement, parallel to the longitudinal axis of the cell, requiring contact with a solid surface.

Cell shape

Bacteria are subdivided into the following groups:

- Cocci (singular, coccus): spherical, or almost spherical, cells, sometimes growing in pairs (diplococci), chains or clumps.
- Rods: straight, cylindrical cells of variable length, with flattened, tapered or rounded ends; sometimes termed bacilli. Short rods are sometimes called cocco-bacilli.
- Curved rods: the curvature varies according to the organism, from short curved rods, sometimes tapered at one end, to spiral shapes.
- Branched filaments: characteristic of actinomycete bacteria.

Gram staining

This is the most important differential staining technique in bacteriology (Box 39.1 gives details). It enables us to divide bacteria into two distinct groups, Gram-positive and Gram-negative, according to a particular staining procedure (the technique is given a capital letter, since it is named after its originator, H.C. Gram). The basis of the staining reaction is the different structure of the cell walls of Gram-positive and Gram-negative bacteria. Heat fixation of air-dried bacteria causes some shrinkage, but cells retain their shape: to measure cell dimensions use a chemical fixative (p. 246).

Gram staining should be carried out using light smears of young, active cultures, since older cultures may give variable results. In particular, certain Gram-positive bacteria may stain Gram-negative if older cultures are used.

Using the hanging drop technique – place a drop of bacterial suspension on a coverslip and invert over a cavity slide so that the drop does not make contact with the slide: motile aerobes are best observed at the edge of the droplet, where oxygen is most abundant.

Assessing motility – if you have not seen bacterial motility before, it is worth comparing your unknown bacterium with a positive and a negative control.

Using cell shape in microbial identification – many bacteria are pleomorphic, varying in size and shape according to the growth conditions and the age of the culture: thus, other characteristics are required for identification.

 SAFETY NOTE The Gram-staining procedure involves toxic dyes and flammable solvents: avoid skin contact and extinguish any naked flames (e.g. Bunsens).

Assessing the Gram status of an unknown bacterium – if a pure culture gives both Gram-positive and Gram-negative cells, identical in size and shape, it can be regarded as a Gram-positive organism that is demonstrating Gram-variability.

Box 39.1   Preparation of a heat-fixed, Gram-stained smear

Preparation of a heat-fixed smear

The following procedure will provide you with a thin film of bacteria on a microscope slide, for staining.

1. **Take a clean microscope slide and pass it through a Bunsen flame twice**, to ensure it is free of grease. Allow to cool.
2. **Using a sterile inoculating loop, place a single drop of water in the centre of the slide and then mix in a small amount of sample** from a single bacterial colony with the drop, until the suspension is slightly turbid. Smear the suspension over the central area of the slide, to form a thin film. For liquid cultures, use a single drop of culture fluid, spread in a similar manner.
3. **Allow to air-dry at room temperature**, or high above a Bunsen flame: air-drying must proceed gently, or the cells will shrink and become distorted.
4. **Fix the air-dried film by passage through a Bunsen flame.** Using a slide holder or forceps, pass the slide, film side up, rapidly through the hottest part of the flame (just above the blue cone). The temperature of the slide should be just too hot for comfort on the back of your hand: note that you must not overheat the slide or you may burn yourself (you will also ruin the preparation).
5. **Allow to cool**: the smear is now ready for staining.

Gram-staining procedure

The version given here is a modification of the Hucker method, since acetone is used to decolorise the smear. Note that some of the staining solutions used are flammable, especially the acetone decolorising solvent: you must make sure that all Bunsens are turned off during staining. The procedure should be carried out with the slides suspended over a sink, using a staining rack.

1. **Flood a heat-fixed smear with 2% w/v crystal violet in 20% v/v ethanol : water** and leave for 1 min.
2. **Pour off the crystal violet and rinse briefly with tap water. Flood with Gram's iodine** (2 g KI and 1 g I_2 in 300 ml water) for 1 min.
3. **Rinse briefly with tap water** and leave the tap running gently.
4. **Tilt the slide and decolorise with acetone** for 2–3 s: acetone should be added dropwise to the slide until no colour appears in the effluent. This step is critical, since acetone is a powerful decolorising solvent and must not be left in contact with the slide for too long.
5. **Immediately immerse the smear in a gentle stream of tap water**, to remove the acetone.
6. **Pour off the water and counterstain for 10–15 s using 2.5% w/v safranin** in 95% v/v ethanol : water.
7. **Pour off the counterstain, rinse briefly with tap water, then dry the smear** by blotting gently with absorbent paper: all traces of water must be removed before the stained smear is examined microscopically.
8. **Place a small drop of immersion oil on the stained smear: examine directly** (without a coverslip) using an oil-immersion objective (p. 255).

Gram-positive bacteria retain the crystal violet (primary stain) and appear purple while Gram-negative bacteria are decolorised by acetone and counterstained by the safranin, appearing pink or red when viewed microscopically.

Other decolorising solvents are sometimes used, including ethanol : water, ethyl ether : acetone and acetone : alcohol mixtures. The time of decolorisation must be adjusted, depending upon the strength of the solvents used, e.g. 95% v/v ethanol : water is less powerful than acetone, requiring around 30 s to decolorise a smear.

This Gram-variability is due to autolytic changes in the cell wall of Gram-positive bacteria. Developing spores are often visible as unstained areas within older vegetative cells of *Bacillus* and *Clostridium*. Other stains are required to demonstrate spores, capsules or flagella (p. 249).

Basic laboratory tests

At least two simple biochemical tests are usually performed:

Oxidase test

This identifies cytochrome *c* oxidase, an enzyme found in obligate aerobic bacteria. Soak a small piece of filter paper in a fresh solution of 1% (w/v) N-N-N′-N′-tetramethyl-p-phenylenediamine dihydrochloride on a clean microscope slide. Rub a small amount from the surface of a young, active

Performing the oxidase test – never use a nichrome wire loop, as this will react with the oxidase reagent, giving a false positive result.

colony on to the filter paper using a glass rod, a *plastic* loop or a wooden applicator stick: a purple-blue colour within 10 s is a positive result.

Catalase test

This identifies catalase, an enzyme found in obligate aerobes and in most facultative anaerobes, which catalyses the breakdown of hydrogen peroxide into water and oxygen ($2 H_2O_2 \rightarrow 2H_2O + O_2$). Transfer a small sample of your unknown bacterium on to a coverslip using a disposable plastic loop or glass rod. Invert on to a drop of hydrogen peroxide: the appearance of bubbles within 30 s is a positive reaction. This method minimises the dangers from aerosols formed when gas bubbles burst.

The oxidase and catalase tests effectively allow us to subdivide bacteria on the basis of their oxygen requirements, without using agar shake cultures (p. 213) and overnight incubation, since, for the most part:

- obligate aerobes will be oxidase and catalase positive;
- facultative anaerobes will be oxidase negative and catalase positive;
- microaerophilic bacteria, aerotolerant anaerobes and strict (obligate) anaerobes will be oxidase and catalase negative – the latter group will grow only under anaerobic conditions (p. 213).

Once you have reached this stage (colony characteristics, motility, shape, Gram reaction, oxidase and catalase status) it may be possible to make a tentative identification, at least for certain Gram-positive bacteria, at the generic level. To identify Gram-negative bacteria, particularly the oxidase-negative, catalase-positive rods, further tests are required.

Identification tables: further laboratory tests

Bacteria are asexual organisms and strains of the same species may give different results for individual biochemical/physiological tests. This variation is allowed for in identification tables (multi-access keys, p. 223), based on the results of a large number of tests. Identification tables are often used for particular subgroups of bacteria, after Gram staining and basic laboratory tests have been performed: an example is shown in Table 39.1.

SAFETY NOTE The catalase and oxidase reagents are irritants and could be harmful if swallowed. Avoid skin contact and ingestion.

Avoiding false negatives – ensure you use sufficient material during oxidase and catalase testing, otherwise you may obtain a false negative result: a clearly visible 'clump' of bacteria should be used.

Carbohydrate utilisation tests and isolation media – many diagnostic agar-based media incorporate one or more specific carbohydrates and pH indicator dyes, thereby providing additional information as part of the isolation procedure (p. 232).

Table 39.1   Identification table for selected Gram-negative rods

Bacterium	Biochemical test								
	1	2	3	4	5	6	7	8	9
Escherichia coli	v	+	−	+	−	v	v	+	−
Proteus mirabilis	−	−	v	−	+	−	+	−	+
Morganella morganii	−	−	−	−	+	−	+	+	−
Vibrio parahaemolyticus	−	+	v	−	−	+	+	+	−
Salmonella spp.	−	+	v	−	−	+	+	−	+

Key to biochemical tests and symbols:
1. sucrose utilisation
2. mannitol utilisation
3. citrate utilisation
4. β-galactosidase activity
5. urease activity
6. lysine decarboxylase activity
7. ornithine decarboxylase activity
8. indole production
9. H_2S production
+, >90% of strains tested positive
−, <10% of strains tested positive
v, 10–90% of strains tested positive

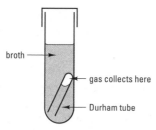

Fig. 39.3 Durham tube in carbohydrate utilisation broth. Air within the Durham tube is replaced by broth during the autoclaving procedure.

A large number of specific biochemical and physiological tests are used in bacterial identification including:

- Carbohydrate utilisation tests. Some bacteria can use a particular carbohydrate as a carbon and energy source. Acidic end products can be identified using a pH indicator dye (p. 147) while CO_2 is detected in liquid culture using an inverted small test tube (Durham tube, Fig. 39.3). Aerobic breakdown (via respiration) is termed oxidation, and anaerobic breakdown is known as fermentation. Identification tables usually incorporate tests for several different carbohydrates, e.g. Table 39.1.
- Enzyme tests. Most of these incorporate a substance that changes colour if the enzyme is present, e.g. a pH indicator, or a chromogenic substrate.
- Tests for specific end products of metabolism, e.g. the production of indole due to the metabolic breakdown of the amino acid tryptophan, or H_2S from sulphur-containing amino acids.

Identification kits

Some biochemical tests are now supplied in kit form, e.g. the API 20E system incorporates 20 tests within a sterile plastic strip (Fig. 39.4). After inoculation and overnight incubation, the results of the tests are converted into a seven-digit code, for comparison with known bacteria using either a reference book (the Analytical Profile Index), or a computer program. While kit identification systems save time and labour, they are more expensive and less flexible than conventional biochemical tests.

Immunological tests

Tests used in diagnostic microbiology include:

- Agglutination tests: based on the reaction between specific antibodies and a particular bacterium (p. 310). These tests are particularly useful for subdividing biochemically similar bacteria.
- Fluorescent antibody tests: the reaction between a labelled antibody and a particular bacterium can be visualised using UV microscopy. The direct fluorescent antibody test uses fluorescein isothiocyanate as the label.
- Enzyme-linked immunoassay tests using antibodies labelled with a particular enzyme, e.g. the double antibody sandwich ELISA or competitive ELISA tests (p. 313).

While such tests can give specific and accurate confirmation of the identity of a bacterium under controlled laboratory conditions, they are often too expensive and time-consuming for routine identification purposes, especially when large numbers of tests are required.

Molecular approaches to microbial identification – *several novel methods of detection and identification are based on nucleic acid techniques, including use of the PCR and Southern blotting to detect particular microbes (pp. 335–6).*

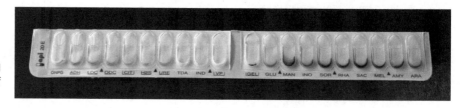

Fig. 39.4 Example of a bacterial identification kit (API 20E) property of bioMérieux S.A / Andrea Bannuscher.

Typing methods

The identification of bacteria at subspecies level is known as typing: this is usually done in a specialist laboratory, e.g. as part of an epidemiological study to establish the source of an infection. Various methods are used:

- Antigen typing or serotyping is based on immunological tests.
- Phage typing is based on the susceptibility of different strains to certain bacterial viruses (phages).
- Biotyping is based on biochemical differences between different strains e.g. enzyme profiles or antibiotic resistance screening ('antibiograms').
- Bacteriocin typing: bacteriocins are proteins released by bacteria that inhibit the growth of other members of the same species.

Practical applications of bacterial typing – E. coli O157:H7 is a serotype of this bacterium that is capable of causing severe human disease: it can be identified on the basis of an agglutination reaction with an appropriate antiserum. Typing is also discussed on p. 225.

Sources for further study

Barrow, G.I. and Feltham, R.K.A. (1993) *Cowan and Steel's Manual for the Identification of Medical Bacteria*, 3rd edn. Cambridge University Press, Cambridge.
[A standard reference work for the identification of clinically-important bacteria.]

Cullimore, D.R. (2000) *Practical Atlas for Bacterial Identification*. CRC Press, Boca Raton, Florida.
[Detailed coverage of practical methods applicable to a wide range of bacteria.]

Fisher, F., Cook, N.B., Fisher, F.W. and Kaszczuk, S. (1998) *Fundamentals of Diagnostic Mycology*. Saunders, Philadelphia.
[Covers identification techniques for medically important fungi.]

Fox, A. (ed.) *Journal of Microbiological Methods*, Elsevier, London (available through Science Direct at: http://www.sciencedirect.com/)
[Provides information on novel developments in all aspects of microbiological methods, including microbial culture/isolation and molecular approaches.]

Holt, J.G. (1994) *Bergey's Manual of Determinative Bacteriology*, 9th edn. Williams & Wilkins, Baltimore.

Macfaddin, J.F. (2000) *Biochemical Tests for the Identification of Medical Bacteria*, 3rd edn. Williams & Wilkins, Philadelphia.
[Explains the operating principles underlying most of the biochemical tests in routine use in diagnostic bacteriology.]

Study exercises

39.1 Describe the colonial characteristics of selected microbes. Either research the features of the following microbes (e.g. via the Web), or look at well-isolated individual colonies in the laboratory, following overnight growth on a suitable medium:

(a) *E. coli*;
(b) *Pseudomonas aerogenes*;
(c) *Staphylococcus aureus*;
(d) *Streptococcus pneumoniae*;
(e) *Bacillus subtilis*;
(f) *Saccharomyces cerevisiae*.

39.2 Find out how some of the tests used in microbial identification work. Research the operating principles that underpin the following tests, using microbiological textbooks or the Web, and prepare brief notes explaining how each of the following tests works:

(a) oxidase test;
(b) indole test;

(c) β-galactosidase test;
(d) urease test;
(e) lysine decarboxylase test;
(f) H_2S production.

39.3 Identify the following oxidase-negative, catalase-positive, Gram-negative rod-shaped bacteria, using Table 39.1.

(a) Positive for citrate, urease, ornithine decarboxylase and H_2S only.
(b) Positive for sucrose, mannitol, β-galactosidase, lysine decarboxylase and indole only.
(c) Positive for mannitol, citrate, lysine decarboxylase, ornithine decarboxylase and H_2S only.
(d) Positive for urease, ornithine decarboxylase and indole only.
(e) Positive for mannitol, β-galactosidase and indole only.

Manipulating and observing specimens

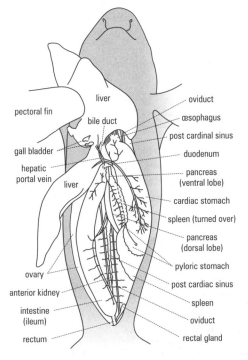

liver

pectoral fin

bile duct

gall bladder

hepatic
portal vein

liver

ovary

anterior kidney

intestine
(ileum)

rectum

oviduct

œsophagus

post cardinal sinus

duodenum

pancreas
(ventral lobe)

cardiac stomach

spleen (turned over)

pancreas
(dorsal lobe)

pyloric stomach

post cardiac sinus

spleen

oviduct

rectal gland

Fig. 40.1 Labelled diagram of abdomen of young female dogfish. Good drawings are essential to get the most out of dissections (see Chapter 28).

Dissection in zoology involves the display or removal of parts of any dead animal, whereas vivisection is an operation on a living animal. Dissection is usually associated with teaching animal structure, but the skills involved are used widely within the life sciences for:

- investigating anatomy and morphology;
- making physiological preparations of nerves, muscles and other organs;
- investigating parasites in various body organs;
- removing specific body organs/tissues for chemical analysis;
- investigating reproductive status;
- removing organs/tissues for histological/histochemical investigation;
- manipulating living material, as in grafting processes and preparing diagrams to show selected features (Fig. 40.1).

There is considerable debate about the ethics of using animals for dissection (p. 116) but this must be distinguished from that of vivisection. It is part of a complex and emotive issue, but one viewpoint is that dissection is required for effective zoology teaching: experience of dissection at an *appropriate* stage of the curriculum is both enlightening and teaches an essential technical skill. However, if dissection is definitely not for you, it is best to discover this as early as possible in your career.

KEY POINT *The primary objectives of dissection are:*
- *personal exploration of animal structure and function;*
- *development of manipulative skills;*
- *production of reports based upon personal observation.*

The ground rules of dissection

There are some important rules to be considered if dissection is to be an acceptable procedure in zoological teaching.

Humane treatment

The use of any animal, whatever its level of organisation, for experimentation or dissection must be a considered act with due regard for humane treatment and killing (see Chapter 19). Remember there are very specific regulations for the use of vertebrate species and some higher invertebrates: check with your supervisor.

Maximum benefit

Any animal should be used for as many investigations as possible.

Preparation

Prepare for the exercise by ensuring that practical schedules and relevant texts are consulted before attempting a dissection. If you have to miss a practical, inform the organiser in advance to prevent animals being killed or prepared unnecessarily.

Types of dissections

Dissection can be carried out at three levels of sophistication, related directly to the size of the organism or structure being investigated. Gross dissection of large organisms requires equipment very different from that

Your responsibility – *because of the ethical aspect to dissection, you have a particular responsibility to attend to and make careful and effective use of a dissection specimen.*

used for 'normal' dissection in that large knives replace scalpels, etc. Normal-scale dissection involves creatures from a dog down to an earthworm, and requires equipment typical of commercial dissection kits/ instruments. Fine-scale dissection for the removal or display of organs, glands, etc., from small animals usually requires only mounted needles, fractured glass edges for cutting and a dissecting microscope or magnifier. Fine-scale dissection requires extensive practice, and the preparation of material (relaxation and fixation, p. 206) is much more critical than for larger specimens.

> ⚠️ **SAFETY NOTE** *Taking care with dissection equipment – sharp blades should be used with extreme caution and disposed of immediately after use in a sharps container.*

Equipment

Table 40.1 lists basic equipment required for normal dissection: some comments on use are given below (see also Fig. 40.2). Commercial dissection kits often contain inappropriate components and you should buy your instruments individually from specialist suppliers, if possible. Equipment for fine-scale dissections can be assembled easily to your own specifications.

Table 40.1 List of basic equipment recommended for dissection.

Quantity	Description
1	All-metal scalpel, stainless steel, 45 mm blade
1 each	Swann–Morton scalpel handles, sizes 3 and 4
1 each	Swann–Morton blades, packets of nos. 10, 11, 12, 15, 22 and 24
2	Dissecting needles, straight, stainless steel
2	Blunt seekers, stainless steel: metal handles
1	Fine forceps, blunt points, stainless steel, 112 mm length
1	Coarse forceps, blunt points, stainless steel, 112 mm length
1	Coarse scissors, open shanks, straight points, stainless steel, 150 mm length
1	Fine scissors, open shanks, straight fine points, stainless steel, 110 mm length
1	Section lifter

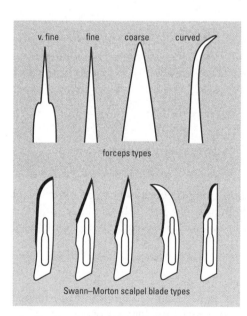

Fig. 40.2 Variations in design of dissection equipment

Scalpels

There are two basic types of scalpels, fixed blade and replaceable blade. Buy at least one 45-mm blade length solid forged scalpel, preferably made of stainless steel: use this for coarser cutting procedures as it can be resharpened using an oil-stone. The Swann–Morton scalpel comprises a handle and a disposable blade: blades come in a variety of shapes and sizes (Fig. 40.2), curved-edge ones being the most useful. Blades fit only specific handle sizes so make sure that they match.

Forceps

Buy at least one each of coarse and fine stainless steel forceps. The latter are very delicate and you should check that the points meet precisely before purchasing. Look after them very carefully as the points are easily damaged: use only on soft tissues. Use large and small forceps for general purposes but keep fine ones for delicate work only.

Dissecting scissors

Buy two pairs of stainless steel, pointed scissors, one medium-large for coarse work and cutting small bones and one fine pair for delicate work. Points and edges are easily damaged and require professional sharpening when repair is needed, so use carefully. A pair of bone cutters is optional but do *not* use scissors to cut large bones.

Dissecting (mounted) needles
Buy at least two with metal handles and protect the points; use them for dissecting membranes in areas where damage is acceptable. Needles have many non-dissection functions associated with other fine manipulative techniques.

Box 40.1 Basic stages of an animal dissection

The basic sequence of steps is outlined in Fig. 40.3.

1. **The animal should have been killed as humanely and as recently as possible**: this will probably be done for you by the class supervisor or technician according to the rules appropriate to the type of animal. The method of killing should be chosen carefully to keep the specimen relaxed. If preserved material is used, wash out excess fixative thoroughly before dissecting.

2. **Orient the specimen carefully**; determine the dorsal/ventral, anterior/posterior or other oral/aboral axes (see Fig. 44.4) and work out the correct orientation of the specimen for dissection. Invertebrates are usually dissected from the dorsal surface (the nerve cord being ventral in position) and vertebrates from the ventral surface (the nerve cord being dorsal in position): however, special objectives may require a different orientation.

3. **Open the body cavity carefully.** This is usually done using forceps to lift the skin away from underlying organs while using a fine pair of scissors to make an initial opening; the scissors are then used to extend this opening in anterior–posterior and lateral directions until the skin flaps can be pinned back. Pin out by placing tension on the skin flaps, breaking down any restricting membranes using a seeker.

4. **Subsequent dissection procedure depends upon your objectives.** To display the system being investigated as clearly and neatly as possible requires:

   (a) identification of organs, blood vessels and nerves initially visible;

   (b) separation of these structures from the membranes that hold them in place. This is best achieved using a blunt instrument such as a seeker to avoid damage;

   (c) removal or displacement of organs that obscure parts of the system you wish to display. Displacement using pins to hold the organ in position is the preferred option when possible;

   (d) if your objective is to display a blood system or a nervous system, you must follow individual vessels/nerves from their point of origin to their destination organ; do this using a blunt seeker to remove covering membranes by working carefully along the structure. Do not pull sideways during this procedure as this often results in breakages; such tissues are usually much stronger in the direction of their length than when pulled laterally.

   (e) tidy up loose pieces and wash away residual blood, etc.; use more pins to finalise the display of the system and then make notes and drawings as necessary (see e.g. Figs 28.3 and 40.1).

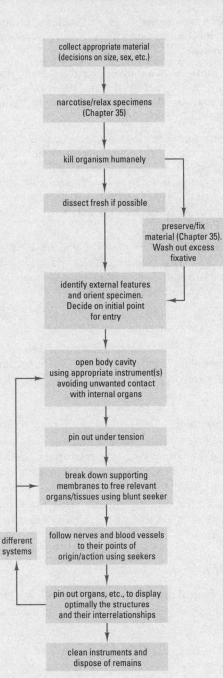

Fig. 40.3 Flowchart for dissection

Box 40.2 Tips for improving dissection technique

The following tips should help you develop your dissection technique:

- **Use a blunt seeker and controlled tension on the tissue to remove the connective tissue** that binds tissues and organs together: use scalpels sparingly.

- **Insert pins/awls obliquely** so that they do not interfere with further dissection. In segmented animals such as earthworms, fine pins can be used to mark the position of specific segments.

- **Dissect most invertebrates under water:** the water buoys up the tissues/organs and assists dissection. Change the water if it becomes clouded but do not allow flowing water to run directly on to the specimen as it will damage delicate structures.

- **Dissect vertebrates and larger invertebrates in air** on a dissection board, using cotton-wool swabs to remove excess blood and other body fluids.

- **Keep tissues under tension while dissecting,** but avoid damaging them with forceps/fingers.

- **Cut away from the organs when using scalpels or scissors** and keep scissor points away from structures you are attempting to free. This is essential when opening the body cavities of small animals.

- **Use appropriately sized equipment:** e.g. do not attempt to use large scissors to open an earthworm. Never use delicate instruments for coarse work.

- ***Never* remove anything until you know what you are removing.**

- **Dissect along structures and not across them,** particularly for tubular structures such as nerves and blood vessels.

- **Use fresh material whenever possible.** Fixatives make tissues more brittle and inelastic: alcohol storage tends to harden skin, muscle and connective tissue.

- **Keep your instruments clean and sharp:** you can't dissect with blunt or dirty equipment. Dry instruments after washing and wipe with an oily cloth. Cut nothing but tissues with scalpels and scissors. Do not sharpen pencils with scalpels or stick mounted needles into the bench or the dissecting board.

- **Be hygienic.** Wear rubber gloves if you have cuts or lesions on your hands. Wash hands and equipment thoroughly when finished. Dispose of animal remains carefully as instructed.

Dissecting seekers

These have blunt points and should have metal handles. Use for breaking down membranes holding delicate organs together. Again, they have multiple uses.

Recommended accessory equipment

A teat (Pasteur) pipette is valuable for washing delicate organs and tissues. A camel-hair brush is useful for removing material from delicate structures. Pins and awls are usually provided by your department and are used for fixing the dissection specimen to the board (awls) or wax dish (pins): small pins are invaluable for pinning organs aside for display purposes. For measurement, and as an aid to drawing, a pair of dividers is very useful, as is a small steel ruler.

Carrying out the dissection

The basic sequence of steps in a typical dissection is shown in Box 40.1 and Fig. 40.3, while Box 40.2 gives tips to help you improve your dissection technique.

Sources for further study

About Inc. *About Biology: Online dissections*. Available: http://biology.about.com/od/onlinedissections/Online_Dissections.htm
Last accessed 09/04/07
[A compendium of Web-based dissections.]

Fishbeck, D. and Sebastini, A. (2001) *Comparative Anatomy: Manual of Vertebrate Dissection*. Morton Publishing Co., Englewood, NJ.

Kardong, K.V. and Zalisko, E.J. (1998) *Comparative Vertebrate Anatomy: A Laboratory Dissection Guide*. McGraw-Hill, Boston.

Morgan, M. *Earthworm Dissection*. Available: http://www.microscopy-uk.org.uk/mag/articles/worm.html

Last accessed 09/04/07.
[A simple, descriptive account of an earthworm dissection.]

Walker, W.F. and Homberger, D. (1997) *Anatomy and Dissection of the Rat,* 3rd edn. W.H. Freeman, New York.

Whitehouse, R.H. and Grove, A.J. (1947) *The Dissection of the Crayfish*. University Tutorial Press, London.
[Part of a classic series including manuals on frog, rabbit, dogfish, earthworm and cockroach. Includes lots of information on how to carry out detailed dissections.]

Study exercises

40.1 Make a list of situations where dissection is used. There are many areas of zoology (and botany) where dissection technique is required. List up to ten situations where you would require to use some form of dissection technique to acquire material to work with, or to carry out manipulations before or after experiments.

40.2 Decide on appropriate dissection instruments. What instruments should you use for the following operations:

(a) following a nerve or blood vessel from source to ending?

(b) opening up the body cavity of the earthworm?

(c) removing or severing muscle tissue in the body wall of a vertebrate?

40.3 Research general rules of dissection. What is the general rule for deciding on which side a general dissection should be commenced: (a) for invertebrates; and (b) for vertebrates?

Many features of interest in biological systems are too small to be seen by the naked eye and can only be observed with a microscope. All microscopes consist of a coordinated system of lenses arranged so that a magnified image of a specimen is seen by the viewer (Figs. 41.1 and 41.2). The main differences are the wavelengths of electromagnetic radiation used to produce the image, the nature and arrangement of the lens systems and the methods used to view the image.

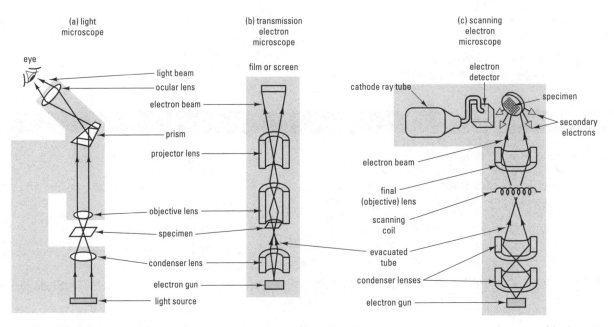

Fig. 41.1 Simplified diagrams of light and electron microscopes. Note that the electron microscopes are drawn upside-down to aid comparison with the light microscope.

Table 41.1 Comparison of microscope types. Resolution is that obtained by a skilled user. LM, light microscope; SEM, scanning electron microscope; TEM, transmission electron microscope

Property	Type of microscope		
	LM	TEM	SEM
Resolution	200 nm	1 nm	10 nm
Depth of focus	Low	Medium	High
Field of view	Good	Limited	Good
Specimen preparation (ease)	Easy	Skilled	Easy
Specimen preparation (speed)	Rapid	Slow	Quite rapid
Relative cost of instrument	Low	High	High

Microscopes allow objects to be viewed with increased resolution and contrast. Resolution is the ability to distinguish between two points on the specimen – the better the resolution, the 'sharper' the image. Resolution is affected by lens design and inversely related to the wavelength of radiation used. Contrast is the difference in intensity perceived between different parts of an image. This can be enhanced (a) by the use of stains, and (b) by adjusting microscope settings, usually at the expense of resolution.

The three main forms of microscopy are light microscopy, transmission electron microscopy (TEM) and scanning electron microscopy (SEM). Their main properties are compared in Table 41.1 and their suitability for observing cells and organelles is shown in Table 41.2.

Light microscopy

Two forms of the standard light microscope, the binocular (compound) microscope and the dissecting microscope, are described in detail in Chapter 43. These are the instruments most likely to be used in routine practical work. Figure 41.2(a) shows a typical image from a light microscope. In more advanced project work, you may use one or more

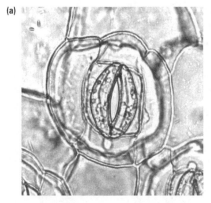

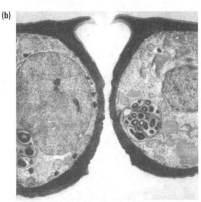

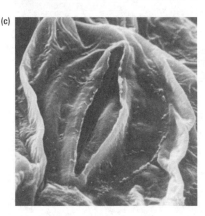

Fig. 41.2 Examples of images of a similar specimen (the stomatal complex of *Commelina communis* L.) obtained using different microscopic techniques: (a) light microscopy (surface view); (b) transmission electron microscopy (transverse section through guard cell pair at mid pore); and (c) scanning electron microscopy (surface view). As an indication of scale, the width of a single guard cell is about 10 µm.

of the following more sophisticated variants of light microscopy to improve image quality:

- Dark-field illumination involves a special condenser that causes reflected and diffracted light from the specimen to be seen against a dark background. The method is particularly useful for near-transparent specimens and for delicate structures like flagella. Care must be taken with the thickness of slides used – air bubbles and dust must be avoided and immersion oil must be used between the dark-field condenser and the underside of the slide.
- Ultraviolet microscopy uses short-wavelength UV light to increase resolution. Fluorescence microscopy uses radiation at UV wavelengths to make certain fluorescent substances (e.g. chlorophyll or fluorescent dyes that bind to specific cell components) emit light of visible wavelengths. Special light sources, lenses and mountants are required for UV and fluorescence microscopy, and filters must be used to prevent damage to users' eyes.
- Phase-contrast microscopy is useful for increasing contrast when viewing transparent specimens. It is superior to dark-field microscopy because a better image of the interior of specimens is obtained. Phase contrast operates by causing constructive and destructive interference effects in the image, visible as increased contrast. Adjustments must be made, using a phase telescope in place of the eyepiece, for each objective lens and a matching phase condenser, and the microscope must be set up carefully to give optimal results.
- Nomarski or Differential Interference Contrast (DIC) microscopy gives an image with a three-dimensional quality. However, the relief seen is optical rather than morphological, and care should be taken in interpreting the result. One of the advantages of the technique is the extremely limited depth of focus that results: this allows 'optical sectioning' of a specimen.
- Polarised-light microscopy can be used to reveal the presence and orientation of optically active components within specimens (e.g. starch grains, cellulose fibres), showing them brightly against a dark background.
- Confocal microscopy allows three-dimensional views of cells or thick sections. A finely focused laser is used to create electronic images of layered horizontal 'slices', usually after fluorescent staining. Images can be viewed individually or reconstructed to provide a 3-D computer-generated image of the whole specimen.

Electron microscopes

Electron microscopes offer an image resolution up to 200 times better than light microscopes (Table 41.1) because they utilise radiation of shorter wavelength in the form of an electron beam. The electrons are produced by a tungsten filament operating in a vacuum and are focused by electromagnets. TEM and SEM differ in the way in which the electron beam interacts with the specimen: in TEM, the beam passes through the specimen (Fig. 41.1(b)), while in SEM the beam is scanned across the specimen and is reflected from the surface (Fig. 41.1(c)). In both cases, the beam must fall on a fluorescent screen before the image can be seen. Permanent images ('electron micrographs') are produced after focusing the beam on photographic film (Figs. 41.2(b) and (c)).

Introduction to microscopy

Table 41.2 Dimensions of some typical cells and organelles with an indication of suitable forms of microscopy for observing them. LM = light microscope; SEM = scanning electron microscope; TEM = transmission electron microscope. Column 2 data after Rubbi (1994).

Cell or organelle	Approximate diameter or width (µm)	Suitable form of microscopy
Prokaryote cell	0.15–5	LM, SEM, TEM
Eukaryote cell	10–100	LM, SEM, TEM
Fungal hypha	5–20	LM, SEM, TEM
Nucleus	5–25	LM, SEM, TEM
Mitochondrion	1–10	SEM, TEM
Chloroplast	2–8	LM, SEM, TEM
Golgi apparatus	1	SEM, TEM
Lysosome/ peroxisome	0.2–0.5	SEM, TEM
Plant cell wall	0.1–10	LM, SEM, TEM

You are unlikely to use either type of electron microscope as part of undergraduate practical work because of the time required for specimen preparation and the need for detailed training before these complex machines can be operated correctly. However, electron microscopy is extremely important in understanding cellular and subcellular structures, and you may be shown electron micrographs with one or more of the following objectives:

- to demonstrate cell ultrastructure (TEM);
- to show surface features of organisms, e.g. when surface features are coated or when cells are 'freeze-fractured', then coated (SEM);
- to investigate changes in the number, size, shape and condition of cells and organelles (TEM);
- to carry out quantitative studies of cell and organelle disposition (TEM).

Aspects of the interpretation of electron micrographs are dealt with in Chapter 44.

Preparative procedures

Without careful preparation of the material being studied, the biological structures viewed with any microscope can be rendered meaningless. Figure 41.3 summarises the processes involved for the main types of microscopy discussed above. The processes involved in preparing material for light microscopy are outlined in Chapter 42.

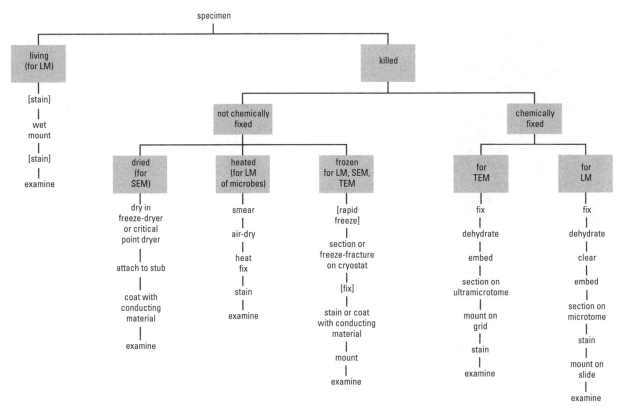

Fig. 41.3 Flowchart of procedures necessary to prepare specimens for different forms of microscopy. Steps enclosed in brackets are optional. LM, light microscope; SEM, scanning electron microscope; TEM, transmission electron microscope.

Text reference

Rubbi, C.P. (1994) *Light Microscopy Essential Data.* Wiley, Chichester.

Sources for further study

Bradbury, S. (1976) *The Optical Microscope in biology.* Edward Arnold, London.

Bradbury, S. (1984) *An Introduction to the Optical Microscope.* Oxford University Press, Oxford.

Davidson, M.W. and Abramowitz, M. *Molecular Expressions. Exploring the World of Optics and Microscopy.* Available: http://micro.magnet.fsu.edu/index.html
Last accessed 09/04/07.
[Covers many areas of basic knowledge underlying microscopy. Includes a microscopy primer.]

Jeffries, C. *Microscopy WWW Sites – by Organisation.* Available: http://www.ou.edu/research/electron/mirror/web-org.html
Last accessed 09/04/07.
[Comprehensive set of links to microscopy websites.]

Murphy, D.B. (2001) *Fundamentals of Light Microscopy and Electronic Imaging.* Wiley-Liss, New York.

van Egmond, W. and Walker, D. *A Virtual Pond-dip.* Available: http://www.microscopy-uk.org.uk/ponddip/x_index.html
Last accessed 09/04/07.
[Introduction to things you can see in pond water using a standard light microscope.]

Study exercises

41.1 Test your microscopy knowledge. Indicate whether the following statements about light microscopy, scanning electron microscopy (SEM) or transmission electron microscopy (TEM) are true or false.

(a) TEM allows you to see at finer resolution than light microscopy.
(b) TEM allows you to see surface features of specimens.
(c) SEM always requires staining of specimens.
(d) The resolution of TEM is about 200 times better than that of light microscopy.
(e) The resolution of a microscope is linked to the wavelength of electromagnetic radiation employed.
(f) The specimen in both TEM and SEM is viewed under near-vacuum conditions.
(g) Specimens for light microscopy can be living or dead.
(h) SEM provides better resolution than TEM.
(i) The depth of focus in light microscopy is greater than that in SEM.
(j) Light microscopy, SEM and TEM all involve the use of a condenser lens within the microscope.

41.2 Fill in the blanks in the following paragraph. Dark-field microscopy involves shining reflected and _____ light on the specimen against a dark background. It is particularly useful for _____ specimens. UV microscopy uses short-wavelength UV light in order to increase image _____. Phase-contrast microscopy utilises constructive and destructive _____ effects to increase image _____. Nomarski microscopy provides a pseudo ____ image, with a very small depth of _____, allowing _____ _____ to be carried out. _____ light microscopy allows visualisation of optically active components in the specimen. Confocal microscopy involves the use of a _____ light source and can yield computer-generated 3-D images.

41.3 Identify the missing preparative procedures. In each sequence below, one or two steps have been missed out. Using Fig. 41.3, identify the missing procedures.

(a) For light microscopy on a killed and fixed specimen: fix – dehydrate – clear – _____ – section – _____ – mount – examine.
(b) For light microscopy on a heat-fixed microbial specimen: smear – _____ – heat fix – _____ – examine.
(c) For TEM on a killed and fixed specimen: fix – _____ – embed – section – mount – stain – examine.

Fixing specimens *– tissues must be fixed as soon as possible after death. If this is not possible, specimens should be stored at low temperature (4°C) and for as short a time as possible.*

Decalcifying specimens *– this may be necessary if calcareous structures remain after fixation: this is usually done using a 5–10 per cent solution of EDTA followed by thorough washing.*

Preparative techniques are crucial to successful microscopical investigation because the chemical and physical processes involved have the potential for making the material difficult to work with and for producing artefacts. The basic steps (outlined in Fig. 41.3) are similar in most cases, but the exact details (e.g. timing, chemicals used and their concentrations) differ according to the material being examined and the purpose of the investigation. It is usually best to follow a recipe that has worked in the past for your material (see Grimstone and Skaer, 1972, Kiernan, 1999).

Chemical fixation

The main purpose of fixation is to preserve material in a lifelike manner. The process of fixation for microscopy is much more critical than for whole specimens (see Chapter 35) and only small pieces of tissue should be used. The fixation solutions used for microscopy are intended to:

- penetrate rapidly to prevent post-mortem changes in the cells;
- coagulate the cell contents into insoluble substances;
- protect tissues against shrinkage and distortion during subsequent processing;
- allow cell parts to become selectively and clearly visible when stained.

Fixative solutions are usually mixtures of chemicals selected for their combined properties (see Chapter 35). Your choice of fixative from the numerous recipes available in reference texts will depend upon both the type of investigation and the nature of the material.

KEY POINT *Poor fixation can produce artefacts, particularly where coagulant fixatives are used.*

When using a fixative for microscopy, observe the following points:

- Use fresh solutions: some of the fluids are unstable and do not keep well. Do not reuse fixative.
- Always use plenty of fixing fluid compared with the volume of material to be fixed (not less than a 10 : 1 fixative : sample volume ratio).
- Avoid underfixation or overfixation: in general the optimum time will be a function of several factors including:
  (a) The penetration capacity of the fixative.
  (b) The size of the piece of tissue: this should always be small and have as large a surface : volume ratio as possible.
  (c) The type of tissue to be fixed: uniform tissues fix more quickly than complex tissues, where one component may form a barrier to others. The presence of chitin usually means a slow rate of penetration. Tissues filled with air can be difficult to submerge and infiltrate; this can be overcome by fixation in a partial vacuum.
  (d) The temperature: increased temperature results in increased penetration rate, but also tends to make tissue brittle.
- Wash the specimen thoroughly after fixation: residues of fixative can interfere with subsequent processes. The washing may be in water or another appropriate solution.

Dehydration and clearing

A high water content in tissues will usually hinder subsequent processing so they must be dehydrated. This is done with an organic solvent using a series of solutions graded from pure water to pure solvent. Ethanol/water mixtures are often used in histology. Dehydration must be carried out carefully, using prescribed time schedules to avoid distortion and hardening. Protect delicate specimens by mixing solutions in the container rather than by transferring the specimen as this is when damage will occur.

The chemicals used for dehydration are not usually soluble in the embedding medium and tissues must therefore be infiltrated with an intermediate fluid, miscible with the waxes and resins normally used: such fluids make the tissues transparent and are termed clearing agents. The most widely used ones are hydrocarbons such as xylene; many are volatile, pose significant health risks, and tend to harden tissues rapidly. Clearing oils such as clove oil and cedarwood oil are safer, but slower in action. Terpineol is useful since it does not require such complete dehydration and can be used straight from 90% v/v ethanol. Note that every trace of oil must be removed during the infiltration (embedding) stage, or specimens will not embed properly.

Embedding and sectioning

Embedding involves infiltrating the specimen with a medium that will solidify and support it when sectioned. The specimen is either passed through a series of gradually increasing concentrations of the embedding material (e.g. wax or epoxy resin) dissolved in the dehydrating or clearing agent, or placed straight into pure embedding agent. Several changes of embedding material are made to remove the last traces of solvent, then the embedding agent is solidified by cooling or by a polymerisation treatment. Precise protocols can be found in specialist books (e.g. Kiernan, 1999). The specimen must be oriented carefully during the solidification process to aid section-cutting in known planes.

The aim of sectioning is to provide a thin slice through the tissues suitable for observing cellular details at maximum resolution. After trimming, blocks of embedded material are sectioned using a microtome. Section-cutting techniques are complex and require time for familiarisation. Sections are attached to slides after flattening the sections by a flotation procedure carried out either directly on the glass slide or in a water bath. The sections are attached to the slide by coating the latter with a very thin layer of albumen before drying them down on to this sticky surface.

For some procedures, specimens are frozen in isopentane, freon or dry-ice/ethanol mixtures and then maintained below −20 °C until sectioned. Sectioning is performed in a cold chamber (cryostat) where there is a microtome; sections are removed directly to glass slides. If storage is necessary, slides are placed over a desiccant in a refrigerator, but this is possible only for short periods.

Hand-cut sections can be made through certain relatively stiff materials, notably stems and roots of herbaceous plants:

1. Grasp the object firmly between your index finger and thumb.
2. Brace your elbows against your ribs to steady your hands.
3. Rest the side of a fresh and sharp razor blade on the index finger holding the specimen (Fig. 42.1).

Allowing time for dehydration – never be tempted to rush, because incomplete dehydration will do more than anything else to ruin a preparation.

 SAFETY NOTE The sharp blades use in hand sectioning are an obvious hazard. Identify beforehand the location of the nearest first aid kit. Take care to carry out such procedures with care and away from others who may inadvertently brush against you.

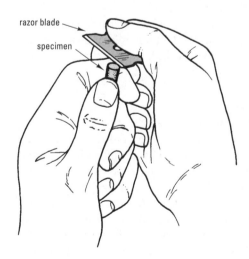

razor blade

specimen

Fig. 42.1 Preparation of a hand-cut section

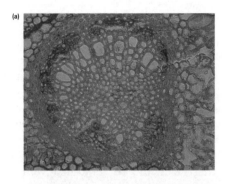

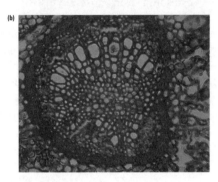

Fig. 42.2 Simulation of the introduction of contrast into a biological specimen by staining. Leaf midrib transverse section showing in greyscale the effects 'before' (a) and 'after' (b) staining xylem tissue red with phloroglucinol. In colour, the effects of staining are even more distinct.

Definitions

Metachromic stains – have the capacity to stain different structures different colours.

Orthochromic stains – never change colour whatever they stain.

Mordants – chemicals (salts and hydroxides of divalent and trivalent metals) that increase the efficiency of stains usually by forming complexes with the stain.

Counterstains – stains that apply a background colour to contrast with stained structures.

Negative staining – where the background is stained rather than the structure of interest.

4. Pull the blade towards your body so that you cut the specimen with a slicing action.
5. Repeat this action quickly, pushing the object slowly upwards with your thumb.
6. Float off the cut sections on to a watch glass of water.

This is a relatively tricky procedure, requiring practice. You may find it helps to lubricate the razor and object by wetting them before cutting. Thin material, such as a leaf, is best sectioned when supported between the halves of a longitudinally split cylinder of pith or fresh carrot. Always cut lots of sections, because only a small proportion will be thin enough to use – the best ones are often wedge-shaped, tapering off to thin edges.

Staining

The purpose of staining in microscopy is to:

- add contrast to the image;
- identify chemical components of interest;
- locate particular tissues, cells or organelles.

This is achieved in different ways for different types of microscopy. In standard light microscopy, contrast is achieved by staining the structure of interest with a coloured dye (e.g. Fig. 42.2); in UV microscopy, contrast is obtained using fluorescent stains. Physico-chemical properties of the stain cause it to attach to certain structures preferentially or be taken up across cell membranes.

Stains for light microscopy are categorised according to the charge on the dye molecule. Stains like haematoxylin, whose coloured part is a cation (i.e. basic dyes), stain acidic, anionic substances like nucleic acids: such structures are termed basophilic. Stains like eosin, whose coloured part is an anion (i.e. acid dyes), stain basic, cationic substances: such structures are termed acidophilic. Acid dyes tend to stain all tissue components, especially at low pH, and are much used as counterstains. Staining is progressive if it results in some structures taking up the dye preferentially. Staining is regressive if it involves initial over-staining followed by decolorisation (differentiation) of those structures that do not bind the dye tightly (e.g. Gram staining, p. 229).

Certain 'vital' stains (e.g. neutral red) are used to determine cell viability or the pH of cell compartments such as plant vacuoles. 'Mortal' stains (e.g. Evans' blue) are excluded from living cells but diffuse into dead ones and are used to assay cell mortality.

Stains and staining procedures

There is a huge range of stains for light microscopy but the features of those used commonly for cytology in botany, microbiology and zoology are given in Table 42.1. Consult appropriate texts for full details of (a) how to make up stains and (b) the protocol to use. Results depend on technique: follow the recommended procedures carefully. There are many stains used in histochemistry for identifying various classes of macromolecules such as DNA, RNA, proteins, lipids, carbohydrates (chitin, cellulose, starch, callose, pectins, glycogen) and heteropolymers (e.g. lipopolysaccharides, peptido-glycans, proteoglycans). Consult specialist texts for methods (e.g. Grimstone and Skaer, 1972; Horobin and Kiernan, 2002).

Table 42.1   A selection of stains for light microscopy of sections

	Stain	What it stains	Comments
Plant cells	Chlorazol black	Cell walls: black Nuclei: black, yellow or green Suberin: amber	The solvent used (70 per cent ethanol in water or water alone) affects colours developed
	Neutral red	Living cells: pink (pH < 7)	A 'vital' stain used to determine cell viability or to visualise plant protoplasts in plasmolysis experiments; best used at neutral external pH
	Phloroglucinol/HCl	Lignified cell walls: red	Care is required because the acid may damage microscope lenses
	Ruthenium red	Pectins: red	Shows up the middle lamella
	Safranin + Fast green	Nuclei, chromosomes, cuticle and lignin: red Other components: green	Stain in safranin first, then counterstain with fast green (light green will substitute). A differentiation step is required
	Toluidine blue	Lignified cell walls: blue Cellulose cell walls: purple	Best to apply dilute and allow progressive staining to occur
Fungi and bacteria	Giemsa	Bacterial chromosome: purple Bacterial cytoplasm: colourless	Also used in zoology to stain protozoa
	Gram	Gram-positive bacteria: violet/purple Gram-negative bacteria: red/pink Yeasts: violet/purple	See p. 230 for procedure
	Gray	Bacterial flagella: red	Uses toxic chemicals: mercuric chloride and formaldehyde. Leifson's stain is an alternative
	Lactophenol cotton blue	Fungal cytoplasm: blue (hyphal wall unstained)	Shrinkage may occur
	Nigrosin or India ink	Background: grey–black	Negative stains for visualisation of capsules: requires a very thin film
	Shaeffer and Fulton	Bacterial endospores: green Vegetative cells: pink/red	Malachite green is primary stain, heated for 5 min. Counterstained with safranin
	Ziehl–Neelsen	Actinomycetes: red Bacterial endospores: red Other microbes: blue	Requires heat treatment of fuchsin primary stain, decolorisation with ethanol–HCl and a methylene blue counterstain (acid-fast structures remain red)
Animal cells	Azure A/eosin B	Nuclei, RNA: blue Basophilic cells: blue–violet Most other cells: pale blue Muscle cells: pink Necrosing cells: pink Cartilage matrix: red–violet Bone: pink Red blood cells: orange–red Mucins: green–blue/blue–violet	Used in pathology – shows up bacteria as blue; must be fresh; care required over pH: Mann's methyl blue/eosin gives similar results
	Chlorazol black	Chitin: greenish-black Nuclei: black, yellow or green Glycogen: pink or red	Solvent (70% v/v ethanol in water or water alone) affects colours formed
	Iron haematoxylin	Nuclei, chromosomes and red blood cells: black Other structures: grey or blue–black	Good for resolving fine detail; iron alum used as mordant before haematoxylin to differentiate
	Mallory	Nuclei: red Nucleoli: yellow Collagen, mucus: blue Red blood cells: yellow Cytoplasm: pink or yellow	Simple, one-stage stain; fades within a year; not to be used with osmium-containing fixatives. Heidenhain's azan gives similar results but does not fade. Cason's one-step Mallory is a rapidly applied stain that is particularly good for connective tissue
	Masson's trichrome	Collagen, mucus: green Cytoplasm: orange or pink	Used as a counterstain after, for example, iron haematoxylin, which will have stained nuclei black. Not to be used after osmium fixation
	Mayer's (haemalum and eosin; 'H&E')	Nuclei: blue/purple Cytoplasm: pink	Alum used as mordant for haematoxylin; eosin is the counterstain. To show up collagen, use van Gieson's stain as counterstain

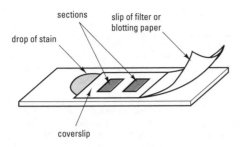

drop of stain
sections
slip of filter or blotting paper
coverslip

Fig. 42.3 How to irrigate a section with stain by drawing it through with a piece of filter paper.

Stains for light microscopy are normally applied by one of four methods:

- Floating the sections on the stain.
- Applying the stain to a smear fixed on to a slide, e.g. when staining bacteria and blood.
- Drawing the stain through under a coverslip as shown in Fig. 42.3.
- Immersing slides with sections attached into a staining trough. This is best for bulk staining.

Most stains need to act in an aqueous medium, so sections that have been embedded in wax must be rehydrated before staining. The wax is dissolved, e.g. in Histo-clear for 1–3 min, the Histo-clear replaced by 100 per cent ethanol (1 min) then 70% (v/v) ethanol : water for 1 min., finally transferring to distilled water. Fresh sections of plants or heat-fixed smears of microbes generally require no pretreatment before staining if they are not to be retained after examination.

Mounting sections

Wet mounts

These are used for observing fresh specimens. The following steps are involved:

1. Isolate the specimen.
2. Place the specimen in a small droplet of the relevant fluid (fresh water, sea water, etc.) on a microscope slide.
3. Gently lower a coverslip on to the droplet, using forceps or two needles and avoiding bubbles.
4. Remove any excess water on or around the coverslip with absorbent paper.

Entire specimens can be examined under the light microscope providing they are small enough to be mounted on a glass slide. They may be mounted in cavity slides or by using ring mounts.

Temporary mounts

These essentially involve wet mounting in a mountant with a short useful life, e.g. for identification purposes. It may be desirable to clear the specimen first and a dual-purpose substance such as lactophenol, which will clear from 70% (v/v) ethanol, is recommended.

Permanent mounts

These protect sections during examination and allow storage without deterioration. A permanent mount involves sealing your section under a coverslip in a mountant. The mountants used are clear resins dissolved in a slowly evaporating solvent. A good mountant has a similar refractive index to the tissue being mounted, remains clear through time, is chemically inert and will harden quickly. Natural resins like Canada balsam take a long time to dry, are variable in quality and tend to colour up and crack in time. The newer synthetic resins and plastics such as DPX

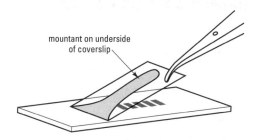

Fig. 42.4 How to lower a coverslip painted with mountant on its underside on to sections on a slide.

mountant on underside of coverslip

__Making squash preparations__ – these may be required for any type of mount. The smallest specimens can be squashed after mounting by applying gentle pressure on the coverslip with your forceps. Larger specimens can be squashed between two slides before fixing and mounting – this ensures that higher pressures are applied evenly.

mountant are superior: they dry quickly, are available in a range of refractive indices, and do not yellow with age. For tissue components soluble in organic solvents, aqueous mounting media based on, for example, gelatine or glycerol should be used. Find out which mountant is recommended for your particular sections.

The recommended procedure when mounting sections on a slide is:

1. Apply a little mountant to a coverslip of appropriate size.
2. Turn the coverslip over and place on its edge to one side of the sections as in Fig. 42.4.
3. Lower the coverslip slowly down on to the sections so as to displace all the air and sandwich the sections between the slide and the coverslip.
4. Press firmly from the centre outwards to distribute the mounting medium evenly.
5. Allow the solvent to evaporate – best results come from slow drying when time allows, but many synthetic mountants will tolerate brief heating when speed is essential.

Text references and sources for further study

Davidson, M.W. and Abramowitz, M. *Molecular Expressions. Exploring the World of Optics and Microscopy.* Available: http://micro.magnet.fsu.edu/index.html
Last accessed 09/04/07.
[Covers many areas of basic knowledge underlying microscopy, including preparative procedures.]

Grimstone, A.V. and Skaer, R.J. (1972) *A Guidebook to Microscopical Methods.* Cambridge University Press, Cambridge.

Horobin, R.W. and Kiernan, J.A. (eds) (2002) *Conn's Biological Stains. A Handbook of Dyes, Stains and Fluorochromes for use in biology and Medicine*, 10th edn. Bios Scientific Publishers, Oxford.

Kiernan, J.A. (1999) *Histological and Histochemical Methods: Theory and Practice*, 3rd edn. Scion Publishing, Bloxham.

Study exercises

42.1 Select appropriate stains. From Table 42.1, identify a stain you could use to help indicate the presence of the following: (a) glycogen in a liver section; (b) woody (lignified) cells in a plant stem section; (c) a fungal pathogen in a leaf section; (d) living cells in an onion epidermal peel; (e) mucus in a lung section; (f) bacteria in a food sample.

42.2 Explain the processes involved in preparing a section for light microscopy. Prepare one-sentence summaries of the reasons for carrying out the following procedures: (a) fixation; (b) dehydration; (c) embedding; (d) sectioning; (e) staining; (f) permanent mounting.

42.3 Investigate toxicity and safety aspects of some of the staining reagents used for light microscopy. Many chemicals used in staining are toxic or otherwise dangerous. As examples, investigate the specific hazards associated with the use of the following stains: (a) chlorazol black; (b) neutral red; (c) malachite green; (d) safranin.

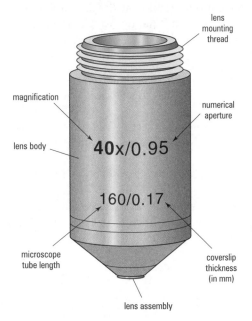

Fig. 43.1 Objective lens parameters. Most lenses are inscribed to show the details labelled above. The numerical aperture is a measure of the light-gathering power of the lens.

The light microscope is probably the most important instrument used in biology practicals and its correct use is one of the basic and essential skills of biology. A standard undergraduate binocular microscope consists of three main types of optical unit: eyepiece, objective (Fig. 43.1) and condenser. These are attached to a stand that holds the specimen on a stage (Fig. 43.2). A monocular microscope is constructed similarly but has one eyepiece lens rather than two.

Setting up a binocular light microscope

Before using any microscope, familiarise yourself with its component parts.

KEY POINT *Never assume that the previous person to use your microscope has left it set up correctly: apart from differences in users' eyes, the microscope needs to be properly set up for each lens combination used.*

The procedures outlined below are simplified to allow you to set up microscopes like those of the Olympus CX series (Fig. 43.2). For monocular microscopes, disregard instructions for adjusting eyepiece lenses in (5).

1. Place the microscope at a convenient position on the bench. Adjust your seating so that you are comfortable operating the focus and stage controls. Unwind the power cable, plug in and switch on after first ensuring that the lamp setting is at a minimum. Then adjust the lamp setting to about two-thirds of the maximum.
2. Select a low-power (e.g. ×10) objective. Make sure that the lens clicks home.
3. Set the eyepiece (ocular) lenses to your interpupillary distance; this can usually be read off a scale on the turret. You should now see a single circular field of vision. If you do not, try adjusting in either direction.
4. Put a prepared slide on the stage. Examine it first against a light source and note the position, colour and rough size of the specimen. Place the slide on the stage (coverslip up!) and, viewing from the side, position it with the stage adjustment controls so that the specimen is illuminated.
5. Focus the image of the specimen using first the coarse and then the fine focusing controls (Fig. 43.3). The image will be reversed and upside down compared with that seen by viewing the slide directly.
   (a) If both eyepiece lenses are adjustable, set your interpupillary distance on the scale on each lens. Close your left eye, look through the right eyepiece with your right eye and focus the image with the normal controls. Now close your right eye, look through the left eyepiece with your left eye and focus the image by rotating the eyepiece holder. Take a note of the setting for future use.
   (b) If only the left eyepiece is adjustable, close your left eye, look with the right eye through the static right eyepiece and focus the image with the normal controls. Now close your right eye, look through the left eyepiece with your left eye and focus the image by rotating the eyepiece holder. Take a note of the setting for future use.

Using binocular eyepieces – if you do not know your interpupillary distance, ask someone to measure it with a ruler. You should stare at a fixed point in the distance while the measurement is taken. Take a note of the value for future use.

Issues for spectacle and contact lens wearers – those who wear glasses can remove them for viewing, as microscope adjustments will accommodate most deficiencies in eyesight (except astigmatism). This is more comfortable and stops the spectacle lenses being scratched by the eyepiece holders. However, it may create difficulties in focusing when drawing diagrams. Those with contact lens should simply wear them as normal for viewing.

Fig. 43.2 Diagram of the Olympus binocular microscope model CX41.

- The lamp in the base of the stand (1) supplies light; its brightness is controlled by an on–off switch and voltage control (2). Never use maximum voltage or the life of the bulb will be reduced – a setting two-thirds to three-quarters of maximum should be adequate for most specimens. A field–iris diaphragm may be fitted close to the lamp to control the area of illumination (3).
- The condenser control focuses light from the condenser lens system (4) on to the specimen and projects the specimen's image on to the front lens of the objective. Correctly used, it ensures optimal resolution.
- The condenser–iris diaphragm (5) controls the amount of light entering and leaving the condenser; its aperture can be adjusted using the condenser–iris diaphragm lever (6). Use this to reduce glare and enhance image contrast by cutting down the amount of stray light reaching the objective lens.
- The specimen (normally mounted on a slide) is fitted to a mechanical stage or slide holder (7) using a spring mechanism. Two controls allow you to move the slide in x and y planes. Vernier scales (see p. 134) on the slide holder can be used to return to the same place on a slide. The fine and coarse focus controls (8) adjust the height of the stage relative to the lens systems. Take care when adjusting the focus controls to avoid hitting the lenses with the stage or slide.
- The objective lens (9) supplies the initial magnified image; it is the most important component of any microscope because its qualities determine resolution, depth of field and optical aberrations. The objective lenses are attached to a revolving nosepiece (10). Take care not to jam the longer lenses on the stage or slide as you rotate the nosepiece. You should feel a distinct click as each lens is moved into position. The magnification of each objective is written on its side; a normal complement would be $\times 4$, $\times 10$, $\times 40$ and $\times 100$ (oil immersion).
- The eyepiece lens (11) is used to further magnify the image from the objective and to put it in a form and position suitable for viewing. Its magnification is written on the holder (normally $\times 10$). By twisting the holder for one or both of the eyepiece lenses you can adjust their relative heights to take account of optical differences between your eyes. The interpupillary distance scale (12) and adjustment knob allow compensation to be made for differences in the distance between users' pupils.

From a photograph of Olympus Model CX41 binocular microscope, supplied by Olympus Microscopes, Olympus Optical Company (UK) Ltd, published courtesy of Olympus Optical Company (UK) Ltd.

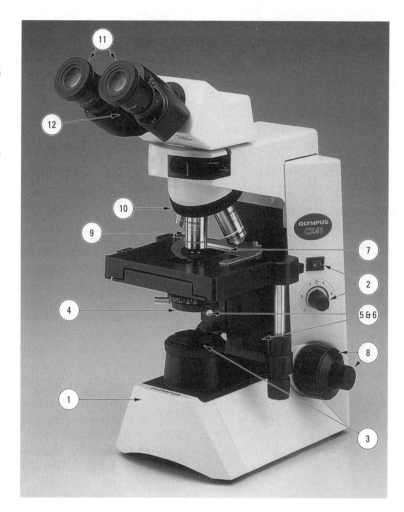

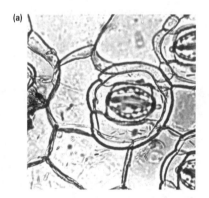 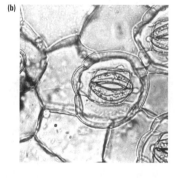

Fig. 43.3 Importance of correct focus in light microscopy. Stomatal complex of *Commelina communis* L., a specimen that is a monolayer of cells approximately 30–50 μm thick. (a) Focal plane is on 'internal' walls of the cells; (b) focal plane is on the 'external' walls and stomatal pore. The two images are different, and while it would not be possible to measure the stomatal pore in (a), it would not be possible to see the vacuolar crystals in (b). When looking at specimens, always use the fine-focus control to view different focal planes.

6. Close the condenser–iris diaphragm (aperture–iris diaphragm), then open it to a position such that further opening has no effect on the brightness of the image (the 'threshold of darkening'). The edge of the diaphragm should not be in view. Turn down the lamp if it is too bright.

7. Focus the condenser. Place an opaque pointed object (the tip of a mounted needle or a sharp pencil point) on the centre of the light source. Adjust the condenser setting until both the specimen and needle tip/pencil point are in focus together. Check that the condenser–iris diaphragm is just outside the field of view.

8. For higher magnifications, swing in the relevant objective (e.g. ×40), carefully checking that there is space for it. Adjust the focus using the fine control only. If the object you wish to view is in the centre of the field with the ×10 objective, it should remain in view (magnified, of course) with the ×40. Adjust the condenser–iris diaphragm and condenser as before – the correct setting for each lens will be different.

9. When you have finished using the microscope, remove the last slide and clean the stage if necessary. Turn down the lamp setting to its minimum, then switch off. Clean the eyepiece lenses with lens tissue. Check that the objectives are clean. Unplug the microscope from the mains and wind the cable round the stand and under the stage. Replace the dust cover.

If you have problems in obtaining a satisfactory image, refer to Box 43.1; if this doesn't help, refer the problem to the class supervisor.

Procedure for observing transparent specimens

Some stained preparations and all colourless objects are difficult to see when the microscope is adjusted as above (Fig. 43.4). Contrast can be improved by closing down the condenser–iris diaphragm. Note that when you do this, diffraction haloes appear round the edges of objects. These obscure the image of the true structure of the specimen and may result in loss of resolution. Nevertheless, an image with increased contrast may be easier to interpret.

Adjusting a microscope with a field–iris diaphragm – adjust this before the condenser–iris diaphragm: close it until its image appears in view as a circle of light, if necessary focusing on the edge of the circle with the condenser controls and centring it with the centring screws. Now open it so the whole field is just illuminated.

High-power objectives – never remove a slide while a high power objective lens (i.e. ×40 or ×100) is in position. Always turn back to the ×10 first. Having done this, lower the stage and remove the slide.

Box 43.1   Problems in light microscopy and possible solutions

No image; very dark image; image dark and illuminated irregularly
- Microscope not switched on (check plug and base)
- Illumination control at low setting or off
- Objective nosepiece not clicked into place over a lens
- Diaphragm closed down too much or off-centre
- Lamp failure

Image visible and focused but pale and indistinct
- Diaphragm needs to be closed down further (see Fig. 43.4)
- Condenser requires adjustment

Image blurred and cannot be focused
- Dirty objective
- Dirty slide
- Slide upside down
- Slide not completely flat on stage
- Eyepiece lenses not set up properly for user's eyes
- Fine focus at end of travel
- Oil-immersion objective in use, without oil

Dust and dirt in field of view
- Eyepiece lenses dirty
- Objective lens dirty
- Slide dirty
- Dirt on lamp glass or upper condenser lens

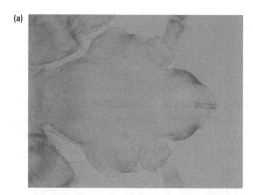

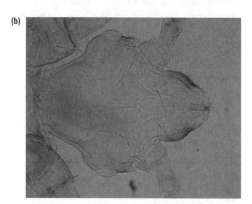

Fig. 43.4 Effect of closing the condenser–iris diaphragm on contrast. Head of human head louse, *Pediculus humanus capitus* DeGeer; (a) with condenser – iris diaphragm open; (b) with condenser – iris diaphragm closed (all other settings the same). Note difference in detail that can be seen in (b), but also that image (b) is darker – when using the condenser–iris diaphragm in this way you may need to compensate by increasing the light setting.

SAFETY NOTE *Take care when moving microscopes, not only because of the cost of replacement, but also because they weigh several kilograms and could cause injury if dropped. Always carry a microscope using two hands.*

Measuring specimens using a dissecting microscope – *because of the low magnification, sizes can generally be estimated by comparison with a ruler placed alongside the specimen. If accurate measurements are required, eyepiece graticules can be used.*

Procedure for oil-immersion objectives

These provide the highest resolution of which the light microscope is capable. They must be used with immersion oil filling the space between the objective lens and the top of the slide. The oil has the same refractive index as the glass lenses, so loss of light by reflection and refraction at the glass/air interface is reduced. This increases the resolution, brightness and clarity of the image and reduces aberration. Use oil-immersion objective(s) as follows:

1. Check that the object of interest is in the field of view using e.g. the ×10 or ×40 objective.
2. Apply a single small droplet of immersion oil to the illuminated spot on the top of the slide, having first swung the ×40 objective away. Never use too much oil: it can run off the slide and mess up the microscope.
3. Move the high-power (×100) oil-immersion objective into position carefully, checking first that there is space for it. Focus on the specimen using the fine control only. You may need a higher brightness setting.
4. Perform condenser–iris diaphragm and condenser focusing adjustments as for the other lenses.
5. When finished, clean the oil immersion lens by gently wiping it with clean lens tissue. If the slide is a prepared one, wipe the oil off with lens tissue.

You should take great care when working with oil immersion lenses as they are the most expensive to replace. Because the working distance between the lens and coverslip is so short (less than 2 mm), it is easy to damage the lens surface by inadvertently hitting the slide or coverslip surface. You must also remember that they need oil to work properly. If working with an unfamiliar microscope, you can easily recognise oil-immersion lenses. Look for a white or black ring on the lens barrel, near the lens, or for 'oil' clearly marked on the barrel.

Care and maintenance of your microscope

Microscopes are delicate precision instruments. Handle them with care and never force any of the controls. Never touch any of the glass surfaces with anything other than clean, dry lens tissue. Bear in mind that a replacement would be very expensive.

If moving a microscope, hold the stand above the stage with one hand and rest the base of the stand on your other hand. Always keep the microscope vertical (or the eyepieces may fall out). Put the microscope down gently.

Clean lenses by gently wiping with clean, dry lens tissue. Use each piece of tissue once only. Try not to touch lenses with your fingers as oily fingerprints are difficult to clean off. Do not allow any solvent (including water) to come into contact with a lens; sea water is particularly damaging.

The dissecting (stereoscopic) microscope

The dissecting microscope (Fig. 43.5) is a form of stereoscopic microscope used for observations at low total magnification (×4 to ×50) where a large

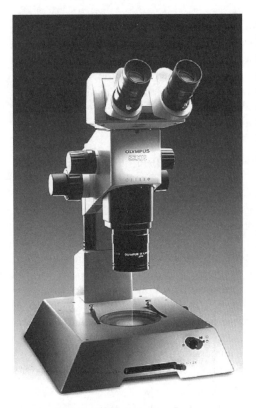

working distance between objectives and stage is required, perhaps because the specimen is not flat or dissecting instruments are to be used. A stereoscopic microscope essentially consists of two separate lens systems, one for each eye. Some instruments incorporate zoom objectives. The eyepiece–objective combinations are inclined at about 15° to each other and the brain resolves the compound image in three dimensions as it does for normal vision. The image is right side up and not reversed, which is ideal for dissections. Specimens are often viewed in a fresh state and need not be placed on a slide – they might be in a Petri dish or on a white tile. Illumination can be from above or below the specimen, as desired.

Most of the instructions for the binocular microscope given above apply equally well to dissecting microscopes, although the latter do not normally have adjustable condensers or diaphragms. With stereoscopic microscopes, make specially sure to adjust the eyepiece lenses to suit your eyes so that you can take full advantage of the stereoscopic effect.

Fig. 43.5 From a photograph of Olympus Model SZX12 dissecting microscope, supplied by Olympus Microscopes, Olympus Optical Company (UK) Ltd, published courtesy of Olympus Optical Company (UK) Ltd.

Sources for further study

Bradbury, S. and Bracegirdle, B. (1998) *Introduction to Light Microscopy*. Bios Scientific Publishers, Oxford.

Davidson, M.W. and Abramowitz, M. *Nikon MicroscopyU: Introduction to Microscope Objectives*. Available: http://www.microscopyu.com/articles/optics/objectiveintro.html.
Last accessed 09/04/07.
[Covers the optics of objective lenses.]

Olympus Microscopy Resource Centre *Introduction to Optical Microscopy, Digital Imaging, and Photomicrography* Available:
http://www.olympusmicro.com/primer/index.html
Last accessed 09/04/07.
[A primer on optical microscopy, including FAQs and links to further resources.]

Study exercises

43.1 Test your knowledge of the parts of a binocular light microscope. Cover up the legend on the left of Fig. 43.2 with a piece of paper or card. Now identify the parts of the light microscope numbered on the diagram. Check your answers from the legend.

43.2 Identify roles of parts of a binocular light microscope. State briefly the *primary* role of each component of a standard binocular light microscope:

(a) condenser
(b) objective lens
(c) condenser–iris diaphragm
(d) interpupillary distance scale
(e) Vernier scales on the mechanical stage.

43.3 Identify the correct sequence of adjustments when setting up a light microscope.

(a) Focus specimen – set interpupillary distance – adjust condenser–iris diaphragm – make individual eyepiece adjustment – focus condenser – focus specimen.

(b) Make individual eyepiece adjustment – set interpupillary distance – focus condenser – focus specimen – adjust condenser–iris diaphragm – focus specimen.

(c) Set interpupillary distance – focus specimen – make individual eyepiece adjustment – adjust condenser–iris diaphragm – focus condenser – focus specimen.

(d) Make individual eyepiece adjustment – set interpupillary distance – focus specimen – adjust condenser–iris diaphragm – focus condenser – focus specimen.

(e) Focus specimen – make individual eyepiece adjustment – set interpupillary distance – focus condenser – adjust condenser–iris diaphragm.

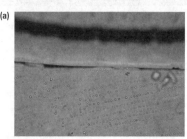

(a)

Microscope images, whether viewed directly or as photomicrographs, need to be interpreted with extreme care. This is not only because of the potentially damaging procedures involved in specimen preparation, but also because scale, section orientation and staining combine to create images that can be misinterpreted by the human mind.

Eliminating non-biological artefacts

(b)

There are several types of non-biological artefacts (e.g. those shown in Fig. 44.1). It is only through experience that you can learn to recognise these. Sometimes they are mistaken for specimens, because the specimen itself is hard to see. A useful tip is to look at your slide carefully before placing it on the stage; if you do this against a light background, you may be able to see the specimen and judge its location as you move the slide under the objective. For slides where specimens are tiny and spread out, such as blood or bacterial smears, it can be useful to focus on the edge of the coverslip (as in Fig. 44.1(a)) to determine the correct focal plane before searching. Correctly setting up the microscope is crucial in such cases.

(c)

Establishing scale and measuring objects

Working out the length of objects on photomicrographs

On a micrograph, the scale will usually be provided as a magnification factor (e.g. ×500) or in the form of a bar of defined length (e.g. 100 nm). If you need to estimate the dimensions of an object in the micrograph, follow the steps below:

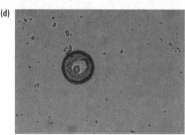

(d)

1. Measure the object as it appears on the micrograph with a ruler or set of Vernier calipers (see p. 134). It may be difficult to decide exactly where the boundary of a structure lies; rather than a discrete boundary, you may be dealing with shades of grey. It is essential to be consistent!
2. If the scale is given as a bar, measure the bar too and find the object's size by proportion. For example, if the object measures 32 mm and a bar representing 100 nm is 20 mm long, then the size of the object is $32/20 \times 100$ nm = 160 nm.
3. If the scale is given as a magnification factor, divide your measurement by this number to obtain the object's size, taking care to enter the correct units in your calculator. For example, if the object measures 32 mm and the print magnification is stated as ×200 000, then its size is 32×10^{-3} m$/200\,000 = 1.6 \times 10^{-7}$ m = 160 nm.

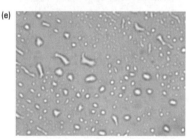

(e)

Avoid putting too many significant figures in any estimates of dimensions: there may be quite large errors in estimating print magnifications that could make the implied accuracy meaningless (p. 408).

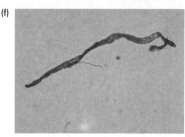

(f)

Fig. 44.1 Artefacts in microscopy. Do not mistake any of the following for biological specimens: (a) edge of coverslip; (b) dust on slide; (c) shard of glass; (d) air bubble; (e) grease spots; (f) cellulose fibre.

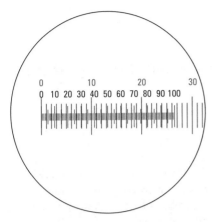

Fig. 44.2 Calibrating an eyepiece micrometer (graticule). Align the two scales and read the number of stage micrometer divisions (shown in colour) for a particular number of eyepiece micrometer divisions (shown in black). In this case 26.5 scale divisions of 0.01 mm (= 2.65 mm) are equivalent to 100 eyepiece divisions, so each eyepiece division 0.265 ÷ 100 = 0.00265 mm = 2.65 μm. This is a typical value for a total magnification of ×400 (e.g. ×40 objective and ×10 eyepiece).

***Using an eyepiece graticule** – choose the eyepiece lens corresponding to your stronger eye and check that you have made the correct adjustments to the eyepiece lenses as detailed on p. 252.*

Examples An object occupying one-third of the area of a field of diameter 0.45 mm would have a cross-sectional area of 0.159/3 (= 0.053 mm²).

Seven objects in a field of area 0.159 mm² would be present at a density of 7 ÷ 0.159 (= 44 objects mm⁻²).

Example For a field area of 0.159 mm², a droplet volume of 30 mm³ and a coverslip of area 484 mm², the field volume would be 30 × (0.159 ÷ 484) = 0.01 mm³, so if you saw 25 objects per field, their density would be 25 ÷ 0.01 = 2 500 mm⁻³.

Adding linear scales to drawings

The magnification of a light-microscope image is calculated by multiplying the objective magnification by the eyepiece magnification. However, the magnification of the image bears no certain relation to the magnification of any drawing of the image – you may equally well choose to draw the same image 10 mm or 10 cm long. For this reason, *it is essential to add a scale to all your diagrams.* You can provide either a bar giving the estimated size of an object of interest, or a bar of defined length (e.g. 100 μm).

The simplest method of estimating linear dimensions is to compare the size of the image to the diameter of the field of view. You can make a rough estimate of the field diameter by focusing on the millimetre scale of a transparent ruler using the lowest power objective. Estimate the diameter of this field directly, then use the information to work out the field diameters at the higher powers *pro rata*. For example, if the field at an overall magnification of ×40 is 4 mm, at an overall magnification of ×100 it will be: 40/100 × 4 mm = 1.6 mm (1600 μm).

Greater accuracy can be obtained if an eyepiece micrometer (graticule) is used. This carries a fine scale and fits inside an eyepiece lens. The eyepiece micrometer is calibrated using a stage micrometer, basically a slide with a fine scale on it. Figure 44.2 shows how to calibrate an eyepiece micrometer, along with a worked example. Once you have calibrated your eyepiece micrometer for each objective lens used, you can use it to measure objects: in the example shown in Fig. 44.2, the scale reading is multiplied by 2.65 μm to give the value in micrometres. So, if you measured the width of a human hair at 34 eyepiece micrometer units, then this will be equal to 34 × 2.65 = 90.1 μm. An alternative approach is to put a scale bar on a diagram, e.g. a 100 μm scale bar would be equivalent to the length of almost 38 eyepiece micrometer divisions.

Estimating the area and area density of objects

You can calculate the area of the field at each magnification using a field diameter estimate obtained as above (obtain the area from πr^2, but don't forget that radius r = diameter ÷ 2). The units you use to measure the diameter will be squared; thus, a field diameter of 0.45 mm converts to a field area of 0.159 mm². Use the field area to estimate the cross-sectional area of a specimen by proportion.

If several objects appear in a field, you can express their frequency on an area basis by dividing the number seen by the field area. It is nearly always best to take an average from several fields of view before calculating.

Volume–density estimates

The density of particles in a liquid suspension can be *roughly* estimated as follows: place a drop of well-mixed suspension on to a slide with a Pasteur pipette. You can estimate the drop volume by weighing drops of water and assuming 1 mg = 1 mm³. Alternatively, use a pipettor (see p. 123). Quickly cover the drop with a coverslip (see p. 251), avoiding air bubbles – a 22 × 22 mm coverslip should allow a 30 mm³ drop to spread out completely. Now count the numbers of the object of interest within each of several fields and calculate a mean value. The volume of each field will be:

$$\text{(drop volume} \times \text{field area)} \div \text{coverslip area} \qquad [44.1]$$

and the density will be:

$$\text{(mean no. of objects per field)} \div \text{field volume} \qquad [44.2]$$

Where accurate estimates of volume–density are required, you should use a haemocytometer (see p. 275).

Determining the position and orientation of the section within the specimen

When interpreting microscope images, you need to determine and take into account what position and orientation the section was taken from.

Always remember that a section is a two-dimensional slice through a three-dimensional structure. Thus, a circular profile in two dimensions could arise from sectioning numerous three-dimensional shapes (Fig. 44.3); you cannot determine which shape without further information (e.g. from serial sections or sections taken in another plane). Likewise, the size of a profile in a section may reveal little about the size of the structure from which it arose. Also bear in mind that very thin sections (say 75 nm thick) may not provide a representative sample of all the organelles present in a cell (which may be 1000–100 000 nm in width).

Always relate what you see to what you know; at a minimum, you should take account of your basic knowledge of animal, microbial or plant cell structure. Note that, in the absence of a defining characteristic, you cannot be sure about the nature of a specimen, because it may simply be due to chance that the section does not include the feature.

The terminology used for sections taken in different planes is shown in Figs. 44.4 and 44.5.

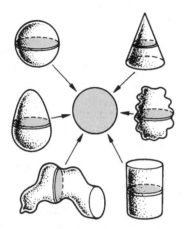

Fig. 44.3 A two-dimensional section may be representative of many different three-dimensional structures.

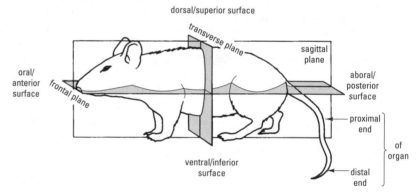

Fig. 44.4 The terms used for various parts and sections through the animal body.

(a)

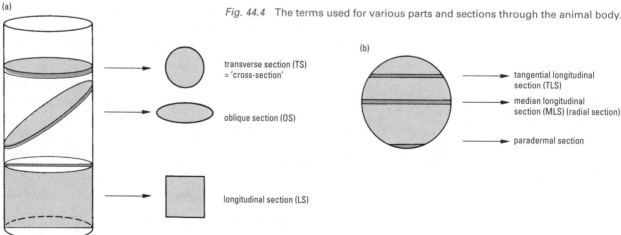

Fig. 44.5 The main section planes: (a) sections through a cylindrical object; (b) various types of longitudinal sections as seen in transverse orientation.

Table 44.1 The main stains used in electron microscopy

Stain	What it stains
Lead citrate	Lipoproteins Ribonucleoproteins Glycogen
Osmium tetroxide	Lipids and proteins
Phosphotungstic acid	Proteins (at pH >3) Polysaccharides Glycoproteins (at pH <3)
Uranyl acetate	Proteins Nucleic acids (at pH 3.5)

Comparing methods – *it is informative to compare the same tissue prepared and viewed by different methods.*

Accounting for methodology

As discussed on pp. 244, 246–51, a great deal of preparative work is required before specimens can be viewed under LM or TEM. Under normal circumstances, it is hoped that the stain has the greatest effect on the image. The colours produced in various cell types and structures by the main stains used in light microscopy are tabulated on p. 249. You should bear in mind that colours are subjective and that dyes may exhibit different colours in different chemical environments (e.g. at different pH values).

Stains for TEM are generally compounds or ions containing elements of high atomic number, such as the heavy metals, because these scatter electrons strongly. On a positive micrograph from a positively stained specimen, dark regions show where electrons have been scattered, i.e. where the stain has attached. Light regions show where electrons have passed relatively unaffected through the specimen, i.e. where the stain has not been taken up. Table 44.1 shows the main stains used in TEM and the specific cell components to which they attach.

Images from SEM are relatively easy to interpret because the brain is used to interpreting surface features. However, it is possible for optical illusions to occur – for instance, it may be difficult to decide whether a shadowed feature projects out of a surface or represents a depression in it (Chapter 27).

Artefacts
The following is a list of the main types of artefact occurring during the preparation process:

- Fixation: swelling or shrinkage may occur, often due to the use of hypotonic or hypertonic fixative solutions. Glutaraldehyde-fixed tissue is particularly likely to shrink if this fixative is used at high concentration.
- Dehydration: shrinkage may occur as water is withdrawn from the tissue. Lipids may dissolve in the solvent used and be lost from the tissue.
- Embedding: difficulties in sectioning may be caused by incorrect hardness of embedding resins.
- Sectioning: too thick a section can result in a blurred image. Compression effects, tearing and knife chattering may occur.
- Mounting: folds may be introduced into the section. These can be detected by the presence of discontinuities in features that cross the fold. In TEM, the pattern of the metal grid may sometimes obscure the image (as a solid black area).
- The stain(s) used on the section: if staining is not carried out correctly, precipitates of stain may appear, often in crystalline form.

 KEY POINT *No preparative procedure gives a 'perfect' image. Artefacts can be introduced at many stages.*

Stereological studies

Stereology allows the study of the three-dimensional organisation of specimens based on two-dimensional information. The subject is a complex one, relying on geometric and statistical principles, and specialist texts should be consulted for detailed background. There are three main approaches:

- Cells or organelles are assumed to have a regular geometric shape, whose surface area or volume can be estimated from measurements of the principal axes (see Table 64.3). Cell and organelle shapes are generally so irregular as to make such estimates very approximate.
- A physical or computer-generated model of structures in three dimensions is built up from a set of serial sections. Considerable expertise is required to obtain the basic information required.
- Representative sections are sampled and information derived from the profiles seen. This method can be based on simple count data and is surprisingly informative. Three parameters of interest are:
  (a) Numerical density, N_V (the numbers of a component found per unit volume). For spherical components, this can be estimated from:

$$N_V = N_A/\bar{D} \qquad [44.3]$$

where N_A is the number of components per unit area and $\bar{D}$ is their mean diameter. N_A is simply obtained by counting component profiles in defined areas. The mean diameter (of spheres) will be approximately equal to the largest diameters seen in the sample, or may be calculated from $\bar{D} = 4/\pi\bar{d}$, where $\bar{d}$ is the mean component diameter of the sample. The situation for non-spherical components is more complex.
  (b) Surface density, S_V (the surface area of the component per unit containing volume). For any shape of object, this can be estimated from:

$$S_V = 4/\pi B_A \qquad [44.4]$$

where B_A is the length of profile boundaries per unit area of section. Profile boundary lengths can be measured by fitting a thread or by using a map-measuring wheel or a planimeter. An even simpler method is to count the number of intersections of the bounding membrane I_i with test lines of total length L_T on an overlay (Fig. 44.6). Then S_V is obtained from:

$$S_V = 2I_i/L_T \qquad [44.5]$$

Note that this method depends on the components being randomly oriented with respect to the lines on the overlay; this can be tested by comparing results with the lines placed in different orientations.
  (c) Volume density, V_V (the volume of the component as a proportion of the total volume). This is numerically equivalent to A_A, the component density on the test area. A_A can be most simply estimated:
  (i) Destructively, by the cut-and-weigh method. Here, the total area to be considered is cut out, weighed, then the component profiles are cut out and weighed: the ratio of weights equals A_A.
  (ii) Non-destructively, by point counting, where a patterned overlay or lattice (e.g. Fig. 44.6) is placed over the micrograph and counts made of the number of points overlaying the feature of interest and the total area. A_A is estimated by the ratio of these point counts.

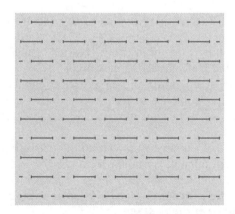

Fig. 44.6 Weibel's multipurpose test overlay for stereological studies. The number of intersections of a membrane with the solid lines can be used when estimating surface densities, while point counts for estimates of profile density on test areas can be made using the ends of the lines.

Both of these methods of estimating A_A require that a reasonably large representative area is sampled. Other methods for estimating A_A include planimetry and linear integration.

Sources for further study

Anon. *Size and Scale: on the Relative Size of Things in the Universe*. Available: http://invsee.asu.edu/nmodules/sizes/sizescalemod/microscopy.htm Last accessed 09/04/07.
[Scale in relation to the history of microscopy.]

Crang, R.F.E. and Klomparens, K.L. (1988) *Artefacts in Biological Electron Microscopy*. Plenum Press, New York.

Russ, J.C. and Dehoff, R.T. (2000) *Practical Stereology*, 2nd edn. Plenum Publishing Corporation, New York.

Study exercises

44.1 Calculate the size of an object seen down a microscope using graticule readings. At ×100 overall magnification, you find that 1 stage micrometer unit of 0.1 mm = 11 eyepiece micrometer units. The length of a parasitic worm on a slide viewed at this magnification appears to be 25 eyepiece units. What is its estimated true length to the nearest μm?

44.2 Calculate the area density of objects seen down a microscope. A light microscope at ×400 overall magnification has a field diameter estimated as 0.425 mm. Observing a sample of leaf surface at this magnification, you count 25 hairs in a single field of view. What is your estimate of the area density (frequency) of hairs, expressed on a per mm^2 basis, to three significant figures?

44.3 Calculate the volume density of objects seen down a microscope. A diluted blood sample viewed at the same magnification as in Study exercise 43.2 gives a mean count of 32 white blood cells per field of view. If the thickness of the liquid layer below the coverslip is 0.1 mm, what is the volume density of white blood cells (express on a per mm^3 basis, to four significant figures)?

44.4 Use point count data to estimate relative volumes of cellular components. A research team fixed and embedded leaf mesophyll cells, sectioned them randomly, stained the sections appropriately and used a TEM to obtain photomicrographs. Then, using an overlay like the one illustrated in Fig. 44.6, they made point counts within each of 25 cell outlines and took a mean of the results (see table below). From their data, answer the following questions, to two decimal places:

(a) What percentage of the cell volume did the cell wall account for?

(b) What percentage of the cell volume was taken up by the cytoplasm?

(c) What is the estimated cyoplasm : vacuole ratio for this cell?

(d) What percentage of the cytoplasm was taken up by chloroplasts?

Point counts over random mesophyll sections

Organelle	Mean point count
Cell wall	34.0
Cytosol	35.0
Nucleus	16.9
Mitochondria	5.1
Chloroplasts	160.8
Other cytoplasmic bodies	2.6
Vacuole	264.7
Total	519.1

Sterile technique (aseptic technique) is the name given to the procedures used in cell culture. Although the same general principles apply to all cell types, you are most likely to learn the basic procedures using bacteria and most of the examples given in this section refer to bacterial culture.

Sterile technique serves two main purposes:

1. To prevent accidental contamination of laboratory cultures due to microbes from external sources, e.g. skin, clothing or the surrounding environment.
2. To prevent microbial contamination of laboratory workers, in this instance you and your fellow students.

 KEY POINT All *microbial cell cultures should be treated as if they contained potentially harmful organisms. Sterile technique forms an important part of safety procedures, and must be followed whenever cell cultures are handled in the laboratory.*

Care is required because:

- You may accidentally isolate a harmful microbe as a contaminant when culturing a relatively harmless strain.
- Some individuals are more susceptible to infection and disease than others – not everyone exposed to a particular microbe will become ill.
- Laboratory culture involves purifying and growing large numbers of microbial cells – this represents a greater risk than small numbers of the original microbe.
- A microbe may change its characteristics, perhaps as a result of gene exchange or mutation.

The international biohazard symbol, shown in Fig. 45.1, is used to indicate a significant risk due to a pathogenic microbe (pp. 267–8).

Fig. 45.1 International symbol for a biohazard. Usually red on a yellow background, or black on a red background.

Sterilisation procedures

Given the ubiquity of microbes, the only way to achieve a sterile state is by their destruction or removal. Several methods can be used to achieve this objective:

Heat treatment

This is the most widespread form of sterilisation and is used in several basic laboratory procedures including the following:

- Red-heat sterilisation. Achieved by heating metal inoculating loops, forceps, needles, etc. in a Bunsen flame (Fig. 45.2). This is a simple and effective form of sterilisation as no microbe will survive even a brief exposure to a naked flame. Flame sterilisation using alcohol is used for glass rods and spreaders (see below).
- Dry-heat sterilisation. Here, a hot-air oven is used at a temperature of at least 160 °C for at least 2 hours. This method is used for the routine sterilisation of laboratory glassware. Dry-heat procedures are of little value for items requiring repeated sterilisation during use.

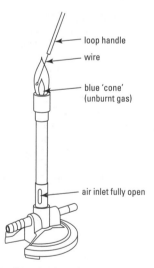

loop handle
wire
blue 'cone' (unburnt gas)
air inlet fully open

Fig. 45.2 'Flaming' a wire loop. Keep the loop in the hottest part of the Bunsen flame (just outside the blue 'cone') until the wire is red-hot.

Fig. 45.3 Autoclave tape – the bottom sample is untreated while the upper sample (with dark diagonal lines) has been autoclaved.

- Moist-heat sterilisation. This is the method of choice for many laboratory items, including most fluids, apart from heat-sensitive media. It is also used to decontaminate liquid media and glassware after use. The laboratory autoclave is used for these purposes. Typically, most items will be sterile after 15 minutes at 121 °C, although large items may require a longer period. The rapid killing action results from the latent heat of condensation of the pressurised steam, released on contact with cool materials in the autoclave. Although special heat-sensitive tape (Fig. 45.3) is sometimes used to check that the autoclave is operating correctly, a better approach is to use spores of *Bacillus stearothermophilus*.

Radiation

Many disposable plastic items used in microbiology and cell biology are sterilised by exposure to UV or ionising radiation. They are supplied commercially in sterile packages, ready for use. Ultraviolet radiation has limited use in the laboratory, while ionising radiation (e.g. γ-rays) requires industrial facilities and cannot be operated on a laboratory scale.

Filtration

Heat-labile solutions (e.g. complex macromolecules, including proteins, antibiotics, serum) are particularly suited to this form of sterilisation. The filters come in a variety of shapes, sizes and materials, usually with a pore size of either $0.2\,\mu m$ or $0.45\,\mu m$. The filtration apparatus and associated equipment is usually sterilised by autoclaving, or by dry heat. Passage of liquid through a sterile filter of pore size $0.2\,\mu m$ into a sterile vessel is usually sufficient to remove bacteria but not viruses, so filtered liquids are not necessarily virus-free.

Using a sterile filter *– most filters are supplied as pre-sterilised items. Make sure you follow a procedure that does not contaminate the filter on removal from its protective wrapping.*

Chemical agents

These are known as disinfectants, or biocides, and are most often used for the disposal of contaminated items following laboratory use, e.g. glass slides and pipettes. They are also used to treat spillages. The term 'disinfection' implies destruction of disease-causing bacterial cells, although spores and viruses may not always be destroyed. Remember that disinfectants require time to exert their killing effect – any spillage should be covered with an appropriate disinfectant and left for at least 10 minutes before mopping up.

 SAFETY NOTE *When working with biocides, take care to avoid skin contact or ingestion, as most are toxic and irritant. If contact does occur, rinse with plenty of water.*

Use of laboratory equipment

Working area

One of the most important aspects of good sterile technique is to keep your working area as clean and tidy as possible. Start by clearing all items from your working surface, wipe the bench down with disinfectant and then arrange the items you need for a particular procedure so that they are close at hand, leaving a clear working space in the centre of your bench.

Media

Cells may be cultured in either a liquid medium (broth), or a solidified medium (p. 271). The gelling agent used in most solidified media is agar, a

Using molten agar *– a water bath (at 45–50°C) can be used to keep an agar-based medium in its molten state after autoclaving. Always dry the outside of the container on removal from the water bath, to reduce the risk of contamination from microbes in the water, e.g. during pour plating (p. 268).*

complex polysaccharide from red algae that produces a stiff transparent gel when used at 1–2% (w/v). Agar is used because it is relatively resistant to degradation by most bacteria and because of its rheological properties – an agar medium melts at 98 °C, remaining solid at all temperatures used for routine laboratory culture. Once melted however, it does not solidify again until the temperature falls to about 44 °C. This means that heat-sensitive constituents (e.g. vitamins, blood, cells, etc.) can be added aseptically to the medium after autoclaving.

Inoculating loops

The initial isolation and subsequent transfer of microbes between containers can be achieved by using a sterile inoculating loop. Most teaching laboratories use nichrome wire loops in a metal handle. A wire loop can be repeatedly sterilised by heating the wire, loop downwards and almost vertical, in the hottest part of a Bunsen flame until the whole wire becomes red-hot. Then the loop is removed from the flame to minimise heat transfer to the handle. After cooling for 8–10 seconds (without touching any other object), it is ready for use.

When re-sterilising a contaminated wire loop in a Bunsen flame after use, do not heat the loop too rapidly, as the sample may spatter, creating an aerosol: it is better to soak the loop for a few minutes in disinfectant than to risk heating a fully charged (contaminated) inoculating loop.

Containers

There is a risk of contamination whenever a sterile bottle, flask or test tube is opened. One method that reduces the chance of airborne contamination is quickly to pass the open mouth of the glass vessel through a flame. This destroys any microbes on the outer surfaces nearest to the mouth of the vessel. In addition, by heating the air within the neck of the vessel, an outwardly directed air flow is established, reducing the likelihood of microbial contamination.

It is general practice to flame the mouth of each vessel immediately after opening and then repeat the procedure just before replacing the top. Caps, lids and cotton wool plugs must not be placed on the bench during flaming and sampling: they should be removed and held using the smallest finger of one hand, to minimise the risk of contamination. This also leaves the remaining fingers free to carry out other manipulations. With practice, it is possible to remove the tops from two tubes, flame each tube and transfer material from one to the other while holding one top in each hand.

Laminar-flow cabinets

These are designed to prevent airborne contamination, e.g. when preparing media or subculturing microbes or tissue cultures. Sterile air is produced by passage through a high-efficiency particulate air (HEPA) filter: this is then directed over the working area, either horizontally (towards the operator) or downwards. The operator handles specimens, media, etc., through an opening at the front of the cabinet. Note that standard laminar-flow cabinets do *not* protect the worker from contamination and must not be used with pathogenic microbes: special safety cabinets are used for work with ACDP hazard group 3 and 4 microbes (Table 45.1) and for samples that might contain such pathogens.

Plastic disposable loops – *these are used in many research laboratories: pre-sterilised and suitable for single use, they avoid the hazards of naked flames and the risk of aerosol formation during heating. Discard into a disinfectant solution after use.*

Using a Bunsen burner to reduce airborne contamination – *working close to the updraught created by a Bunsen flame reduces the likelihood of particles falling from the air into an open vessel.*

Using glass pipettes – *these are plugged with cotton wool at the top before being autoclaved inside a metal can. Flame the open end of the can on removal of a pipette, to prevent contamination of the remaining pipettes. Autopipettors and sterile disposable tips (p. 123) offer an alternative approach.*

Table 45.1 Classification of microbes on the basis of hazard. The following categories are recommended by the UK Advisory Committee on Dangerous Pathogens (ACDP).

Hazard group	Comments
1	Unlikely to cause human disease
2	May cause disease: possible hazard to laboratory workers, minimal hazard to community
3	May cause severe disease: may be a serious hazard to laboratory workers, may spread to community
4	Causes severe disease: is a serious hazard to laboratory workers, high risk to community

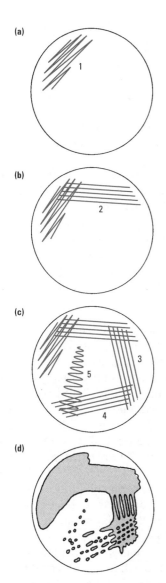

Fig. 45.4 Preparation of a streak plate for single colonies. (a) Using a sterile metal loop, take a small sample of the material to be streaked. Distribute the sample over a small sector of the plate (area 1), then flame the loop and allow to cool (approximately 8–10 s). (b) Make several small streaks from the initial sector into the adjacent sector (area 2), taking care not to allow the streaks to overlap. Flame the loop and allow to cool. (c) Repeat the procedure for areas 3 and 4, re-sterilising the loop between each step. Finally, make a single, long streak, as shown for area 5. (d) The expected result after incubation at the appropriate temperature (e.g. 37 °C for 24 h): each step should have diluted the inoculum, giving individual colonies within one or more sectors on the plate. Further subculture of an individual colony should give a pure (clonal) culture.

Microbiological hazards

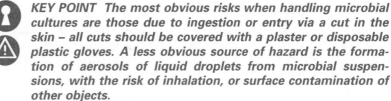

KEY POINT The most obvious risks when handling microbial cultures are those due to ingestion or entry via a cut in the skin – all cuts should be covered with a plaster or disposable plastic gloves. A less obvious source of hazard is the formation of aerosols of liquid droplets from microbial suspensions, with the risk of inhalation, or surface contamination of other objects.

The following steps will minimise the risk of aerosol formation:

- Use stoppered tubes when shaking, centrifuging or mixing microbial suspensions.
- Pour solutions gently, keeping the difference in height to a minimum.
- Discharge pipettes on to the side of the container.

Other general rules that apply in all laboratories include:

- Take care with sharp instruments, including needles and glass Pasteur pipettes.
- Do not pour waste cultures down the sink – they must be autoclaved.
- Put other contaminated items (e.g. slides, pipettes) into disinfectant after use.
- Wipe down your bench with disinfectant when practical work is complete.
- Always wash your hands before leaving the laboratory.

Plating methods

Many culture methods make use of a solidified medium within a Petri plate. A variety of techniques can be used to transfer and distribute the organisms prior to incubation. The three most important procedures are described below.

Streak-dilution plate

Streaking a plate for single colonies is one of the most important basic skills in microbiology, since it is used in the initial isolation of a cell culture and in maintaining stock cultures, where a streak-dilution plate with single colonies all of the same type confirms the purity of the strain. A sterile inoculating loop is used to streak the organisms over the surface of the medium, thereby diluting the sample (Fig. 45.4). The aim is to achieve single colonies at some point on the plate: ideally, such colonies are derived from single cells (e.g. in the case of unicellular bacteria, animal and plant cell lines) or from groups of cells of the same species (in filamentous or colonial forms). Single colonies, containing cells of a single species and derived from a single parental cell, form the basis of all pure culture methods (p. 212).

Note the following:

- Keep the lid of the Petri plate as close to the base as possible to reduce the risk of aerial contamination.
- Allow the loop to glide over the surface of the medium. Hold the handle at the balance point (near the centre) and use light, sweeping movements, as the agar surface is easily damaged and torn.
- Work quickly, but carefully. Do not breathe directly on to the exposed agar surface and replace the lid as soon as possible.

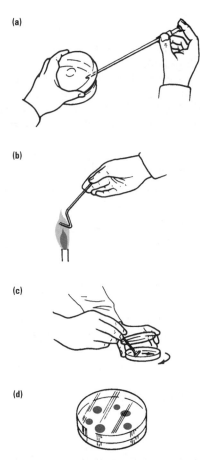

(a)

(b)

(c)

(d)

Fig. 45.5 Preparation of a spread plate. (a) Transfer a small volume of cell suspension (0.05–0.5 ml) to the surface of a solidified medium in a Petri plate. (b) Flame sterilise a glass spreader and allow to cool (8–10 s). (c) Distribute the liquid over the surface of the plate using the sterile spreader. Make sure of an even coverage by rotating the plate as you spread: allow the liquid to be absorbed into the agar medium. Incubate under suitable conditions. (d) After incubation, the microbial colonies should be distributed over the surface of the plate.

Spread plate

This method is used with cells in suspension, either in a liquid growth medium or in an appropriate sterile diluent. It is one method of quantifying the number of viable cells (or colony forming units) in a sample, after appropriate dilution (p. 131).

An L-shaped glass spreader is sterilised by dipping the end of the spreader in a beaker containing a small amount of 70% v/v alcohol, allowing the excess to drain from the spreader and then igniting the remainder in a Bunsen flame. After cooling, the spreader is used to distribute a known volume of cell suspension across the plate (Fig. 45.5). *There is a significant fire risk associated with this technique*, so take care not to ignite the alcohol in the beaker, e.g. by returning an overheated glass rod to the beaker. The alcohol will burn with a pale blue flame that may be difficult to see, but will readily ignite other materials (e.g. a laboratory coat). Another source of risk comes from small droplets of flaming alcohol shed by an overloaded spreader on to the bench and this is why you *must* drain excess alcohol from the spreader *before* flaming. Some laboratories now provide plastic disposable spreaders for student use, to avoid the risk of fire.

Pour plate

This procedure also uses cells in suspension, but requires molten agar medium, usually in screw-capped bottles containing sufficient medium to prepare a single Petri plate (i.e. 15–20 ml), maintained in a water bath at 45–50 °C. A known volume of cell suspension is mixed with this molten agar, distributing the cells throughout the medium. This is then poured without delay into an empty sterile Petri plate and incubated, giving widely spaced colonies (Fig. 45.6). Furthermore, as most of the colonies are formed within the medium, they are far smaller than those of the surface streak method, allowing higher cell numbers to be counted (e.g. up to 1000 colonies per plate): some workers pour a thin layer of molten agar on to the surface of a pour plate after it has set, to ensure that no surface colonies are produced. Most bacteria and fungi are not killed by brief exposure to temperatures of 45–50 °C, though the procedure may be more damaging to microbes from low temperature conditions, e.g. psychrophilic bacteria.

One disadvantage of the pour-plate method is that the typical colony morphology seen in surface-grown cultures will not be observed for those colonies that develop within the agar medium. A further disadvantage is that some of the suspension will be left behind in the screw-capped bottle. This can be avoided by transferring the suspension to the Petri plate, adding the molten agar, then swirling the plate to mix the two liquids. However, even when the plate is swirled repeatedly and in several directions, the liquids are not mixed as evenly as in the former procedure.

Working with phages

Bacterial viruses ('bacteriophages', or simply 'phages') are often used to illustrate the general principles involved in the detection and enumeration of viruses. They also have a role in genome mapping of bacteria. Individual phage particles (virions) are too small to be seen by light microscopy, but are detected by their effects on susceptible host cells:

- Virulent phages will infect and replicate within actively growing host cells, causing cell lysis and releasing new infective phages – this 'lytic cycle' takes ≈30 min for T-even phages of *E. coli*, e.g. T4.

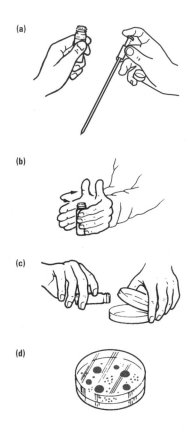

Fig. 45.6 Preparation of a pour plate. (a) Add a known volume of cell suspension (0.05–1.0 ml) to a small bottle of molten agar medium from a 45 °C water bath. (b) Mix thoroughly, by rotating between the palms of the hands: do not shake or this will cause frothing of the medium. (c) Pour the mixture into an empty, sterile Petri plate and allow to set. Incubate under suitable conditions. (d) After incubation, the microbial colonies will be distributed throughout the medium: any cells deposited at the surface will give larger, spreading colonies.

● Temperate phages are a specialised group, capable either of lytic growth or an alternative cycle, termed lysogeny – the phage becomes latent within a host cell (lysogen), typically by insertion of its genetic information into the host cell genome, becoming a 'prophage'. At a later stage, termed induction, the prophage may enter the lytic cycle. A widely used example is λ phage of *E. coli*.

The lytic cycle can be used to detect and quantify the number of phages in a sample. A known volume of sample is mixed with susceptible bacterial cells in molten soft agar medium (45–50 °C), then poured on top of a plate of the same medium, creating a thin layer of 'top agar'. The upper layer contains only half the normal amount of agar, to allow phages to diffuse through the medium and attach to susceptible cells. On incubation, the bacteria will grow throughout the agar to produce a homogeneous 'lawn' of cells, except in those parts of the plate where a phage particle has infected and lysed the cells to create a clear area, termed a plaque (Fig. 45.7). Each plaque is due to a single functional phage (i.e. a plaque-forming unit, or PFU). A count of the number of plaques can be used to give the number of phages in a particular sample (e.g. as $PFU\,ml^{-1}$), with appropriate correction for dilution and the volume of sample counted in an analogous manner to a bacterial plate count (p. 276). When counting plaques in phage assays you should view them against a black background to make them easier to see: mark each plaque with a spirit-based marker to ensure an accurate count. Temperate phages often produce cloudy plaques, because many of the infected cells will be lysogenised rather than lysed, creating turbidity within the plaque. Samples of material from within the plaque can be used to subculture the phage for further study, perhaps in a broth culture where the phages will cause widespread cell lysis and a decrease in turbidity. Alternatively, phages can be stored by adding chloroform to aqueous suspensions – this will prevent contamination by cellular micro-organisms. A similar approach can be used to detect and count animal or human viruses, using a monolayer of susceptible host cells.

Electron microscopy (EM, p. 243) provides an alternative approach to the detection of viruses, avoiding the requirement for culture of infected host cells, and giving a faster result. However, it requires specialised equipment and expertise. EM counts are often higher than culture-based methods, for similar reasons to those described for bacteria (p. 274).

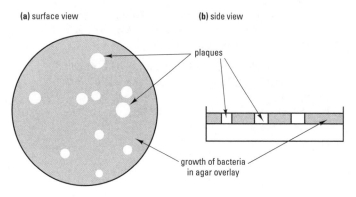

(a) surface view   **(b)** side view

plaques

growth of bacteria in agar overlay

Fig. 45.7 Phage plaques in a 'lawn' of susceptible host bacterium

Labelling Petri plates – *the following information should be recorded on the base:*

- *date*
- *the growth medium*
- *your name or initials*
- *brief details of the experimental treatment.*

Labelling your plates and cultures

Petri plates should always be labelled on the *base*, rather than the lid. Restrict your labelling to the outermost region of the plate, to avoid problems when counting colonies, assessing growth, etc. After labelling, Petri plates usually are incubated upside down in a temperature-controlled incubator (often at 37 °C) for an appropriate period (usually 18–72 hours). Plates are also usually kept upside down on the lab bench – following incubation, the base (containing medium and microbes) can then be lifted from the lid and examined.

Sources for further study

Anon. (1995) *Advisory Committee on Dangerous Pathogens: Categorisation of Biological Agents According to Hazard and Categories of Containment*, 4th edn. HSE Books, London.

Anon. (2004) *Advisory Committee on Dangerous Pathogens: Approved List of Biological Agents.* Available: http://www.hse.gov.uk/pubns/misc208.pdf Last accessed 09/04/07.

Collins, C.H., Lyne, P.M., Grange, J.M. and Falkinham, J. (2004) *Collins and Lyne's Microbiological Methods*, 8th edn. Hodder-Arnold, London.

Hawkey. P. and Lewis, D. (2004) *Medical Bacteriology: A Practical Approach,* 2nd edn. Oxford University Press, Oxford.

Rhodes, P.M. and Stanbury, P.F. (1997) *Applied Microbial Physiology: a Practical Approach.* Oxford University Press, Oxford.

Study exercises

45.1 Decide on the best method of sterilisation. What would be the most appropriate method of sterilisation for the following items?
   (a) A box of 100 plastic tips to be used with a pipettor.
   (b) A 50 ml batch of blood, for use in 5% v/v blood agar plates.
   (c) A 1 litre batch of MacConkey agar.
   (d) Ten 5 ml glass pipettes.
   (e) A microbiological wire, used for 'stab' cultures.
   (f) A 10 ml sample of a heat-sensitive solution of an antibiotic, to be used as a component of a selective isolation medium.

45.2 Find out the biohazard classification (UK ACDP categorisation – Table 45.1) for the following microbes:
   (a) *Salmonella typhimurium*
   (b) *Leptospira interrogans*
   (c) *Shigella dysenteriae* (type 1)
   (d) *E. coli* K12
   (e) *E. coli* O157
   (f) Human immunodeficiency virus (HIV)
   (g) *Cryptococcus neoformans* var. *neoformans*
   (h) *Mycobacterium tuberculosis*
   (i) *Lactobacillus plantarum*
   (j) Marburg virus.

45.3 Consider the advantages and disadvantages of spread-plating and pour-plating methods. Having read through this chapter, list up to six pros/cons of each plating method and compare your answers either with the list we have provided on p. 453, or with those of other students, as a group exercise.

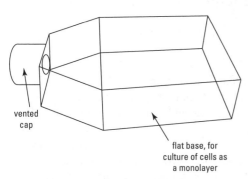

vented
cap

flat base, for
culture of cells as
a monolayer

Fig. 46.1 Plastic flask for animal cell culture – this design provides a large surface area for growth of an adherent monolayer of cells.

Microbial, animal and plant cell culture methods are based on the same general principles, requiring:

- a pure culture (also known as an axenic culture), perhaps isolated as part of an earlier procedure, or from a culture collection;
- a suitable nutrient medium to provide the necessary components for growth. This medium must be sterilised before use;
- satisfactory growth conditions including temperature, pH, atmospheric requirements, ionic and osmotic conditions;
- sterile technique (p. 264) to maintain the culture in pure form.

Heterotrophic animal cells, fungi and many bacteria require appropriate organic compounds as sources of carbon and energy. Non-exacting bacteria can utilise a wide range of compounds and they are often grown in media containing complex natural substances (including meat extract, yeast extract, soil, blood). Animal cells have more stringent growth requirements (p. 281).

Photoautotrophic bacteria, cyanobacteria and algae are grown in a mineral medium containing inorganic ions including chelated iron, with a light source and CO_2 supply. Plant cells may require additional vitamins and hormones (p. 284). For chemoautotrophic bacteria, the light source is replaced by a suitable inorganic energy source, e.g. H_2S for sulphur-oxidising bacteria, NH_3/NH_4^+ for nitrifying bacteria, etc.

Growth on solidified media

Many organisms can be cultured on an agar-based medium (p. 265).

 KEY POINT *An important benefit of agar-based culture systems is that an individual cell inoculated onto the surface can develop to form a visible colony: this is the basis of most microbial isolation and purification methods, including the streak dilution, spread plate and pour plate procedures (p. 267).*

Animal cells are often grown as an adherent monolayer on the surface of a plastic or glass culture vessel (Fig. 46.1), rather than on an agar-based medium (p. 281).

Several types of culture vessel are used:

- Petri plates (Petri dishes): usually the pre-sterilised, disposable plastic type, providing a large surface area for growth.
- Glass bottles or test tubes: these provide sufficient depth of agar medium for prolonged growth of bacterial and fungal cultures, avoiding problems of dehydration and salt crystallisation. Inoculate aerobes on the surface and anaerobes by stabbing down the centre, into the base (stab culture).
- Flat-sided bottles: these are used for animal cell culture, to provide an increased surface area for attachment and allow growth of cells as a surface monolayer. Usually plastic and disposable.

The dynamics of growth are usually studied in liquid culture, apart from certain rapidly growing filamentous fungi, where increases in colony diameter can be measured accurately, e.g. using Vernier calipers (p. 134).

Cell culture

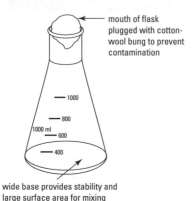

mouth of flask plugged with cotton-wool bung to prevent contamination

1000
800
1000 ml
600
400

wide base provides stability and large surface area for mixing

Fig. 46.2 Conical (Erlenmeyer) flask

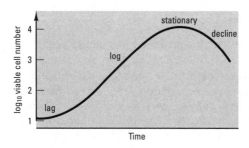

Fig. 46.3 Population growth curve for cells in batch culture (liquid medium).

Example Suppose you counted 2×10^3 cells ($\log_{10} = 3.30$) per unit volume at t_0 and 6.3×10^4 cells ($\log_{10} = 4.80$) after 2 h (t_x).

Substitution into eqn [46.1] gives $[2.303(4.8 - 3.3)] \div 2 = 1.727 \, h^{-1}$ (or $0.0288 \, min^{-1}$).

Substituting the same values into eqn [46.2] gives:
$[0.301 \times 2] \div [4.8 - 3.3] = 0.40 \, h$ (or 24 min).

Working with logarithms – *note that there is no log value for zero, so you cannot plot zero on a log–linear growth curve or on a death curve.*

Growth in liquid media

Many cells, apart from primary cultures of animal cells, can be grown as a homogeneous unicellular suspension in a suitable liquid medium, where growth is usually considered in terms of cell number (population growth) rather than cell size. Most liquid culture systems need agitation, to ensure adequate mixing and to keep the cells in suspension. An Erlenmeyer flask of 100–2000 ml capacity (Fig. 46.2) can be used to grow a batch culture on an orbital shaker, operating at 20–250 cycles per minute. For aerobic organisms, the surface area of such a culture should be as large as possible: restrict the volume of medium to not more than 20 per cent of the flask volume. Larger cultures may need to be gassed with sterile air and mixed using a magnetic stirrer rather than an orbital shaker. The simplest method of air sterilisation is filtration, using glass wool, non-absorbent cotton wool or a commercial filter unit of appropriate pore size (usually $0.2 \, \mu m$). Air is introduced via a sparger (a glass tube with many small holes, so that small bubbles are produced) near the bottom of the culture vessel to increase the surface area and enhance gas exchange. More complex systems have baffles and paddles to further improve mixing and gas exchange.

Liquid culture systems may be subdivided under two broad headings:

Batch culture

This is the most common approach for routine liquid culture. Cells are inoculated into a sterile vessel containing a fixed amount of growth medium. Your choice of vessel will depend upon the volume of culture required: larger-scale vessels (e.g. 1 litre and above) are often called 'fermenters' or 'bioreactors', particularly in biotechnology. Growth within the vessel usually follows a predictable S-shaped (sinusoidal) curve when plotted in log–linear format (Fig. 46.3), divided into four components:

1. Lag phase: the initial period when no increase in cell number is seen. The larger the inoculum of active cells the shorter the lag phase will be, provided the cells are transferred from similar growth conditions.
2. Log phase, or exponential phase: where cells are growing at their maximum rate. This may be quantified by the growth rate constant, or specific growth rate (μ), where:

$$\mu = \frac{2.303 \, (\log N_x - \log N_0)}{(t_x - t_0)} \qquad [46.1]$$

where N_0 is the initial number of cells at time t_0 and N_x is the number of cells at time t_x. For times specified in hours, μ is expressed as h^{-1}.

Prokaryotes grow by binary fission, whereas eukaryotes grow by mitotic cell division; in both cases each cell divides to give two identical offspring. Consequently, the doubling time or generation time (g, or T_2) is:

$$g = \frac{0.301 \, (t_x - t_0)}{\log N_x - \log N_0} \qquad [46.2]$$

Cells grow at different rates, with doubling times ranging from under 20 min for some bacteria to 24 h or more for animal and plant cells. Exponential phase cells are often used in laboratory experiments, since growth and metabolism are nearly uniform.

3. Stationary phase: growth decreases as nutrients are depleted and waste products accumulate. Any increase in cell number is offset by death. This phase is usually termed the 'plateau' in animal cell culture.
4. Decline phase, or death phase: this is the result of prolonged starvation and toxicity, unless the cells are subcultured. Like growth, death often shows an exponential relationship with time, which can be characterised by a rate constant (death rate constant), equivalent to that used to express growth or, more often, as the decimal reduction time (d, or T_{90}), the time required to reduce the population by 90 per cent:

$$d = \frac{t_x - t_0}{\log N_x - \log N_0} \quad [46.3]$$

Some cells undergo rapid autolysis at the end of the stationary period while others show a slower decline.

Batch-culture methods can be used to maintain stocks of particular organisms; cells are subcultured on to fresh medium before they enter the decline phase. However, primary cultures of animal cells have a finite life unless transformed to give a continuous cell line, capable of indefinite growth (p. 281).

Continuous culture

This is a method of maintaining cells in exponential growth for an extended period by continuously adding fresh growth medium to a culture vessel of fixed capacity. The new medium replaces nutrients and displaces some of the culture, diluting the remaining cells and allowing further growth.

After inoculating the vessel, the culture is allowed to grow for a short time as a batch culture, until a suitable population size is reached. Then medium is pumped into the vessel: the system is usually set up so that any increase in cell number due to growth will be offset by an equivalent loss due to dilution, i.e. the cell number within the vessel is maintained at a steady state. The cells will be growing at a particular rate (μ), counterbalanced by dilution at an equivalent rate (D):

$$D = \frac{\text{flow rate}}{\text{vessel volume}} \quad [46.4]$$

where D is expressed per unit time (e.g. h^{-1}). In a chemostat, the growth rate is limited by the availability of some nutrient in the inflowing medium, usually either carbon or nitrogen (see Fig. 46.4). In a turbidostat, the input of medium is controlled by the turbidity of the culture, measured using a photocell. A turbidostat is more complex than a chemostat, with additional equipment and controls.

To determine the specific growth rate (μ) of a continuous culture:

1. Measure the flow of medium through the vessel over a known time interval (e.g. connect a sterile measuring cylinder or similar volumetric device to the outlet), to calculate the flow rate.
2. Divide the flow rate by the vessel volume (eqn [46.4]) to give the dilution rate (D).
3. This equals the specific growth rate, since $D = \mu$ at steady state.

Example Suppose you counted 5.2×10^5 cells ($\log_{10} = 5.716$) per unit volume at t_o and 3.7×10^3 cells ($\log_{10} = 3.568$) after 60 min (t_x). Substitution into eqn [46.3] gives $60 \div [2.148] = 27.9$ min. To the nearest minute, this gives a value for d of 28 min.

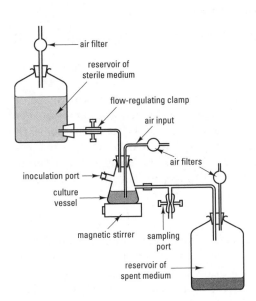

Fig. 46.4 Example of a two-dimensional lab equipment diagram of components of a chemostat

Example Suppose a continuous culture system of 2000 ml volume had a flow of 600 ml over a period of 40 min (flow rate $600 \div 40 = 15$ ml min^{-1}. Substitution into eqn [46.4] gives a dilution rate D of $15 \div 2000 = 0.00075$ min^{-1} or $0.00075 \times 60 = 0.45$ h^{-1}.

4. If you want to know the doubling time (g), calculate using the relationship:

$$g = \frac{0.693}{\mu} \qquad [46.5]$$

(Note that eqn [46.5] also applies to exponential phase cells in batch culture and is useful for interconverting g and μ.)

Continuous culture systems are more complex to set up than batch cultures. They are prone to contamination, having additional vessels for fresh medium and waste culture: strict aseptic technique is necessary when the medium reservoir is replaced, and during sampling and harvesting. However, they offer several advantages over batch cultures, including the following:

- The physiological state of the cells is more clearly defined, since actively growing cells at the same stage of growth are provided over an extended time period. This is useful for biochemical and physiological studies.
- Monitoring and control can be automated and computerised.
- Modelling can be carried out for biotechnology/fermentation technology.

Measuring growth in cell cultures

The most widely used methods of measuring growth are based on cell number.

Direct microscopic counts

One of the simplest methods is to count the cells in a known volume of medium using a microscope and a counting chamber or haemocytometer (Box 46.1). While this gives a rapid assessment of the total cell number, it does not discriminate between living and dead cells. It is also time-consuming as a large number of cells must be counted for accurate measurement. It may be difficult to distinguish individual cells, e.g. for cells growing as clumps.

Electronic particle counters

These instruments can be used to give a direct (total) count of a suspension of microbial cells. The Coulter counter detects particles due to change in electrical resistance when they pass through a small aperture in a glass tube (Fig. 46.5). It gives a rapid count based on a larger number of cells than direct microscopy. It is well suited for repeat measurements or large sample numbers and can be linked to a microcomputer for data processing. If correctly calibrated, the counter can also measure cell sizes. A major limitation of electronic counters is the lack of discrimination between living cells, dead cells, cell clumps and other particles (e.g. dust). In addition, the instrument must be set up and calibrated by trained personnel. Flow cytometry is a more specialised alternative, since particles can be sorted as well as counted.

Culture-based counting methods

A variety of culture-based techniques can be used to determine the number of microbes in a sample. A major assumption of such methods is

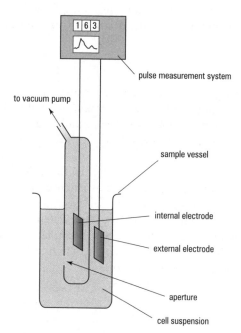

Fig. 46.5 Components of an electronic particle counter. During operation, the cell suspension is drawn through the aperture by the vacuum, creating a 'pulse' of resistance between the two electrodes as each cell passes through the aperture.

to vacuum pump
pulse measurement system
sample vessel
internal electrode
external electrode
aperture
cell suspension

Box 46.1　How to use a counting chamber or haemocytometer

A counting chamber is a specially designed slide containing a chamber of known depth with a grid etched on to its lower surface. When a flat coverslip is placed over the chamber, the depth is uniform. Use as follows:

1. **Place the special coverslip over the chamber.** Press the edges firmly, to ensure that the coverslip makes contact with the surface of the slide, but take care that you do not break the slide or coverslip by using too much force. When correctly positioned, you should be able to see interference rings (Newton's rings) at the edge of the coverslip.

2. **Add a small amount of your cell suspension to fill the central space above the grid.** Place on the microscope stage and allow the cells to settle (2–3 min).

3. **Examine the grid microscopically.** using the ×10 objective lens first, since the counting chamber is far thicker than a standard microscope slide. Then switch to the ×40 objective: take care not to scratch the surface of the objective lens, as the special coverslip is thicker than a normal coverslip. For a dense culture, the small squares are used, while the larger squares are used for dilute suspensions. You may need to dilute your suspension if it contains more than thirty cells per small square.

4. **Count the number of cells in several squares:** at least 600 cells should be counted for accurate measurements. Include those cells that cross the upper and left-hand boundaries, but not those that cross the lower or right-hand rulings (p. 181). A hand tally may be used to aid counting. Motile cells must be immobilised prior to counting (e.g. by killing with a suitable biocide).

5. **Divide the total number of cells (C) by the number of squares counted (S),** to give the mean cell count per square.

6. **Determine the volume (in ml) of liquid corresponding to a single square (V),** e.g. a Petroff–Hausser chamber has small squares of linear dimension 0.2 mm, giving an area of 0.04 mm²; since the depth of the chamber is 0.02 mm, the volume is $0.04 \times 0.02 = 0.0008$ mm³; as there are 1000 mm³ in 1 ml, the volume of a small square is 8×10^{-7} ml; similarly, the volume of a large square (equal to 25 small squares) is 2×10^{-5} ml. Note that other types of counting chamber will have different volumes: check the manufacturer's instructions. For example, the improved Neubauer chamber (Fig. 46.6) has small squares of volume 0.00025 mm³ $= 2.5 \times 10^{-7}$ ml.

Fig. 46.6 Haemocytometer grid (improved Neubauer rulings) viewed microscopically. The large square (delimited by triple etched lines) has a volume of 1/250th mm³ (0.004 mm³ = 4 μl) while each small square (16 contained within the large square) has a volume of 1/4000 mm³ (0.00025 mm³ = 0.25 μl). Note that the boundary line for squares delimited by triple-etched lines is the *middle* line, so this line must be used when counting (see Fig. 30.4, p. 181, for how to deal with objects straddling the gridlines).

7. **Calculate the cell number per ml by dividing the mean cell count per square by the volume of a single square (in ml).**

8. **Remember to take account of any dilution of your original suspension** in your final calculation by multiplying by the reciprocal of the dilution (M), e.g. if you counted a one in twenty dilution of your sample, multiply by twenty, or if you diluted to 10^{-5}, multiply by 10^5.

The complete equation for calculating the total microscopic count is:

$$\text{Total cell count (per ml)} = (C \div S \div V) \times M \quad [46.6]$$

For example, if the mean cell count for a hundred-fold dilution of a cell suspension, counted using a Petroff–Hausser chamber, was 12.4 cells in ten small squares, the total count would be

$$(12.4 \div 10 \div 8 \times 10^{-7}) \times 10^2 = 1.55 \times 10^8 \text{ ml}^{-1}.$$

A simpler, less accurate approach is to use a known volume of sample under a coverslip of known area on a standard glass slide, counting the number of cells per field of view using a calibrated microscope of known field diameter, then multiplying up to give the cell number per ml (see p. 259).

Box 46.2 How to make a plate count of bacteria using an agar-based medium

1. **Prepare serial decimal dilutions of the sample in a sterile diluent (p. 131).** The most widely used diluents are 0.1% w/v peptone water or 0.9% w/v NaCl, buffered at pH 7.3. Take care that you mix each dilution before making the next one. For soil, food, or other solid samples, make the initial decimal dilution by taking 1 g of sample and making this up to 10 ml using a suitable diluent. Gentle shaking or homogenisation may be required for organisms growing in clumps. The number of decimal dilutions required for a particular sample will be governed by your expected count: dilute until the expected number of viable cells is around 100–1 000 ml^{-1}.

2. **Transfer an appropriate volume (e.g. 0.05–0.5 ml) of the lowest dilution to an agar plate** using either the spread-plate method or the pour-plate procedure (p. 268). At least two, and preferably more, replicate plates should be prepared for each sample. You may also wish to prepare plates for more than one dilution, if you are unsure of the expected number of viable cells.

3. **Incubate under suitable conditions for 18–72 h, then count the number of colonies on each replicate plate at the most appropriate dilution.** The most accurate results will be obtained for plates containing 30–300 colonies. Mark the base of the plate with a spirit-based pen each time you count a colony. Determine the mean colony count per plate at this dilution (C).

4. **Calculate the colony count per ml of that particular dilution** by dividing by the volume (in ml) of liquid transferred to each plate (V).

5. **Now calculate the count per ml of the original sample** by multiplying by the reciprocal of the dilution: this is the multiplication factor (M); e.g. for a dilution of 10^{-3}, the multiplication factor would be 10^3. For soil, food or other solid samples, the count should be expressed per g of sample.

The complete equation for calculating the viable count is:

$$\text{Count per ml (or per g)} = (C \div V) \times M \quad [46.7]$$

For example, for a sample with a mean colony count of 5.5 colonies per plate for a volume of 0.05 ml at a dilution of 10^{-7}, the count would be:

$$(5.5 \div 0.05) \times 10^7 = 1.1 \times 10^9 \text{ CFU ml}^{-1}$$

The count should be reported as colony-forming units (CFU) per ml, rather than as cells per ml, since a colony may be the product of more than one cell, particularly in filamentous microbes or in organisms with a tendency to aggregate. You should also be aware of the problems associated with counts of zero – these are best recorded as '<1', and you should then apply the appropriate correction factors for dilution and volume to obtain the detection limit. For example, a zero count (<1) of 100 µl of a five-fold dilution gives a detection limit of $(<1 \div 0.1) \times 5 = <50$ CFU ml^{-1}.

Definition

CFU – colony-forming unit: a cell or group of cells giving rise to a single colony on a solidified medium.

Alternative approaches in plate counting – when large numbers of samples have to be counted, a single agar plate can be divided into segments and a single droplet of each dilution placed into the appropriate segment ('Miles and Misra' droplet counting).

that, under suitable conditions, an individual viable microbial cell will be able to multiply and grow to give a visible change in the growth medium, i.e. a colony on an agar-based medium, or turbidity ('cloudiness') in a liquid medium. You are most likely to gain practical experience using bacterial cultures, counted by one or more of the following methods:

- Spread- or pour-plate methods ('plate counts', p. 268). The most widespread approach is to transfer a suitable amount of the sample to an agar medium, incubate under appropriate conditions and then count the resulting colonies (Box 46.2).
- Membrane filtration. For bacterial samples where the expected cell number is lower than 10 CFU ml^{-1}, pass the sample through a sterile filter (pore size 0.2 µm or 0.45 µm). The filter is then incubated on a suitable medium until colonies are produced, giving a count by dividing the mean colony count per filter by the volume of sample filtered.
- Multiple tube count, or most probable number (MPN). A bacteriological technique where the sample is diluted and known volumes are transferred to several tubes of liquid medium (typically, five tubes at three volumes), chosen so that there is a low probability of the smallest volumes containing a viable cell. After incubation, the

number of tubes showing growth (turbidity) is compared to tabulated values to give the most probable number (MPN per ml).

The principal advantage of culture-based counting procedures is that dead cells will not be counted. However, for such techniques, the incubation conditions and media used may not allow growth of all cells, underestimating the true viable count. Further problems are caused by cell clumping and dilution errors. In addition, such methods require sterile apparatus and media and the incubation period is lengthy before results are obtained. An alternative approach is to use direct microscopy, combined with 'vital' or 'mortal' staining. For example, the direct epifluorescence technique (DEFT) uses acridine orange and UV epifluorescence microscopy to separate living and dead bacteria, while neutral red is a vital stain used for plant cells. Chapter 42 gives examples of vital/mortal stains for other cell types. A further approach is to use DNA-based methods, such as the polymerase chain reaction (PCR, p. 335).

Counting injured or stressed microbes – a resuscitation stage may be required, to allow cells to grow under selective conditions, p. 213.

Sources for further study

Ball, A.S. (1997) *Bacterial Cell Culture: Essential Data.* Wiley, New York.

Cann, A.J. (1999) *Virus Culture: A Practical Approach.* Oxford University Press, Oxford.

Cartledge, T.G. (ed.) (1992) *In Vitro Cultivation of Micro-organisms.* Butterworth-Heinemann, Oxford. [Includes in-depth coverage of the mathematical principles underlying growth in batch and continuous culture.]

Jennings, D.H. and Isaac, S. (1995) *Microbial Culture.* Bios, Oxford.

Rhodes, P.M. and Stanbury, P.F. (1997) *Applied Microbial Physiology: A Practical Approach.* Oxford University Press, Oxford. [Provides details of the growth requirements and culture methods applicable to a range of different microbes.]

Study exercises

46.1 Calculate the specific growth rate and doubling time of cells in culture. What are the specific growth rates and doubling times of the following (give all answers to three significant figures)?
  (a) A broth culture of the yeast *Saccharomyces cerevisiae* growing in exponential phase and containing 5.2×10^4 cells at 10 a.m. and 3.4×10^6 cells at 7 p.m.
  (b) A log phase culture of *E. coli* containing 3.0×10^4 CFU ml^{-1} at the start of the experiment and 6.7×10^7 CFU ml^{-1} at the end of the experiment, 200 minutes later.
  (c) An actively growing culture of *Bacillus subtilis* containing 32 bacteria in 25 squares of a haemocytometer chamber at 13.00 hours and 250 bacteria in 25 squares of the same haemocytometer chamber at 15.30 hours.

46.2 Practise the calculations involved in using a haemocytometer to make a direct microscopic count. Express your answers to three significant figures in all cases.

  (a) The mean cell count of a yeast suspension per small square of an improved Neubauer counting chamber was 6.42. If the volume of a small square is 2.5×10^{-7} ml, what is the cell count per ml?
  (b) The following counts were obtained for bacteria in 20 individual small squares of a Petroff–Hauser counting chamber: 26, 36, 42, 35, 27, 16, 29, 50, 24, 43, 41, 35, 18, 36, 33, 47, 25, 46, 32, 57. If the volume of each small square is 8×10^{-7} ml, what is the cell count per ml?
  (c) A 10^{-2} dilution of a dense suspension of *E. coli* was examined microscopically using an improved Neubauer counting chamber, giving a total of 78 cells in a total of 25 small squares. If the volume of each small square is 2.5×10^{-7} ml what is the cell count per ml of the original (undiluted) suspension.
  (d) A dilute suspension of yeast cells was concentrated ten-fold by centrifuging 10 ml and

(continued)

Study exercises (continued)

resuspending the pellet in 1 ml. The mean cell count of the concentrated yeast suspension per large square of a Petroff–Hauser counting chamber was 6.8. If the volume of the large square is 2×10^{-5} ml, what is the cell count per ml of the original suspension?

46.3 Practise the calculations involved in making a plate count. Express your answers to three significant figures.

(a) The mean spread-plate count for 100 μl of a 10^{-5} dilution of a culture of *E. coli* was 54.4 CFU. What is the mean plate count per ml of the original suspension?

(b) Three replicate plates of nutrient agar were each spread with 200 μl of a 10^{-3} dilution of a bacterial suspension, giving colony counts of 34, 40 and 37 after 24 h incubation. What is the mean plate count per ml of the original suspension?

(c) A twenty-fold dilution of a yeast suspension was used to prepare four replicate pour plates, each containing 500 μl of this dilution and giving counts of 211, 186, 194 and 202 after incubation. What is the mean plate count per ml of the original suspension?

(d) A sample of 50 g of raw seafood was homogenised and diluted to 2% w/v in sterile saline solution. Three replicate pour plates were prepared using 1 ml of the diluted sample, giving counts for *E. coli* of 35, 41 and 32. What is the average count of *E. coli* per 100 g of raw seafood?

(e) Duplicate samples of 250 ml of river water were filtered through separate sterile membranes of pore size 0.2 μm. The membranes were then transferred to the surface of an agar-based medium and incubated for 48 h at 37°C, giving colony counts of 45 and 32 respectively. What is the mean count per 100 ml of river water?

(f) A sample of bottled drinking water was processed by serial decimal dilution and 500 μl samples of each dilution were pour plated and incubated at 22°C for 72 h. The colony counts obtained for the three replicate plates of the 10^{-1} dilution were 28, 32 and 39. Does this water meet the EU regulations for mineral water, which specify a maximum average plate count at 22°C of 100 CFU ml^{-1}?

47 Working with animal and plant tissues and cells

Although the aim of many studies is to isolate, quantify or characterise individual molecules from a biological system, e.g. the purification of a particular enzyme (p. 318), this is not always the most appropriate course of action. Depending on the purpose of the investigation, it may be more relevant to study the functioning of biomolecules within more complex systems, in order to understand their role in a particular biological process. At one extreme this may be carried out *in vivo*, using whole multicellular organisms (e.g. individual animals or plants), while the other extreme is represented by *in vitro* studies, using subcellular 'cell-free' extracts. Between these two extremes, a range of tissue and cell culture techniques offers some of the biological complexity of the intact organism combined with a degree of experimental control that may not be obtainable *in vivo*. The ethics and costs of whole animal experimentation have provided an additional stimulus to the development of *in vitro* methods, e.g. toxicity tests using mammalian cell culture, while developments in molecular genetics have led to further applications of cell and tissue culture.

Animal tissues and organs

Physiological experiments are carried out using either whole organisms, or a range of animal organs and tissues, including heart, liver, muscle, etc.

 KEY POINT *A major practical consideration is that the tissue should be studied as soon as possible after the death of the animal, typically under laboratory conditions that mimic the* in vivo *environment as closely as possible.*

Working with vertebrate animals and their organs/tissues – remember that procedures must be consistent with the law, i.e. in the UK, the Animals Scientific Procedures Act 1986.

In most instances, the experiments are relatively short term (<24 h) and the aim is to maintain the tissue in a physiological state similar to that within the living organism. For metabolic studies, the whole organ or a tissue slice (typically 1–10 mm thick) will be bathed in an appropriate perfusion fluid, supplied either by gravity or by peristaltic pump. Practical considerations include:

- Inorganic solute requirements – the chemical composition of the perfusion fluid is usually chosen to reflect the major inorganic ion requirements of the tissue. For short-term studies, a number of so-called 'physiological salt solutions' may be used, e.g. Ringer's solution, one formulation of which is given in Table 47.1.

Using human tissues – in the UK, the Human Tissue Act 2004 applies and work is licensed and overseen by the Human Tissue Authority. Note that established cell lines (p. 281) are excluded from the regulations.

- Oxygen requirements – it may be necessary to increase the O_2 content of the perfusion fluid by bubbling with air, in order to meet the oxygen demand of the innermost parts of the tissue. However, this can lead to oxygen toxicity in the outermost parts and an alternative approach is simply to increase the rate of perfusion.
- Physico-chemical conditions – including temperature (usually controlled to $\pm 1\,^\circ C$ of normal body temperature), water status (the perfusion fluid and the tissue should be isotonic, p. 144), pH and buffering capacity (e.g. some perfusion fluids have elevated an $NaHCO_3$ concentration, to mimic the buffering capacity of mammalian serum).

Table 47.1 Composition of Ringer's solution (simplified formulation, for amphibians, etc., pH 7.4–7.6)

Compound	Amount per litre (g)
NaCl	6.0
KCl	0.075
CaCl$_2$	0.1
NaHCO$_3$	0.1

Definitions

Apoplasm – that part of the plant body outside the symplasm.

Light compensation point – the amount of photosynthetically active radiation (PAR) where photosynthetic CO_2 uptake is balanced by CO_2 production due to respiration and photorespiration.

Plasmodesmata – transverse connections through the cell wall, linking the cytoplasm of adjacent plant cells and creating a symplasm.

Working with plant tissues and organs – the ethical problems associated with animal and human tissues are avoided, and there is a decreased risk of infection to the laboratory worker; practical manipulations are often carried out in horizontal laminar-flow cabinets.

Table 47.2 Components of Long Ashton medium (nitrate version)

Stock solution: mass of component required per litre of solution (g)	Volume of stock solution to make 1 litre of medium (ml)
Major nutrients	
KNO$_3$: 50.60	8
Ca(NO$_3$)$_2$: 80.25	8
MgSO$_4$·7H$_2$O: 46.00	8
NaH$_2$PO$_4$·2H$_2$O: 52.00	4
Micronutrients	
FeKEDTA: 3.30	5
MnSO$_4$·4H$_2$O: 2.23	1
ZnSO$_4$·7H$_2$O: 0.29	1
CuSO$_4$·5H$_2$O: 0.25	1
H$_3$BO$_3$: 3.10	1
Na$_2$MoO$_4$·2H$_2$O: 0.12	1
NaCl: 5.85	1
CoSO$_4$·7H$_2$$n$O: 0.056	1

Effects of humans on plants – remember that your exhaled breath will be nearly saturated with water vapour and will contain CO_2 at 3–4% v/v, some 100 times more concentrated than atmospheric CO_2: your breath can thus affect rates of transpiration and photosynthesis.

- Organic nutrient requirements – for longer-term studies, suitable nutrients will be required: these may be chemically defined additives, e.g. vitamins, amino acids, proteins, etc., or biological fluids such as plasma or serum. Glucose is often added as a carbon and energy source (Freshney, 2005).

Plant tissues and organs

Individual plant components (e.g. leaves, leaf slices and epidermal strips) can be isolated from the main plant body for study under controlled conditions. Since photosynthetic plant parts are autotrophic, they may be maintained *in vitro* for longer than animal organs, given adequate light and CO_2. However, most plant cells are joined via plasmodesmata, and the separation of such connections when the component is removed from the plant often leads to death when these connections are broken.

KEY POINT **Plants show wound responses that may affect the metabolic processes under study. For these reasons, the most suitable systems for longer-term studies are often whole organs, e.g. whole leaves or entire root systems.**

Water culture (hydroponics), e.g. in Long Ashton medium (Table 47.2), is an alternative approach, offering greater control over the root environment (Dodds and Robert, 1995).

It is best to use vigorous, healthy stock plants and to follow a well-established procedure, taking account of the following:

- Sterility – strict attention to sterile technique can be essential to the success of many longer-term experiments. Decontamination of plant organs may be especially difficult where specimens are obtained from soil: to achieve this, use a surface wash with disinfectant (e.g. 10% w/v sodium hypochlorite), followed by several rinses with sterile water.
- Gaseous environment – in general, the experimental system should be well ventilated. Actively photosynthetic tissues will rapidly deplete the atmospheric CO_2 in a closed vessel: plant parts may also produce physiologically active gases, such as ethylene, especially at wound sites. Turgor loss may occur in isolated plant parts unless a high humidity is maintained.
- Nutrition – plant tissues may benefit from a supply of inorganic ions, including K^+, SO_4^{2-}, etc., and may require certain vitamins, micronutrients and plant hormones for prolonged studies.
- Physico-chemical conditions – light is the most important environmental requirement for green plant parts. At atmospheric CO_2 concentrations, the light compensation point is about 5–10 µmol photons PAR m^{-2} s^{-1} (p. 343) and photosynthesis is usually saturated between 500 and 2000 µmol photons PAR m^{-2} s^{-1}, depending on the plant type. Light quality and photoperiod (daylength) are also important: fluorescent tubes and incandescent bulbs that mimic the photosynthetic spectrum of sunlight are available (see p. 343), while photoperiod can be controlled using a timer. The water potential of aqueous media can be adjusted using an impermeant osmoticum such as mannitol, and pH values are often kept close to those of the apoplasm ($\approx$pH 6) using appropriate buffers, if necessary (p. 151).

Cell and tissue cultures

Many of the basic principles involved in culturing animal and plant cells are broadly similar to those described for microbial cell culture (p. 271). Definitions for several key terms are given in the margin.

Applications of cell and tissue culture

The main uses of animal and plant cell culture systems include:

- Experimental model systems in biochemistry, pharmacology and physiology: cell culture offers certain advantages over whole organism studies, with greater control over environmental conditions and biological variability. The use of genetically defined clones of cells may simplify the analysis of experimental data. Conversely, results obtained with specialised cell-based systems might be unrepresentative of a broader range of cell types and may be more difficult to interpret in terms of the whole organism.
- Studies of the growth requirements of particular cells: including studies of the positive effects of growth factors or growth-promoting substances, and the negative effects of xenobiotics or cytotoxic compounds. The use of cell culture in bioassays and mutagenicity testing is considered in Chapter 49.
- Studies of cell development and differentiation: including aspects of the cell cycle and gene expression. Cell cultures retaining their ability to differentiate *in vitro* are particularly interesting to researchers, while the lack of differentiation and unlimited growth of many animal cell lines makes them useful models of tumour development.
- Pathological studies: including culture of foetal cells for karyotyping and detection of genetic abnormalities, e.g. trisomy, etc.
- Genetic manipulation: cell culture techniques have played an essential role in the development of molecular biology, including the production of transgenic animals and plants by techniques such as transfection, etc.
- Biotechnology: including the industrial production of therapeutic proteins, vaccines and monoclonal antibodies using large-scale hybridoma culture techniques similar to those used in microbiology (p. 272).
- Stem cell technology: increasingly important in biomedical research and development.

 KEY POINT *One of the main differences between organ and tissue incubation techniques and those used in cell/tissue culture is that the former aim simply to maintain metabolic and physiological activity for a limited period, while the latter provide conditions suitable for cell growth, division and development* in vitro *over an extended timescale, from a few days to several months.*

Animal cell culture systems

These may be established either from whole organisms (e.g. chick embryo), discrete organs (e.g. rat liver) or from blood (e.g. lymphocytes), typically using *mild* enzymic tissue disruption techniques, where necessary.

KEY POINT *Although, in theory, it is possible to culture nucleated cells from virtually any source, in practice, the highest rates of success are most often achieved with young, actively growing tissues.*

The principal considerations in animal cell culture are:

- Safety: it is important to be aware of the potential dangers of infection from cell cultures. Although avian and rodent cells present a reduced risk of disease transmission compared with human cells, all cell cultures must be regarded as a potential source of pathogenic microbes, and appropriate sterile technique (Box 47.1) must be used at all times (see also p. 264). Work involving human tissues and cell cultures must be carried out in a biosafety cabinet by trained, experienced personnel – you are unlikely to gain practical experience with such cultures in the early stages of your course. For other cell cultures, you will need to follow the code of safe practice of your department – consult your departmental safety officer if you have any doubts about safe working procedures.

- Whether to use a primary culture or a cell line: freshly isolated cells are more likely to reflect the biochemical activities of cells *in vivo*, though they will have a limited lifespan in culture, requiring repeated isolation for longer-term projects. Continuous cell lines are more easily cultured and offer the advantage that their growth requirements in culture may be known in some detail, especially for the more widely used cell lines (e.g. BHK, HeLa).

- The requirements for a solid substratum: some cells must be attached to a solid surface in order to grow. Anchorage dependence is a typical feature of primary cultures and finite cell lines – such cultures show density-dependent growth inhibition once the cells have formed a confluent monolayer on the surface of the substratum. An alternative approach is to grow such cells on a particulate support using 'microcarrier' beads. In contrast, many continuous cell lines can be maintained in suspension culture, as individual cells or aggregates.

- The physico-chemical conditions, including pH (typically 7.2–7.5) and buffering capacity, osmolality (usually $300 \pm 20 \, \text{mosmol kg}^{-1}$) and temperature (e.g. 35–37 °C for mammalian cells).

- The requirements of the culture medium: these will include the provision of inorganic ions (as a balanced salt solution), a carbon/energy source plus other organic nutrients, and in some cases a supplement containing antimicrobial agents to counter the risks of contamination. For example, Dulbecco's modified Eagle's medium is used for many mammalian cell types that grow as adherent monolayers, while suspension cultures of continuous cell lines can be maintained using less stringent media. To support growth, the basal medium is usually supplemented with serum (usually foetal calf serum, at up to 20% v/v), or a chemically defined serum-like supplement containing a mixture of proteins, polypeptides, hormones, lipids and trace components. The *in vitro* level of CO_2 and O_2 must also be considered: many cell cultures are buffered using bicarbonate, and must be maintained in an atmosphere of elevated CO_2, either in a sealed culture vessel or in a CO_2 incubator, to maintain pH balance. In some cases, a pH-sensitive dye (e.g. phenol red, p. 147) may be incorporated into the growth medium, to provide a visual check on pH

 SAFETY NOTE *When working with tissue cultures safely, do not confuse laminar-flow hoods with biosafety cabinets (p. 266).*

SAFETY NOTE *When sterilising during tissue culture (Box 47.1), take care when using 70% v/v alcohol near a naked flame (e.g. Bunsen or spirit lamp) – it is easily ignited and burns with a weakly visible flame.*

Example Dulbecco's modified Eagle's medium contains Ca^{2+}, Fe^{3+}, Mg^{2+}, K^+, Na^+, Cl^-, SO_4^{2-}, PO_4^{2-}, glucose, 20 amino acids, 10 vitamins, inositol and glutathione. Foetal calf serum is usually added at up to 20% v/v.

Using antibiotics in cell culture – *a typical antimicrobial supplement might include antibacterial agents, e.g. penicillin and streptomycin, an antifungal agent, e.g. griseofulvin, and an antimycoplasmal agent, e.g. gentamicin. Use sparingly to treat contamination and discontinue once the contaminant has been eradicated.*

Box 47.1 Sterile technique and its application to animal and plant cell culture

Although Chapter 45 gives general advice on the basic principles of sterile technique with microbial cultures, animal and plant cell cultures need additional precautions due to the complexity of some of the procedures and the likelihood of rapid overgrowth of any contaminating microbes. For routine work on the open bench, you should be aware of the following points:

1. **Consider personal clothing and hygiene**: long hair should be tied back, or retained by a net/cap. Avoid pendulous earrings and similar items. Wash your hands at the outset, to remove any loose skin flakes: surgical gloves may be worn, though there is loss of tactile sensitivity and comfort. Hands/ gloves should be swabbed with 70% v/v alcohol to further reduce the risk of contamination. If you have a cold/cough, consider wearing a face mask.

2. **Work in a designated quiet area**: there should be no air currents (avoid open windows) and no 'through traffic' or other activity that might give rise to contamination (e.g. microbes should not be cultured in the same location). The term 'quiet area' also highlights that no talking should occur, to reduce the risk of aerosol contamination from the oral microflora.

3. **Organise your work surface**: clear everything from the bench, swab with 70% v/v alcohol, then position all items around a central area, so you can reach every item without reaching across anything. A well-organised workspace reduces the risk of accidental contact between sterile and non-sterile items. Swab bottles and flasks with 70% v/v alcohol before positioning them at the edge of your workspace. Also, swab your work surface between procedures and at the end of the session.

4. **Work close to a Bunsen flame**, positioned centrally within your workspace where convection currents create an upward airflow, reducing the likelihood of particles falling from the air into an open vessel. Always flame the tops of glass bottles (p. 266), but not plastic items, for 2–3 seconds both *before* and *after* opening and closing, rotating the bottles during exposure to the flame. Flame glass pipettes, as described on p. 266.

5. **Tilt flasks and bottles during use**: uncapped culture flasks and media bottles are best kept at a shallow angle, to minimise the risk of airborne contamination.

6. **Work without delay, but do not hurry**: always keep in mind that the air contains contaminant microbes

and the longer you leave a vessel open to the air, the more likely it is that it will become contaminated.

When preparing to work in a laminar-flow cabinet, the following additional aspects should be considered:

1. **Note whether the airflow is horizontal or vertical**: horizontal flow gives greatest protection to the work area and provides the most stable air flow, while vertical-flow cabinets reduce exposure of the operator. A biosafety cabinet should be used for work with potentially hazardous cultures (e.g. human cells/tissues, or with any cells known to be infected with a virus).

2. **Prepare the cabinet**: switch on the cabinet and leave running for at least 10 minutes, then swab the work surface and other interior surfaces with 70% v/v alcohol. Swab the outsides of bottles, flasks, etc., before bringing them into the cabinet. Arrange items around a crescent-shaped central work area.

3. **Carry out your work with due regard for the airflow within the cabinet**: use of a Bunsen burner is often discouraged, as it disrupts the correct airflow, creates heat and can ignite flammable items within the restricted interior of the cabinet. Always try to keep your hands/arms further away from the sterile airstream than the items within the cabinet, e.g. avoid working *behind* an open vessel in a horizontal-flow cabinet and *above* an open vessel in a vertical-flow cabinet.

4. **Take particular care when using pipettes**: a laminar-flow hood will have a restricted interior volume, and it is easy for the novice to touch a pipette tip on the hood, or on another item during use. When removing a pipette from its packaging and connecting to a pipette filler (p. 123) you should point the tip of the pipette away from the user and into the air flow, holding it well above the graduated scale markings to avoid contamination. The pipette and filler are then held horizontally until required: as this can be a tricky procedure for the novice, you should not take a pipette out of its container until you are ready to use it.

5. **Work carefully with open bottles and flasks**: caps can be either placed top-down on the work surface, or held in a crooked finger (Fig. 45.6(a), p. 269). Since the airflow is sterile, there is less need to work swiftly when compared to the open bench, but you should still re-cap all containers as soon as you have finished a particular procedure.

Using a horizontal laminar-flow hood – note that this is designed to minimise contamination of the culture, rather than the worker.

Definitions

Callus – an aggregation of undifferentiated plant cells in culture.

Embryogenic callus – tissue with the capacity to differentiate under defined laboratory conditions, typically in response to plant growth regulators in the medium.

Explant – a fragment of tissue used to initiate a culture (the term is also used in animal culture).

Protoplasts – cells lacking their cell walls.

Sphaeroplasts – cells with attached fragments of their cell walls: osmotically sensitive.

Totipotency – the ability (of any plant cell) to de-differentiate and re-differentiate into any of the cell types found in the mature plant.

status during growth – a colour change indicates acidification of the medium and the growth medium should be renewed.

- The equipment required: this may include a laminar-flow hood or biosafety cabinet to reduce the possibility of microbial contamination, suitable culture vessels (typically pre-sterilised, disposable polystyrene dishes, bottles and flasks, treated to create a negatively charged, hydrophilic surface), a supply of high-purity water (typically distilled, deionised and carbon filtered), a suitable incubator with temperature control of $\pm 0.5\,^{\circ}\text{C}$ or better, often with CO_2 control and mechanical mixing, and an inverted microscope to examine adherent cell monolayers during growth.

The successive stages of isolation of animal components are shown in Fig. 47.1(a). Box 47.1 gives advice on key aspects of sterile technique in cell culture; additional guidance on practical procedures in animal cell culture is given by Reed *et al.* (2007).

Plant tissue and cell culture systems

Plant tissue cultures can be established by growing explants of sterilised tissue on the surface of an agar-based growth medium to give a callus of undifferentiated cells. Initial sterilisation is usually achieved by incubation for 15–20 min in 10% w/v sodium hypochlorite.

KEY POINT *For most plants, cell cultures can be established from a broad range of tissue types, reflecting the totipotency of many plant cells.*

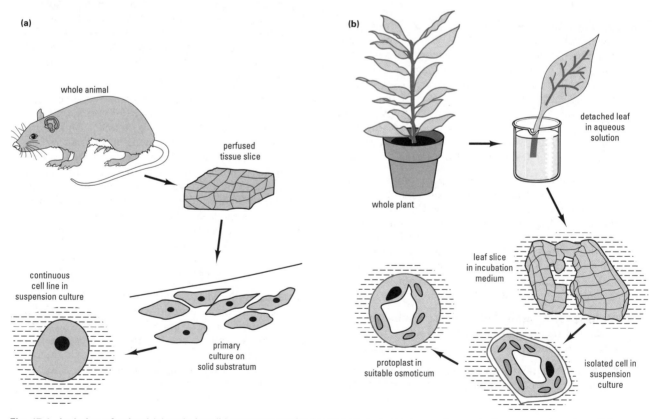

Fig. 47.1 Isolation of animal (a) and plant (b) components for *in vitro* study (note decreasing scale, from organisms to cells).

Embryogenic callus may be induced to differentiate, forming tissues and organs on a medium containing appropriate plant hormones: in many cases, these cultures will develop to form plantlets that can be grown on to mature plants, or encapsulated to produce so-called 'artificial seeds'. This approach can be used to study the conditions necessary for differentiation and development, or to propagate rare plants and other valuable stock (e.g. virus-free stock, or genetically altered plants). Callus derived from anthers can be used to provide haploid cell cultures and haploid plants – these are often useful for experimental genetics and breeding purposes.

Plant cell suspension cultures (Fig. 47.1(b)) are usually obtained by transferring fragments of actively growing callus to a 'shake' flask containing liquid medium in an orbital incubator: gentle agitation causes fragmentation of the callus tissue, to give a suspension culture that will contain individual plant cells and cell aggregates. This culture can be maintained by repeated subculturing of material from the upper layers of the liquid, encouraging the growth of small aggregates. However, in contrast to animal cell suspension cultures, it is rare for plant suspension cultures to be entirely unicellular, due to the presence of plasmodesmata and plant cell walls. Suspension cultures often require a minimum inoculum size on subculture – an inoculum volume of 10% v/v may be necessary to ensure successful subculture. It may also be helpful to add a small amount of 'conditioned' medium from a previous culture. Commercial-scale suspension cultures are used to produce certain plant pigments and secondary metabolites such as flavourings and high-value pharmaceutical compounds.

The growth media used for callus and suspension cultures are more complex than those required for intact plant organs and tissues, with organic nutrient supplements in addition to a balanced salt solution. Such organic supplements often include a major carbon and energy source, e.g. sucrose, plus various vitamins and growth regulators (typically, at least one auxin and a cytokinin), together with undefined components such as yeast extract and hydrolysed casein in some instances. Otherwise, the techniques are broadly similar to those described for microbial systems (Chapters 36 and 46).

Plant protoplasts

Some experimental procedures using plant cells require the enzymatic removal of their cell walls, creating protoplasts that can be manipulated *in vitro*, then maintained under conditions that allow the regeneration of cell walls, with subsequent growth and differentiation to give genetically modified plants. Protoplast isolation often involves pretreatment in a concentrated osmoticum (e.g. sucrose or mannitol, at $300-500\,\mathrm{mmol\,l^{-1}}$) to plasmolyse the cells, weakening the linkage between cell wall and plasma membrane. Enzymatic treatment is often prolonged, taking several hours in a suitable mixture of enzymes, e.g. cellulase and Macerozyme. The resulting material can be sieved through fine nylon mesh, centrifuged at low speed and then resuspended, to remove sub-cellular debris and cell aggregates. The protoplast preparation can be checked for fragments of cell wall using a suitable stain, e.g. 0.1% w/v calcofluor white and a fluorescence microscope.

The fusion of protoplasts from different plants can be used to produce a somatic hybrid: this process can be used to circumvent interspecies reproductive barriers, creating novel plants. Protoplast fusion can be induced by chemical 'fusogens', e.g. using polyethylene glycol (PEG) at high concentration, or by electrofusion (incubation under low alternating current to encourage aggregation, then brief exposure to a high-voltage electrical field – typically $1\,000\,V\,cm^{-1}$ for 1–2 ms – causing protoplast fusion). Similar techniques can be used with animal cells. The successive stages of isolation of plant components are illustrated in Fig. 47.1(b).

Text references

Dodds, J.H. and Robert, L.W. (1995) *Experiments in Plant Tissue Culture*, 3rd edn. Cambridge University Press, Cambridge.

Freshney, R.I. (2005) *Culture of Animal Cells: A Manual of Basic Technique*, 5th edn. Wiley, New York.

Reed, R.H., Holmes, D., Weyers, J.D.B. and Jones, A.M. (2007) *Practical Skills in Biomolecular Sciences*, 3rd edn. Prentice Hall, Harlow.

Sources for further study

Anon. (2006) *Fundamental Techniques in Cell Culture – A Laboratory Handbook*. Sigma-Aldrich/European Collection of Cell Cultures. Available at: http://www.sigmaaldrich.com/Area_of_Interest/Life_Science/Cell_Culture/ Key_Resources/ECACC_Handbook.html. Last accessed: 09/04/07.

Butler, M. (2003) *Animal Cell Culture and Technology: The Basics*, 2nd edn. Garland Science Press, Oxford.

Doyle, A., Griffiths, J.B. and Newell, D.G. (1998) *Cell and Tissue Culture: Laboratory Procedures*. Wiley, New York.

Evans, D., Coleman, J. and Kearns, A (2003) *Plant Cell Culture: The Basics*. Garland Science Press, Oxford.

Helgason, C.D. (2004) *Basic Cell Culture Protocols*, 3rd edn. Humana Press, New Jersey.

Loyola-Vargas, V. M. and Vazques-Flota, F. (2005) *Plant Cell Culture Protocols*, 2nd edn. Humana Press, New Jersey.

Masters, J. (2000) *Animal Cell Culture: A Practical Approach*, 3rd edn. Oxford University Press, Oxford.

Razdan, M.K. (2003) *Introduction to Plant Tissue Culture*, 2nd edn. Intercept, Andover.

Study exercises

47.1 Calculate the osmolarity of Ringer's solution from its individual constituents. Use the data in Table 47.1 and the information in Chapter 23, and assume that each constituent behaves according to ideal thermodynamic principles. Note the following M_r values: NaCl 58.44; KCl 74.55; $CaCl_2$ 110.99; $NaHCO_3$ 84.01, required to convert the data in Table 47.1 to molar concentrations (see p. 141 if you are unsure about this conversion). Give your final answer in $mosmol\,kg^{-1}$, to one decimal place.

47.2 Find out about the availability of cell cultures using the World Wide Web. Research the culture collections of particular countries via their websites. How useful is each site in terms of features such as: ease of use; access to catalogues; information on individual strains, e.g. culture media; details of costs?

47.3 Research the origins of the code names used for individual cell lines. In many instances, the code names given to individual cell lines are derived from their original source. Use the Web to find out the origin of the following descriptors for cell lines: (a) HeLa; (b) Vero; (c) BHK; (d) CHO; (e) TBY-2.

Definitions

The following terms are used when discussing the quality and features of images:

Colour balance – the way in which the perceived natural colours of a subject are reproduced in an image: this can be affected by image capture, lighting and printing.

Colour depth – in digital images, the amount of information used to define the colour of each pixel: at 1 bit per pixel, images are black and white; at 24 bits per pixel, images are displayed as over 16 million colours, beyond the colour-discerning ability of the human eye.

Contrast – the degree of gradation between colours or the number of grey shades in black-and-white film: the higher the contrast, the sharper is the gradation.

Depth of field – the range of shot that is in focus for a given camera setting: a low value may be required to highlight a specific subject or you may wish the whole field to be in focus, as in a panoramic view.

Focal length – the distance (in mm) at which a lens system will focus an image when set to 'infinity': this affects the magnification of the image and the angle of view, which are inversely related.

Grain – flaws in a film image, caused by the method of image capture.

Image stabilisation (antishake) – a feature available on digital cameras that compensates for camera shake on long exposures.

Noise – flaws, often white pixels, in a digital image caused by sensor errors.

Resolution – the sharpness and quality of an image, often expressed in 'dots per inch' (a non-SI unit).

Saturation – the term used to describe the intensity of a colour: a saturated colour is an intense colour.

Zoom – lens feature that magnifies an image by altering focal length. Optical zoom accomplishes this using the lens configuration, without great loss of image quality; however, camera shake may become a significant factor at high magnifications. Digital zooms achieve this electronically, and there is always loss of image quality.

Photography and imaging techniques are used primarily for:

- documentation and validation of observations for projects or coursework;
- recording visual information for later measurement, counting and/or analysis;
- preparation of material for illustrating talks;
- preparation of poster displays;
- producing 'hard copy' of microscope images.

Professional technical assistance may be available for this type of work but there is an important role for you in defining your requirements precisely. A basic understanding of the key factors that affect the final product will help you do this.

Types of photography

Photographs can be taken using two main systems: the digital camera and the film camera. Although the basic operation of both types is similar, they differ in their light-capturing and image-recording systems.

Digital photography and imaging

This is the production of an electronically stored image, using either a digital camera (still or video) or a scanning device from a pre-existing image. The image is held as a series of discrete picture elements called pixels (Fig. 48.1); the greater the number of pixels per unit surface area, the better the potential resolution of the image and the greater the amount of detail captured. The image may be in black/white, greyscale or colour. Image resolution is constrained by the nature of the sensor array and the amount of memory available for recording, while size and quality will also depend on the display or printer used.

Film photography

Black-and-white or colour-print photography involves capturing images on a film of plastic coated with light-sensitive silver halide salts. When exposed to light and subsequently chemically treated (developed), silver grains are produced to form a negative image, later converted to a positive image during printing. Transparency (slide or positive) films are based on colour dyes and are effectively grainless. Table 48.1 summarises the relationship between speed, grain and resolution for print films. The choice of film is a compromise between speed and resolution. Slower film is best for fine detail, but requires high light levels for effective results. Faster film is more suitable for moving subjects or when light levels are low, but there is a loss of contrast and detail.

Comparing digital and film systems

Despite advances in digital image technology and reductions in the price of digital cameras and associated memory systems, even the best digital images are of somewhat lower quality (resolution) than high-quality film images (see Table 48.2). However, digitised images have the advantages of

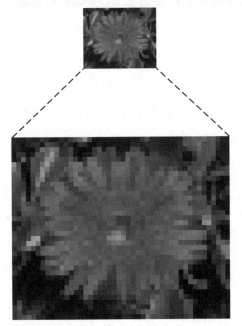

Fig. 48.1 Pixel structure in a digital image

Table 48.1 The speed and grain relationship in film

Speed of film	ISO number	DIN number	Grain	Resolution
Slow	25–64	15–19	Very fine	Very sharp
Medium	64–200	19–24	Fine	Sharp
Fast	200–400	24–27	Medium	Medium
Extra-fast	400–1600	27–32	Coarse	Poor to medium

speed and ease of production, since a developing stage is avoided, and they are also more readily manipulated. The preview facility on digital cameras is extremely useful when it is important to confirm that a successful image has been taken, e.g. when an event is unrepeatable. On the other hand, digital images are stored electronically and are not secure from computer failure, or damage from strong magnetic fields (see Box 10.1).

Other points of comparison include:

- the need for precision in exposure for digital systems is higher than for colour-print films: this means that the exposure timing is more critical to the quality achieved;
- digital systems are weak at recording highlights but tend to provide better shadow detail than film;
- digital systems tend to offer greater depth of field (the range of shot that is in focus), but this is not always an advantage as it can then be

Table 48.2 A comparison of digital and film photography. This table assumes you would be using a 6 megapixel SLR (single-lens reflex) digital camera and a 35 mm SLR film camera with 100 ASA film. Although it is not the only factor determining image quality, results would in general be poorer for a compact digital camera or one offering fewer megapixels.

Aspect	SLR digital camera	SLR film camera
Resolution	Primarily limited by number of pixels, i.e. 6 megapixels.	Limited by grain density of film and generally better. For scanned images, probably higher than 6 megapixels (up to 25 megapixels, depending on scanner).
Image quality	Better for shadow detail. Long-exposure images may be of poor quality. Delays in image capture may make it difficult to capture moving subjects. Depth of field longer with compact digital cameras.	Better for highlights. Grain is present, and will show if exposure time is long and/or image is magnified. Good-quality long exposures are possible. Easier to control depth of field.
Convenience	Easier to take repeated shots of same subject; easier to send images to others digitally; easier to incorporate into word-processed documents.	Results not known until film is developed – hence use of bracketing (p. 290), which is wasteful of film. Unless scanned, less easy to send or incorporate into word-processed documents.
Costs	Costs of equipment, operation and printing similar. No film or processing costs, but memory card required.	Costs of equipment, operation and printing similar, but film must be bought and subsequently processed, with additional cost. Similar cost of printing.
Post-exposure image manipulation	Easy using commercially available software.	Possible at developing stage but only after scanning thereafter.
Storage and copying of images	Electronic. Digital images can be stored without loss of quality and occupy little physical space. Copies of files are exact facsimiles. Back-up easier.	Physical (hard copy). Colour-film images may fade though time and negatives and prints occupy significant space. There is always loss of quality when they are copied. Back-up less easy, unless scanned.

Understanding the light sensitivity of photographic systems – both digital and film media are rated in terms of their 'speed', measured in either ISO (= ASA) or DIN units. A good range for general purposes is 100–160 ISO (21–23 DIN). Film of 1600 ISO is available, as are digital cameras with settings as high as 3200 ISO. These are useful for low light situations, but with some loss of image quality.

Capturing moving images – this may be valuable in biology, for example, to record animal behaviour, speed up plant movements or study fungal growth patterns.

Focusing images correctly – the autofocus feature on cameras may result in lack of focus on the parts of the image you are interested in because the algorithm used (generally spot or area) may not be suitable for your purpose. Better results may be obtained using manual focusing.

Tips for better photography:

- Use your camera on a tripod whenever possible, and use high shutter speed to minimise the effect of movements. Otherwise, adopt a posture with your elbows firmly locked against your body to reduce camera shake.
- Use electronic flash wherever possible as it provides uniformity of colour balance. Unless you want shadowless lighting, do not place the flash on the camera (hot-shoe) connection.
- Shadowless lighting of smaller objects is usefully obtained through a ring-flash system.

Storing colour film – colour film is highly sensitive to environmental factors such as heat and humidity, which cause changes in film speed and colour rendition. Make sure that your film has a sufficiently long expiry date. Store film for extended periods in a refrigerator or freezer: if stored in a freezer, allow a 24 h thawing period before use.

difficult to focus on critical features against a blurred background, in contrast to film;
- moving subjects can be easier to capture using film as there is often a significant delay between pressing the shutter and image capture in digital cameras.
- digital cameras frequently offer a 'movie' option, and most digital movie cameras provide a 'still' facility allowing you to capture an event at a critical point.

Capturing images

Camera and lens types

For both digital and film systems, the primary difference is whether you preview the image through the lens (TTL), often referred to as single-lens reflex (SLR), or whether a separate viewfinder is used. Digital cameras also incorporate a monitor that allows you to preview and review images. Lenses are either fixed, i.e. have a set focal length, or zoom, allowing a range of focal lengths, and hence magnification and angle of view. In general, higher-quality images can be taken with an SLR camera; these systems have better-quality lenses and allow greater control over the final image.

Setting up for photography

In the biosciences, this is usually carried out in one of three general situations:

1. laboratory or studio environments – where conditions are under your control and problems are relatively easy to overcome;
2. field situations – where many factors may be difficult to control, and you may need to use artificial means to enhance the shot (e.g. flash photography, or fast film speed);
3. working with specialised equipment – such as photo-microscopes or electron microscopes, where control of the photographic process is largely managed by the instrument's operating system.

The amount and type of light available to create an image is very important, especially for colour photography. Colour-image creation requires the appropriate colour temperature balance to be selected. Colour temperature is measured in Kelvin (K: see Fig. 48.2). In digital cameras this is usually automatic, but can be manually controlled via the 'white balance' settings. Filters are used with film cameras, to control overall colour balance, but the most important factor is the film type selected:

- daylight film is balanced for daylight conditions (5400 K) and electronic flash;
- artificial light types A (3400 K) or B (3200 K), are designed for studio lighting conditions.

Digital images are also modifiable after being taken, unlike film images where minimal adjustment is possible after taking the photograph.

The quantity of light can be measured by a photographic light meter, which may be external or built-in. The more light there is, the smaller the lens aperture you can select (larger f-number), and the greater will be the depth of field (= depth of focus). With the camera on a tripod, you can select slow shutter speeds and larger aperture settings (f-numbers), to

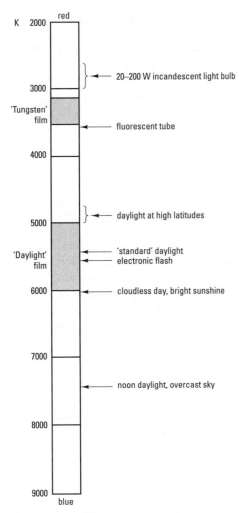

Fig. 48.2 Colour temperature and the Kelvin scale in relation to sensitivity of film type

> **Backing up digital image files** – since no physical copy exists, you must remember to download and back up digital image files to appropriate media (hard drive, network drive, CD or USB memory stick) as soon as possible.

maximise depth of focus. Where an electronic flash system is employed to provide some or all of the lighting this is even easier, since the effective shutter speed with electronic flash is extremely short (1 ms). Remember that your choice of lighting arrangement will affect the quality of the picture: shadowless lighting is appropriate in some situations, but often shadows help to give three-dimensional form to subjects. In general, the use of more than one light source is advisable to prevent hard shadows.

 KEY POINT *To use an SLR camera properly, you need to understand the relationship between aperture (f-number), shutter speed and depth of field. For moving objects, give priority to fast shutter speed by opening the lens aperture (small f-number). Where depth of field is required, close the lens aperture (high f-number); this will result in slower shutter speeds, so take care to avoid camera shake. Bracket your exposures (make extra exposures at plus and minus 1 f-number at same shutter speed) to ensure good results.*

Scanning

Scanners can be useful to digitise a printed image. The quality of the scanned image is defined in terms of the number of dots per inch (dpi). Dedicated facilities may be available in your university; however, many modern printers also have a scan feature, though often at lower resolution. A scanner with a value of 600 dpi is adequate for most purposes.

Image processing

Images are recorded in digital cameras as files on various forms of memory cards and the data are later transferred either to a computer for manipulation or directly to a printer. An appropriate card reader or USB cable may be required. For critical work, the rule is to capture images using the largest possible file size, as reductions (e.g. compression of file size) of digital information are possible, but you can never add information.

Digital image files can be stored in various formats (e.g. .bmp, .jpg) and can be manipulated using programs such as Adobe Photoshop or Paint Shop Pro. Aspects of image quality such as contrast and colour balance can be adjusted and images can be cropped and resized to suit your application (e.g. small file sizes are recommended for websites). Digital images are valuable for use with word processors and Web-authoring software and can be transferred readily between such programs.

Film images require chemical development, usually a professional task. A film image may be impaired irreversibly by faulty development and the resulting image can only be modified to a limited extent during subsequent printing. This is a key reason for selecting digital image systems.

Photomicrography

This requires the use of a camera system mounted on a microscope. The important operations for successful photomicrography are outlined below:

1. Carefully prepare the specimen: ensure thorough cleanliness of any slides or coverslips used.
2. Choose the correct film type and any filters required for alteration of colour balance, depending upon the type of light source available.

3. Decide on the magnification to be used: make sure that you know how this relates to the magnification of the negative/transparency. Always include a photograph of a stage micrometer (p. 259) so the final magnification can be calculated and given in the legend to the photograph.
4. Carefully focus the specimen onto the film plane: there will be a system-specific method for doing this (consult the manual, as this is a critical aspect).
5. Bracket extra exposures above and below the automatic settings, especially for colour photography.

Printing images

The most appropriate choice of print type will depend on your purpose. Glossy-finish prints generally appear sharper than matt or other finishes. Moreover, the addition of lettering and scales is often only possible on the smooth surface of a glossy print. However, when preparing prints for display on a poster, matt/velvet finishes are often preferable, as there is less glare for the viewer. Printing from computers using ink-jet and similar printers requires care, since colours viewed on monitors do not always reproduce on paper as expected. You may need to experiment with settings and paper quality to obtain the appropriate results.

Adding scales and labels to images
Having acquired a suitable print, it is often necessary to add information to it:

- When digital images are copied into word-processor or presentation software, additional information such as lettering, scales and labels can be added by overlaying a 'text box', or equivalent. You may need to make the background transparent, e.g. using the 'no fill' option in PowerPoint. Lettering can be added to printed media using transfer letters (e.g. Letraset).
- Choose a simple font type that will not distract and a size that is legible but not too dominant, e.g. 12 point in final reproduction. Always use the same font and type size for a related sets of images.
- Choose an appropriate part of the image for the lettering, i.e. a black area for a white letter. If the background is variable, create a coloured or white area and place the letters on this (see Fig. 48.3).

Storing and mounting images

Digital images
Indexing software can be used to organise and catalogue your images (e.g. Picasa, available for free download at: http://picasa.google.com/). These programs allow you to add appropriate information and store this electronically along with the image, facilitating viewing and selection when your collection of images is large.

Prints and transparencies
Prints, negatives and transparencies are easily damaged and should be stored with care:

- Prints should be stored flat in boxes or in albums. When used for display, mount on stiff board either with modern photographic adhesives or dry-mounting tissue.

Using scales with images – these are vital to the interpretation of any biological image (see Chapter 44 for information on different types and how to calculate them).

Fig. 48.3 Use of white areas for labelling dark prints, illustrated by an SEM micrograph of a stellate hair on a leaf surface.

Recording information about images – this should include subject, location, image-capture details and processing procedures. Bear in mind that these data may be of use when the image is used in a different context from the one you initially planned.

- Negative strips should be stored in transparent or translucent paper filing sheets. Obtain a set of contact prints for each film and store this with the negatives for easy reference.
- Transparencies (slides) should be mounted in plastic mounts but not between glass, as this often causes more problems than it solves. Labelling of transparencies is best done on the mount, using small, self-adhesive labels.

Beware of cheap plastic filing materials as some contain residual chemicals that can cause damage over time. Avoid dampness: silica gel desiccant can be used where necessary. Both physical copies and records should be well organised, including all potentially relevant information.

Sources for further study

Adobe Systems Inc. *Digital Imaging and Photo Editing Software for Digital Photography*. Available: http://www.adobe.com/digitalimag/ Last accessed: 09/04/07

Ang, T. (2002) *Digital Photographer's Handbook*. Dorling Kindersley, London.

Daney, C. *Photography and Digital Imaging Resources – a Collection of Links*. Available: http://www.mbay.net/~cgd/photo/pholinks.htm Last accessed: 09/04/07

Gookin, D. (2000) *Digital Scanning and Photography*. Microsoft Press International, Redmond, Washington.

McCartney, S. (2001) *Mastering the Basics of Photography*. Allworth Press, New York.

Oldfield, R., Dickes, F.W. and Rost, F.W.D. (2000) *Photography with a Microscope*. Cambridge University Press, Cambridge.

Study exercises

48.1 Choose an appropriate image-recording method. You wish to record a stage in an experiment as an image. What are the relative merits of doing this using a film camera relative to using a digital camera?

48.2 Make a list of situations where an image is often required. When in biology project work might you require to use images?

48.3 Choose an appropriate combination of film/ lighting/image for specific purposes. What film, lighting and image types would you select to record the following items *in situ*?
(a) a flower in a field
(b) a field site
(c) plants in a greenhouse
(d) mice in a laboratory experiment.

The ability of organisms to grow, develop and respond to stimuli is fundamental to life. The aims of studies concerned with these phenomena include:

- investigating developmental processes;
- comparing developmental patterns;
- understanding behavioural responses;
- characterising physiological processes;
- measuring the levels of a particular compound (bioassays);
- assessing sensitivity to a stimulus.

> **KEY POINT** *The growth and development of a multicellular organism is a complex phenomenon that cannot be easily measured by a single variable. An overall picture emerges from changes in size, weight, number, complexity and function evaluated at cell, tissue and organ levels.*

Growth and development

Length and area are relatively easy to measure (see p. 134 for techniques) though curved organisms may present difficulties. Volume measurements can be made from the displacement of water employing Archimedes' principle, or derived from fresh weight if the density of the organism can be estimated. Growth may not occur smoothly through time (e.g. plants increase in height mainly at night, when cell turgor is highest). For rapidly growing organisms it is desirable to have a continuous photographic or video record, though this may only be feasible with sedentary organisms. Care is required to ensure that focus is maintained on the points that will be measured, or your data will be inaccurate. If you are interested in growth of different parts of the organism, use reference points, e.g. recognisable morphological features such as hairs or marks made with ink or paint.

Fresh weight might at first sight appear a useful measure of growth: however, water content may vary according to availability and consumption. Dry weight (p. 135) does not have this disadvantage, but it may decrease when respiration dominates metabolism (e.g. during seed germination), despite the fact that growth is obviously occurring. Protein or nitrogen content may be better correlated with 'mass of protoplasm', but they are both complex to measure and are 'destructive' measurements.

Whichever of the above variables is measured, growth generally shows a sigmoidal increase in time (Fig. 49.1(a)) that is reminiscent of the population growth curve of cells in batch culture (p. 272). The slope of this curve at any given time represents the absolute growth rate, with units of increment in size per unit time. Plotting this against time generally shows a bell-shaped curve (Fig. 49.1(b)). The relative growth rate or rate of increase in size per unit of size measures the 'efficiency' of growth (Fig. 49.1(c)).

Cell number is a common measure of the growth of unicellular suspensions (p. 274). For multicellular organisms, this can be estimated from measurements of average cell size and total volume. Numbers of organs (e.g. leaves), specialised cells (e.g. hairs) or organelles (e.g. mitochondria) can indicate the progress of differentiation and development, but in general these

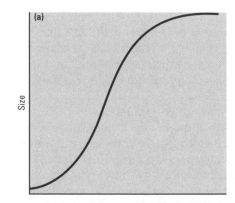

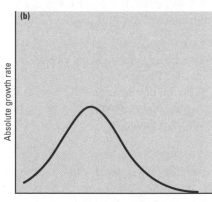
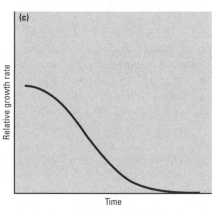

Fig. 49.1 Different ways of analysing growth: (a) simple plot of size against time; (b) plot of absolute growth rate for the same data; (c) plot of relative growth rate for the same data.

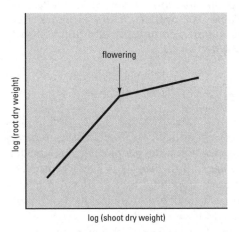

Fig. 49.2 Example showing changes in an allometric relationship based on a published analysis of the growth of a grass. Both root and shoot dry weight increase as the plant ages. The onset of flowering causes a clear change in their relationship.

Analysing allometric data – when fitting eqn [49.2] to allometric data, you should use Model II regression as the variables are interdependent (see p. 435). Note that for monophasic log–log relationships, **a** can be determined from the intercept (log **a**) and **b** from the slope (log **b**).

are difficult processes to characterise quantitatively. Changes in numbers can also be expressed as absolute and relative rates, as above.

Display and analysis of growth data

Care should be taken when plotting a growth curve. When plotting change in size against time for a single organism, a smooth curve passing through all measurement points is appropriate (i.e. measurement error is assumed to be negligible). If dealing with a sample of organisms, the approach depends on whether (a) the same sample is measured each time (as with non-destructive harvesting, e.g. for linear measurement), when the curve should pass through each mean point, or (b) the samples are independent (as with destructive harvesting, e.g. for dry-weight measurement), when a smooth trend line should be drawn that need not pass through all of the points. Regarding statistical analysis, note that time-series measurements can only be treated as independent if the experimental design is as in (b) above; otherwise, in comparing changes in size with time, you should use tests for paired data. Note that variance may increase with size, so non-parametric tests (Chapter 66) are likely to be appropriate.

When presenting growth rates, use histograms if the data are obtained from differences in mean values divided by the time interval between measurements. A smooth curve can be used if it represents the slope (i.e. first derivative) of a function that has been fitted to the growth curve. Growth data are often log-normally distributed, so curves should be fitted to logarithmically transformed data. Fitting functions to growth curves is complex and specialist advice should be sought.

Allometry

This is the study of relationships between the dimensions of organisms. It can be used for describing changes in the allocation of resources within a single species (e.g. root-to-shoot ratio in plants, Fig. 49.2) or among species (e.g. brain size in relation to body mass). The underlying relationship between dimensions x and y is generally:

$$y = ax^b \qquad [49.1]$$

where a is a positive constant. The coefficients a and b can be estimated from a log–log plot (e.g. Fig. 49.2), which may show one or more linear phases of the form:

$$\log y = \log a + b \log x \qquad [49.2]$$

Care should be taken in interpreting data of this type: remember that eqn [49.1] is an experimentally derived relationship rather than a theoretical one. Consult a specialist text (e.g. Causton and Venus, 1981) if you wish to fit a line or lines to such data.

Responses to stimuli

Responses to stimuli are involved in many activities essential for life. For example, they are required to capture resources, escape or deter predators, move towards or away from a given set of conditions or interact with other individuals in a community. Understanding these responses requires answers to the following questions:

- What is the precise response and what causes it?
- How does the response develop?

- What is the function of the response?
- How might the response have evolved?
- How much of the response is inherited and how much learned?

The first stage when investigating a response is to describe it fully (preferably under conditions as near to those in nature as possible) and determine the precise stimuli that cause it. For instance, a study of plant phototropism might start with research into which cells perceive the light stimulus and which respond; effective wavelengths and doses of light might then be investigated. When working with animals, avoid anthropomorphism and describe objectively what happens. All circumstances of potential relevance (e.g. weather, time of day) should be recorded until a clear pattern emerges. A full description of most responses obviously requires physiological and biochemical techniques whose scope is too wide to be covered here.

You may wish to follow an individual animal in space or time and this will require either recognising its features or marking it in some way that does not interfere with the response (e.g. leg-ringing birds or painting symbols on insects). When recording the elements of a behavioural pattern, a coding system may be helpful (e.g. 1 = flaps wings, 2 = drops neck, etc.). As with growth, photography and video techniques are good aids for recording movements and analysing them later. Tape recorders can be used to record a description of responses so long as your voice does not interfere with the response. They can also record the sound component of a response and might be used later to elicit a response under controlled conditions (e.g. for a suspected alarm cry).

The second question listed above requires a study of the response throughout the developmental stages of the organism. The third and fourth questions require a certain amount of interpretive skills but, as with the fifth question, are amenable to experimentation. You may be presented with a natural experiment if conditions change in the right way in the course of your observations. Experiments in a laboratory give you control over conditions but this may well be at the expense of realism (see p. 185).

In designing experiments to investigate animal behaviour, you must take account of the fact that an animal's behaviour may change according to its experience. If not actually investigating learning *per se*, you need to ensure that the opportunity for a subject to learn is the same for each treatment, or bias will occur. Ways round this problem are either to use naïve animals for every replicate or to adopt a Latin square design (p. 189) so that experiences are shared evenly among treatments (fully randomised or randomised block designs may not achieve this).

Bioassays and their applications

A bioassay is a method of quantifying a chemical substance (analyte) by measuring its effect on a biological system under controlled conditions. The hypothetical underlying phenomena are summarised by the relationship:

$$A + Rec \rightleftharpoons ARec \longrightarrow R \qquad [49.3]$$

where A is the analyte, Rec the receptor, ARec the analyte–receptor complex and R the response. This relation is analogous to the formation

---

Distinguishing between innate and learned behaviour – innate behaviour is genetically determined; it is always executed perfectly and in its entirety the first and every subsequent time the stimulus occurs. Learned behaviour is modified by experience.

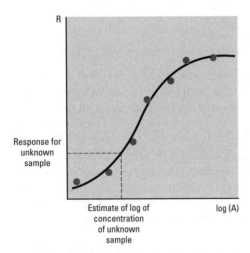

Fig. 49.3 Typical bioassay response curve, showing estimation of an unknown sample. Closed circles represent responses with standard samples. R = response; A = analyte.

Examples The following are typical bioassays:

- measuring the amount of a plant growth substance in a plant extract using the coleoptile straight growth assay;
- measuring the effect of a drug on the dilation of a person's pupils;
- measuring the mutagenic properties of a chemical compound in the Ames test (see Box 49.1).

Legal use of bioassays – in the UK, where bioassays involving 'higher' (vertebrate) animals are controlled by the Animals Scientific Procedures Act (1986), they can only be carried out under the direct supervision of a scientist licensed by the Home Office.

of product from an enzyme–substrate complex and, using similar mathematical arguments to those of enzyme kinetics (p. 322), it can be shown that the expected relationship between [A] and rate of response is hyperbolic (sigmoidal in a log–linear plot like Fig. 49.3). This pattern of response is usually observed in practice if a wide enough range of [A] is tested.

To carry out the assay, the response elicited by the unknown sample is compared with the response obtained for differing concentrations of the substance, as shown in Fig. 49.3. When fitting a curve to standard points and estimating unknowns, the available methods, in order of increasing accuracy, are:

1. fitting by eye;
2. using linear regression on a restricted 'quasi-linear' portion of the assay curve;
3. linearisation followed by regression (e.g. by probit transformation);
4. non-linear regression (e.g. to the Morgan–Mercer–Flodin equation).

In general, bioassay techniques have more potential faults than physico-chemical assay techniques. These may include the following:

- A greater level of variability: error in the estimate of the unknown compound will result because no two organisms will respond in exactly the same way. Assay curves vary through time, and because they are non-linear, a full standard curve is required each time the assay is carried out.
- Lack of chemical information: bioassays provide information about *biological* activity; they say little about the chemical structure of an unknown compound. The presence of a specific compound may need to be confirmed by a physico-chemical method (e.g. mass spectrometry).
- Possibility of interference: though many bioassays are very specific, it is possible that different chemicals in the extract may influence the results.

Despite these problems, bioassays are still much used. They are 'low tech' and generally cheap to set up. They often allow detection at very low concentrations. Bioassays also provide the means to assess the biological activity of chemicals and to study changes in sensitivity to a chemical, which physico-chemical techniques cannot do. Changes in sensitivity may be evident in the shape of the dose–response curve and its position on the concentration axis.

Bioassays can involve responses of whole organisms or parts of organisms. 'Isolated' responding systems (e.g. excised tissues or cells, Chapter 47) decrease the possibility of interference from other parts of the organism. Disadvantages include disruption of nutrition and wound damage during excision. Isolation can continue down to the molecular level, as in immunoassays (Chapter 51).

Bioassays are the basis for characterising the efficacy of drugs and the toxicity of chemicals. Here, response is often treated as a quantal (all-or-nothing) event. The E_{50} is defined as that concentration of a compound causing 50 per cent of the organisms to respond. Where death is the observed response, the LD_{50} describes the concentration of a chemical that would cause 50 per cent of the test organisms to die within a specified period under a specified set of conditions. Box 49.1 presents details of the Ames test, a widely used bioassay to assess the mutagenicity of chemicals.

Box 49.1 Mutagenicity testing using the Ames test – an example of a widely used bioassay

Chemical carcinogens can be identified by the formation of tumours in laboratory animals exposed to the compound under controlled conditions. However, such animal bioassays are time-consuming and expensive. Dr B.N. Ames and co-workers have shown that most carcinogens are also mutagens, i.e. they will induce mutational changes in DNA. The Ames test makes use of this correlation to provide a simple, rapid and inexpensive bioassay for the initial screening of potential carcinogens. The test makes use of particular strains of *Salmonella typhimurium* with the following characteristics:

- histidine auxotrophy – the tester strains are unable to grow on a minimal medium without added histidine: this characteristic is the result of specific mutational changes to the DNA of these strains, including base substitutions and frame shifts;
- increased cell envelope permeability, to permit access of the test compound to the cell interior;
- defects in excision repair systems and enhanced error-prone repair systems, to reduce the likelihood of DNA repair after treatment with a potential mutagen.

When grown in the presence of a chemical mutagen, the bacteria may revert to prototrophy as a result of back mutations that restore the wild-type phenotype: such revertants grow independently of external histidine and are able to form colonies on minimal medium, unlike the original test strains. The extent of reversion can be used to assess the mutagenic potential of a particular chemical compound. Since many chemicals must be activated *in vivo*, the test incorporates a rat liver homogenate (so-called 'S-9 activator', containing microsomal enzymes) to simulate the metabolic events within the liver. The tester strains are mixed with S-9 activator and a small amount of molten soft agar, then poured as a thin agar overlay (top agar) on a minimal medium plate. The top agar layer contains a very small amount of histidine, to allow the tester strains to divide a few times and express any mutational changes (i.e. prototrophy).

The test can be performed in one of two ways:

1. **Spot test:** a concentrated drop of the test compound is placed at the centre of the plate, either directly on the agar surface, or on a small filter paper disc. The test compound will diffuse into the agar and revertants appear as a ring of colonies around the site of inoculation, as shown in Fig. 49.4. The distance of the ring of colonies from this site provides a measure of the toxicity of the compound, while the number of colonies within the ring gives an indication of the relative mutagenicity of the test substance. The spot test is often carried out in a simplified form, without added S-9 activator, as a rapid preliminary test prior to quantitative analysis.

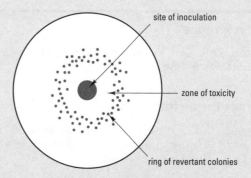

Fig. 49.4 Typical outcome of spot test (Ames test)

2. **Agar incorporation test:** known amounts of the test compound are mixed separately with molten top agar and the other constituents and poured on to separate minimal agar plates. After incubation for 48 h, revertants will appear as evenly dispersed colonies throughout the agar overlay. The number of colonies reflects the relative mutagenicity of the test compounds, with a direct relationship between colony count and the amount of mutagen. Agar incorporation tests can be used to generate dose–response curves similar to that shown in Fig. 49.3. The simplicity, sensitivity and reproducibility of the Ames test has resulted in its widespread use for screening potential carcinogens in many countries, though it is not an infallible test for carcinogenicity.

 SAFETY NOTE Correct handling procedures must be followed at all times, as the test substances may be carcinogenic – testers should wear gloves and avoid skin contact or ingestion.

Considerations when setting up a bioassay
- The response should be easily measured and as metabolically 'close' to the initial binding event as possible.
- The experimental conditions should mimic the *in vivo* environment.
- The standards should be chemically identical to the compound being measured and spread over the expected concentration range being tested.

- The samples should be purified if interfering compounds are present and diluted so the response will be on the 'linear' portion of the assay curve.

To check for interference, the bioassay may be standardised against another method (preferably physico-chemical). Related compounds known to be present in the analyte solution should be shown to have minimal activity in the bioassay. If an interfering compound is present, this may show up if a known amount of standard is added to sample vials – the result will not be the sum of independently determined results for the standard and sample.

Text reference

Causton, D.R. and Venus, J.C. (1981) *The Biometry of Plant Growth*. Edward Arnold, London.

Sources for further study

Drickamer, L.C., Vessey, S.H. and Jakob, E.M. (2001) *Animal Behavior: Mechanisms, Ecology, Evolution*. McGraw-Hill, Dubuque, Iowa.

Hunt, R. (1990) *Basic Growth Analysis*. Unwin Hyman, London.

Martin, P. and Bateson, P. (1993) *Measuring Behaviour: An Introductory Guide*, 2nd edn. Cambridge University Press, Cambridge.

Price, E.O. and Stokes, A.W. (1975) *Animal Behavior in Laboratory and Field*, 2nd edn. W.H. Freeman, San Francisco.

Salanki, J., Jeffrey, D. and Hughes, G.M. (1994) *Biological Monitoring of the Environment. A Manual of Methods*. International Union of Biological Sciences Methodology Series, CAB International, Oxford.

Study exercises

49.1 Estimate relative growth rate. A baby hamster weighed at the start of a week has a mass of 35 g. By the end of one week it has a mass of 38.5 g.
(a) What is its absolute growth rate on a daily basis, expressed in mg day^{-1}?
(b) What is its relative growth rate on a weekly basis, expressed as a percentage?

49.2 Use bioassay data to estimate the amount of analyte in unknown samples. Bioassay data obtained by exposing water fleas (*Daphnia* sp.) to a purified algal toxin are shown in the table opposite.
(a) Graph the data and use the graph to estimate the toxin contents of two samples obtained from two lochs that gave (i) a heart rate of 130 beats per minute and (ii) a heart rate of 25 beats per minute.

(b) Which one of the two estimates would be the less reliable and why?
(c) How could you improve the reliability of this estimate, if you had more of the sample?
(d) What unmeasured factors might limit the usefulness of this assay?

Responses of *Daphnia* to toxin

Toxin concentration (mmol l^{-1})	Mean heart rate of *Daphnia* sp. (beats per minute)
1.00×10^{-3}	20
3.16×10^{-4}	23
1.00×10^{-4}	26
3.16×10^{-5}	50
1.00×10^{-5}	180
3.16×10^{-6}	192
1.00×10^{-6}	195

(continued)

Study exercises (continued)

49.3 Use agar diffusion bioassay data to quantify a growth inhibitor in a sample. Agar diffusion bioassay is a simple means of estimating the amount of an inhibitory substance in a test sample, by adding the substance, in aqueous solution, to wells cut in a Petri plate of agar medium seeded with a microbe sensitive to the substance and then incubating the plate at an appropriate growth temperature. Diffusion of the inhibitory substance gives a zone of growth inhibition proportional to $\log_{10}$ of the concentration of substance. In the following example, the amount of the antibiotic nisin in a test sample can be determined by comparing the degree of growth inhibition of *Bacillus subtilis* for the test solution and a series of standards containing known concentrations of nisin. Prepare a calibration curve and determine the nisin concentration of solution 5 (unknown). Express your answer in $\mu g\,ml^{-1}$, to one decimal place.

Growth inhibition of *B. subtilis* in solutions containing different amounts of nisin

Solution	Nisin content ($\mu g\,ml^{-1}$)	Zone of growth inhibition (mm)
1	40	26.5
2	20	23
3	10	19.5
4	5	17
5	unknown (test solution)	22.5

Advanced laboratory techniques

There are many instances where it is necessary to measure the quantity of a test substance using a calibrated procedure. You are most likely to encounter this approach in one or more of the following practical exercises:

- Quantitative spectrophotometric assay of biomolecules (Chapter 59).
- Flame or atomic absorption spectroscopic analysis of metal ions in biological solutions (p. 369).
- Using a chromogenic or fluorogenic substrate to determine the activity of an enzyme (p. 319).
- Quantitative chromatographic analysis, e.g. GC or HPLC (Chapter 60).
- Using a bioassay system (p. 296) to quantify a test substance: examples include immunodiffusion (Chapter 51) and radioimmunoassay (p. 312).

Understanding quantitative measurement – Chapter 25 contains details of the basic principles of valid measurement, while Chapters 51–60 deal with some of the specific analytical techniques used in biology. The use of internal standards is covered on p. 379.

 KEY POINT *In most instances, calibration involves the establishment of a relationship between the measured response (the 'signal') and one or more 'standards' containing a known amount of substance.*

In some instances, you can measure a signal due to an inherent property of the substance, e.g. the absorption of UV light by nucleic acids (p. 369), whereas in other cases you will need to react it with another substance to see the result (e.g. molecular weight measurements of DNA fragments after electrophoresis, visualised using ethidium bromide, p. 335), or to produce a measurable response (e.g. the reaction of cupric ions and peptide bonds in the Biuret assay for proteins, p. 319).

Calibrating laboratory apparatus – this is important in relation to validation of equipment, e.g. when determining the accuracy and precision of a pipettor by the weighing method: see p. 124.

The different types of calibration curve

By preparing a set of solutions (termed 'standards'), each containing either (i) a known *amount* or (ii) a specific *concentration* of the substance, and then measuring the response of each standard solution, the underlying relationship can be established in graphical form as a 'calibration curve', or 'standard curve'. This can then be used to determine either (i) the amount or (ii) the concentration of the substance in one or more test samples. Alternatively, the response can be expressed solely in mathematical terms: an example of this approach is the determination of chlorophyll pigments in plant extracts by measuring absorption at particular wavelengths, and then applying a formula based on previous (published) measurements for purified pigments (p. 369).

There are various types of standard curve: in the simplest cases, the relationship between signal and substance will be linear, or nearly so, and the calibration will be represented best by a straight-line graph (see Box 50.1). In some instances (Fig. 50.1(a)), you will need to transform either the x values or y values (Chapter 61), in order to produce a linear graph (e.g. in radial immunodiffusion bioassay, where the y values are squared, pp. 310–12. In other instances, the straight-line relationship may only hold up to a certain value (the 'linear dynamic range') and beyond this point the graph may curve (e.g. in quantitative spectrophotometry, the Beer–Lambert relationship often becomes invalid at high absorbance, giving a curve, Fig. 50.1(b)). Some calibration curves are

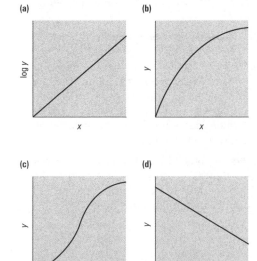

Fig. 50.1 Calibration curves: (a) log–linear; (b) curvilinear; (c) sigmoid, or S-shaped; (d) inverse

Box 50.1 The stages involved in preparing and using a calibration curve

1. **Decide on an appropriate test method** – for example, in a project, you may need to research the best approach to the analysis of a particular metabolite in your biological material.

2. **Select either (a) amount or (b) concentration, and an appropriate range and number of standards** – in practical classes, this may be given in your schedule, along with detailed instructions on how to make up the standard solutions. In other cases, you may be expected to work this out from first principles (Chapters 21–23 give worked examples) – aim to have evenly spaced values along the x axis.

3. **Prepare your standards very carefully** – due attention to detail is required: for example, you should ensure that you check the calibration of pipettors beforehand, using the weighing method (p. 124). Don't forget the 'zero standard' plus any other controls required, e.g. to test for interference due to other chemical substances. Your standards should cover the range of values expected in your test samples.

4. **Assay the standards and the unknown (test) samples** – preferably all at the same time, to avoid introducing error due to changes in the sensitivity or drift in the zero setting of the instrument with time. It is a good idea to measure all of your standard solutions at the outset, and then measure your test solutions, checking that the 'zero standard' and 'top standard' give the same values after, say, every six test measurements. If the re-measured standards do not fall within a reasonable margin of the previous value, then you will have to go back and recalibrate the instrument, and repeat the last six test measurements. If your test samples lie outside the range of your standards, you may need to repeat the assay using diluted test samples (extrapolation of your curve may not be valid, see p. 394).

5. **Draw the standard curve, or determine the underlying relationship** – Figure 50.2 gives an example of a typical linear calibration curve, where the spectrophotometric absorbance of a series of standard solutions is related to the amount of substance. When using a spreadsheet (Box 50.2) or graphic package (Chapter 12), it is often appropriate to use Model I linear regression (p. 435)

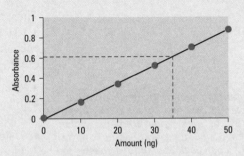

Fig. 50.2 Typical calibration curve for spectrophotometric analysis

to produce a linear trend line (also termed the 'line of best fit') and you can then quote the value of r^2, which is a measure of the 'fit' of the measurements to the line (see p. 434). However, you should take care not to use a linear plot when the underlying relationship is clearly non-linear (Fig. 62.19) and you must consider whether the assumptions of the regression analysis are valid (e.g. for transformed data, pp. 384–5).

6. **Determine the amount or concentration in each unknown sample** – either by reading the appropriate value from the calibration curve, or by using the underlying mathematical relationship, i.e. $y = a + bx$ (p. 411). Make sure you draw any horizontal and vertical construction lines very carefully – students often lose marks unnecessarily by submitting poorly drawn construction lines within practical reports.

7. **Correct for dilution or concentration, where appropriate** – for example, if you diluted each test sample by ten-fold, then you would need to multiply by 10 to determine the value for the undiluted test sample. As another example, if you assayed 0.2 ml of test sample, you would need to multiply the value obtained from the calibration curve by 5, to give the value per ml.

8. **Quote your test results to an appropriate number of significant figures** – this should reflect the accuracy of the method used (see pp. 406–8), not the size of your calculator's display.

sigmoid (Fig. 50.1(c)). Finally, the signal may *decrease* in response to an increase in the substance (Fig. 50.1(d)), e.g. radioimmunoassay, where an inverse sigmoid calibration curve is obtained (p. 312). In some practical classes, you may be told that the relationship is expected to be linear, curvilinear, or whatever, while in others you may be expected to decide the form of the standard curve as part of the exercise.

Practical considerations

Amount or concentration?

This first step is often the most confusing for new students. It is vital that you understand the difference between *amount* of substance (e.g. mg, ng, etc.), and *concentration* (the amount of substance per unit volume, e.g. $mmol\,l^{-1}$, $mol\,m^{-3}$, % w/v, etc.) before you begin your practical work.

 KEY POINT *Essentially, you have to choose whether to work in terms of either (i) the total amount of substance in your assay vessel (e.g. test tube, or cuvette) or (ii) the final concentration of the substance in your assay vessel, which is independent of the volume used.*

The interconvession of amount and concentration is covered in more detail on p. 139. Either way, this is usually plotted on the x (horizontal) axis and the measured response on the y (vertical) axis.

Choice of standards

In your early practical classes, you may be provided with a stock solution (p. 130), from which you then have to prepare a specified number of standard solutions. In such cases, you will need to understand how to use dilutions to achieve the required amounts or concentrations (p. 130). In later work and projects, you may need to prepare your standards from chemical reagents in solid form, where the important considerations are purity and solubility (p. 128). For professional analysis (e.g. in forensic science, or clinical biochemistry), it is often important to be able to trace the original standard or stock solution back to national or international standards or to certified reference materials.

How many standards are required?

This may be given in your practical schedule, or you may have to decide what is appropriate (e.g. in project work and research). If the form and working range of the standard curve is known in advance, this may influence your choice – for example, linear calibration curves can be established with fewer standards than curvilinear relationships. In some instances, analytical instruments can be calibrated using a single standard solution, often termed a 'calibrator'. Replication of each standard solution is a good idea, since it will give you some information on the variability involved in preparing and assaying the standards. Consider whether you should plot mean values on your standard curve, or whether it is better to plot the individual values (if one value appears to be well off the line, you have made an error, and you may need to check and repeat).

Example A test tube cntaining 8 ml of water plus 2 ml of 1 % w/v NaCl (M_r) = 58.44) would have a *mass concentration* of 0.2 % w/v NaCl (a five-fold dilution of the original NaCl solution), which can also be expressed as $2\,gl^{-1}$; in terms of *molar concentration* (p. 138), this would be equivalent to $2 \div 58.44 = 0.0342\,moll^{-1}$ (to three significant figures). Expressed in terms of the *amount* of NaCl test tuble, this would be 0.02 g in *mass*, or $0.02 \div 58.44 = 3.42 \times 10^{-4}$ mol (342 μmol in *moles*.

Organisations providing national/international standards – these include:

- *Laboratory of the Government Chemist (http://www.lgc.co.uk/), for UK standards and European Reference Materials.*
- *National Institute for Science and Technology (http://www.cstl.nist.gov/), for biochemical/chemical standards in the USA.*
- *Institute for Reference Materials and Measurements (http://www.irmm.jrc.be/), for European standards, including BCR and ERM materials.*
- *OIE Biological Standards Commission (http://www.oie.int/bsc/eng/en_bsc.htm), for international animal standards.*

Plotting a standard curve – do not force your calibration line to pass through zero if it clearly does not. There is no reason to assume that the zero value is any more accurate than any other reading you have made.

Box 50.2 How to use a spreadsheet (Microsoft Excel) to produce a linear regression plot

Two approaches are possible: either using the trendline feature or using the regression analysis function within the data analysis tool pack. Both make the assumption that the criteria for Model I linear regression are met (see p. 435). In the example shown below, the following simple data set has been used:

Amount (ng)	Absorbance
0	0.00
10	0.19
20	0.37
30	0.56
40	0.63
50	0.78

Using the *Trendline* feature. This quick method provides a line of best fit on an Excel chart and can also provide a set of equation values for predictive purposes.

1. **Create a graph (*chart*) of your data.** Enter the data in two columns within your spreadsheet, select the data array (highlight using left mouse button) and then, using the *Chart Wizard* icon, select *Chart type > XY (Scatter)*.
2. **Add a trend line.** Right-click on any of the data points on your graph, and select the *Add Trendline* menu. Choose the *Linear* trend-line option, but do not click *OK* at this stage. Rather, from the *Options* menu, select: (i) *Display equation on chart* and (ii) *Display R-squared value on chart*. Now click OK. The equation (shown in the form $y = bx + a$) gives the slope and intercept of the line of best fit, while the R-squared value (coefficient of determination, p. 435) gives the proportional fit to the line (the closer this value is to 1, the better the fit of the data to the trend line).
3. **Modify the graph to improve its effectiveness.** For a graph that is to be used elsewhere (e.g. in a lab write-up or project report), adjust the display to remove the default background and gridlines and change the symbol shape (see Box 62.2, p. 391 for more advice and examples). Right-click on the trend line and use the *Format Trendline > Custom*

menu to adjust the *Weight* of the line to make it thinner. Drag and move the equation panel if you would like to alter its location on the chart, or delete it, having noted the values. Figure 50.3 shows a calibration curve produced in this way for the data presented above.

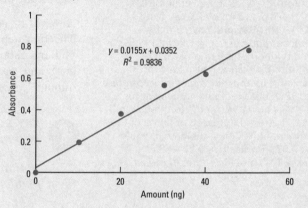

$$y = 0.0155x + 0.0352$$
$$R^2 = 0.9836$$

Fig. 50.3 Calibration curve showing line of best fit and details of linear regression equation

4. **Use the regression equation to estimate unknown (test) samples.** By rearranging the equation for a straight line and substituting a particular *y*-value, you can predict the amount/concentration of substance (*x*-value) in a test sample. This is more precise than simply reading the values from the graph using construction lines (Box 50.1). If you are carrying out multiple calculations, the appropriate equation, $x = (y - a)/b$, can be entered into a spreadsheet, for convenience.

Using the Regression Analysis tool. This requires the 'data analysis tool pack' to be loaded beforehand (using *Tools > Add-Ins > Analysis ToolPak*) and provides summary output that contains details of slope, intercept, and coefficient of determination along with an analysis of variance (ANOVA) table (Chapter 66 gives further details of these aspects, including an example of output in Study exercise 66.4).

Dealing with interfering substances – one approach is to use the method of 'standard additions', where the standards all contain a fixed additional amount of the sample (see e.g. Dean, 1997). Internal standards (p. 379) can also be used to detect such problems.

Preparing your standards

It is extremely important to take the greatest care to measure out all chemicals and liquids very accurately, to achieve the best possible standard curve. The grade of volumetric flask used and temperature of the solution also affect accuracy (grade A apparatus is best). You may also consider what other additives might be required in your standard solutions. For example, do your test samples have high levels of potentially interfering substances, and should these also be added to your standards? Also consider what controls and blank solutions to prepare.

KEY POINT *The validity of your standard curve depends upon careful preparation of standards, especially in relation to accurate dispensing of the volumes of any stock solution and diluting liquid – the results for your test samples can only be as good as your standard curve.*

Preparing the calibration curve and determinining the amount of the unknown (test) sample(s)

This is described in stepwise fashion within Box 50.1. Check you understand the requirements of graph drawing, especially in relation to plotted curves (p. 387) and the mathematics of straight-line graphs (p. 411). Spreadsheet programs such as Microsoft Excel can be used to produce a regression line for a straight-line calibration plot (pp. 434–7). Examples of how to do this are provided in Box 50.2.

Text reference

Dean, J.D. (1997) *Atomic Absorption and Plasma Spectroscopy*, 2nd edn. Wiley, Chichester.

[Chapter 1 deals with calibration, and covers the principle of standard additions.]

Sources for further study

Brown, P.J. (1994) *Measurement, Regression, and Calibration*. Clarendon Press, Oxford.
[Covers advanced methods, including curve fitting and multivariate methods.]

Mark, H. (1991) *Principles and Practice of Spectroscopic Calibration*. Wiley, New York.

Miller, J.N. and Miller, J.C. (2005) *Statistics and Chemometrics for Analytical Chemistry*, 5th edn. Prentice Hall, Harlow.
[Gives detailed coverage of calibration methods and the validity of analytical measurements.]

Study exercises

50.1 Determine unknowns from a hand-drawn calibration curve. The following data are for a set of calibration standards for Zn, measured by atomic absorption spectrophotometry.

Absorbance measurements for a series of standard solutions containing different amounts of zinc

Zinc concentration ($\mu g\,ml^{-1}$)	Absorbance
0	0.000
1	0.082
2	0.174
3	0.257
4	0.340
5	0.408
6	0.463
7	0.511
8	0.543
9	0.561
10	0.575

Draw a calibration curve by hand using graph paper and estimate the concentration of zinc in the following water samples:
(a) an undiluted sample, giving an absorbance of 0.157
(b) a twenty-fold dilution, giving an absorbance of 0.304
(c) a five-fold dilution, giving an absorbance of 0.550.
Give your answer to three significant figures in each case.

50.2 Determine unknowns from a calibration curve produced in Excel. The following data are for a set of calibration standards for protein content, measured by the Lowry (Folin–Ciocalteau) method (p. 320).

(continued)

Study exercises (continued)

Absorbance measurements for a series of protein standards

Protein (mg per tube)	Absorbance
0.00	0.000
0.02	0.161
0.04	0.284
0.06	0.438
0.08	0.572
0.10	0.762

Using PC-based software (e.g. Excel), fit a trend line (linear regression) and determine the protein content of solutions with the following absorbances:
(a) 0.225   (b) 0.465   (c) 0.682
Give your answers to three significant figures in each case.

50.3 **Identify the errors in a calibration curve.** The figure shows a calibration curve of the type that might be submitted in a practical write-up. List the errors and compare your observations with the list given on p. 455.

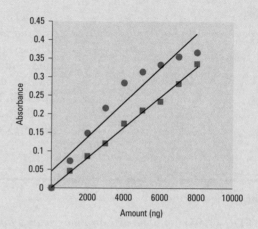

Calibration graph

Antibody – a protein produced in response to an antigen (an *anti*body-*gen*erating foreign macromolecule).

Epitope – a site on the antigen that determines its interaction with a particular antibody.

Hapten – a substance that contains at least one epitope, but is too small to induce antibody formation unless it is linked to a macromolecule.

Ligand – a molecule or chemical group that binds to a particular site on another molecule.

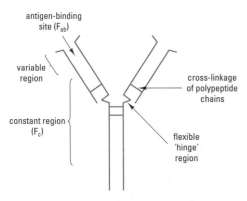

antigen-binding site (F_ab)

variable region

constant region (F_c)

cross-linkage of polypeptide chains

flexible 'hinge' region

Fig. 51.1 Diagrammatic representation of IgG (antibody)

Producing polyclonal antibodies – in the UK, this is controlled by government regulations, since it involves vertebrate animals: personnel must be licensed by the Home Office and must operate in accordance with the Animals Scientific Procedures Act (1986) and with the Code of Practice for the Housing and Care of Animals (2005).

Antibodies are an important component of the immune system, which protects animals against certain diseases (see Delves et al., 2006). They are produced by B lymphocytes in response to foreign macromolecules (antigens). A particular antibody will bind to a site on a specific antigen, forming an antigen–antibody complex (immune complex). Immunological bioassays use the specificity of this interaction for:

- identifying macromolecules, cellular components or whole cells;
- quantifying a particular substance.

Antibody structure

An antibody is a complex globular protein, or immunoglobulin (Ig). Although there are several types, IgG is the major soluble antibody in vertebrates and is used in most immunological assays. Its main features are:

- Shape: IgG is a Y-shaped molecule (Fig. 51.1), with two antigen-binding sites.
- Specificity: variation in amino acid composition at the antigen-binding sites explains the specificity of the antigen–antibody interaction.
- Flexibility: each IgG molecule can interact with epitopes that are different distances apart, including those on different antigen molecules.
- Labelling: regions other than the antigen-binding sites can be labelled, e.g. using a radioisotope or enzyme (p. 312).

KEY POINT *The presence of two antigen-binding sites on a single flexible antibody molecule is relevant to many immunological assays, especially the agglutination and precipitation reactions.*

Antibody production

Polyclonal antibodies

These are commonly used at undergraduate level. They are produced by repeated injection of antigen into a laboratory animal. After a suitable period (3–4 weeks) blood is removed and allowed to clot, leaving a liquid phase (polyclonal antiserum) containing many different IgG antibodies, resulting in:

- cross-reaction with other antigens or haptens;
- batch variation, as individual animals produce slightly different antibodies in response to the same antigen;
- non-specificity, as the antiserum will contain many other antibodies.

Standardisation of polyclonal antisera therefore is difficult. You may need to assess the amount of cross-reaction, interbatch variation or non-specific binding using appropriate controls, assayed at the same time as the test samples.

Monoclonal antibodies

These are specific to a single epitope and are produced from individual clones of cells (hybridomas), grown using cell culture techniques (p. 281).

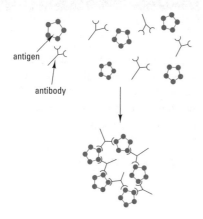

Fig. 51.2 Formation of an antigen–antibody complex

Such cultures provide a stable source of antibodies of known, uniform specificity. Although monoclonal antibodies are likely to be used increasingly in future years, polyclonal antisera are currently employed for many routine immunological assays.

Agglutination tests

When antibodies interact with a suspension of a particulate antigen, e.g. cells or latex particles, the formation of immune complexes (Fig. 51.2) causes visible clumping, termed agglutination. A positive haemagglutination reaction gives an even 'carpet' of red cells over the base of the tube, while a negative reaction gives a tightly packed 'button' of red cells at the bottom of the tube. Agglutination tests are used in several ways:

- Microbial identification: at the species or subspecies level (serotyping), e.g. mixing an unknown bacterium with the appropriate antiserum will cause the cells to agglutinate.
- Latex agglutination (bound antigens): by coating soluble (non-particulate) antigens on to microscopic latex spheres, their reaction with a particular antibody can be visualised.
- Latex agglutination (bound antibodies): antibodies can be bound to latex microspheres, leaving their antigen-binding sites free to react with soluble antigen.
- Haemagglutination: red blood cells can be used as agglutinating particles. However, in some instances, such reactions do not involve antibody interactions (e.g. some animal viruses may haemagglutinate unmodified red blood cells).

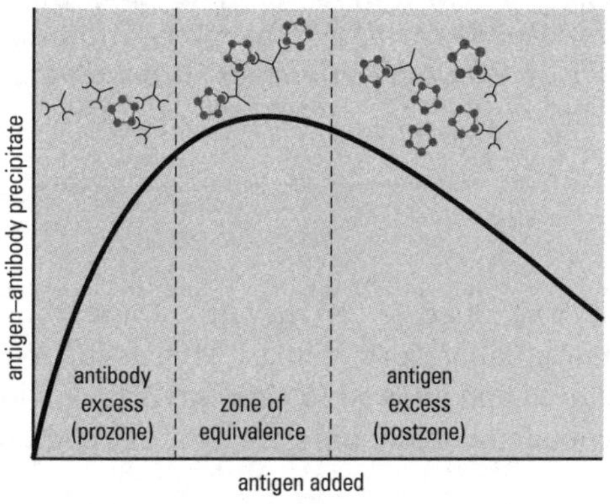

Precipitin tests

Immune complexes of antibodies and soluble antigens (or haptens) usually settle out of solution as a visible precipitate: this is termed a precipitin test, or precipitation test. The formation of visible immune complexes in agglutination and precipitation reactions only occurs if antibody and antigen are present in an optimal ratio (Fig. 51.3). It is important to appreciate the shape of this curve: cross-linkage is maximal in the zone of equivalence, decreasing if either component is present in excess. The quantitative precipitin test can be used to measure the antibody content of a solution (Clausen, 1989, and Lefkovits, 1997, give details). Visual assessment of precipitation reactions forms the basis of several other techniques, described below.

Fig. 51.3 Precipitation curve for an antigen titrated against a fixed amount of antibody

Immunodiffusion assays

These techniques are easier to perform and interpret than the quantitative precipitin test. Precipitation of antibody and antigen occurs within an agarose gel, giving a visible line corresponding to the zone of equivalence (Fig. 51.3). Details of the main techniques are given in Box 51.1.

Immunoelectrophoretic assays

These methods combine the precipitin reaction with electrophoretic migration, providing sensitive, rapid assays with increased separation and resolution.

Box 51.1   How to carry out immunodiffusion assays

The two most widespread approaches are (i) single radial immunodiffusion (Mancini technique) and (ii) double-diffusion immunoassay (Ouchterlony technique).

Single radial immunodiffusion (RID)

This is used to quantify the amount of antigen in a test solution, as follows:

1. **Prepare an agarose gel** (1.5% w/v), containing a fixed amount of antibody: allow to set on a glass slide or plate, on a level surface.
2. **Cut several circular wells in the gel.** These should be of a fixed size between 2 and 4mm in diameter (see Fig. 51.4(a)). Cut your wells carefully. They should have straight sides and the agarose must not be torn or lifted from the glass plate. All wells should be filled to the top, with a flat meniscus, to ensure identical diffusion characteristics. Non-circular precipitin rings, resulting from poor technique, should not be included in your analysis.
3. **Add a known amount of the antigen or test solution to each well.**
4. **Incubate on a level surface at room temperature in a moist chamber:** diffusion of antigen into the gel produces a precipitin ring. This is usually measured after 2–7 days, depending on the molecular mass of the antigen.
5. **Examine the plates against a black background** (with side illumination), or stain using a protein dye (e.g. Coomassie blue).
6. **Measure the diameter of the precipitin ring,** e.g. using Vernier calipers (p. 134).

7. **Prepare a calibration curve** (Chapter 50) from the samples containing known amounts of antigen (Fig. 51.4(b)): the squared diameter of the precipitin ring is directly proportional to the amount of antigen in the well.
8. **Quantify the amount of antigen in your test solutions,** using a calibration curve prepared from standards assayed at the same time.

Double-diffusion immunoassay (Ouchterlony technique)

This technique is widely used to detect particular antigens in a test solution, or to look for cross-reaction between different antigens.

1. **Prepare an agarose gel** (1.5% w/v) on a level glass slide or plate: allow to set.
2. **Cut several circular wells in the gel.**
3. **Add test solutions of antigen or polyclonal antiserum to adjacent wells.** Both solutions diffuse outwards, forming visible precipitin lines where antigen and corresponding antibody are present in optimal ratio (Fig. 51.5).

The various reactions between antigen and antiserum are:

- **Identity:** two wells containing the same antigen, or antigens with identical epitopes, will give a fused precipitin line (identical interaction between the antiserum and the test antigens, Fig. 51.5(a)).
- **Non-identity:** where the antiserum contains antibodies to two different antigens, each with its own distinct epitopes, giving two precipitin lines that intersect without any interaction (no cross-reaction, Fig. 51.5(b)).
- **Partial identity:** where two antigens have at least one epitope in common, but where other epitopes are present, giving a fused precipitin line with a spur (cross-reaction, Fig. 51.5(c)).

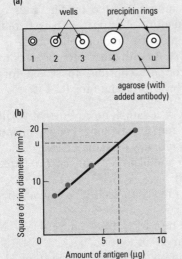

(a)

wells   precipitin rings

1   2   3   4   u

agarose (with added antibody)

(b)

Fig. 51.4 Single radial immunodiffusion (RID). (a) Assay: four standards are shown (wells 1 to 4, each one double the strength of the previous standard), and an unknown (u), run at the same time, (b) Calibration curve. The unknown contains 6.25 µg of antigen. Note the non-zero intercept of the calibration curve, corresponding to the square of the well diameter: do not force such calibration lines through the origin.

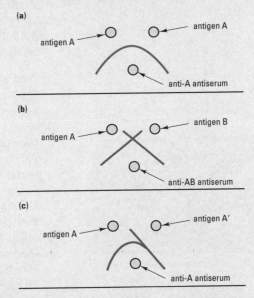

Fig. 51.5 Precipitin reactions in double-diffusion immunoassay: (a) identity; (b) non-identity; (c) partial identity

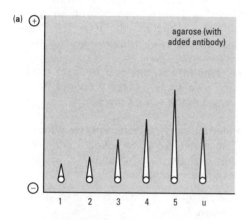

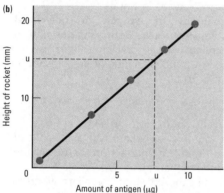

Fig. 51.6 Laurell rocket immunoelectrophoresis. (a) Assay: precipitin rockets are formed by electrophoresis of five standards of increasing concentration (wells 1 to 5) and an unknown (u). (b) Calibration curve: the unknown sample contains 7.7 μg of antigen.

Crossover electrophoresis (countercurrent electrophoresis)

Similar to the Ouchterlony technique, since antigen and antibody are in separate wells. However, the movement of antigen and antibody towards each other is driven by a voltage gradient (p. 334): most antigens migrate towards the anode, while IgG migrates towards the cathode. This method is faster and more sensitive than double immunodiffusion, taking 15–20 minutes to reach completion.

Quantitative immunoelectrophoresis (Laurell rocket immunoelectrophoresis)

Similar to RID, as the antibody is incorporated into an agarose plate while the antigen is placed in a well. However, a voltage gradient moves the antigen into the gel, usually towards the anode, while the antibody moves towards the cathode, giving a sharply peaked, rocket-shaped precipitin line, once equivalence is reached (within 2–10 hours). The height of each rocket shape at equivalence is directly proportional to the amount of antigen added to each well. A calibration curve for samples containing known amount of antigen can be used to quantify the amount of antigen present in test samples (Fig. 51.6).

Radioimmunological methods

These methods use radioisotopes to detect and quantify the antigen–antibody interaction, giving improved sensitivity over agglutination and precipitation methods. The principal techniques are:

Radioimmunoassay (RIA)

This is based on competition between a radioactively labelled antigen (or hapten) and an unlabelled antigen for the binding sites on a limited amount of antibody. The quantity of antigen in a test solution can be determined using a known amount of radiolabelled antigen and a fixed amount of antibody (Fig. 51.7). As with other immunoassay methods, it is important to perform appropriate controls, to screen for potentially interfering compounds. The basic procedure for RIA is:

1. Add appropriate volumes of a sample to a series of small test tubes. Prepare a further set of tubes containing known quantities of the substance to be assayed to provide a standard curve.
2. Add a known amount of radiolabelled antigen (or hapten) to each tube (sample and standard).
3. Add a fixed amount of antibody to each tube (the antibody must be present in limited quantity).
4. Leave at constant temperature for a fixed time (usually 24 hours), to allow antigen–antibody complexes to form.
5. Precipitate the antibody and bound antigen using saturated ammonium sulphate, followed by centrifugation.
6. Determine the radioactivity of the supernatant or the precipitate (p. 348).
7. Prepare a calibration curve of radioactivity against $\log_{10}$ antigen (Fig. 51.7). The curve is most accurate in the central region, so adjust the amount of antigen in your test sample to fall within this range.

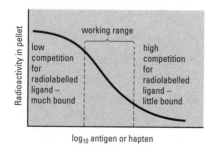

Fig. 51.7 Radioimmunoassay calibration curve. Note that the assay is insensitive at very low and very high antigen levels.

Note the following:

- You must be registered to work with radioactivity: check with the departmental safety officer, if necessary (p. 352).
- Measure all volumes as accurately as possible as the end result depends on the volumetric quantities of unlabelled (sample) antigen, radiolabelled antigen and antibody: an error in any of these reagents will invalidate the assay.
- Incorporate replicates, so that errors can be quantified.
- Seek your supervisor's advice about fitting a curve to your data: this can be a complex process (see p. 296).

Immunoradiometric assay (IRMA)

This technique uses radiolabelled antibody, rather than antigen, for direct measurement of the amount of antigen (or hapten) in a sample. Most immunoradiometric assays are similar to the double-antibody sandwich method described below, except that the second antibody is labelled using a radioisotope. The important advantages over RIA are:

- linear relationship between amount of radioactivity and test antigen;
- wider working range for test substance;
- improved stability/longer shelf life.

Enzyme immunoassays (EIA)

These techniques are also known as enzyme-linked immunosorbent assays (ELISA). They combine the specificity of the antibody–antigen interaction with the sensitivity of enzyme assays using either an antibody or an antigen conjugated (linked) to an enzyme at a site that does not affect the activity of either component. The enzyme is measured by adding an appropriate chromogenic substrate, which yields a coloured product. Enzymes offer the following advantages over radioisotopic labels:

- Increased sensitivity: a single enzyme molecule can produce many product molecules, amplifying the signal.
- Simplified assay: enzyme assays are usually easier than radioisotope assays (p. 319).
- Improved stability of reagents: components are generally more stable than their radiolabelled counterparts, giving them a longer shelf life.
- No radiological hazard: no requirement for specialised containment/disposal facilities.
- Automation is straightforward: using disposable microtitre plates and an optical scanner.

The principal techniques are:

Double-antibody-sandwich ELISA

This is used to detect specific antigens, involving a three-component complex between a capture antibody linked to a solid support, the antigen, and a second, enzyme-linked antibody (Fig. 51.8). This can be used to detect a particular antigen, e.g. a virus in a clinical sample, or to quantify the amount of antigen.

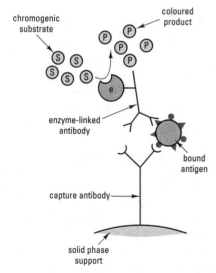

Fig. 51.8 Double-antibody sandwich ELISA

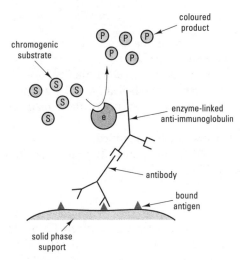

Fig. 51.9 Indirect ELISA

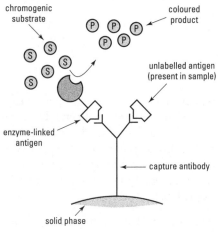

Fig. 51.10 Competitive ELISA

Indirect ELISA

This is used for antibody detection, with a specific antigen attached to a solid support. When the appropriate antibody is added, it binds to the antigen and will not be washed away during rinsing. Bound antibody is then detected using an enzyme-linked anti-immunoglobulin, e.g. a rabbit IgG antibody raised against human IgG (Fig. 51.9). One advantage of the indirect assay is that a single enzyme-linked anti-immunoglobulin can be used to detect several different antibodies, since the specificity is provided by the bound antigen.

Competitive ELISA

Here, any antigen present in a test sample competes with added enzyme-labelled antigen for a limited number of binding sites on the capture antibody (Fig. 51.10). Most commercial systems use 96-well microplates (12 columns by 8 rows), where each well is coated with the appropriate antibody. Following addition of known volumes of (i) sample and (ii) enzyme-labelled antigen, the plates are incubated (typically, up to 1 hour), then washed thoroughly to remove all unbound material. Bound enzyme-labelled antigen is then detected using a suitable substrate. Quantitative results can be obtained by measuring the absorbance of each well using a spectrophotometric microplate reader: the absorbance at a particular wavelength is *inversely* proportional to the amount of antigen present in the test sample. Calibration standards (p. 305) are required to convert the readings to an amount or concentration. Alternatively, the test can be carried out in positive/negative format. Box 51.2 gives practical details for a sandwich-type ELISA.

Dipstick immunoassays

Improvements in technology have led to the development of a range of diagnostic kits that can be used either in a medical setting 'near-patient testing' or in the home, including immunoassays for cancer screening and drug testing. The first of these available to the general public was the home pregnancy test kit, which measures the hormone human chorionic gonadotrophin (hCG) in urine. The level of this hormone is raised substantially during pregnancy. The principle of this test, which is similar to other home test kits, is that coloured plastic microbeads are conjugated to an anti-hCG monoclonal antibody and then coated on to an absorbent plastic or cellulose strip. When this strip is dipped into urine any hCG present will bind to the anti-hCG antibody and both hormone–antibody complexes and free antibody will then move up the strip by capillary action until they reach a second hCG-specific antibody that is coupled chemically to the strip so that it cannot move (see Fig. 51.11). Any hormone–antibody complex will bind to this 'fixed' antibody (through the hCG component); due to the presence of the coloured plastic microbeads, a distinct line will then appear at this location on the strip, signifying the presence of the hormone. If no hormone–antibody complex is present the first antibody will simply continue up the strip until it reaches an antibody raised against the anti-hCG antibody itself, coupled to the strip at a higher point. This will bind any free anti-hCG antibody in the sample and a coloured line will appear at this point. Since there will always be an excess

Box 51.2 How to perform an ELISA assay

While the following example is for a sandwich (capture) ELISA assay in microplate format, the same general principles apply to the other types of ELISA:

1. **Prepare the apparatus.** Switch on all equipment required:
    (i) The microplate reader – used to measure the absorbance of the solution in each well: set the reader to the required wavelength.
    (ii) The microplate washer (where used) – each well must be washed at various stages during the procedure. When using an automated washer, first check that the wash bottle contains sufficient diluent and then test using an old microplate, to check that all wells are being washed correctly. Where required, use a wire needle to clean any blocked wash delivery tubes and repeat. For manual washing, use a wash bottle or multichannel pipettor – make sure you fill each well and empty out all of the wash solution at the end of each wash stage.
    (iii) The computer – this will contain the software required to label the wells, draw the calibration curve and calculate the results for test samples: fill out the ELISA template with details of the assays to be carried out.
2. **Prepare the various solutions to be analysed.** These include:
    (i) Test samples – make sure that each sample is identified with a code that enables you to record what each test well contains.
    (ii) Calibrators/standards – including 'cutoff' calibrators and known positive standards.
    (iii) Controls – positive and negative controls and blanks.
3. **Coat the wells with the capture antibody.** Typically 100 μl of a solution of the appropriate monoclonal or polyclonal antibody is added to each well and microplates are then incubated overnight at 4 °C to allow binding to the well.
4. **Wash the wells.** Transfer the microplate to the washer (or wash manually) – wash six times to remove excess coating (capture) antibody. The final rinse should be programmed so that the washer leaves the wells empty of diluent.
5. **Add blocking solution to each well.** Typically 100 μl of an inert protein solution (bovine serum albumin) is added to each well to block any free binding sites on the well. Microplates are incubated at room temperature for 30 minutes and then washed, as in step 4.
6. **Add test samples, calibrators and controls to wells.** Typically 100 μl of appropriately diluted test sample, control, etc. is added to each well. Microplates are incubated at room temperature for 90 minutes and then washed as step 4.
7. **Add the detection antibody to each well.** Add 100 μl of monoclonal or polyclonal detection antibody labelled with a suitable enzyme (e.g. horseradish peroxidise HRP) to each well. The microplate is then incubated at room temperature for 30–60 min and then washed, as in step 4.
8. **Add chromogenic substrate to each well.** For example, with HRP-labelled antigen add 100 μl of a standard solution of tetramethylbenzidine (TMB) and hydrogen peroxide to each well and re-incubate in darkness for 30 min, to allow colour development. The TMB is oxidised in the presence of hydrogen peroxide to produce a blue colour.

 SAFETY NOTE *TMB and hydrogen peroxide are harmful by inhalation – use a fume hood.*

9. **Stop the reactions.** For example, by adding 100 μl of 2 mol l^{-1} sulphuric acid to denature the enzyme. The colour of the oxidised TMB will change from blue to yellow as a result of the pH shift. While the human eye can readily distinguish different shades of blue, it is more difficult to visually assess different shades of yellow, once the reaction has been stopped.
10. **Measure the absorbance of each sample/calibrator/control well.** Transfer the microplate to the reader and assay at an appropriate wavelength: for TMB, use 450 nm.
11. **Interpret the results.** Check that the absorbance values of calibrators and control are within the required range. Then, for each sample, either record the absolute value (convert to concentration or amount e.g. using a calibration curve, pp. 303–5) or record as 'positive' or 'negative' (based on values for 'cutoff' calibrators), as appropriate.

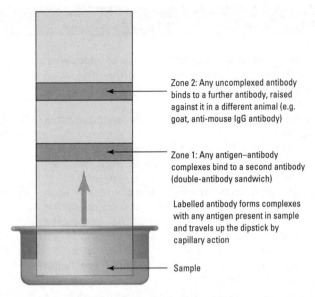

Zone 2: Any uncomplexed antibody binds to a further antibody, raised against it in a different animal (e.g. goat, anti-mouse IgG antibody)

Zone 1: Any antigen–antibody complexes bind to a second antibody (double-antibody sandwich)

Labelled antibody forms complexes with any antigen present in sample and travels up the dipstick by capillary action

Sample

Fig. 51.11 Two-site dipstick immunoassay. For a positive test, as shown here, coloured bands will appear at zones 1 and 2 while a negative test will give colour in zone 2 only (any other colour options indicate a failed test).

of the original anti-hCG antibody conjugated to the plastic beads, the appearance of two coloured lines means that the hormone is present and the test result is positive, whereas if only the second coloured line appears, it means that the test is negative (this also serves as a control to confirm that the test is working – lack of any coloured lines indicates that the test has failed).

Text references

Clausen, J. (1989) *Immunochemical Techniques for the Identification and Estimation of Macromolecules*, 3rd edn. Elsevier, Amsterdam.

Delves, P.J., Martin, S., Burton, D. and Roitt, I. (2006) *Roitt's Essential Immunology*, 11th edn. Blackwell, Oxford.

Lefkovits, I. (1997) *Immunology Methods Manual: The Comprehensive Sourcebook of Techniques*. Academic Press, New York.

Sources for further study

Gosling, J.R.G. (2000) *Immunoassays: A Practical Approach*. Oxford University Press, Oxford.

Hay, F.C., Westwood, O.M.R. and Nelson, P.N. (2002) *Practical Immunology*. Blackwell, Oxford.

Study exercises

51.1 Determine the amount of antigen in a test solution using Laurell rocket immunoelectrophoresis. The figure below represents the results from a rocket immunoelectrophoresis assay for a series of standards. Determine the amount of antigen in a test solution giving a rocket (precipitin line) height of 18.5 mm.

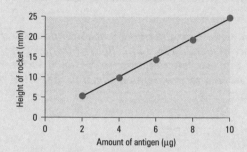

51.2 Determine the amount of antigen in a test solution using single radial immunodiffusion (RID). Using the following data for a series of standards, determine the amount of antigen in a test solution giving a ring diameter of 6.5 mm.

Single radial immunodiffusion results for a series of standards containing 4–20 μg antigen

Amount of antigen (μg)	Ring diameter (mm)
4	3.5
8	5.5
12	7.0
16	8.0
20	9.0

51.3 Interpret data from a haemagglutination test. In some conditions, such as Hashimoto's disease, autoantibodies to the thyroid protein thyroglobulin are produced in large amounts. The level of these antibodies in serum can be measured by indirect haemagglutination: red blood cells coated with thyroglobulin are mixed with serial doubling dilutions of the test serum. The titre of antithyroglobulin autoantibodies is the lowest dilution at which haemagglutination occurs. The figure below represents the outcome of an antithyroglobulin haemagglutination test on a serum sample. The first well – numbered 1 – represents a ten-fold dilution of the serum sample, with serial doubling dilutions in subsequent wells.

(a) Write a brief explanation of the test results shown in the figure.

(b) Determine the autoantibody titre.

(c) Score the sample as 'positive' if the titre is more dilute than 1 in 5000 (i.e. 1 : >5000) and 'negative' if it is less dilute than 1 in 5000 (i.e. 1 : <5000).

Enzymes are globular proteins that increase the rate of specific biochemical reactions. Each enzyme operates on a limited number of substrates of similar structure to generate products under well-defined conditions of concentration, pH, temperature, etc. In metabolism, groups of enzymes work together in sequential pathways to carry out a particular process, e.g. the multi-reaction conversion of glucose to lactate (glycolysis).

Example Enzyme EC 1.1.1.1 is usually known by its trivial name, alcohol dehydrogenase.

🔑 *KEY POINT Enzymes are categorised according to the chemical reactions involved, leading to a four-figure Enzyme Commission code number and a systematic name for each enzyme. Most enzymes also have a recommended trivial name, often denoted by the suffix 'ase'.*

Measuring enzyme reactions

Activity

This is measured in terms of the rate of enzyme reaction. Activity may be expressed directly as amount of substrate utilised per unit time (e.g. nmol min^{-1}, etc.), or in terms of the non-SI international unit (U, or sometimes IU), defined as the amount of enzyme that will convert 1 µmol of substrate to product(s) in 1 minute under specified conditions. However, the recommended (SI) unit of enzyme activity is the katal (kat), which is the amount of enzyme that will convert 1 mol of substrate to product(s) in 1 second under optimal conditions, determined from the following equation:

Example Using eqn [52.1], the conversion of 4.5 µmol (4.5×10^{-6} mol) of substrate in 5 min (300 s) would give an enzyme activity of $4.5 \times 10^{-6} \div 300 = 1.5 \times 10^{-8}$ kat = 15 nkat.

$$\text{enzyme activity (kat)} = \frac{\text{substrate converted (mol)}}{\text{time (s)}} \qquad [52.1]$$

This unit is relatively large ($1 \text{ kat} = 6 \times 10^7 \text{ U}$) so SI prefixes are often used, e.g. nkat or pkat (p. 159). Note that the units are *amount of substrate* (mol), not concentration ($mol\,l^{-1}$ or $mol\,m^{-3}$).

For enzymes with macromolecular substrates of unknown molecular weight (e.g. deoxyribonuclease, amylase), activity can be expressed as the mass of substrate consumed (e.g. ng DNA min^{-1}), or amount of product formed (e.g. nmol glucose min^{-1}). You must ensure that your units clearly specify the substrate or product used, especially when the enzyme transformations involve different numbers of substrate or product molecules.

Example The hydrolysis of 1 molecule of maltose to give 2 glucose molecules by α-glucosidase means that enzyme activity specified in terms of substrate consumption (nmol maltose) would be half the value expressed with respect to product formation (nmol glucose).

The turnover number of an enzyme is the *amount of substrate* (mol) converted to product in 1 second by 1 mol of enzyme operating at maximum rate under optimum conditions. In practice, this requires information on the molecular weight of the enzyme, the amount of enzyme present and its maximum activity (p. 323).

Purity

The specific activity of an enzyme preparation expresses the enzyme activity in terms of the quantity of protein present:

Example For an enzyme preparation with an activity of 1.5×10^{-8} kat in 24 mg ($= 2.4 \times 10^{-8}$ kg), eqn [52.2] gives a specific activity of $1.5 \times 10^{-8} \div 2.4 \times 10^{-8} = 0.625$ kat kg^{-1}.

$$\text{specific activity} = \frac{\text{enzyme activity (kat)}}{\text{mass of protein (kg)}} \qquad [52.2]$$

Alternative units may be used, e.g. $U\,mg^{-1}$, $ng\,min^{-1}\,mg^{-1}$. The specific activity is a useful way of comparing the purity of different enzyme

Preparing a calibration curve for protein (enzyme) assay – Box 50.1 (p. 304) gives practical details and study exercise 50.2 (p. 307) gives typical data for a protein assay.

preparations (purified enzyme preparations have high specific activities), comparing the stages in purification of an enzyme (specific activity increases as other proteins are eliminated), and assessing enzyme stability (an unstable enzyme will show a decrease in specific activity with time).

Enzyme activity and purity are often expressed in terms of the protein content of the enzyme preparation. Most assays for protein content do not give an absolute value, but require standard solutions, containing appropriate amounts of a particular protein, to be analysed at the same time, enabling a standard curve to be constructed. Bovine serum albumin (BSA) is commonly used as a general-purpose protein standard. However, you may need to use an alternative standard if your enzyme has an amino acid composition that is markedly different from BSA, depending on the specific method you use (Box 52.1).

Box 52.1 Methods of determining the amount of protein in an aqueous solution

For all of the following methods, the amounts are appropriate for semi-micro cuvettes (1.5 ml volume, path length 1 cm). Appropriate controls (blanks) must be analysed, to assess possible interference (e.g. due to buffers, etc.).

Biuret method

This is based on the specific reaction between cupric ions (Cu^{++}) in alkaline solution and two adjacent peptide bonds, as found in proteins. As such, it is not significantly affected by differences in amino acid composition between proteins.

1. **Prepare protein standards over an appropriate range** (typically, between 1–20 mg ml^{-1}).

2. **Add 1 ml of each standard solution to separate test tubes. Prepare a reagent blank, using 1 ml of distilled water, or an appropriate solution.**

3. **Add 1 ml of each unknown solution to separate test tubes.**

4. **Add 1 ml of Biuret reagent (1.5 g CuSO$_4$·5H$_2$O, 6.0 g sodium potassium tartrate in 300 ml of 10 w/v NaOH) to all standard and unknown tubes and to the reagent blank.**

5. **Incubate at 37 °C for 15 min.**

6. **Read the absorbance of each solution at 520 nm against the reagent blank.** The colour is stable for several hours.

The main limitation of the Biuret method is its lack of sensitivity – it is unsuitable for solutions with a protein content of less than 1 mg ml^{-1}.

Direct measurement of UV absorbance (Warburg–Christian method)

Proteins absorb EMR maximally at 280 nm (due to the presence of aromatic amino acids) and this forms the basis of the method. The principal advantages of this approach are its simplicity and the fact that the assay is non-destructive. The most common interfering substances are nucleic acids, which can be assessed by measuring the absorbance at 260 nm: a pure solution of protein will have a ratio of absorption (A_{280}/A_{260}) of approximately 1.8, decreasing with increasing nucleic acid contamination. Note also that any free aromatic amino acids in your solution will absorb at 280 nm, leading to an overestimation of protein content. The simplest procedure, which includes a correction for small amounts of nucleic acid, is as follows (use quartz cuvettes throughout):

1. **Measure the absorbance of your solution at 280 nm (A_{280}): if A_{280} is greater than 1, dilute by an appropriate amount and re-measure.**

2. **Repeat at 260 nm (A_{260}).**

3. **Estimate the approximate protein content using the following relationship:**

$$[\text{protein}] \, \text{mg ml}^{-1} = 1.45 \, A_{280} - 0.74 \, A_{260} \qquad [52.3]$$

This equation is based on the work of Warburg and Christian (1942), for enolase. For other proteins, it should not be used for quantitative work, since it gives only a rough approximation of the amount of protein present, due to variations in aromatic amino acid composition.

(continued)

Box 52.1 (continued)

Lowry (Folin–Ciocalteau) method

This is a colorimetric assay, based on a combination of the Biuret method, described above, and the oxidation of tyrosine and tryptophan residues with Folin and Ciocalteau's reagent to give a blue–purple colour. The method is extremely sensitive (down to a protein content of $20\,\mu g\,ml^{-1}$), but is subject to interference from a wide range of non-protein substances, including many organic buffers (e.g. TRIS, HEPES, etc.), EDTA, urea and certain sugars. Choice of an appropriate standard is important, as the intensity of colour produced for a particular protein is dependent on the amount of aromatic amino acids.

1. **Prepare protein standards within an appropriate range for your samples** (the method can be used from 0.02–$1.00\,mg\,ml^{-1}$).

2. **Add 1 ml of each standard solution to separate test tubes. Prepare a reagent blank, using 1 ml of distilled water, or an appropriate solution.**

3. **Add 1 ml of each of your unknown solutions to separate test tubes.**

4. **Then, add 5 ml of 'alkaline solution' (prepared by mixing 2% w/v Na_2CO_3 in $0.1\,mol\,l^{-1}$ NaOH, 1% w/v aqueous NaKtartrate in the ratio 100:1:1).** Mix thoroughly and then allow to stand for 3 min.

5. **Add 0.5 ml of Folin–Ciocalteau reagent (commercial reagent, diluted 1:1 with distilled water on the day of use).** Mix rapidly and thoroughly and then allow to stand for 30 min.

6. **Read the absorbance of each sample at 600 nm.**

Dye-binding (Bradford) method

Coomassie brilliant blue combines with proteins to give a dye–protein complex with an absorption maximum of 595 nm. This provides a simple and sensitive means of measuring protein content, with few interferences. However, the formation of dye–protein complex is affected by the number of basic amino acids within a protein, so the choice of an appropriate standard is important. The method is sensitive down to a protein content of approximately $5\,\mu g\,ml^{-1}$ but the relationship between absorbance and concentration is often non-linear, particularly at high protein content.

1. **Prepare protein standards over an appropriate range (between 5 and $100\,\mu g\,ml^{-1}$).**

2. **Add $100\,\mu l$ of each standard solution to separate test tubes. Prepare a reagent blank, using $100\,\mu l$ of distilled water, or an appropriate solution** (note that these small volumes must be accurately dispensed, e.g. using a calibrated pipettor, p. 123).

3. **Add $100\,\mu l$ of your unknown solutions to separate test tubes.**

4. **Add 5.0 ml of Coomassie brilliant blue G250 solution ($0.1\,g\,l^{-1}$).**

5. **Mix and incubate for at least 5 min: read the absorbance of each solution at 595 nm.**

Other methods are less widely used. They include determination of the total amount of nitrogen in solution (e.g. using the Kjeldahl technique) and calculating the protein content, assuming a nitrogen content of 16 per cent. An alternative approach is to precipitate the protein (e.g. using trichloracetic acid, tannic acid, or salicylic acid) and then measure the turbidity of the resulting precipitate (using a nephelometer, or a spectrophotometer, p. 368).

Types of assay

The rate of substrate utilisation or product formation must be measured under controlled conditions, using some characteristic that changes in direct proportion to the concentration of the test substance.

Spectrophotometric assays

Many substrates and products absorb visible or UV light, and the change in absorbance at a particular wavelength provides a convenient assay method (p. 368). Artificial substrates are used where no suitable spectrophotometric assay is available for the natural substrates. In most cases, these are chromogenic analogues of the natural substrate, producing coloured products. In other cases, a product may be measured by a colorimetric chemical reaction.

Example α-Glucosidase activity can be measured using the artificial substrate p-nitrophenol-α-D-glucose, which liberates p-nitrophenol (yellow) with an absorption maximum at 404 nm.

Example Phosphoenolpyruvate carboxylase can be assayed by coupling to malate dehydrogenase: the first enzyme converts phosphoenolpyruvate to oxaloacetate which is then oxidised to malate by the second enzyme (present in excess) with a stoichiometric (1 : 1) reduction of NAD^+. The coupled assay is monitored spectrophotometrically (p. 368) as an increase in A_{340} against time.

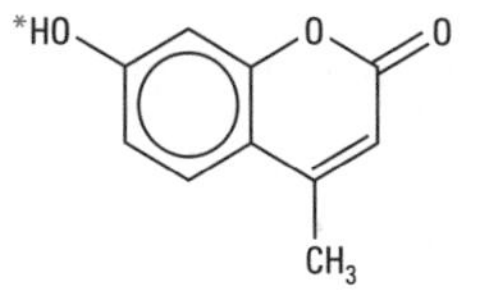

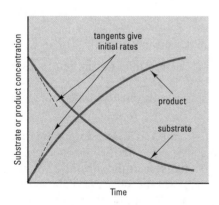

Fig. 52.1 Structure of 4-methylumbelliferone (4-MU). The free hydroxlyl group* is used to create fluorogenic substrates, e.g. coupled to galactose (4-MU-GAL) it can be used to asay β-galactosidase activity, since the galactoside derivative is non-fluorescent and hydrolysis can be detected by the fluorescence of liberated 4-MU.

Several assays are based on interconversion of the nicotinamide adenine dinucleotide coenzymes NAD^+ or $NADP^+$, which are reversibly reduced by many enzymes. The reduced form (either NADH or NADPH) can be detected at 340 nm, where the oxidised form has negligible absorbance (p. 368). An alternative approach is to use a coupled-enzyme assay, where the product of the test enzyme is used as a substrate for a second enzyme reaction that involves oxidation/reduction of nucleotide coenzymes. Such assays are particularly useful for continuous monitoring of enzyme activity (see below) and for reactions where the product from the test substance is too low to detect by other methods, since coupled assays are more sensitive.

Fluorimetric assays

Certain substrates liberate fluorescent products as a result of enzyme activity, providing a highly sensitive assay method. Fluorogenic substrates include fluorescein and methylumbelliferone derivatives (Fig. 52.1). Care is required, since impurities in the enzyme preparation may produce background fluorescence, or quenching (reduction) of the signal.

Radioisotopic assays

These are useful where the substrate and product can be easily separated, e.g. in decarboxylase assays using a ^{14}C-labelled substrate, where gaseous $^{14}CO_2$ is produced.

Electrochemical assays

Enzyme reactions involving acids and bases can be monitored using a pH electrode (p. 148), though the change in pH will affect the activity of the test enzyme. An alternative approach is to measure the amount of acid or alkali required to maintain a constant pre-selected pH in a pH-stat.

An oxygen electrode can be used if O_2 is a substrate or a product (p. 355). Other ion-specific electrodes can monitor ammonia, nitrate, etc.

Monitoring substrate utilisation/product formation

Continuous assays (kinetic assays)

The change in substrate or product is monitored as a function of time, to provide a progress curve for the reaction (Fig. 52.2). These curves start off in a near-linear manner, decreasing in slope as the reaction proceeds and substrate is used up. The initial velocity of the reaction (v_0) is obtained by drawing a tangent to the curve at zero time and measuring its slope. Continuous monitoring can be used when the test substance can be assayed rapidly and non-destructively, e.g. using a chromogenic substrate.

Discontinuous assays (fixed-time assays)

It is sometimes necessary to measure the amount of substrate consumed or product formed after a fixed time period, e.g. where the test substance is assayed by a (destructive) colorimetric chemical method. It is vital that the time period is kept as short as possible, with the change in substrate concentration limited to around 10 per cent, so that the assay is within the linear part of the progress curve (Fig. 52.2). This may need to be established in a preliminary experiment, e.g. using a continuous assay.

Fig. 52.2 Enzyme reaction progress curve: substrate utilisation/product formation as a function of time.

Enzyme purification

The purification of an enzyme usually involves several stages:

1. Tissue/cell disruption, often using mechanical/ultrasonic homogenisation.
2. Differential centrifugation, removing larger particulate components (p. 361).
3. Ammonium sulphate fractionation: selective precipitation by the stepwise addition of ammonium sulphate at low temperature ($<10\,°C$). The precipitated protein is then redissolved in a fresh buffer solution.
4. Chromatographic and/or electrophoretic separation, including ion exchange and gel filtration chromatography and polyacrylamide gel electrophoresis or isoelectric focusing. This stage may involve several individual steps and can lead to a fully purified enzyme (purification to homogeneity), with maximum specific activity.

At each step in the procedure the percentage yield of enzyme in an individual fraction can be determined:

$$\text{yield} = \frac{\text{enzyme in fraction (kat or U)}}{\text{enzyme in original preparation (kat or U)}} \times 100 \qquad [52.4]$$

Note that the yield equation does not use a concentration-based measure of enzyme activity, as this would be affected by the volumes of the solutions involved. The relative purification of an enzyme is usually expressed as 'n-fold purification', where:

$$n = \frac{\text{specific activity of fraction}}{\text{specific activity of original preparation}} \qquad [52.5]$$

Enzyme kinetics

For most enzymes, when the initial velocity (v_0) using a fixed amount of an enzyme is plotted as a function of the concentration of a single substrate [S] with all other substrates present in excess, a rectangular hyperbola is obtained (Fig. 52.3). At low substrate concentrations v_0 is directly proportional to [S], with a decreasing proportional response as substrate concentration is increased until saturation is achieved. The shape of this plot can be described by a mathematical relationship, known as the Michaelis–Menten equation:

$$v_0 = \frac{V_{max}\,[S]}{K_m + [S]} \qquad [52.6]$$

This equation makes use of two kinetic constants:

- V_{max}, the maximum velocity of the reaction (at infinite substrate concentration).
- K_m, the Michaelis constant, is the substrate concentration where $v_0 = \frac{1}{2}V_{max}$.

V_{max} is a function of the amount of enzyme and is the appropriate rate to use when determining the specific activity of a purified enzyme (p. 318). The Michaelis constant can provide a measure of the substrate affinity of an enzyme and is an important characteristic of a particular enzyme. Thus, an enzyme with a large K_m usually has a low affinity for the substrate whereas an enzyme with a small K_m usually has a high affinity.

Your first step in determining the kinetic constants for a particular enzyme is to measure the rate of reaction at several substrate concentrations,

Example If an original preparation of 50 ml had an activity of 100 pkat ml^{-1} (total 5 nkat) and a subsequent fraction of 5 ml volume had an activity of 200 pkat ml^{-1} (total 1 nkat), using eqn [52.4] this would represent a percentage yield of 20 per cent.

Example Using eqn [52.5] an increase in specific activity from 10 nkat (mg protein)$^{-1}$ to 120 nkat (mg protein)$^{-1}$ represents a twelve-fold purification.

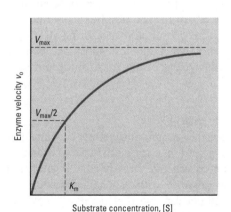

Fig. 52.3 Effect of substrate concentration on enzyme activity

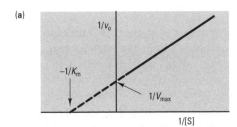

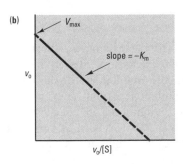

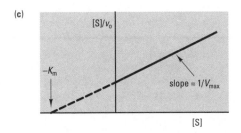

Fig. 52.4 Graphical transformations for determining the kinetic constants of an enzyme. (a) Lineweaver–Burk plot. (b) Eadie–Hofstee plot. (c) Hanes–Woolf plot.

as in Fig. 52.3. There are various ways to obtain K_m and V_{max} from such data, mostly involving drawing a graph representing a linear transformation of eqn [52.6]:

- The Lineweaver–Burk plot: a graph of the reciprocal of the reaction rate $(1/v_0)$ against the reciprocal of the substrate concentration $(1/[S])$ gives $-1/K_m$ as the intercept of the x axis and $1/V_{max}$ as the intercept of the y axis (Fig. 52.4(a)). Note that the slope of the plot is most affected by the least-accurate values, i.e. those measured at low substrate concentration.
- The Eadie–Hofstee plot: v_0 against $v_0/[S]$, where the intercept on the y axis gives V_{max} and the slope equals $-K_m$ (Fig. 52.4(b)).
- The Hanes–Woolf plot: $[S]/v_0$ against $[S]$, giving $-K_m$ as the intercept of the x axis and $1/V_{max}$ from the slope (Fig. 52.4(c)).

There are several computer packages that will plot the above relationships and calculate the kinetic constants from a given set of data using linear regression analysis (p. 435). While the Eadie–Hofstee and Hanes–Woolf plots distribute the data points more evenly than the Lineweaver–Burk plot, the best approach to such data is to use non-linear regression on untransformed data. This is usually outside the scope of the simpler computer programs, though tailor-made commercial packages can carry out such analyses. Note also that some enzymes do not show Michaelis–Menten kinetics, e.g. those with more than one active site per molecule (allosteric enzymes), which often give sigmoid curves of v_0 against [S]. Such enzymes are usually involved in the control of metabolism.

Factors affecting enzyme activity

If you want to measure the maximum rate of a particular enzyme reaction, you will need to optimise the following:

Substrate concentration
The substrates must be present in excess, to ensure maximum reaction velocity. Equation [52.6] shows that a substrate concentration equivalent to $10 \times K_m$ will give 91% of V_{max}, while a concentration of $100 \times K_m$ will give 99 per cent of the maximum rate.

Cofactors
Many enzymes require appropriate concentrations of specific cofactors for maximum activity. These are subdivided into coenzymes (soluble, low molecular weight organic compounds that are actively involved in catalysis by accepting or donating specific chemical groups, i.e. they are co-substrates of the enzyme; examples include NAD^+ and ADP); and activators (inorganic metal ions, required for maximal activity, e.g. Mg^{2+}).

Temperature
Enzyme activity increases with temperature, until an optimum is reached. Above this point, activity decreases as a result of protein denaturation (Fig. 52.5). Note that the optimum temperature for enzyme *activity* may not be the same as that for maximum *stability* (enzymes are usually stored at temperatures near to or below $0\,°C$, to maximise stability).

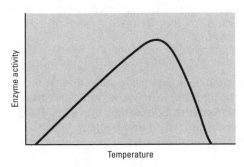

Fig. 52.5 Effect of temperature on enzyme activity

Enzyme studies

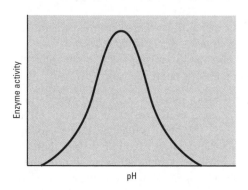

Fig. 52.6 Effect of pH on enzyme activity

pH

Enzymes work best at a particular pH, due to changes in ionisation of the substrates or of the amino acid residues within the enzyme (Fig. 52.6). Most enzyme assays are performed in buffer solutions (p. 150), to prevent changes in pH during the assay.

 KEY POINT Note that temperature and pH optima are dependent upon reaction conditions – you should therefore specify the experimental conditions under which such optima are determined.

Inhibitors

Many compounds can reduce the rate of an enzyme reaction, e.g. substances that compete for the active site of the enzyme due to similarities in chemical structure with the natural substrate (competitive inhibitors), or substances that bind to the enzyme at other sites, inhibiting normal function (non-competitive inhibitors). Most non-competitive inhibitors are not chemically related to the natural substrate, e.g. heavy-metal ions, such as Hg^{2+} and Cd^+, acting as irreversible, non-competitive inhibitors. Competitive inhibitors reduce K_m but have no effect on V_{max} whereas non-competitive inhibitors reduce V_{max} but have no effect on K_m.

Text reference

Warburg, O. and Christian, W. (1942) 'Isolierung und Kristallisation des Garungsferments Enolase', *Biochemische Zeitschrift*, **310**, pp. 384–421.

Sources for further study

Copeland, R.A. (2000) *Enzymes: A Practical Introduction to Structure, Mechanism, and Data Analysis*. Wiley, New York.
[Covers basic principles, kinetics and experimental systems]

Purich, D.L. and Allison, R.D. (2000) *Handbook of Biochemical Kinetics*. Academic Press, New York.

Study exercises

52.1 **Determine enzyme activity and purity (specific activity).** An extract of α-glucosidase containing 2.5 mg of protein was incubated with *p*-nitrophenol-α-D-glucose at 1.0 mmol l^{-1} and assayed by following the liberation of *p*-nitrophenol at 404 nm. This gave a progress curve with an initial velocity equal to an absorbance change of 0.25 min^{-1}. Given that the conversion of 10 nmol of substrate to product results in an absorbance change of 0.122:

   (a) Determine the amount of product formed in nmol min^{-1}.

   (b) Calculate the enzyme activity in kat.

   (c) Determine the specific activity, in terms of (i) nmol min^{-1} (mg protein)$^{-1}$ and (ii) kat (mg protein)$^{-1}$.

   Quote all answers to three significant figures.

52.2 **Calculate yield and relative purification of an enzyme.** The table opposite shows representative data for an enzyme purification involving five stages. Calculate the yield and *n*-fold purification at each stage (give all answers to three significant figures).

   (continued)

Study exercises (continued)

Volume, protein concentration and enzyme activity at several stages during a protein purification procedure

Stage	Total volume of extract (ml)	Enzyme activity ($U\,ml^{-1}$)	Protein concentration ($mg\,ml^{-1}$)
1	78	1.2	4.8
2	70	1.2	1.7
3	10	6.4	2.5
4	5	10.8	1.4
5	0.6	65.7	3.4

52.3 Estimate K_m and V_{max} for an enzyme. The table below shows representative data for enzyme activity as a function of substrate concentration.

Use this to determine the kinetic constants K_m and V_{max} of the enzyme (give your answers to three significant figures).

Enzyme activity at various substrate concentrations

Substrate concentration ($\mu mol\,l^{-1}$)	Enzyme activity (U)
0	0
5	16.2
7	20.5
10	28.1
15	32.7
25	39.2
50	56.7
100	68.2

Genotype – an individual's genetic make-up, i.e. the organism's genes.

Phenotype – the observable characteristics of an individual organism; the consequence of the underlying genotype and its interaction with the environment.

Essential vocabulary for Mendelian genetics – *make sure you know what the following terms and symbols mean:*
- chromosome, sex chromosome, autosome;
- gene, allele;
- dominant, recessive, lethal;
- haploid, diploid, gamete, zygote;
- heterozygous, homozygous;
- P, F1, F2, ♀, ♂.

Genomes and chromosome numbers – *remember that mitochondria and chloroplasts contain DNA molecules, but these are not included when calculating the chromosome number.*

Meiosis – division of a diploid cell that results in haploid daughter cells carrying half the original number of chromosomes. Occurs during gamete (sperm and egg) formation.

Mitosis – division of a cell into two new cells, each with the same chromosome number. Occurs in somatic cells, e.g. during growth, development, repair, replacement.

Gregor Mendel, an Austrian monk, made pioneering studies of the genetics of eukaryotic organisms in the middle of the nineteenth century. He made crosses between different forms of flowering plants. Through careful examination and numerical analysis of the observable characteristics, or phenotype, of the parents and their progeny, Mendel was able to deduce much about their genetic characteristics, or genotype. The principles derived from these experiments explain the basis of heredity, and hence underpin our understanding of sexual reproduction, biodiversity and evolution. Mendelian genetics is concerned primarily with the transmission of genetic information, as opposed to molecular genetics, which deals with the molecular details of the genome and techniques for altering genes (see Chapter 54).

 KEY POINT *A common initial stumbling block in genetics is terminology. In many cases the definitions are interdependent, so your success in this subject depends on your grasp of all the definitions and underlying ideas explained below.*

Important terms and concepts

Each character in the phenotype is controlled by the organism's genes, the basic units of inheritance. Each gene includes the 'genetic blueprint' (DNA) that usually defines the amino acid sequence for a specific polypeptide or protein – often an enzyme or a structural protein. The protein gives rise to the phenotype through its activity in metabolism or its contribution to the organism's structure. The full complement of genes in an individual is known as its genome. Individual genes can exist in different forms, each of which generally leads to a different form of the protein it codes for. These different gene forms are known as alleles.

In eukaryotes, the genes are located in a particular sequence on chromosomes within the nucleus. The number of chromosomes per cell is characteristic for each organism (its chromosome number, n). For example, the chromosome number for man is 23. In cells of most 'higher' organisms, there are two of each of the chromosomes (2n). This is known as the diploid state. As a result of the process of meiosis, which precedes reproduction, special haploid cells are formed (gametes) that contain only one of each chromosome (1n). In sexual reproduction, haploid gametes from two individuals fuse to form a zygote, a diploid cell with a new genome, which gives rise to a new individual through the process of mitosis. Cell numbers are increased by this process, producing genetically identical cells.

Organisms vary in the span of the diploid and haploid phases. In some 'lower' organisms, the haploid phase is the longer-lasting form; in most 'higher' organisms, the diploid phase is dominant. A life-cycle diagram can be used to show how the phases are organised and what life forms are involved in each case (see e.g. Fig. 28.6, p. 171).

Since each diploid individual carries two of each chromosome, it has two copies of each gene in every cell. The number of alleles of each character present depends on whether the relevant genes are the same – the homozygous state – or whether they are different – the heterozygous state. Hence, while there may be many alleles for any given gene, an individual could have two at most, and might only have one if it were homozygous. This will depend on the alleles present in the parental gametes that fused when the zygote was formed. Offspring that inherit a different combination of alleles at the two loci compared with their parents are known as recombinants.

The basis of Mendel's experiments and of many exercises in genetics are crosses, where individuals showing particular phenotypes are mated and the phenotypes of the offspring, or F_1 generation, are studied (see Box 53.1). If homozygous individuals carrying alternative alleles for a character are crossed, one of the alleles may be dominant, and all the F_1 generation will show that character in their phenotype. The character not evident is said to be recessive.

In describing crosses, geneticists denote each character with a letter of the alphabet, a capital letter being used for the dominant allele and lower case for the recessive. Taking for example a gene with dominant and recessive forms A and a respectively, there are three possibilities for each individual: it can be (a) homozygous recessive aa; (b) homozygous dominant AA; or (c) heterozygous Aa.

The reasons for dominance might relate to the activity of an enzyme coded by the relevant gene; for example, Mendel's yellow pea allele is dominant because the gene involved codes for the breakdown of chlorophyll. In the homozygous recessive case, none of the functional enzyme will be present, chlorophyll breakdown cannot occur and the seeds remain green. Not all alleles exhibit dominance in this form. In some cases, the heterozygous state results in a third phenotype (incomplete or partial dominance); in others, the heterozygous individual expresses both genotypes (codominance). Another possible situation is epistasis, in which one gene affects the expression of another gene that is independently inherited. Note also that many genes, such as those coding for human blood groups, have multiple alleles.

 KEY POINT *Genetics problems may well involve one of the 'standard' crosses shown in Box 53.1. Before tackling the problem, try to analyse the information provided to see if it fits one of these types of cross. Fig. 53.1 is a flowchart detailing the steps you should take in answering genetics problems.*

Unless otherwise stated or obvious from the evidence, you should assume that the genes being considered in any given case are on separate chromosomes. This is important because it means that they will assort independently during meiosis. Thus, the fact that an allele of gene A is present in any individual will not influence the possibility of an allele of gene B being present. This allows you to apply simple probability in predicting the genetic make-up of the offspring of any cross (see p. 330).

Where genes are present on the same chromosome, they are said to be linked genes, and thus it would appear that they would not be able to assort independently during metaphase 1 of meiosis, when homologous

Example In the garden pea, *Pisum sativum*, studied by Mendel, the yellow seed allele Y was found to be dominant over the green seed allele y. A cross of YY x yy genotypes would give rise to Yy in all of the F_1 generation, all of which would thus have the yellow seed phenotype. If the F_1 generation were interbred, this Yy x Yy cross would lead to progeny in the next, F_2, generation with the genotypes YY:Yy:yy in the expected ratio 1:2:1. The expected phenotype ratio would be 3:1 for yellow:green seed.

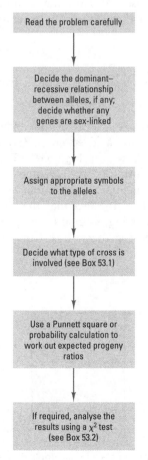

Fig. 53.1 Flowchart for tackling problems in Mendelian genetics

Pedigree notation:

○   normal female

□   normal male

● ■   female or male with an inherited condition

○———□   mating

           offspring

Example of simple family pedigree:

This diagram shows the offspring of a normal male and female. The two daughters are normal, but the son has the inherited condition.

Fig. 53.2 Pedigree notation and family trees

Denoting linked genes – *these are often shown diagramatically, with a double line indicating the chromosome pair. For example, the two possible linkages for the genotype AaBb would be shown as:*

$$\frac{A \quad b}{a \quad B} \quad or \quad \frac{A \quad B}{a \quad b}$$

chromosomes are independently oriented. However, although physically attached to each other, they may become separated when crossing-over occurs between the homologous chromosomes at an early stage of meiosis. Exchange of genetic information between homologous chromosomes is called recombination. Linkage can be detected from a cross between individuals heterozygous and homozygous recessive for the relevant genes, e.g. AaBb × aabb. If the genes A and B are on different chromosomes, we expect the ratio of AaBb, aabb, Aabb and aaBb to be $1:1:1:1$ in the F_1. However, if the dominant alleles of both genes occur on the same chromosome, the last two combinations will occur, but rarely. Just how rarely depends on how far apart they lie on the relevant chromosome – the further apart, the more likely it is that crossing-over will occur. This is the basis of chromosome mapping (see Box 53.1).

Another complication you will come across is sex-linked genes. These occur on one of the X or Y chromosomes that control sex. Because one or other of the sexes – depending on the organism – is determined as XX and the other as XY (see Box 53.1), this means that rare recessive genes carried on the X chromosome may be expressed in XY individuals. Sex-linked genes are sometimes obvious from differences in the frequencies of phenotypes in male and female offspring. Pedigree charts (Fig. 53.2) are codified family trees that are often used to show the inheritance and expression of sex-linked characteristics through various generations.

Analysis of crosses

There are two basic ways of working out the results of crosses from known or assumed genotypes:

- The Punnett square method provides a good visual indication of potential combinations of gametes for a given cross. Lay out your Punnett squares consistently as shown in Fig. 53.3. Then group together the like genotypes to work out the genotype ratio and proceed to work out the corresponding phenotype ratio if required.
- Probability calculations are based on the fact that the chance of a number of independent events occurring is equal to the probabilities of each event occurring, multiplied together. Thus if the probability

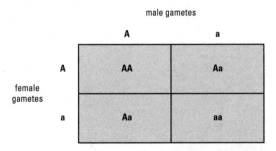

Fig. 53.3 Layout for a simple Punnett square for the cross Aa × Aa. The genotypic ratios for this cross are $1:2:1$ for AA:Aa:aa, and the phenotypic ratio would be $3:1$ for characteristic A to characteristic a. In this simple Punnett square, the allele frequencies are treated as equal ($f = 0.5$); if different from this, the probability of genotypes in each combination will be the relevant frequencies multiplied together (see p. 330).

Box 53.1 Types of cross and what you can (and can't) learn from them

Monohybrid cross – the simplest form of cross, considering two alleles of a single gene.

Example: AA × aa

If only the parental *phenotypes* are known, you can't always deduce the parental genotypes from the phenotype ratio in the F_1. An individual of dominant phenotype in the F_1 could arise from a homozygous dominant or heterozygous genotype. However, crossing the F_1 generation with themselves to give an F_2 generation may provide useful information from the phenotype ratios that are found.

Dihybrid cross – a cross involving two genes, each with two alleles.

Example: AaBB × AaBb

As with a monohybrid cross, you can't always deduce the parental genotype from the phenotype ratio in the F_1 generation alone.

Test cross – a cross of an unknown genotype with a homozygous recessive.

Example: AABb × aabb

A test cross is one between an individual dominant for A and B with one recessive for both genes. The progeny will all be dominant for A, revealing the homozygous nature of the parent for this gene, but the progeny phenotypes will be split approximately 1:1 dominant to recessive for gene B, revealing the heterozygous nature of the parent for this gene. This type of cross reveals the unknown parental genotype in the proportions of phenotypes in the F_1.

Sex-linked cross – a cross involving a gene carried on the X chromosome; this can be designated as dominant or recessive using appropriate superscripts (e.g. X^A and X^a).

Example: $X^A X^a \times X^A Y$

In sex-linked crosses you need to know the basis of sex determination in the species concerned – which of XX or XY is male (for example, the former in birds,

butterflies and moths, the latter in mammals and *Drosophila*). The expected ratios of phenotypes in the offspring will depend on this. Note that the recessive genotype a will be expressed in the $X^a Y$ case.

Crosses with linked genes – genes are linked if they are on the same chromosome. This is revealed from a cross between individuals heterozygous and homozygous recessive for the relevant genes.

Example 1 (genes on separate chromosomes): expected offspring frequency from the cross AaBb × aabb is AaBb, aabb, Aabb and aaBb in the ratio 1:1:1:1.

Example 2 (linked genes): the frequencies of the last two combinations in example 1 might be skewed according to the direction of parental linkage.

Chromosome mapping uses the frequency of crossing-over of linked genes to estimate their distance apart on the chromosome on the basis that crossing-over is more likely, the further apart are the genes. So-called 'map units' are calculated on the following basis:

$$\frac{\text{no. of recombinant progeny}}{\text{total no. of progeny}} \times 100 = \% \text{ crossing-over}$$

[53.1]

By convention, 1% crossing-over = 1 map unit (centimorgan, cM). The order of a number of genes can be worked out from their relative distances from each other. Thus, if genes A and B are 12 map units apart, while A and B are respectively 5 and 7 map units from C, the assumed order on the chromosome is ACB (Fig. 53.4).

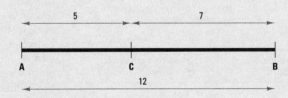

Fig. 53.4 Genetic map showing relative positions of genes A, B and C

Using probability calculations – this can be simpler and faster than Punnett squares when two or more genes are considered.

Carrying out genetic crosses – typically these are performed using organisms that have a large number of offspring, to even out random variation. Other useful attributes include short generation time, ease of maintenance, and wide range of mutations readily observed in the phenotype.

Limitation of the Chi² test – note that the formula cited in Box 53.2 is valid only if expected numbers are greater than 5.

Table 53.1 Values of Chi² (χ^2) for which $P = 0.05$. The value for $(n - 1)$ degrees of freedom (d.f.) should be used, where $n =$ the number of categories (= phenotypes) considered (normally fewer than 4 in genetics problems). If χ^2 is less than or equal to this value, accept the null hypothesis that the observed values arose by chance; if χ^2 is greater than this value, reject the null hypothesis and conclude that the difference between the observed and expected values is statistically significant.

Degrees of freedom	χ^2 value for which $P = 0.05$
1	3.84
2	5.99
3	7.82
4	9.49

P of a child being a boy is 0.5 and the probability of the child of particular parents being blue-eyed is 0.5, then the probability of that couple having a blue-eyed son is $0.5 \times 0.5 = 0.25$, and that of having two blue-eyed boys is:

$$P = (0.5 \times 0.5) \times (0.5 \times 0.5) = 0.0625$$

How do you decide whether the results of an experimental cross fit your expectation from theory as calculated above? This isn't easy, because of the element of chance in fertilisation. Thus, while you might expect to see a $3 : 1$ phenotype ratio of progeny for a given cross, in 500 offspring you might actually observe a ratio of $379 : 121$, which is a ratio of just over $3.13 : 1$. Can you conclude that this is significantly different from $3 : 1$ in the context of random error? The answer to this problem comes from statistics. However, the answer isn't certain, and your conclusion will be based on a balance of probabilities (see Box 53.2 and Table 53.1).

Box 53.2 Example of a Chi² (χ^2) test

This test allows you to assess the difference between observed (O) and expected (E) values and is extremely useful in biology. It is particularly valuable in determining whether progeny phenotype ratios fit your assumptions about their genotypes. The operation of the test is best illustrated by the use of an example. Assume that your null hypothesis (see p. 426) is that the phenotypic ratio is $3 : 1$ and you observe that in 500 offspring the phenotype ratio is $379 : 121$ whereas the expected ratio is $375 : 125$.

Start the test by calculating the test statistic χ^2. The general formula for calculating χ^2 is:

$$\chi^2 = \Sigma \, \frac{(O - E)^2}{E} \qquad [53.2]$$

In this example, this works out as:

$$\chi^2 = \frac{(379 - 375)^2}{375} + \frac{(121 - 125)^2}{125} = \frac{16}{375} + \frac{16}{125} = 0.171$$

The probability associated with this value can be obtained from χ^2 **tables** for $(n - 1)$ degrees of freedom (d.f.), where $n =$ the number of categories = number of phenotypes considered. Here the d.f. value is $2 - 1 = 1$. Since the χ^2 value of 0.171 is lower than the tabulated value for 1 d.f. (3.84, Table 53.1), we therefore accept the null hypothesis and conclude that the difference between observed and expected results is not significant (since $P > 0.05$). Had χ^2 been $\geqslant 3.84$, then $P \leqslant 0.05$ and we would have rejected the null hypothesis and concluded that the difference was significant, i.e. that the progeny phenotype did not fit the expected ratio.

You can carry out Chi² calculations using the CHITEST function in Microsoft Excel. The *Help* function within that program contains useful guidance.

Population genetics

Population genetics is largely concerned with the frequencies of alleles in a population and how these may change in time. The Hardy–Weinberg Principle states that the frequency of alleles f remains the same between generations, unless influenced by some outside factor(s).

To understand why this is the case, consider alleles H and h for a particular gene, which exist in the breeding population at frequencies p and q respectively. If the individuals carrying these alleles interbreed randomly, then the expected phenotype and allele ratios in the F_1 generation can be calculated simply as:

$$f(HH) = p^2;$$
$$f(Hh) = 2pq; \quad \text{and}$$
$$f(hh) = q^2$$

If you wish to confirm this, lay out a Punnett square with appropriate frequencies for each allele. Now, by summation, the frequency of H in the $F_1 = p^2 + pq$ (a similar calculation can be made for allele h); and since in this example there are only two alleles, $p + q = 1$ and so $q = (1 - p)$. Substituting $(1 - p)$ for q, the frequency of H in the F_1 is thus:

$$p^2 + p(1 - p) = p^2 + p - p^2 = p$$

i.e. the frequency of the allele is unchanged between generations. A similar relationship exists for the other alleles.

The Hardy–Weinberg Principle was named after its first, independent, protagonists. It holds so long as the following criteria are satisfied:

1. random mating – so that no factors influence each individual's choice of a mate;
2. large population size – so that the laws of probability will apply;
3. no mutation – so that no new alleles are formed;
4. no emigration, immigration or isolation – so that there is no interchange of genes with other populations nor isolation of genes within the population;
5. no natural selection – so that no alleles have a reproductive advantage over others.

Population geneticists use the Hardy–Weinberg Principle to gain an idea of the rate of evolution and the influences on evolution. By ensuring that criteria 1–4 hold, if there are any changes in allele frequency between generations, then the rate of change of allele frequencies indicates the rate of evolutionary change (natural selection).

Example Cystic fibrosis occurs in 0.04% of Caucasian babies. If this condition results from a double recessive allele *aa*, then following the Hardy–Weinberg Principle, $q^2 = 0.0004$ (0.04% expressed as a fraction of 1) and so $q = \sqrt{0.0004} = 0.02$, or 2%. Since $p = (1 - q)$, $p = 0.98$, or 98%. The frequency of carriers of cystic fibrosis in the Caucasian population (people having the alleles Aa) is given by $2pq$. From the above, $2pq = 2(0.02 \times 0.98) = 0.0392$. Hence 3.92% of the Caucasian population are carriers (roughly one in 25).

Sources for further study

Blumberg, R.B. *MendelWeb*. Available: http://www.mendelweb.org/
Last accessed 09/04/07.
[A resource for students interested in the origins of classical genetics.]

Falconer, D.S. and MacKay, T.F.C. (1996) *Introduction to Quantitative Genetics*, 4th edn. Longman, Harlow.

Klug, W.S. and Cummings, M.R. (2005) *Essentials of Genetics*, 5th edn. Prentice-Hall, Upper Saddle River, New Jersey.

Winter, P.C., Hickey, G.I. and Fletcher, H.L. (1998) *Instant Notes in Genetics*. Bios Scientific Publishers, Oxford.

Study exercises

53.1 Use a Punnett square to predict the outcome of a cross. Lay out a Punnett square for a cross between genotypes RrOO × RrOo, where R is a semi-dominant gene for flower colour such that RR = red, Rr = pink and rr = white; and O is a dominant gene for corolla shape such that OO = closed corolla, Oo = closed corolla and oo = open corolla. From the Punnett square, derive both the genotypic and phenotypic ratios for the cross.

53.2 Use probability to predict the outcome of a cross. Two hazel-eyed parents are heterozygous for the eye-colour gene B. When expressed as bb, the individual is blue-eyed. Mum's hair is (genuine) blonde but Dad's is mousy-brown. In this case she is double recessive (mm) for the hair-colour gene M and he is heterozygous (Mm). What is the probability that they will have a blue-eyed, blonde daughter?

53.3 Work out a likely genetic scenario for a given set of results. Five tailless female mice were crossed with normal males (with tails). There were 31 normal mice and 28 tailless mice in the F$_1$ progeny. When pairs of tailless mice from the F$_1$ generation were crossed, their (F$_2$)

progeny were as follows: normal, 27; tailless, 55; dead on birth, but tailless, 30. In each case the ratio of males to females was roughly 50:50. Provide a logical explanation for these results.

53.4 Predict parental genotypes from the results of crosses involving sex-linked genes. A red-eye gene is known to be sex-linked in *Drosophila*; that is, the alleles R (red-eyed) or r (white-eyed) are carried on the X chromosome, while the Y chromosome does not carry these eye-colour alleles. In the fruit-fly, XX = female and XY = male. Predict the possible parental genotypes from the following F$_1$ progeny ratios:

(a) 35 red-eyed female, 17 red-eyed male, 19 white-eyed male;

(b) 27 red-eyed male, 29 red-eyed female;

(c) 19 red-eyed female, 18 white-eyed female, 22 red-eyed male, 21 white-eyed male.

53.5 Carry out a χ^2 (Chi^2) test. A geneticist expects the results of a test cross to be in the phenotype ratio 1:2:1. He observes 548 progeny from his cross in the ratio 125:303:120. What should he conclude?

Deoxyribonucleic acid (DNA) is the genetic material of all cellular organisms. The sequential arrangement of nucleotide subunits represents a genetic code for the synthesis of specific cellular proteins or functional RNA molecules. A portion of DNA that encodes the information for a single polypeptide, protein or polyribonucleotide is usually referred to as a gene while the entire genetic information of an organism is known as its genome.

Recent advances in the manipulation of nucleic acids in the laboratory have increased our understanding of the structure and function of genes at the molecular level. Additionally, these techniques can be used to alter the genome of an organism (genetic engineering or gene cloning), e.g. to create a bacterium capable of synthesising a foreign protein such as a potentially valuable hormone or vaccine, or a recombinant protein.

Conforming to regulations – in the UK, the Genetically Modified Organisms (Contained Use) Regulations (2000) provide the regulatory framework for all research procedures involving the genetic modification of organisms.

 KEY POINT *Although genetic manipulation must be carried out under strict containment, in accordance with appropriate legislation, the procedures involved in the isolation, amplification, recombination and cloning of DNA are often used at undergraduate level, to illustrate the general features of the techniques.*

Basic principles

Gene cloning involves several steps:

1. Isolation of the DNA sequence (gene) of interest from the genome of an organism. This usually involves DNA purification followed by either enzymic digestion, mechanical fragmentation or PCR amplification, to liberate the target DNA sequence.
2. Creation of an artificial recombinant DNA molecule (sometimes referred to as rDNA), by inserting the gene into a DNA molecule capable of replicating in a host cell, i.e. a 'cloning vector'. Suitable cloning vectors for bacterial cells include plasmids, bacteriophages (bacterial viruses) and cosmids.
3. Introduction of the recombinant DNA molecule into a suitable host, e.g. *Escherichia coli* (transformation when a plasmid is used, or transfection for recombinant viral nucleic acid).
4. Selection and growth of the transformed (or transfected) cell, using the techniques of cell culture (p. 271). Since a single transformed host cell can be grown to give a clone of genetically identical cells, each carrying the gene of interest, the technique is often referred to as 'gene cloning', or molecular cloning.

Definitions

Plasmids – circular molecules of DNA that are capable of autonomous replication. They can be isolated, manipulated and then reintroduced into bacterial cells.

Cosmids – hybrid plasmid vectors containing the cos (cohesive) sites from phage λ, enabling *in-vitro* packaging into phage capsids. Useful for cloning large segments of DNA, typically up to 50 kb.

Transformation – In bacteria: stable incorporation of external DNA, e.g. a plasmid. In eukaryotes: the conversion of a cell culture of finite life to a continuous (immortal) cell line (also occurs in cancer).

Transfection – In bacteria: uptake of viral nucleic acid. In eukaryotes: uptake of any naked, foreign DNA.

Extraction and purification of DNA

Specific details of the steps involved in the isolation of DNA vary, depending upon the source material. However, the following sequence shows the principal stages in the purification of plasmid DNA from bacterial cells.

1. Cell wall digestion: incubation of bacteria in a lysozyme solution will remove the peptidoglycan cell wall. This is often carried out under isotonic conditions, to stop the cells from bursting open and releasing

Preparing glassware – all glassware for DNA purification must be siliconised, to prevent adsorption of DNA. All glass and plastic items must be sterilised before use.

Molecular biology techniques

chromosomal DNA. Note that Gram-negative bacteria are relatively insensitive to lysozyme, requiring additional treatment to allow the enzyme to reach the cell wall layer, e.g. osmotic shock, or incubation with a chelating agent, e.g. ethylenediaminetetra-acetate (EDTA). The latter treatment will also inactivate any bacterial deoxyribonucleases (DNases) in the solution, preventing enzymic degradation of plasmid DNA during extraction.

2. Lysis using strong alkali (NaOH) and a detergent, e.g. sodium dodecyl sulphate (SDS), to solubilise the cellular membranes and partially denature the proteins. Neutralisation of this solution (e.g. using potassium acetate) causes the chromosomal DNA to aggregate as an insoluble mass, leaving the plasmid DNA in solution.

3. Removal of other macromolecules, particularly RNA and proteins by, for example, enzymic digestion using ribonuclease and proteinase. Additional chemical purification steps give further increases in purity, e.g. proteins can be removed by mixing the extract with water-saturated phenol (50% v/v), or a phenol/chloroform mixture. On centrifugation, the DNA remains in the upper aqueous layer, while the proteins partition into the lower organic layer. Repeated cycles of phenol/chloroform extraction can be used to minimise the carry-over of these macromolecules. Additional purification can be obtained using isopycnic density gradient centrifugation in CsCl (p. 361).

4. Precipitation of DNA using around 70% v/v ethanol : water (produced by adding two volumes of 95% v/v ethanol to one volume of aqueous extract), followed by centrifugation, to recover the DNA pellet. Further rinsing with 70% v/v ethanol : water will remove any salt contamination from the previous stages. The extracted DNA can then be redissolved in buffer solution and frozen for future use.

Separation of DNA using agarose gel electrophoresis

Electrophoresis is the term used to describe the movement of ions in an applied electrical field, as shown for DNA in Fig. 54.1. DNA molecules are negatively charged, migrating through an agarose gel towards the anode at a rate that is dependent upon molecular size – smaller, compact DNA molecules can pass through the sieve-like agarose matrix more easily than large, extended fragments. Electrophoresis of plasmid DNA is usually carried out using a submerged agarose gel. The amount of agarose is adjusted, depending on the size of the DNA molecules to be separated, e.g. 0.3% w/v agarose is used for large fragments (>20 000 bases) while 0.8% is used for smaller fragments. Note the following:

- Individual samples are added to preformed wells using a pipettor. The volume of sample added to each well is usually less than 25 μl so a steady hand and careful dispensing are needed to pipette each sample.

- The density of the samples is usually increased by adding a small amount of glycerol or sucrose, so that each sample is retained within the appropriate well.

- A water-soluble anionic tracking dye (e.g. bromophenol blue) is also added to each sample, so that migration can be followed visually.

- DNA fragment-size markers ('molecular weight' standards) are added to one or more wells. After electrophoresis, the relative position of

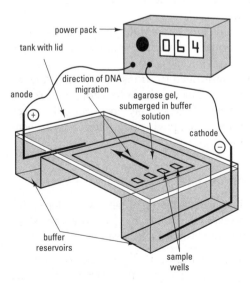

power pack

tank with lid

anode

direction of DNA migration

agarose gel, submerged in buffer solution

cathode

buffer reservoirs

sample wells

Fig. 54.1 Agarose gel electrophoresis of DNA

bands of known size (length) can be used to prepare a calibration curve (usually, by plotting $\log_{10}$ of the molecular weight of each band against the distance travelled).

- The gel should be run until the tracking dye has migrated across 80 per cent of the gel (see manufacturer's instructions for appropriate voltages/times).

- After electrophoresis, the bands of DNA can be visualised by soaking the gel for around 5 minutes in $0.1-0.5\,\mathrm{mg\,l^{-1}}$ ethidium bromide, which binds to DNA by intercalation between the paired nucleotides of the double helix.

- Under UV light, bands of DNA are visible due to the intense orange–red fluorescence of the ethidium bromide. The limit of detection using this method is around 10 ng DNA per band. The migration of each band from the well can be measured using a ruler. Alternatively, a photograph can be taken, using a digital camera and adaptor or a dedicated system such as GelDoc.

- If a particular band is required for further study (e.g. a plasmid), the piece of gel containing that band is cut from the gel using a scalpel and the DNA extracted, e.g. using a silica membrane 'spin column'.

Identification of specific DNA molecules using Southern blotting

After electrophoretic separation in agarose gel, the fragments of DNA can be immobilised on a filter using a technique named after E.M. Southern:

1. The fragments are first denatured using concentrated alkali, giving single-stranded DNA (ssDNA); this is necessary to allow hybridisation with 'probe' DNA after blotting.

2. A nitrocellulose filter is then placed directly on to the gel, followed by several layers of absorbent paper. The DNA is 'blotted' on to the filter as the buffer solution is pulled through.

3. Specific sequences can be identified by incubation with labelled complementary probes of single stranded DNA, which will hybridise with a particular sequence, followed by visualisation, often using an enzyme-based system.

DNA amplification using the polymerase chain reaction (PCR)

The polymerase chain reaction is particularly useful for amplifying small amounts of DNA, e.g. from a virus in a clinical specimen, or from a small tissue sample in forensic science. The three basic steps are:

1. Denaturation of double-stranded DNA by heating to 94–98 °C, to separate the individual DNA strands.
2. Annealing of oligonucleotide primers (short, synthetic DNA sequences that will hybridise at a specific position on the target DNA, when the temperature is reduced to 37–65 °C.
3. Primer extension by a thermostable DNA polymerase (e.g. *Taq* polymerase from *Thermus aquaticus*) at 72 °C.

Molecular biology techniques

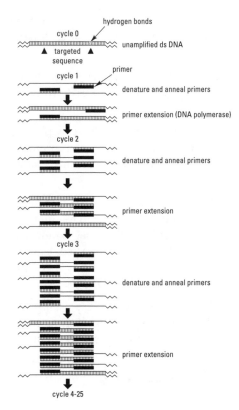

Fig. 54.2 The polymerase chain reaction (PCR)

Repeated cycling generates an exponentially increasing number of DNA fragments corresponding to the sequence between the two primers (see Fig. 54.2). Box 54.1 gives further practical details.

KEY POINT *PCR is so sensitive that one of the main problems associated with the technique is contamination with 'foreign' DNA. Great care is required to avoid sample contamination during in vitro amplification. It is good practice to carry out PCR amplification in a designated laboratory, to minimise the risk of contamination.*

DNA assay

The relative amount of DNA in an agarose gel can be quantified by visual comparison of stained bands with 'marker' bands of known quantity. A more accurate method is to use band densitometry, e.g. using the GelDoc system. A simple approach to quantifying the amount of nucleic acid in an aqueous solution is to measure the absorbance of the solution at 260 nm using a spectrophotometer, as detailed on p. 369. Note that the A_{260} value applies to purified DNA, whereas a plasmid DNA extract prepared using the protocol on p. 333 will contain a substantial amount of contaminating RNA, with similar absorption characteristics to DNA. Any contaminating protein would also invalidate the calculation; the protein can be detected by measuring the absorbance of the solution at 280 nm. Purified DNA has a value for A_{260}/A_{280} of around 1.8 and contaminating protein will give a lower ratio. If your solution gives a ratio substantially lower than this (<1.7), you should repeat the later steps of your extraction procedure (p. 334).

Enzymic cleavage and ligation of DNA

Type II restriction endonucleases (commonly called restriction enzymes) recognise and cleave a particular sequence of double-stranded DNA (usually, four or six nucleotide pairs), known as the restriction site. Each enzyme is given a code name derived from the name of the bacterium from which it is isolated, e.g. *Hin* dIII was the third restriction enzyme isolated from *Haemophilus influenzae* strain Rd (Fig. 54.3). Most enzymes cleave each strand at a different position, producing short, single-stranded regions known as cohesive ends, or 'sticky ends'. A few enzymes cleave DNA to give blunt-ended fragments. Restriction enzymes that cleave DNA to give 'sticky ends' are widely used in genetic engineering, since two DNA molecules cut with the same restriction enzyme will have complementary single-stranded

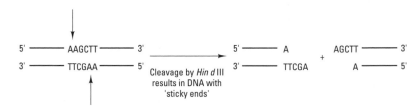

Fig. 54.3 Recognition site for the restriction enzyme *Hin* dIII. This is the conventional representation of double-stranded DNA, showing the individual bases, where A is adenine, C cytosine, G guanine and T thymine. The cleavage site on each strand is shown by an arrow.

Box 54.1 How to carry out the polymerase chain reaction (PCR)

The protocol below shows the main stages in a typical PCR: note that temperatures, incubation times and the number of cycles will vary with the particular application.

1. **Check you have the required apparatus and reagents to hand**, including (i) thermal cycler; (ii) template DNA (at least 50 ng μl^{-1}); (iii) stock solutions of deoxyribonucleoside triphosphates (dNTPs; at 5 mmol l^{-1} each); (iv) thermostable DNA polymerase (5 U μl^{-1}); (v) primers (e.g. at 30 pmol μl^{-1}); (vi) stock buffer solution, e.g. containing 100 mmol l^{-1} TRIS (pH 8.4), 500 mmol l^{-1} KCl, 15 mmol l^{-1} MgCl$_2$, 1% w/v gelatine, 1% v/v Triton X0100 (the stock buffer is often termed 10x PCR buffer).

2. **Prepare a reaction mixture for your template DNA**: for example, a mixture containing (i) 1.0 μl of template DNA; (ii) 2.5 μl of stock buffer solution; (iii) 1.0 μl of each of the two primers; 1.0 μl of each of the four dNTPs; (iv) 0.1 μl of DNA polymerase; (v) 15.4 μl of sterile ultrapure water, to give a total reaction volume of 25.0 μl.

3. **Prepare appropriate positive and negative controls**: a positive control is a PCR template that is known to work under the reaction conditions used. The most commonly used negative control is the reaction mixture minus the template DNA, though negative controls can be set up lacking any one of the reaction components.

4. **Cycle in the thermal cycler**: for example, an initial period of 5 min at 94 °C, followed by 30 cycles of 94 °C for 1 min (denaturation), then 50 °C for 1 min (primer annealing), then 72 °C for 1 min (chain extension).

5. **Assess the effectiveness of the PCR**: for example, by agarose gel electrophoresis (Fig. 54.1) and ethidium bromide (or SYBR Safe) staining.

To avoid contamination in PCR:

- Use a laminar-flow cabinet (p. 266) as a dedicated workstation for PCR use, located in a separate lab from that used to carry out other DNA manipulation. For very sensitive work, use a short-wavelength UV light inside the workstation for 20–30 min before starting work, to degrade any contaminating DNA. Make sure that you do not expose your skin or eyes to the short-wavelength UVC source (see Safety Note, p. 335).
- Keep exclusive supplies of pipettors, tips, tubes and reagents for sample preparation, reactions and product analysis.
- Autoclave all buffer solutions, pipette tips and tubes.
- Wear disposable gloves at all times and change them frequently.

Troubleshooting – common problems include:

- Lack of a PCR product (one or more components of the reaction mixture may have been left out; annealing temperature may be too high; denaturation temperature may be too low).
- Too many bands present (one or more components of the reaction mixture may be present in excess; primers may not be specific enough to the target sequence; annealing temperature may be too low).
- Bands corresponding to primer-dimers present (the primers may show partial complementarity – check sequences; primer concentration may be too high).

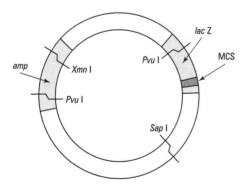

Fig. 54.4 Restriction map of the plasmid pUC 19. The position of some individual restriction sites is shown together with the genes for ampicillin resistance (*bla*) and β-galactosidase (*lac Z*). MCS = multiple cloning site for 40 restriction enzymes within the *lac Z* gene.

regions, allowing them to anneal (base pair), due to the formation of hydrogen bonds between individual bases within this region.

Restriction enzymes have two important applications in molecular genetics:

- Mapping. A DNA molecule can be cleaved into several restriction fragments whose number, size and orientation relative to one another can be determined using agarose gel electrophoresis. The position of individual restriction sites can be used to create a restriction enzyme map for a particular molecule, e.g. a plasmid such as pUC19 (Fig. 54.4).
- Genetic engineering. Two restriction fragments cut using the same enzyme (Fig. 54.5) and annealed by complementary base pairing can be permanently joined together using another bacterial enzyme (DNA ligase), which forms covalent bonds between the annealed strands, creating the recombinant molecule (Fig. 54.6). When the two molecules involved are the cloning vector and the target DNA, the size of the

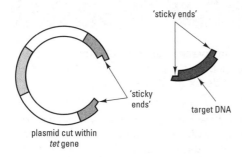

Fig. 54.5 Restriction of plasmid and foreign DNA e.g. with *Hin* dIII

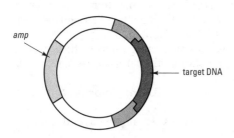

Fig. 54.6 Annealing and ligation of plasmid and target DNA to give a recombinant plasmid that confers resistance to ampicillin. As target DNA has been inserted within the *lac Z* gene, the gene is now discontinuous and inactive.

recombinant plasmid can be predicted (e.g. a plasmid of 4500 base pairs, plus a target DNA fragment of 2500 base pairs will give a recombinant molecule of 7000 base pairs), allowing separation by electrophoresis. Note that most of the plasmids used in genetic engineering code for two or more easily detectable marker genes (e.g. antibiotic resistance), with single restriction sites on each plasmid (Fig. 54.4).

Transformation/transfection of a suitable host cell

Once a recombinant plasmid has been produced *in vitro*, it must be introduced into a suitable host cell (e.g. *E. coli*). Several procedures can be used:

- Pretreatment with $CaCl_2$ at low temperature: exponential phase cells (p. 272) are incubated with hypotonic $CaCl_2$ at 4 °C for around 30 min, followed by a brief heat shock (e.g. 42 °C for 2 min). The low-temperature incubation allows DNA to adhere to the cells, while the heat shock promotes DNA uptake.
- Electroporation: cells or protoplasts are subjected to electric shock treatment (typically, $>1\,kV\,cm^{-1}$) for short periods ($<10\,ms$).
- For animal and plant cells a range of techniques can be used, including electroporation, various microinjection treatments, either using a microsyringe or microprojectiles ('biolistics'), or cationic lipids (Fig. 54.7), which form lipid–DNA complexes that fuse with cell membranes and allow DNA to enter the cell.

These treatments mostly cause a temporary increase in membrane permeability, leading to the uptake of plasmid DNA from the external medium. Such systems are often inefficient, with fewer than 0.1 per cent of all cells showing stable transformation. However, this is not usually a significant problem, since a single viable transformant can be grown to give a large number of identical cells, using standard microbiological plating techniques (p. 267).

To maximise the transformation efficiency of *E. coli* in the heat shock/$CaCl_2$ procedure, use the minimum volume of solution in a thin-walled glass tube, so that the cells experience a rapid change in temperature.

Selection and detection of transformants

As the efficiency of transformation is often very low, many of the plasmids used in genetic manipulation code for antibiotic resistance and other selectable traits, e.g. pUC19 (Fig. 54.4) carries the gene for the β-lactamase

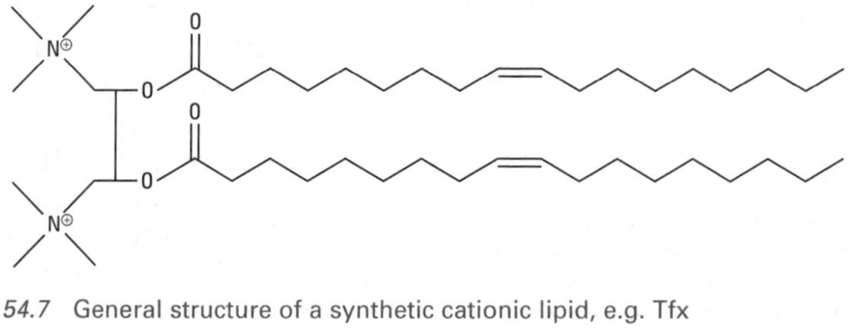

Fig. 54.7 General structure of a synthetic cationic lipid, e.g. Tfx

Recognising transformants – after plating bacteria on to medium containing ampicillin, you may notice a few small 'feeder' or 'satellite' colonies surrounding a single larger (transformant) colony. The feeder colonies are derived from non-transformed cells that survive due to the breakdown of antibiotic in the medium around the transformant colony, and they should not be selected for subculture.

gene (*bla*) and also the gene for the enzyme β-galactosidase (*lac Z*). These genes act as markers in a single-step selection procedure. Expression of *bla* confers resistance to ampicillin, which can be demonstrated by including this antibiotic in the growth medium, thus selecting for transformants (non-transformed cells would be unable to grow on a medium containing ampicillin). The other gene is used as a marker to distinguish transformants that have acquired the recombinant plasmid from those that have taken up the original plasmid, as follows: an intact *lac Z* gene can be detected using a combination of (i) an inducer (e.g. isothiopropylgalactoside, IPTG) plus (ii) a suitable chromogenic substrate, e.g. 5-bromo-4-chloro-3-indolyl-β-D-galactoside (XGAL), which is broken down to release a blue-green indigo derivative. Thus a colony of cells containing the native plasmid would express *lac Z* and would appear blue, while the equivalent recombinant colony would have a discontinuous *lac Z* gene (Fig. 54.5) and would appear off-white, due to insertional inactivation of this gene. The multiple cloning site ('polylinker' region) within the *lac Z* gene enables use of any one of a number of different restriction enzymes, depending on the sites present in the target DNA sequence. Another useful feature of pUC19 and similar plasmids is that they are present in high copy number within a transformed cell, with several thousand identical copies per cell, giving improved yield of plasmid DNA on extraction (p. 333).

Other plasmids make use of different markers; for example, the luciferase gene can be detected by bioluminescence in the presence of the substrate luciferin, or the green fluorescent protein (GFP) from the jellyfish *Aequorea victoria*. More specialised vectors are available for specialised functions, such as M13mp19, a single-stranded phage vector for DNA sequencing.

Box 54.2 provides an overview of Web-based resources for bioinformatics, while Reed et al. (2007) gives further practical information on additional procedures in molecular biology (including DNA sequencing, probes and hybridisation, and various forms of PCR including real-time PCR), bioinformatics (databases and tools) and proteomics (studying the expressed proteins in an organism).

Text reference

Reed, R.H., Holmes, D., Weyers, J.D.B. and Jones, A.M. (2007) *Practical Skills in Biomolecular Sciences,* 3rd edn. Prentice Hall, Harlow.

Sources for further study

Brown, T.A. (2006) *Gene Cloning and DNA Analysis,* 5th edn. Blackwell, Oxford.

Hardin, C., Edwards, J., Riell, A. et al. (2001) *Cloning, Gene Expression, and Protein Purification: Experimental Procedures and Process Rationale.* Oxford University Press, Oxford.
[A practical handbook and laboratory manual, with basic experimental protocols together with underlying theoretical principles.]

McPherson, M.J. and Møller, S.G. (2006) *PCR: the Basics,* 2nd edn. Taylor & Francis, London.

Russell, P. (2005) *iGenetics: A Molecular Approach.* Addison-Wesley, Harlow.

Twyman, R.H. and Primrose, S.B. (2004) *Principles of Gene Manipulation and Genomics,* 7th edn. Blackwell, Oxford.

Box 54.2 Bioinformatics – Internet resources

Bioinformatics is a term used to describe the application of computers in biology and, in particular, to the analysis of sequence data for biopolymers such as proteins and nucleic acids. These complex molecules contain a large amount of information within their structures, and the only practical approach to understanding this is to use a computer – generally, by comparing sequence data for all or part of a specific biomolecule with that in a database. For ease of access, many bioinformatics databases and programs have been made freely accessible via the Internet.

The major databases holding primary sequence information for nucleic acids and proteins are operated by the European Bioinformatics Institute (http://www.ebi.ac.uk/Databases/) and the National Center for Biotechnology Information (http://www.ncbi.nlm.nih.gov). These allow you to:

- **find and retrieve a nucleotide or amino acid sequence** from the database;
- **translate a nucleotide sequence** into an amino acid sequence and vice versa;
- **search for similarity** between a particular sequence or sequences within the database, e.g. by comparing and aligning the sequences for several proteins (or nucleic acids), to identify regions of sequence similarity (see Fig. 54.8); and
- **carry out phylogenetic analysis,** constructing 'ancestry trees' to show the most likely evolutionary relationships between sequences from various organisms.

Secondary databases have been created using primary sequence data, to provide information on the patterns identified within particular types of biomolecules, e.g. PROSITE (http://www.expasy.org/prosite/) and PRINTS (http://www.bioinf.man.ac.uk/dbbrowser/index/index.html) databases, which identify protein families by their diagnostic 'signature' motifs, or the ExPaSy proteomics resource (http://www.expasy.org/), for analysis of protein structure. Other websites have been constructed to bring together information for a particular organism, e.g. The *Saccharomyces* Genome Database (SGD, http://www.yeastgenome.org), The *Arabidopsis* Information Resource (http://www.arabidopsis.org/) or The Caenorhabditis elegans Genome Project (http://www.sanger.ac.uk/Projects/C_elegans/) and the Human Genome Project (http://www.sanger.ac.uk/). Many of these sites are rather complex and newcomers can find them a little difficult to navigate, especially if you are just browsing. You are most likely to make use of bioinformatics databases in computer-based exercises: for example, where you are given a particular sequence and you then have to 'interrogate' a particular database, e.g. to identify similar sequences. Alternatively, you could try out the tutorials available at many of these sites.

```
H. sapiens     atgtatggcattgagaatgaagtcttcctgagccttccatgtatcctcaatgcccggggg
               ||||| ||||| ||||| |||||||||| || || || ||||| ||||| |||||
R. norvegicus  atgtacggcatcgagaacgaagtcttcctcagtctcccgtgcatccttaatgctcgggg
```

Fig. 54.8 Representative output from a BLAST (Basic Local Alignment Search Tool) alignment search for DNA sequences from part of the lactate dehydrogenase genes of a human (*Homo sapiens*) and a laboratory rat (*Rattus norvegicus*). Here, identical bases are shown in green, and non-identical bases are shown in black. This region shows 49 identical bases out of 59, i.e. $49 \div 59 \times 100 = 83\%$ identity (to the nearest integer).

Study exercises

54.1 Explain the purpose of the various stages involved in the extraction and purification of DNA. Having read through this chapter, briefly describe the main purpose of each of the following reagents within a DNA extraction protocol, and rearrange them into their order of use within the procedure:

(a) phenol : chloroform extraction;
(b) incubation in 70% v/v ethanol : water;
(c) ribonuclease and proteinase digestion;
(d) incubation with lysozyme/EDTA;
(e) sodium dodecyl sulphate incubation;
(f) NaOH treatment, followed by neutralisation with potassium acetate.

54.2 Construct a restriction map. The figure overleaf represents an electrophoretic separation of plasmid pARJ, digested with different restriction enzymes, both individually and in combination. All of the enzymes cut the plasmid only once. Use this information to construct a simple restriction map of the plasmid. Key to lanes:

1 = DNA standards (0.5 – 8 kbp);
2 = *Eco* RI;
3 = *Hin* dIII;
4 = *Pvu* II;
5 = *Eco* RI and *Hin* dIII;
6 = *Eco* RI and *Pvu* II;
7 = *Hin* dIII and *Pvu* II;
8 = *Eco* RI, *Hin* dIII and *Pvu* II;
9 = DNA standards (0.5 – 8 kbp).

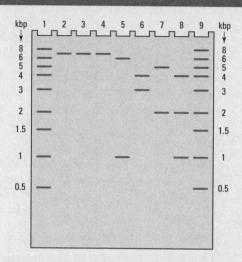

54.3 Calculate transformation efficiency. A volume of 100 µl of *E. coli* cells was transformed using 20 µl of plasmid DNA (prepared at 0.1 ng µl^{-1}) and then made up to a final volume of 500 µl in buffer. A ten-fold dilution of this suspension was prepared and 200 µl of this dilution was surface-spread (p. 268) on to a suitable medium containing an antibiotic to select for transformants, giving 180 colonies on the spread plate after overnight incubation. What is the transformation efficiency, expressed as the number of transformants (colony-forming units, CFU) per µg DNA? (Give your answer in exponential notation, to three significant figures.)

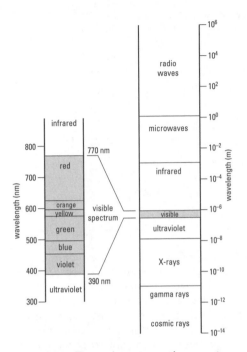

Fig. 55.1 The electromagnetic spectrum, *Methods for Physical and Chemical Analysis of Frest Waters*, 2nd edition, International Biological Programme, No. 8, Blackwell Scientific (Golterman, H. L., Clymo, R. S. and Ohnstad, M.A.M., 1978), reproduced by kind permission of Fisher Scientific UK Ltd.

Light measurement is directly relevant to several aspects of biology, including: photosynthesis, photomorphogenesis and photoperiodism in plants; perception and thermoregulation in animals; aquatic biology.

The nature of light

Light is most strictly defined as that part of the spectrum of electromagnetic radiation detected by the human eye. However, the term is also applied to radiation just outside that visible range (e.g. UV and infrared 'light'). Electromagnetic radiation is emitted by the sun and by other sources (e.g. an incandescent lamp) and the electromagnetic spectrum is a broad band of radiation, ranging from cosmic rays to radio waves (Fig. 55.1). Most biological experiments involve measurements within the UV, visible and infrared regions (generally, within the wavelength range 200–1000 nm).

Radiation has the characteristics of a particle and of a vibrating wave, travelling in discrete particulate units, or 'packets', termed photons. A quantum is the amount of energy contained within a single photon (it is important not to confuse these two terms, although they are sometimes used interchangeably in the literature). In some circumstances, it is appropriate to measure light in terms of the number of photons, usually expressed directly in moles (6.02×10^{23} photons = 1 mol); older textbooks may use the redundant term Einstein as the unit of measurement, where 1 Einstein = 1 mol photons. Alternatively, the energy content (power) may be measured (e.g. in $W\,m^{-2}$). Radiation also behaves as a vibrating electrical and magnetic field moving in a particular direction, with the magnetic and electrical components vibrating perpendicular to one another and perpendicular to the direction of travel. The wave nature of radiation gives rise to the concepts of wavelength (λ, usually measured in nm), frequency (v, measured in s^{-1}), speed (c, the speed of electromagnetic radiation, which is $3 \times 10^8\,m\,s^{-1}$ in a vacuum), and direction. In other words, radiation is a vector quantity, where

$$c = \lambda v \qquad [55.1]$$

Photometric and radiometric measurements

Photometric measurements

These are based on the energy perceived by a 'standard' human eye, with maximum sensitivity in the yellow–green region, around 555 nm (Fig. 55.2). The unit of measurement is the candela, a base unit in the SI system, defined in terms of the visual appearance of a specific quantity of platinum at its freezing point. Derived units are used for the luminous flow (lumen) and luminous flow per unit area (lux). These units were once used in photo-biology and you may come across them in older literature. However, it is now recognised that such measurements are of little direct relevance to biologists, including even those who may wish to study visual responses, because they are not based on fundamental physical principles.

Table 55.1 Wavelength ranges for UV, visible and IR radiation, compiled from several sources.

Type	Wavelength range (nm))
UVC	100–290*
UVB	290–320
UVA	320–400
Violet	390–450
Blue	450–500
Green	500–600
Yellow	560–600
Orange	600–620
Red	620–770
IR	>770

* Note the UVC is absorbed by the Earth's ozone layer.

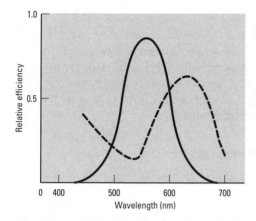

Fig. 55.2 Relative efficiency of vision (solid line) and photosynthesis (dotted line) as a function of wavelength.

Radiometric measurements

The radiometric system is based on physical properties of the electro-magnetic radiation itself, expressed either as the number of photons, or as their energy content. The following terms are used (units of measurement in parentheses):

- Photon flux (mol photons s^{-1}) is the number of photons arriving at an object within the specified time interval.
- Photon exposure (mol photons m^{-2}) is the total number of photons received by an object, usually expressed per unit surface area.
- Photon flux density (mol photons m^{-2} s^{-1}) or PFD is the most commonly used term to describe the number of photons arriving at a particular surface, expressed per unit surface area and per unit time interval.
- Photosynthetically active radiation or PAR is radiation within the waveband 400–700 nm, since the photosynthetic pigments (chlorophylls, carotenoids, etc.) show maximum absorption within this band.
- Photosynthetic photon flux density (mol photons m^{-2} s^{-1}) or PPFD is the number of photons within the waveband 400–700 nm arriving at a particular surface, expressed per unit surface area and per unit time interval. Often this term is used interchangeably with PFD.
- Irradiance ($J\,m^{-2}\,s^{-1} = W\,m^{-2}$) is the amount of energy arriving at a surface, expressed per unit surface area and per unit time interval.
- Photosynthetic irradiance ($W\,m^{-2}$) or PI is the energy of radiation within the waveband 400–700 nm arriving at a surface, expressed per unit surface area and per unit time interval.

Choice of measurement scale

Photon flux density

This is often the most appropriate unit of measurement for biological systems where individual photons are involved in the underlying process, e.g. in photosynthetic studies, where PPFD is measured, since each photochemical reaction involves the absorption of a single photon by a pigment molecule. Most modern light-measuring instruments (radiometers) can measure this quantity, giving a reading in μmol photons $m^{-2}\,s^{-1}$.

Irradiance

This is appropriate if you are interested in the energy content of the light, e.g. if you are studying energy balance, or thermal effects. Many radiometers measure photosynthetic irradiance within the waveband 400–700 nm, giving a reading in $W\,m^{-2}$. It is possible to make an approximate conversion between PPFD and PI measurements, providing the spectral properties of the light source are known (see Table 55.2).

Table 55.2 Approximate conversion factors for a photosynthetic irradiance (PI) of $1\,W\,m^{-2}$ to photosynthetic photon flux density (PPFD)

Source	PPFD (μmol photons $m^{-2}\,s^{-1}$)
Sunlight	4.6
'Cool white' fluorescent tube	4.6
Osram 'daylight' fluorescent tube	4.6
Quartz-iodine lamp	5.0
Tungsten bulb	5.0

(*Source*: Lüning, 1981)

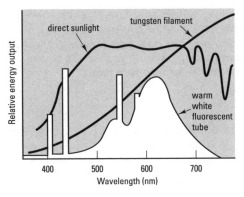

Fig. 55.3 Spectral distribution of energy output from various sources. (Adapted from Golterman et al., 1978.)

Spectral distribution

This can be determined using a spectroradiometer, e.g. to compare different light sources (Fig. 55.3). A spectroradiometer measures irradiance or photon flux density in specific wavebands. This instrument consists of a monochromator (p. 368) to allow separate narrow wavebands (5–25 nm bandwidth) to be measured by a detector; some instruments provide a plot of the spectral characteristics of the source.

Using a radiometer ('light meter')

The main components of a radiometer are:

- Receiver: either flat-plate, hemispherical or spherical, depending upon requirements. Most incorporate a protective diffuser, to reduce reflection.
- Detector: either thermoelectric or photoelectric. Some photoelectric detectors suffer from fatigue, with a decreasing response on prolonged exposure: check the manufacturer's handbook for exposure times.
- Processor and readout device to convert the output from the detector into a visible reading, in analogue or digital form.

Box 55.1. gives practical details of the steps involved in using a radiometer. In your write-up, give full details of how the measurement was made, e.g. the type of light source, instrument used, where the sensor was placed, whether an average was calculated, etc.

Box 55.1 Measuring photon flux density or irradiance using a battery-powered radiometer

1. **Check the battery**. Most instruments have a setting that gives a direct readout of battery voltage. Recharge if necessary before use.

2. **Select the appropriate type of measurement** (e.g. photon flux density or irradiance over the PAR waveband, or an alternative range): the simpler instruments have a selection dial for this purpose.

3. **Place the sensor in the correct location and position** for the measurement: it may be appropriate to make several measurements at different positions, and take an average.

4. **Choose the most appropriate scale** for the readout device: for needle-type meters, the choice of maximum reading is usually selected by a dial, within the range 0.3 to 30 000. Start at a high range and work down until the reading is on the scale.

Your final scale should be chosen to provide the most accurate reading, e.g. a reading of 15 μmol photons m^{-2} s^{-1} should be made using the 0–30 scale, rather than a higher range.

5. **Read the value from the meter**. For needle-type instruments there may be two scales, the upper one marked from 0 to 10 and the lower one from 0 to 3: make sure you use the correct one, e.g. a half-scale deflection on the 0–30 scale is 1.5.

6. **Check that the answer is realistic**, e.g. full sunlight has a PPFD of up to 2 000 μmol photons m^{-2} s^{-1} ($PI \equiv 400\,W\,m^{-2}$), though the value will be far lower on a dull or cloudy day, while the PPFD at a distance of 1 m from a mercury lamp is around 150 μmol photons m^{-2} s^{-1}, and 50 μmol photons m^{-2} s^{-1} at the same distance from a fluorescent lamp.

Text references

Golterman, H.L., Clymo, R.S. and Ohnstad, M.A.M. (1978) *Methods for Physical and Chemical Analysis of Fresh Waters*. Blackwell Scientific, Oxford.

Lüning, K.J. (1981) 'Light', in C.S. Lobban and M.J. Wynne (eds), *The Biology of Seaweeds*, pp. 326–55, Blackwell Scientific, Oxford.

Source for further study

Boyd, R.W. (1983) *Radiometry and the Detection of Optical Radiation*. Wiley, New York.

55.1 Carry out a calculation involving interconversion of photosynthetic photon flux and photosynthetic irradiance. For sunlight with a photosynthetic photon flux density (PPFD) of 1610 mol m^{-2}, what is the total amount of photosynthetic energy (in Joules) falling on a leaf of area 45 cm^2 over a 30-minute experimental period?

55.2 Compare the photosynthetic photon flux densities (PPFD) of different locations. The following values represent typical light levels for a range of different situations, expressed either as photosynthetic irradiance (PI) or PPFD:

(a) outside, on a sunny day: $PI = 300 \text{ W m}^{-2}$;

(b) in a room lit by 'cool white' fluorescent tubes: $PPFD = 6.50 \text{ nmol photons cm}^{-2} \text{ s}^{-1}$;

(c) under the leaf canopy in a forest: $PPFD = 275 \text{ }\mu\text{mol photons m}^{-2} \text{ s}^{-1}$;

(d) in a growth cabinet lit by a bank of 'daylight' fluorescent lights: $PI = 35.2 \text{ W m}^{-2}$;

(e) in a room lit by a tungsten bulb: $PI = 1.15 \text{ mW cm}^{-2}$.

Convert all to PPFD in the same units (expressed to three significant figures) and then rank the locations in order of decreasing PPFD.

55.3 Research the various types of commercial radiometers (light meters) available on the Web. Make a list of the features that you might want to find out about and see how much information you can find.

Examples $^{12}_{6}C$, $^{13}_{6}C$ and $^{14}_{6}C$ are three of the isotopes of carbon. About 98.9 per cent of naturally occurring carbon is in the stable $^{12}_{6}C$ form. $^{13}_{6}C$ is also a stable isotope but it only occurs at 1.1 per cent natural abundance. Trace amounts of radioactive $^{14}_{6}C$ are found naturally; this is a negatron-emitting radioisotope (see Table 56.2).

Examples ^{226}Ra decays to ^{222}Rn by loss of an alpha particle, as follows:

$$^{226}_{88}Ra \rightarrow \, ^{222}_{86}Rn + \, ^{4}_{2}He^{2+}$$

^{14}C shows beta decay, as follows:

$$^{14}_{6}C \rightarrow \, ^{14}_{7}N + \beta^{-}$$

^{22}Na decays by positron emission, as follows:

$$^{22}_{11}Na \rightarrow \, ^{22}_{10}Ne + \beta^{+}$$

^{55}Fe decays by electron capture and the production of an X-ray, as follows:

$$^{55}_{26}Fe \rightarrow \, ^{55}_{25}Mn + X$$

The decay of ^{22}Na by positron emission (β^{+}) leads to the production of a γ ray when the positron is annihilated on collision with an electron.

The isotopes of a particular element have the same number of protons in the nucleus but a different number of neutrons, giving them the same proton number (atomic number) but a different nucleon number (mass number, i.e. number of protons + number of neutrons). Isotopes may be stable or radioactive. Radioactive isotopes (radioisotopes) disintegrate spontaneously at random to yield radiation and a decay product.

Radioactive decay

There are three forms of radioactivity (Table 56.1) arising from three main types of nuclear decay:

- Alpha decay involves the loss of a particle equivalent to a helium nucleus. Alpha (α) particles, being large and positively charged, do not penetrate far in living tissue, but they do cause ionisation damage and this makes them generally unsuitable for tracer studies.
- Beta decay involves the loss or gain of an electron or its positive counterpart, the positron. There are three subtypes:
  (a) Negatron (β^{-}) emission: loss of an electron from the nucleus when a neutron transforms into a proton. This is the most important form of decay for radioactive tracers used in biology. Negatron-emitting isotopes of biological importance include ^{3}H, ^{14}C, ^{32}P and ^{35}S.
  (b) Positron (β^{+}) emission: loss of a positron when a proton transforms into a neutron. This only occurs when sufficient energy is available from the transition and may involve the production of gamma rays when the positron is later annihilated by collision with an electron.
  (c) Electron capture (EC): when a proton 'captures' an electron and transforms into a neutron. This may involve the production of X-rays as electrons 'shuffle' about in the atom (as with ^{125}I) and it frequently involves electron emission.
- Internal transition involves the emission of electromagnetic radiation in the form of gamma (γ) rays from a nucleus in a metastable state and always follows initial alpha or beta decay. Emission of gamma radiation leads to no further change in atomic number or mass.

Note from the above that more than one type of radiation may be emitted when a radioisotope decays. The main radioisotopes used in biology and their properties are listed in Table 56.2.

Table 56.1 Types of radioactivity and their properties

Radiation	Range of maximum energies (MeV*)	Penetration range in air (m)	Suitable shielding material
Alpha (α)	4–8	0.025–0.080	Unnecessary
Beta (β)	0.01–3	0.150–16	Plastic (e.g. Perspex)
Gamma (γ)	0.03–3	1.3–13†	Lead

*Note that $1\,MeV = 1.6 \times 10^{-13}\,J$
† Distance at which radiation intensity is reduced to half

Table 56.2 Properties of some isotopes used commonly in biology. Physical data obtained from Lide (2006).

Isotope	Emission(s)	Maximum energy (MeV)	Half-life	Main uses	Advantages	Disadvantages
^{3}H	β^-	0.01861	12.3 years	Suitable for labelling organic molecules in wide range of positions at high specific activity	Relatively safe	Low efficiency of detection, high isotope effect, high rate of exchange with environment
^{14}C	β^-	0.15648	5715 years	Suitable for labelling organic molecules in a wide range of positions	Relatively safe, low rate of exchange with environment	Low specific activity
^{22}Na	β^+ (90%) + γ, EC	2.842 (β^+)	2.60 years	Transport studies	High specific activity	Short half-life, hazardous
^{32}P	β^-	1.710	14.3 days	Labelling proteins and nucleotides (e.g. DNA)	High specific activity, ease of detection	Short half-life, hazardous
^{35}S	β^-	0.167	87.2 days	Labelling proteins and nucleotides	Low isotope effect	Low specific activity
^{36}Cl	β^-, β^+, EC	0.709 (β^-) 1.142 (β^+, EC)	300000 years	Transport studies	Low isotope effect	Low specific activity, hazardous
^{125}I	EC + γ	0.178 (EC)	59.9 days	Labelling proteins and nucleotides	High specific activity	Hazardous
^{131}I	$\beta^- + \gamma$	0.971 (β^-)	8.04 days	Labelling proteins and nucleotides	High specific activity	Hazardous

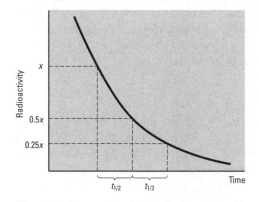

Fig. 56.1 Decay of a radioactive isotope with time. The time taken for the radioactivity to decline from x to $0.5x$ is the same as the time taken for the radioactivity to decline from $0.5x$ to $0.25x$, and so on. This time is the half-life ($t_{1/2}$) of the isotope.

Example For ^{35}S, with a half-life of 87.2 days (Table 56.2), the fraction remaining after 28 days would be worked out as follows:

$x = (-0.693 \times 28) \div 87.2 = -0.222522936$,

then using eqn [56.1],

$f = e^{-0.222522936} = 0.800496646$

(approximately 80.0% of original activity).

Each radioactive particle or ray carries energy, usually measured in electron volts (eV). The particles or rays emitted by a particular radioisotope exhibit a range of energies, termed an energy spectrum, characterised by the maximum energy of the radiation produced, E_{max} (Table 56.2). The energy spectrum of a particular radioisotope is relevant to the following:

- Safety: isotopes with the highest maximum energies will have the greatest penetrating power, requiring appropriate shielding (Table 56.1).
- Detection: different instruments vary in their ability to detect isotopes with different energies.
- Discrimination: some instruments can distinguish between isotopes, based on the energy spectrum of the radiation produced (p. 349).

The decay of an individual atom (a 'disintegration') occurs at random, but that of a population of atoms occurs in a predictable manner. The radioactivity decays exponentially, having a characteristic half-life ($t_{1/2}$). This is the time taken for the radioactivity to fall from a given value to half that value (Fig. 56.1). The $t_{1/2}$ values of different radioisotopes range from fractions of a second to more than 10^{19} years (see also Table 56.2). If the $t_{1/2}$ is very short, as with ^{15}O ($t_{1/2} \approx 2$ min), then it is generally impractical to use the isotope in experiments because you would need to account for the decay during the experiment and counting period.

To calculate the fraction (f) of the original radioactivity left after a particular time (t), use the following relationship:

$$f = e^x, \quad \text{where} \quad x = -0.693 t / t_{\frac{1}{2}} \qquad [56.1]$$

Note that the same units must be used for t and $t_{1/2}$ in the above equation.

Measuring radioactivity

The SI unit of radioactivity is the becquerel (Bq), equivalent to one disintegration per second (d.p.s.), but disintegrations per minute (d.p.m.) are also used. The curie (Ci) is a non-SI unit equivalent to the number of disintegrations produced by 1 g of radium (37 GBq). Table 56.3 shows the relationships between these units. In practice, most instruments are not able to detect all of the disintegrations from a particular sample, i.e. their efficiency is less than 100 per cent and the rate of decay may be presented as counts min^{-1} (c.p.m.) or counts s^{-1} (c.p.s.). Most modern instruments correct for background radiation and inefficiencies in counting, converting count data to d.p.m. Alternatively, the results may be presented as the measured count rate, although this is only valid where the efficiency of counting does not vary greatly among samples.

 KEY POINT *The specific activity is a measure of the quantity of radioactivity present in a known amount of the substance.*

$$\text{specific activity} = \frac{\text{radioactivity (Bq, Ci, d.p.m., etc.)}}{\text{amount of substance (mol, g, etc.)}} \quad [56.2]$$

This is an important concept in practical work involving radioisotopes, since it allows interconversion of disintegrations (activity) and amount of substance (see Box 56.1).

Two SI units refer to doses of radioactivity and these are used when calculating exposure levels for a particular source. The sievert (Sv) is the amount of radioactivity giving a dose in man equivalent to 1 gray (Gy) of X-rays: 1 Gy = an energy absorption of 1 J kg^{-1}. The dose received in most biological experiments is a negligible fraction of the maximum permitted exposure limit. Conversion factors from older units are given in Table 56.3.

The most important methods of measuring radioactivity for biological purposes are described below.

The Geiger–Müller (G–M) tube

This operates by detecting radiation when it ionises gas between a pair of electrodes across which a voltage has been applied. You should use a hand-held Geiger–Müller tube for routine checking for contamination (although it will not pick up ^{3}H activity).

The scintillation counter

This operates by detecting the scintillations (fluorescence 'flashes') produced when radiation interacts with certain chemicals called fluors (Fig. 56.2). In solid (or external) scintillation counters (often referred to as 'gamma counters') the radioactivity causes scintillations in a crystal of fluorescent material held close to the sample. This method is only suitable for radioisotopes producing penetrating radiation.

Liquid scintillation counters are mainly used for detecting beta decay and they are especially useful in biology. The sample is dispersed or dissolved in a suitable solvent containing the fluor(s) – the 'scintillation cocktail'. The radiation first interacts with the solvent, and the energy from this interaction is passed to the fluors, which produce detectable light. The scintillations are measured by photomultiplier tubes (Fig. 56.2) that turn the light pulses into electronic pulses, the magnitude of which is directly related to the energy of the original radioactive event. The spectrum of electronic pulses is thus related to the energy spectrum of the radioisotope.

Table 56.3 Relationships between units of radioactivity. For abbreviations, see text.

1 Bq = 1 d.p.s.	1 Sv = 100 rem
1 Bq = 60 d.p.m.	1 Gy = 100 rad
1 Bq = 27 pCi	1 Gy ≈ 100 roentgen
1 d.p.s. = 1 Bq	1 rem = 0.01 Sv
1 d.p.m. = 0.016 7 Bq	1 rad = 0.01 Gy
1 Ci = 37 GBq	1 roentgen ≈ 0.01 Gy
1 mCi = 37 MBq	
1 µCi = 37 kBq	

Example If 0.4 ml of a ^{32}P-labelled DNA solution at a concentration of 50 µmol l^{-1} (amount = 0.4 × 50 ÷ 1000 = 0.02 µmol) gave a count of 2490 d.p.m. (= 41.5 Bq), using eqn [56.2] this would correspond to a specific activity of 2490 ÷ 0.02 = 124 500 d.p.m. µmol^{-1} (or 2075 Bq µmol^{-1}).

photomultiplier tubes (2)

lead shielding

pulse analysis and readout

1 4 5

sample vial

sample, dispersed or dissolved in 'scintillation cocktail'

Fig. 56.2 Components of a scintillation counter. Note that in most modern instruments, all components are enclosed within a single cabinet.

Box 56.1 How to determine the specific activity of an experimental solution

Suppose you need to make up a certain volume of an experimental solution, to contain a particular amount of radioactivity. For example, 50 ml of a mannitol solution at a concentration of 25 mmol l^{-1}, to contain 5 Bq μl^{-1} – using a manufacturer's stock solution of ^{14}C-labelled mannitol (specific activity = 0.1 Ci/mmol^{-1}).

1. **Calculate the total amount of radioactivity in the experimental solution**, in this example 5 × 1 000 (to convert μl to ml) × 50 (50 ml required) = 2.5 × 10^5 Bq (i.e. 250 kBq).

2. **Establish the volume of stock radioisotope solution required**: for example, a manufacturer's stock solution of ^{14}C-labelled mannitol contains 50 μCi of radioisotope in 1 ml of 90% v/v ethanol: water. Using Table 56.3, this is equivalent to an activity of 50 × 37 = 1 850 kBq. So, the volume of solution required is 250/1850 of the stock volume, i.e. 0.135 1 ml (135 μl).

3. **Calculate the amount of non-radioactive substance required** as for any calculation involving concentration (see pp. 131, 139), e.g. 50 ml (0.05 l) of a 25 mmol l^{-1} (0.025 mol l^{-1}) mannitol (relative molecular mass 182.17) will contain 0.05 × 0.025 × 182.17 = 0.227 7 g.

4. **Check the amount of radioactive isotope to be added**. In most cases, this represents a negligible amount of substance, e.g. in this instance, 250 kBq of stock solution at a specific activity of 14.8 × 10^6 kBq/mmol^{-1} (converted from 0.4 Ci mmol^{-1} using Table 56.3) is equal to 250/14 800 000 = 16.89 nmol, equivalent to approximately 3 μg mannitol. This can be ignored in calculating the mannitol concentration of the experimental solution.

5. **Make up the experimental solution** by adding the appropriate amount of non-radioactive substance and the correct volume of stock solution.

6. **Measure the radioactivity in a known volume of the experimental solution**. If you are using an instrument with automatic correction to Bq, your sample should contain the predicted amount of radioactivity, e.g. an accurately dispensed volume of 100 μl of the mannitol solution should give a corrected count of 100 × 5 = 500 Bq (or 500 × 60 = 30 000 d.p.m.).

7. **Note the specific activity of the experimental solution**: in this case, 100 μl (1 × 10^{-4} l) of the mannitol solution at a concentration of 0.025 mol l^{-1} will contain 25 × 10^{-7} mol (2.5 μmol) mannitol. Dividing the radioactivity in this volume (30 000 d.p.m.) by the amount of substance (eqn [36.2]) gives a specific activity of 30 000/2.5 = 12 000 d.p.m. μmol^{-1}, or 12 d.p.m. nmol^{-1}. This value can be used:

   (a) To assess the accuracy of your protocol for preparing the experimental solution: if the measured activity is substantially different from the predicted value, you may have made an error in making up the solution.

   (b) To determine the counting efficiency of an instrument: by comparing the measured count rate with the value predicted by your calculations.

   (c) To interconvert activity and amount of substance: the most important practical application of specific activity is the conversion of experimental data from counts (activity) into amounts of substance. This is only possible where the substance has not been metabolised or otherwise converted into another form; e.g., a tissue sample incubated in the experimental solution described above with a measured activity of 245 d.p.m. can be converted to nmol mannitol by dividing by the specific activity, expressed in the correct form. Thus 245/12 = 20.417 nmol mannitol.

Correcting for quenching – find out how your instrument corrects for quenching and check the quench indication parameter (QIP) on the printout, which measures the extent of quenching of each sample. Large differences in the QIP would indicate that quenching is variable among samples and might give you cause for concern.

Modern liquid scintillation counters use a series of electronic 'windows' to split the pulse spectrum into two or three components. This may allow more than one isotope to be detected in a single sample, provided their energy spectra are sufficiently different (Fig. 56.3). A complication of this approach is that the energy spectrum can be altered by pigments and chemicals in the sample, which absorb scintillations or interfere with the transfer of energy to the fluor; this is known as quenching (Fig. 56.3). Most instruments have computer-operated quench-correction facilities (based on measurements of standards of known activity and energy spectrum) that correct for such changes in counting efficiency.

Box 56.2 Tips for preparing samples for liquid scintillation counting

Modern scintillation counters are very simple to operate; problems are more likely to be due to inadequate sample preparation than to incorrect operation of the machine. Common pitfalls are the following:

- **Incomplete dispersal of the radioactive compound in the scintillation cocktail.** This may lead to under-estimation of the true amount of radioactivity present:

  (a) Water-based samples may not mix with the scintillation cocktail – change to an emulsifier-based cocktail. Take care to observe the recommended limits, upper and lower, for amounts of water to be added or the cocktail may not emulsify properly.

  (b) Solid specimens may absorb disintegrations or scintillations: extract radiochemicals using an intermediate solvent like ethanol (ideally within the scintillation vial) and then add the cocktail. Tissue-solubilising compounds such as Soluene are effective, particularly for animal material, but extremely toxic, so the manufacturer's instructions must be followed closely. Radioactive compounds on slices of agarose or polyacrylamide gels may be extracted using a product such as Protosol. Agarose gels can be dissolved in a small volume of boiling water.

  (c) Particulate samples may sediment to the bottom of the scintillation vial – suspend them by forming a gel. This can be done with certain emulsifier-based cocktails by adding a specific amount of water.

- **Chemiluminescence.** This is where a chemical reacts with the fluors in the scintillation cocktail causing spurious scintillations, a particular risk with solutions containing strong bases or oxidising agents. Symptoms include very high initial counts that decrease through time. Possible remedies are:

  (a) Leave the vials at room temperature for a time before counting. Check with a suitable blank that counts have dropped to an acceptable level.

  (b) Neutralise basic samples with acid (e.g. acetic acid or HCl).

  (c) Use a scintillation cocktail that resists chemiluminescence such as Hiconicfluor.

  (d) Raise the energy of the lower counts detected to about 8 keV – most chemiluminescence pulses are weak (0–7 keV). This approach is not suitable for weak emitters, e.g. ^{3}H.

Liquid scintillation counting of high-energy β-emitters – β-particles with energies greater than 1 MeV can be counted in water (Čerencov radiation), with no requirement for additional fluors (e.g. ^{32}P).

Many liquid scintillation counters treat the first sample as a 'background', subtracting whatever value is obtained from the subsequent measurements as part of the procedure for converting to d.p.m. If not, you will need to subtract the background count from all other samples. Make sure that you use an appropriate background sample, identical in all respects to your radioactive sample but with no added radioisotope, in the correct position within the machine. Check that the background reading is reasonable (15–30 c.p.m.). Tips for preparing samples for liquid scintillation counting are given in Box 56.2.

Gamma-ray (γ-ray) spectrometry

This is a method by which a mixture of γ-ray-emitting radionuclides can be resolved quantitatively by pulse-height analysis. It is based on the fact that pulse heights (voltages) produced by a photomultiplier tube are proportional to the amounts of γ-ray energy arriving at the scintillant or a lithium-drifted germanium detector. The lithium-drifted germanium detector, which is abbreviated to Ge(Li) – pronounced 'jelly' – provides high resolution (narrow peaks), essential in the analysis of complex mixtures.

Autoradiography

This is a method where photographic film is exposed to the isotope. It is used mainly to locate radioactive tracers in thin sections of an organism or

Fig. 56.3 Energy spectra for three radioactive samples, detected using a scintillation counter. Sample *a* is a high-energy β-emitter while *b* contains a low-energy β-emitter, giving a lower spectral range. Sample *b'* contains the same amount of low-energy β-emitter, but with quenching, shifting the spectral distribution to a lower energy band. The counter can be set up to record disintegrations within a selected range (a 'window'). Here, 'window a' could be used to count isotope *a* while 'window b' could give a value of isotope *b*, by applying a correction for the counts due to isotope *a*, based on the results from 'window a'. Dual counting allows experiments to be carried out using two isotopes (double labelling).

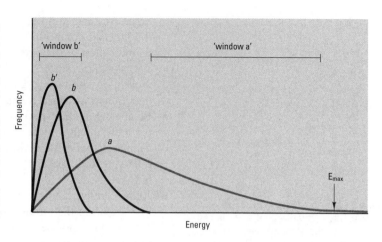

on chromatography papers and gels, but quantitative work is possible. The radiation interacts with the film in a similar way to light, silver grains being formed in the developed film where the particles or rays have passed through. The radiation must have enough energy to penetrate into the film, but if it has too much energy the grain formation may be too distant from the point where the isotope was located to identify precisely the point of origin (e.g. high-energy β-emitters). Autoradiography is a relatively specialised method and individual lab protocols should be followed for particular isotopes/applications.

Biological applications for radioactive isotopes

The main advantages of using radioactive isotopes in biological experiments are:

- Radioactivity is readily detected. Methods of detection are sufficiently sensitive to measure extremely small amounts of radioactive substances.
- Studies can be carried out on intact, living organisms. If care is taken, minimal disruption of normal conditions will occur when radiolabelled compounds are introduced.
- Protocols are simple compared with equivalent methods for chemical analysis.

The main disadvantages are:

- The 'isotope effect'. Molecules containing different isotopes of the same atom may react at slightly different rates and behave in slightly different ways from the natural isotope. The isotope effect is more extreme the smaller the atom and is most important for ^{3}H-labelled compounds of low molecular mass.
- The possibility of mistaken identity. The presence of radioactivity does not tell you anything about the compound in which the radioactivity is present: it could be different from the one in which it was applied, due to metabolism or spontaneous breakdown of a ^{14}C-containing organic compound.

The main types of experiments are:

- Investigations of metabolic pathways: a radioactively labelled substrate is added (often to an *in vitro* experimental 'system' rather than a whole

Metabolism of radiolabelled compounds – *you may need to separate individual metabolites before counting, e.g. using chromatography (Chapter 60), or electrophoresis (p. 334).*

organism) and samples taken at different time intervals. By identifying the labelled compounds and plotting their appearance through time, an indication of the pathway of metabolism can be obtained.

- Translocation studies: radioisotopes are used to follow the fate of molecules within an organism. Uptake and translocation rates can be determined with relative ease.
- Ecological studies: radioisotopic tracers provide a convenient method for determining food web interrelationships and for investigating behaviour patterns, while environmental monitoring may involve following the 'spectral signature' of isotopes deliberately or accidentally released.
- Radio-dating: the age of plant or mineral samples can be determined by measuring the amount of a radioisotope in the sample. The age of the specimen can be found using the $t_{1/2}$ by assuming how much was originally incorporated.
- Mutagenesis and sterilisation: radioactive sources can be used to induce mutations, particularly in micro-organisms. Gamma emitters of high energy will kill microbes and are used to sterilise equipment such as disposable Petri dishes.
- Assays: radioisotopes are used in several quantitative detection methods of value to biologists. Radioimmunoassay is described on p. 312. Isotope dilution analysis works on the assumption that introduced radiolabelled molecules will equilibrate with unlabelled molecules present in the specimen. The amount of substance initially present can be worked out from the change in specific activity of the radioisotope when it is diluted by the 'cold' material. A method is required whereby the substance can be purified from the sample and sufficient substance must be present for its mass to be measured accurately.

Billington et al. (1992) give further details and practical advice on using radioisotopes in biological experiments.

Working practices when using radioactive isotopes

By law, undergraduate work with radioactive isotopes must be very closely supervised. In practical classes, the protocols will be clearly outlined, but in project work you may have the opportunity to plan and carry out your own experiments, albeit under supervision. Some of the factors that you should take into account, based on the assumption that your department and laboratory are registered for radioisotope use, are discussed below:

1. Must you use radioactivity? If not, it may be a legal requirement that you use an alternative method.
2. Have you registered for radioactive work? Normal practice is for all users to register with a local Radiation Protection Supervisor. Details of the project may have to be approved by the appropriate administrator(s). You may have to have a short medical examination before you can start work.
3. What labelled compound will you use? Radioactive isotopes must be ordered well in advance through your department's Radiation Protection Supervisor. Aspects that need to be considered include:
   (a) The radionuclide. With many organic compounds this will be confined to ^{3}H and ^{14}C (but see Table 56.2). The risk of a significant 'isotope effect' may influence this decision (see p. 351).

Example Carbon dating – living organisms have essentially the same ratio of ^{14}C to ^{12}C as the atmosphere; however, when an organism dies, its ^{14}C/^{12}C falls because the radioactive ^{14}C isotope decays. Since we know the half-life of ^{14}C (5 715 years), a sample's ^{14}C/^{12}C ratio will allow us to estimate its age; e.g. if the ratio were exactly 1/8 of that in the atmosphere, the sample is three half-lives old and was formed 17 145 years before present. Such estimates carry an error of the order of 10 per cent and are unreliable for samples older than 50 000 years, for which longer-lived isotopes can be used.

Registering for radioisotope work – in the UK, institutions must be registered for work with specific radioisotopes under the Radioactive Substances Act (1993).

Supervision of work with radioisotopes – in the UK, the Ionizing Radiations Act (1985) provides details of local arrangements for the supervision of radioisotope work.

(b) The labelling position. This may be a crucial part of a metabolic study. Specifically labelled compounds are normally more expensive than those that are uniformly ('generally') labelled.

(c) The specific activity. The upper limit for this is defined by the isotope's half-life but, below this, the higher the specific activity, the more expensive the compound.

4. Are suitable facilities available? You'll need a suitable work area, preferably out of the way of general lab traffic and within a fume cupboard for those cases where volatile radioactive substances are used or may be produced.

In conjunction with your supervisor, decide whether your method of application will introduce enough radioactivity into the system, how you will account for any loss of radioactivity during recovery of the isotope and whether there will be enough activity to count at the end. You should be able to predict approximately the amount of radioactivity in your samples, based on the specific activity of the isotope used, the expected rate of uptake/exchange and the amount of sample to be counted. Use the isotope's specific activity to estimate whether the non-radioactive ('cold') compound introduced with the radiolabelled ('hot') compound may lead to excessive concentrations being administered. Advice for handling data is given in Box 56.1.

 Safety and procedural aspects

Make sure the bench surface is one that can be easily decontaminated by washing (e.g. Formica) and always use a disposable surfacing material such as Benchkote. It is good practice to carry out as many operations as possible within a Benchkote-lined plastic tray so that any spillages are contained. You will need a lab coat to be used exclusively for work with radioactivity, safety spectacles and a supply of thin latex or vinyl disposable gloves. Suitable vessels for liquid waste disposal will be required and special plastic bags for solids – make sure you know beforehand the disposal procedures for liquid and solid wastes. Wash your hands after handling a vessel containing a radioactive solution and again before removing your gloves. Gloves should be placed in the appropriate disposal bag as soon as your experimental procedures are complete.

It is important to comply with the following guidelines:

- Read and obey the local rules for safe usage of radiochemicals.
- Maximise the distance between you and the source as much as possible.
- Minimise the duration of exposure.
- Wear protective clothing (properly fastened lab coat, safety glasses, gloves) at all times.
- Use appropriate shielding at all times (Table 56.1).
- Monitor your working area for contamination frequently.
- Mark all glassware, trays, bench work areas, etc., with tape incorporating the international symbol for radioactivity (Fig. 56.4).
- Keep adequate records of what you have done with a radioisotope – the stock remaining and that disposed of in waste form must agree.
- Store radiolabelled compounds appropriately and return them to storage areas immediately after use.

 SAFETY NOTE *Each new experiment should be planned carefully and experimental protocols laid down in advance so you work as safely as possible and do not waste expensive radioactively labelled compounds.*

Carrying out a 'dry run' – *consider doing this before working with radioactive compounds, perhaps using a dye to show the movement or dilution of introduced liquids, as this will lessen the risks of accident and improve your technique.*

 SAFETY NOTE *The correct way to use Benchkote and similar products is with the waxed surface down (to protect the bench or tray surface) and the absorbent surface up (to absorb any spillage). Write the date in the corner when you put down a new piece. Monitor using a G–M tube and replace regularly under normal circumstances. If you are aware of spillage, replace immediately and dispose of correctly.*

CAUTION RADIOACTIVE MATERIAL

Fig. 56.4 Tape showing the international symbol for radioactivity

- Dispose of waste promptly and with due regard for local rules.
- Make the necessary reports about waste disposal, etc., to your departmental Radiation Protection Supervisor.
- Clear up after you have finished each experiment.
- Wash thoroughly after using radioactivity.
- Monitor the work area and your body when finished.

Text references and sources for further study

Billington, D., Jayson, G.G. and Maltby, P.J. (1992) *Radioisotopes*. Bios Scientific Publishers, Oxford.

L'Annunziata, M.F. (ed.) (2003) *Handbook of Radioactivity Analysis*, 2nd edn. Academic Press, San Diego.

Lide, D.R. (ed.) (2006) *CRC Handbook of Chemistry and Physics*, 87th edn. CRC Press, Boca Raton, Florida.

Slater, J. (ed). (2002) *Radioisotopes in Biology: A Practical Approach,* 2nd edn. Oxford University Press, Oxford.

Study exercises

56.1 Carry out a half-life calculation. A rat dropping found in a pyramid in Egypt had a $^{14}C:^{12}C$ ratio that was 57.25 per cent of a modern-day standard. Use this value to estimate the approximate date when the rat visited the pyramid, to the nearest century.

56.2 Practise radioactivity interconversions. Express the following values in the alternative units indicated, with appropriate prefixes as necessary. Answer to three significant figures.
(a) 72 000 d.p.m. as Bq;
(b) 20 μCi as d.p.m.;
(c) 44 400 Bq as μCi;
(d) 6.3×10^5 d.p.m. mol^{-1} as Bq g^{-1}, for a compound with a relative molecular mass of 350;
(e) 3108 d.p.m. as pmol, for a sample of a standard where the specific activity is stated as 50 Ci mol^{-1}.

56.3 Use the concept of specific activity in calculations. A researcher wishes to estimate the rate of uptake of the sugar galactose by carrot cells in a suspension culture. She prepares 250 ml of the cell culture medium containing 10^7 cells per ml and unlabelled galactose at a concentration of 5 mmol l^{-1}. She then 'spikes' this with 5 μl (regard this as an insignificant volume) of radioactive standard containing 55 MBq of ^{14}C-labelled galactose (regard as an insignificant concentration). Answer to two significant figures.
(a) Calculate the specific activity of the galactose in the culture solution in Bq mol^{-1}.
(b) If the total cell sample takes up 79.2×10^5 Bq in a 2-hour period, calculate the galactose uptake rate in mol s^{-1} cell^{-1}.

Oxygen electrodes measure oxygen in solution using the amperometric principle, i.e. by monitoring the current flowing between two electrodes when a voltage is applied. The most widespread electrode is the Clark type (Fig. 57.1), manufactured in the UK by Rank Bros, Cambridge, which is suitable for measuring O_2 concentrations in cell, organelle and enzyme suspensions. Platinum and silver electrodes are in contact with a solution of electrolyte (normally saturated KCl). The electrodes are separated from the medium by a Teflon membrane, permeable to O_2. When an electrical potential is applied to the electrodes, this generates a current proportional to the O_2 concentration (Wilson and Walker, 2005). The reactions can be summed up as:

$$4Ag \rightarrow 4Ag^+ + 4e^- \qquad \text{(at silver anode)}$$

$$O_2 + 2e^- + 2H^+ \rightleftharpoons H_2O_2 \qquad \text{(in electrolyte solution; } O_2 \text{ replenished by diffusion from test solution)}$$

$$H_2O_2 + 2e^- + 2H^+ \rightleftharpoons 2H_2O \quad \text{(at platinum cathode)}$$

Setting up and using a Clark (Rank) oxygen electrode

Box 57.1 describes the steps involved in setting up this type of electrode, and Box 57.2 explains how to analyse the results you obtain. If you are

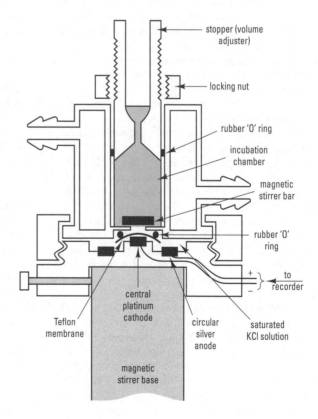

Fig. 57.1 Transverse section through a Clark (Rank) oxygen electrode

setting up from scratch, perform steps 1–13 in Box 57.1. If a satisfactory membrane is already in place, start at step 7.

The temperature of the incubation vessel can be controlled by passing water (e.g. from a water bath) through the outer chamber. Cells or organelles may be present in the solution added to the incubation chamber or can be added via the hole in the stopper using a syringe, as can any chemicals you wish to introduce, such as metabolic substrates or inhibitors. Take care not to introduce air bubbles and remove any that appear by gently raising and lowering the stopper. The electrode can be used repeatedly, providing the membrane is satisfactory: remove solutions

---

Box 57.1 How to set up a Clark (Rank) oxygen electrode

1. **Detach the base of the incubation vessel** (see Fig. 57.1) by unscrewing the locking ring.
2. **Add enough saturated KCl to cover the electrodes.**
3. **Cut a 1 mm square hole in the centre of a 10 × 10 mm square of lens tissue** and place on to the KCl solution with hole over the central cathode.
4. **Cut a 10 × 10 mm square of Teflon membrane and place over the lens tissue;** seal by gently lowering the incubation vessel and tightening the locking ring, making sure that the rubber O-ring is correctly positioned over the membrane.
   (a) Do not overtighten the locking ring.
   (b) Take care not to trap air bubbles beneath the membrane.
   (c) Make sure that the membrane does not become twisted.
5. **Clamp the electrode over the magnetic stirrer base using the clamping screw.**
6. **Connect the electrode leads to the polarising unit/recording device** (silver anode to positive, platinum cathode to negative). You can use either a digital readout system on the control unit or the output can be passed to a chart recorder, giving a readout of oxygen status as a function of time (see Box 57.2). Check that the polarising voltage is set to 0.60 V and adjust, if necessary, using the 'polarising voltage' control (typically, this requires a small screwdriver to adjust).
7. **Add air-saturated experimental solution and a small Teflon-coated magnetic stirrer bar to the chamber.** The volume of the incubation chamber can be adjusted by moving the locking nut on the stopper. To adjust, add the appropriate amount of liquid to the chamber using a pipette, insert the stopper and screw the locking nut until the solution just fills the incubation chamber.

8. **Gently push the stopper (volume adjuster) into position**, making sure that no air bubbles are trapped in the chamber, and switch on the stirrer. Adjust the rate of stirring using the 'stirrer' control, if required.
9. **Set the zero** – remove the stopper and add a few crystals of sodium dithionite – the dithionite ions ($S_2O_4^{2-}$) will be oxidised to sulphite (SO_3^{2-}) and sulphate ions (SO_4^{2-}), thereby consuming all of the O_2 in the solution within 3–5 min (an alternative is to bubble N_2 through the solution for > 10 min). Once the reading has stabilised, if this is not at zero, then adjust using the 'set zero' control (in practice, the zero is usually quite stable and often needs no adjustment).
10. **Adjust the sensitivity** – rinse out the dithionite solution thoroughly/repeatedly, then replace with air-equilibrated water and allow the reading to stabilise (this may take 5–10 min.). Then adjust the 'sensitivity' control of the electrode system to set the oxygen-saturated value at an appropriate point (for a controller with a digital readout, this is best set to 100, and then the values represent percentage oxgen saturation). To check that the instrument is working correctly, switch off the stirrer for a few moments – the reading should drop quickly as oxygen is consumed at the electrode surface and should then rise rapidly when the stirrer is switched back on.
11. **Rinse the incubation chamber thoroughly and add fresh experimental solution.**
12. **Carry out your experiment.**
13. **Remove the solution and check the calibration.** If the reading for air-equilibrated water is different, the electrode's sensitivity or the temperature may have changed and you may need to recalibrate and repeat the measurement.

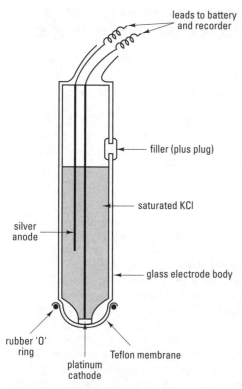

leads to battery and recorder

filler (plus plug)

saturated KCl

silver anode

glass electrode body

rubber 'O' ring

platinum cathode

Teflon membrane

Fig. 57.2 A Clark-type oxygen probe

carefully (e.g. using a pipette, or vacuum line and trap). Keep water in the chamber when not in use. Replace the membrane if:

- the reading becomes noisy;
- the electrode will not zero after adding sodium dithionite;
- the response becomes too slow (check by switching off the stirrer – the reading should drop rapidly as the available O_2 is consumed).

 KEY POINT *Successful operation of a Clark (Rank) oxygen electrode requires an intact Teflon membrane and clean electrodes. When replacing the Teflon membrane, you should always check that the electrode surfaces are clean (shiny): if necessary, use a mild abrasive paste to remove any oxidised material.*

Oxygen probes

Clark-type oxygen electrodes are also available in probe form for immersion in the test solution (Fig. 57.2). Such probes are particularly useful for field studies since they allow direct measurement of oxygen status *in situ*, in contrast to chemical assays (e.g. the Winkler method). The main point to note is that the solution must be stirred during measurement, to replenish the oxygen consumed by the electrode. Golterman et al. (1978) discuss field sampling protocols for measuring dissolved gases in fresh waters.

Table 57.1 Oxygen-saturation values for distilled water and sea water at standard atmospheric pressure and a range of temperatures (derived from Green and Carritt, 1967).

Temperature °C	Distilled water			Sea water		
	$mol\,m^{-3}$ (mmol l^{-1})	$mg\,l^{-1}$ (p.p.m.)	$ml\,l^{-1}$	$mol\,m^{-3}$ (mmol l^{-1})	$mg\,l^{-1}$ (p.p.m.)	$ml\,l^{-1}$
0	0.460	14.7	10.3	0.359	11.5	8.04
2	0.435	13.9	9.75	0.342	10.9	7.65
4	0.413	13.2	9.24	0.326	10.4	7.30
6	0.392	12.5	8.78	0.311	9.95	6.97
8	0.373	11.9	8.35	0.298	9.54	6.66
10	0.355	11.4	7.95	0.285	9.12	6.37
12	0.339	10.8	7.59	0.273	8.74	6.11
14	0.324	10.4	7.25	0.261	8.35	5.85
16	0.310	9.92	6.94	0.251	8.03	5.62
18	0.297	9.50	6.65	0.241	7.71	5.40
20	0.285	9.12	6.38	0.232	7.42	5.19
22	0.274	8.77	6.12	0.224	7.17	5.00
24	0.263	8.42	5.89	0.215	6.88	4.82
26	0.253	8.1	5.67	0.208	6.66	4.64
28	0.244	7.81	5.46	0.200	6.40	4.48
30	0.235	7.52	5.27	0.193	6.18	4.32
37	0.211	6.75	4.71	0.174	5.57	3.88

Note: tabulated values assume atmospheric pressure = 101.3 kPa ($\approx$ 760 mm Hg); for more accurate work, a correction for any deviation can be made by multiplying the appropriate figures from the table by the ratio of the real pressure to the assumed pressure.

Box 57.2 How to convert a chart recorder trace to a rate of O_2 consumption or production

Having set up your oxygen electrode (Box 57.1) or probe, the most common experiments are those in which you measure the rate of oxygen consumption (respiration) or production (photosynthesis) in a fixed volume of solution, e.g. the chamber of a Rank (Clark) oxygen electrode. Typically, this involves attaching the electrode to a chart recorder and following the change in oxygen concentration with time. The principal steps are as follows:

1. **Calculate the amount of oxygen in the electrode chamber** – multiply the appropriate oxygen-saturation concentration, from Table 57.1, by the volume of the chamber. For example, if you are working in mmol l^{-1} ($= \mu mol\ ml^{-1}$) then at 20°C, air-equilibrated (oxygen-saturated) distilled water contains 0.285 μmol ml^{-1}. For an electrode chamber of volume 5 ml, there will be $0.285 \times 5 = 1.425$ μmol oxygen in the water within the chamber at saturation.

2. **Set the zero on the recorder** – following the addition of a small amount of sodium dithionite to the chamber, to remove oxygen (Box 57.1), the trace is monitored until it has stabilised. Then adjust the 'zero' or 'back off' control of the recorder until the chart trace is set to zero on the chart paper: this is marked as 'A' on Fig 57.3.

3. **Set the oxygen-saturated reading on the recorder** – rinse out the dithionite solution thoroughly/ repeatedly, replace with air-equilibrated water and allow the reading to stabilise (this may take 3–5 min). Then adjust the 'sensitivity' control of the chart recorder to set the oxygen-saturated value at an appropriate point on the chart – for respiration measurements, this can be close to the full width of the chart paper, whereas for photosynthesis measurements it is more usual to set it closer to the mid-point of the chart paper, to allow measurements over 100 per cent to be recorded. In the example shown in Fig 57.3, where the chart paper is calibrated in mm divisions, the oxygen saturated reading is set at 80 divisions (80 mm), marked as 'B'. Note that the recorder pen can be lifted between readings, and during rinsing of the chamber, to reduce the amount of 'noise' on the chart.

4. **Calculate the amount of oxygen equivalent to a single division of the chart paper** – divide the amount of oxygen in the chamber by the number of divisions (mm) to give the amount per division

(e.g. $1.425 \div 80 = 0.0178125$ μmol ($= 17.8125$ nmol per division).

5. **Carry out your experiment** – the example shown in Fig. 57.3 represents the consumption of oxygen during respiration of a suspension of yeast cells, marked as 'C'. Continue the recording until you are satisfied that the rate has remained stable for at least 5 min. In many experiments, you will be adding substances that change the rate (e.g. metabolic inhibitors, or substrates – this can be done by injecting a solution containing the substance through the small hole in the stopper of the electrode (keep the volume of this solution to < 1 per cent of the total volume of the chamber, to minimise the effect of the added solution).

6. **Convert the gradient of the trace into a rate of oxygen consumed or produced** – using a transparent ruler, draw the line of best fit through that part of the trace showing a stable relationship (see Fig. 57.3). Use this line of best fit to work out the gradient of the trace, by converting from linear dimensions to an amount per unit time: e.g. the gradient shown in Fig. 57.3 is equivalent to a decrease of 24 chart divisions (mm) in 3 cm. Using the value calculated in step 4, the amount of oxygen consumed is thus $17.8125 \times 24 = 427.5$ nmol. For a chart speed of 0.2 cm min^{-1}, this represents the change occurring over a time interval of $3 \div 0.2 = 15$ min. Thus the rate of oxygen consumed is $427.5 \div 15 = 28.5$ nmol min^{-1}.

7. **Express the rate on an appropriate basis** – in most instances, this will be either per cell (e.g. determined using a haemocytometer, p. 275) or per unit mass of protein (p. 319), chlorophyll (p. 369), or an equivalent marker. For the example shown in Fig. 57.3, if there were 2.4×10^7 yeast cells per ml of suspension, then in the 5 ml chamber of the electrode there would be a total of $2.4 \times 10^7 \times 5 = 1.2 \times 10^8$ yeast cells, giving a rate of $28.5 \div (1.2 \times 10^8) = 2.375 \times 10^{-7}$ nmol min^{-1} $cell^{-1}$, probably better expressed as 237.5 amol min^{-1} $cell^{-1}$. Note that when inhibitors/substrates are used, you should always wait until a new stable rate has been achieved (for 5 min) before using the trace to determine the new rate of oxygen consumption/ production.

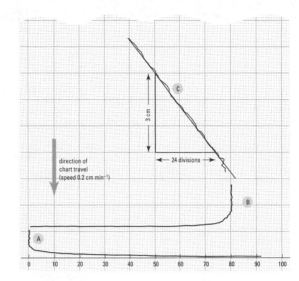

Fig. 57.3 Chart recorder output for a Clark (Rank) oxygen electrode. A = zero oxygen; B = oxygen saturation; C = respiration of yeast.

Text references and sources for further study

Golterman, H.L., Clymo, R.S. and Ohnstad, M.A.M. (1978) *Methods for Physical and Chemical Analysis of Fresh Waters*. Blackwell Scientific, Oxford.

Green, E.J. and Carritt, D.E. (1967) 'New Tables for Oxygen Saturation of Seawater', *Journal of Marine Research*, **25**, pp. 140–7.

Wilson, K. and Walker, J. (eds.) (2005) *Principles and Techniques of Biochemistry and Molecular Biology*, 6th edn. Cambridge University Press, Cambridge.

Study exercises

57.1 Calculate the oxygen content of specified volumes of water at defined temperatures. Using the information in Table 57.1 and assuming air equilibration at 101.3 kPa, work out the amount of oxygen in each of the following (give all answers to three significant figures):

(a) 4 ml of distilled water at 20°C (express your answer in μmol);

(b) 20 ml of sea water at 12°C (express your answer in μmol);

(c) 10 ml of distilled water at 15°C (express your answer in μmol);

(d) 250 ml of distilled water at 37°C (express your answer in mg);

(e) 200 ml of sea water at 25°C (express your answer in ml).

57.2 Calculate respiration using data from an oxygen electrode readout. An oxygen electrode was set up to contain 5.0 ml of air-equilibrated distilled water at 30°C, giving a full-scale reading on a chart recorder (200 divisions on the chart-recorder paper), while a zero reading gave no deflection (0 divisions on the chart recorder paper). A yeast cell suspension (containing 4.1×10^9 cells in total) was added to the same electrode system along with a suitable growth medium, giving an approximately linear trace on the chart recorder with a slope of −25 units per cm, at a chart speed of 0.5 cm min^{-1}. What is the respiratory oxygen consumption, expressed to three significant figures in nmol min^{-1} $(10^9$ cells)$^{-1}$?

57.3 Determine net photosynthetic rate from oxygen-electrode readings. The following data represent oxygen-electrode readings taken every minute for an illuminated suspension of a cyanobacterium in a 5.0 ml chamber of an oxygen electrode at 20°C with readings from 0 min to 8 min of: 70.4; 73.6; 75.3; 78.5; 83.6; 85.9; 88.2; 91.4; 94.1. The system was first calibrated to read 100.0 for air-equilibrated distilled water and 0.0 for anaerobic water, in effect making the scale read in terms of percentage oxygen saturation. The chlorophyll a content of an equivalent amount of cyanobacterial suspension to that used in the electrode chamber was measured at 2.13 μg ml^{-1}. What is the net photosynthetic rate of the cyanobacterial suspension, in μmol min^{-1} (mg chlorophyll a)$^{-1}$?

Particles suspended in a liquid will move at a rate that depends on:

- the applied force – particles in a liquid within a gravitational field, e.g. a stationary test tube, will move in response to the earth's gravity;
- the density difference between the particles and the liquid – particles less dense than the liquid will float upwards while particles denser than the liquid will sink;
- the size and shape of the particles;
- the viscosity of the medium.

For most biological particles (cells, organelles or molecules) the rate of flotation or sedimentation in response to the earth's gravity is too slow to be of practical use in separation.

KEY POINT *A centrifuge is an instrument designed to produce a centrifugal force far greater than the earth's gravity, by spinning the sample about a central axis (Fig. 58.1). Particles of different size, shape or density will thereby sediment at different rates, depending on the speed of rotation and their distance from the central axis.*

Table 58.1 Relationship between speed (r.p.m.) and acceleration (relative centrifugal field, RCF) for a typical bench centrifuge with an average radius of rotation, $r_{av} = 115\,mm$.

r.p.m.	RCF*
500	30
1000	130
1500	290
2000	510
2500	800
3000	1160
3500	1570
4000	2060
4500	2600
5000	3210
5500	3890
6000	4630

*RCF values rounded to nearest 10.

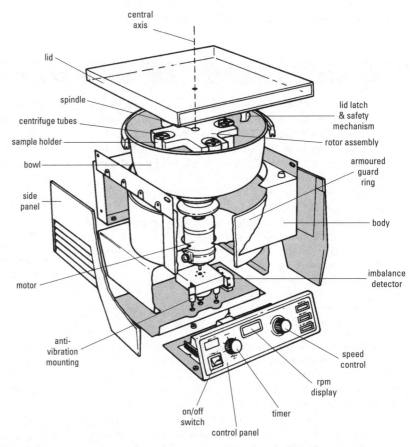

Fig. 58.1 Principal components of a low-speed bench centrifuge, from diagram of low-speed centrifuge *model MSE Centaur 2*, supplied by Fisher Scientific UK Ltd, reproduced by kind permission of Fisher Scientific UK Ltd.

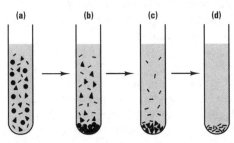

Fig. 58.2 Differential sedimentation. (a) Before centrifugation, the tube contains a mixed suspension of large, medium and small particles of similar density. (b) After low-speed centrifugation, the pellet is predominantly composed of the largest particles. (c) Further high-speed centrifugation of the supernatant will give a second pellet, predominantly composed of medium-sized particles. (d) A final ultracentrifugation step pellets the remaining small particles. Note that all of the pellets apart from the final one will have some degree of cross-contamination.

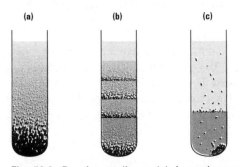

Fig. 58.3 Density gradients. (a) A continuous (linear) density gradient. (b) A discontinuous (stepwise) density gradient, formed by layering solutions of decreasing density on top of each other. (c) A single-step density barrier, designed to allow selective sedimentation of one type of particle.

How to calculate centrifugal acceleration

The acceleration of a centrifuge is usually expressed as a multiple of the acceleration due to gravity (g = 9.80 m s⁻²), termed the relative centrifugal field (RCF, or 'g value'). The RCF depends on the speed of the rotor (n, in revolutions per minute, r.p.m.) and the radius of rotation (r, in mm) where:

$$RCF = 1.118\, r \left(\frac{n}{1\,000}\right)^2 \quad [58.1]$$

This relationship can be rearranged, to calculate the speed (r.p.m.) for specific values of r and RCF:

$$n = 945.7 \sqrt{\left(\frac{RCF}{r}\right)} \quad [58.2]$$

However, you should note that RCF is not uniform within a centrifuge tube: it is highest near the outside of the rotor (r_{max}) and lowest near the central axis (r_{min}). In practice, it is customary to report the RCF calculated from the average radius of rotation (r_{av}), as shown in Fig. 58.5. It is also worth noting that RCF varies relative to the *square* of the speed: thus the RCF will be doubled by an increase in speed of approximately 41 per cent (Table 58.1).

Centrifugal separation methods

Differential sedimentation (pelleting)

By centrifuging a mixed suspension of particles at a specific RCF for a particular time, the mixture will be separated into a pellet and a supernatant (Fig. 58.2). The successive pelleting of a suspension by spinning for a fixed time at increasing RCF is widely used to separate organelles from cell homogenates. The same principle applies when cells are harvested from a liquid medium.

Density-gradient centrifugation

The following techniques use a density gradient, a solution that increases in density from the top to the bottom of a centrifuge tube (Fig. 58.3).

- Rate-zonal centrifugation. By layering a sample on to a shallow pre-formed density gradient, followed by centrifugation, the larger particles will move faster through the gradient than the smaller ones, forming several distinct zones (bands). This method is time-dependent, and centrifugation *must* be stopped before any band reaches the bottom of the tube (Fig. 58.4).
- Isopycnic centrifugation. This technique separates particles on the basis of their buoyant density. Several substances form density gradients during centrifugation (e.g. sucrose, CsCl, Ficoll, Percoll, Nycodenz). The sample is mixed with the appropriate substance and then centrifuged – particles form bands where their density corresponds to that of the medium (Fig. 58.4). This method requires a steep gradient and sufficient time to allow gradient formation and particle redistribution, but is unaffected by further centrifugation.

Bands within a density gradient can be sampled using a fine Pasteur pipette, or a syringe with a long, fine needle. Alternatively, the tube may be punctured and the contents (fractions) collected dropwise in several tubes. For accurate work, an upward displacement technique can be used (see Ford and Graham, 1991).

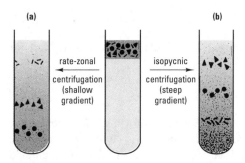

Fig. 58.4 Density-gradient centrifugation. The central tube shows the position of the sample prior to centrifugation, as a layer on top of the density gradient medium. Note that particles sediment on the basis of size during rate-zonal centrifugation (a), but form bands in order of their densities during isopycnic centrifugation (b). ●, large particles, intermediate density; ▲, medium-sized particles, low density; ▬ small particles, high density.

Density-barrier centrifugation

A single-step density barrier (Fig. 58.3(c)) can be used to separate cells from their surrounding fluid, e.g. using a layer of silicone oil adjusted to the correct density using dinonyl phthalate. Blood cell types can be separated using a density barrier of e.g. Ficoll.

Types of centrifuge and their uses

Low-speed centrifuges

These are bench-top instruments for routine use, with a maximum speed of 3 000–6 000 r.p.m. and RCF up to 6 000 g (Fig. 58.1). They are used to harvest cells, larger organelles (e.g. nuclei, chloroplasts) and coarse precipitates (e.g. antibody–antigen complexes, p. 310). Most modern machines also have a sensor that detects any imbalance when the rotor is spinning and cuts off the power supply (Fig. 58.1). However, some of the older models do not, and must be switched off as soon as any vibration is noticed, to prevent damage to the rotor or harm to the operator. Box 58.1 gives details of operation for a low-speed centrifuge.

Microcentrifuges (microfuges)

These are bench-top machines, capable of rapid acceleration up to 12 000 r.p.m. and 10 000 g. They are used to sediment small sample volumes (up to 1.5 ml) of larger particles (e.g. cells, precipitates) over short timescales (typically, 0.5–15 min). They are particularly useful for the rapid separation of cells from a liquid medium, e.g. silicone oil microcentrifugation.

Continuous-flow centrifuges

Useful for harvesting large volumes of cells from their growth medium. During centrifugation, the particles are sedimented as the liquid flows through the rotor.

High-speed centrifuges

These are usually larger, free-standing instruments with a maximum speed of up to 25 000 r.p.m. and RCF up to 60 000 g. They are used for microbial cells, many organelles (e.g. mitochondria, lysosomes) and protein precipitates. They often have a refrigeration system to keep the rotor cool at high speed. You would normally use such instruments only under direct supervision.

Ultracentrifuges

These are the most powerful machines, having maximum speeds in excess of 30 000 r.p.m. and RCF up to 600 000 g, with sophisticated refrigeration and vacuum systems. They are used for smaller organelles (e.g. ribosomes, membrane vesicles) and biological macromolecules. You would not normally use an ultracentrifuge, though your samples may be run by a member of staff.

Rotors

Many centrifuges can be used with tubes of different size and capacity, either by changing the rotor, or by using a single rotor with different buckets/adaptors.

- Swing-out rotors: sample tubes are placed in buckets that pivot as the rotor accelerates (Fig. 58.5(a)). Swing-out rotors are used on many

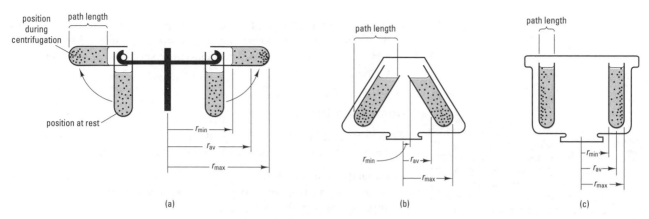

Fig. 58.5 Rotors: (a) swing-out rotor; (b) fixed-angle rotor; (c) vertical-tube rotor

low-speed centrifuges: their major drawback is their extended path length and the resuspension of pellets due to currents created during deceleration.

- Fixed-angle rotors: used in many high-speed centrifuges and microcentrifuges (Fig. 58.5(b)). With their shorter path length, fixed rotors are more effective at pelleting particles than swing-out rotors.

Box 58.1 How to use a low-speed bench centrifuge

1. **Choose the appropriate tube size and material for your application**, with caps where necessary. Most low-speed machines have four-place or six-place rotors – use the correct number of samples to *fill* the rotor assembly whenever possible.

2. **Fill the containers to the appropriate level:** do not overfill, or the sample may spill during centrifugation.

 3. **It is vital that the rotor is balanced during use.** Therefore, *identical* tubes must be prepared, to be placed opposite each other in the rotor assembly. This is particularly important for density-gradient samples, or for samples containing materials of widely differing densities, e.g. soil samples, since the density profile of the tube will change during a run. However, for low-speed work using small amounts of particulate matter in aqueous solution, it is sufficient to counterbalance a sample with a second tube filled with water, or a saline solution of similar density to the sample.

 4. **Balance each pair of sample tubes** (plus the corresponding caps, where necessary) to within 0.1 g using a top-pan balance; add liquid dropwise to the lighter tube, until the desired weight is reached. Alternatively, use a set of scales. For small sample volumes (up to 10 ml) added to disposable, lightweight plastic tubes, accurate pipetting of your solution may be sufficient for low-speed use.

5. **For centrifuges with swing-out rotors**, check that each holder/bucket is correctly positioned in its locating slots on the rotor and that it is able to swing freely. All buckets must be in position on a swing-out rotor, even if they do not contain sample tubes – buckets are an integral part of the rotor assembly.

6. **Load the sample tubes into the centrifuge.** Make sure that the outside of the centrifuge tubes, the sample holders and sample chambers are dry: any liquid present will cause an imbalance during centrifugation, in addition to the corrosive damage it may cause to the rotor. For sample holders where rubber cushions are provided, make sure that these are correctly located. Balanced tubes must be placed opposite each other – use a simple code if necessary, to prevent mix-ups.

7. **Bring the centrifuge up to operating speed** by gentle acceleration. Do not exceed the maximum speed for the rotor and tubes used.

8. **If the centrifuge vibrates at any time during use, switch off** and find the source of the problem.

9. **Once the rotor has stopped spinning, release the lid and remove all tubes.** If any sample has spilled, make sure you clean it up thoroughly using a non-corrosive disinfectant, e.g. Virkon, so that it is ready for the next user.

10. **Close the lid (to prevent the entry of dust) and return all controls to zero.**

- Vertical-tube rotors: used for isopycnic density-gradient centrifugation in high-speed centrifuges and ultracentrifuges (Fig. 58.5(c)). They cannot be used to harvest particles in suspension as a pellet is not formed.

Centrifuge tubes

These are manufactured in a range of sizes (from 1.5 ml up to 1000 ml) and materials. The following aspects may influence your choice:

- Capacity. This is obviously governed by the volume of your sample. Note that centrifuge tubes must be completely full for certain applications, e.g. for high-speed work.
- Shape. Conical-bottomed centrifuge tubes retain pellets more effectively than round-bottomed tubes, while the latter may be more useful for density-gradient work.
- Maximum centrifugal force. Detailed information is supplied by the manufacturers. Standard Pyrex glass tubes can only be used at low centrifugal force (up to $2\,000\,g$).
- Caps. Most fixed-angle and vertical-tube rotors require tubes to be capped, to prevent leakage during use and to provide support to the tube during centrifugation. For low-speed centrifugation, caps must be used for any hazardous samples. Make sure you use the correct caps for your tubes.
- Solvent resistance. Glass tubes are inert, polycarbonate tubes are particularly sensitive to organic solvents (e.g. ethanol, acetone), while polypropylene tubes are more resistant. See manufacturer's guidelines for detailed information.
- Sterilisation. Disposable plastic centrifuge tubes are often supplied in sterile form. Glass and polypropylene tubes can be repeatedly sterilised. Cellulose ester tubes should *not* be autoclaved. Repeated autoclaving of polycarbonate tubes may lead to cracking/stress damage.
- Opacity. Glass and polycarbonate tubes are clear, while polypropylene tubes are more opaque.
- Ability to be pierced. If you intend to harvest your sample by puncturing the tube wall, cellulose acetate and polypropylene tubes are readily punctured using a syringe needle.

Balancing the rotor

For the safe use of centrifuges, the rotor must be balanced during use (Box 58.1), or the spindle and rotor assembly may be damaged permanently; in severe cases, the rotor may fail and cause a serious accident.

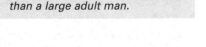 Safe practice

Given their speed of rotation and the extremely high forces generated, centrifuges have the potential to be extremely dangerous, if used incorrectly. You should only use a particular centrifuge if you fully understand the operating principles: if unsure, check with a member of staff. For safety reasons, all centrifuges are manufactured with an armoured casing that should contain any fragments in cases of rotor failure. Machines usually have a safety lock to prevent the motor from

 SAFETY NOTE *When working with centrifuge tubes never be tempted to use a tube or bottle that was not designed to fit the machine you are using (e.g. a general-purpose glass test tube, or a screw-capped bottle), or you may damage the centrifuge and cause an accident.*

Using microcentrifuge tubes *– the integral push-on caps of microcentrifuge tubes must be correctly pushed home before use or they may come off during centrifugation.*

 SAFETY NOTE *Never balance centrifuge tubes 'by eye' – use a balance. Note that a 35 ml tube full of liquid at an RCF of 3 000 g has an effective weight greater than a large adult man.*

 SAFETY NOTE *It is vital that you balance your loaded centrifuge tubes before use. As a general rule, balance all sample tubes to within 1 per cent or better, using a toppan balance or scales. Place balanced tubes opposite each other.*

being switched on unless the lid is closed and to stop the lid from being opened while the rotor is moving. Don't be tempted to use older machines without a safety lock, or centrifuges where the locking mechanism is damaged/inoperative. Be particularly careful to make sure that hair and clothing are kept well away from moving parts.

Text reference

Ford, T.C. and Graham, J.M. (1991) *An Introduction to Centrifugation*. Bios Scientific Publishers, Oxford.

Sources for further study

Boyer, R. F. (2005) *Biochemistry Laboratory: Modern Theory and Techniques*. Benjamin Cummings, San Francisco.
[Also covers other topics, including chromatography and spectroscopy.]

Graham, J. (2001) *Biological Centrifugation*. Bios Scientific Publishers, Oxford.

Rickwood, D., Ford, T. and Steensgard, J. (1994) *Centrifugation: Essential Data*. Wiley, New York.

Study exercises

58.1 **Decide on the type of centrifuge required for a particular application.** What centrifuge would you use for each of the following?

(a) Separating mitochondria from a cell homogenate.

(b) Separating yeast cells from their surrounding medium in an experiment to study the uptake of a radiolabelled amino acid as a function of time.

(c) Harvesting cells from a bioreactor containing 25 litres of growth medium.

58.2 **Determine centrifugal acceleration for a specific centrifuge.** Calculate the centrifugal acceleration (RCF, i.e. the *g* value) of the following (express your answers to three significant figures):

(a) A bench centrifuge with an average radius of rotation of 125 mm operating at 4000 r.p.m.

(b) A bench-top microcentrifuge with an average radius of rotation of 60 mm, operating at 12 000 r.p.m.

(c) An ultracentrifuge with an average radius of rotation of 186 mm, operating at 30 000 r.p.m.

58.3 **Determine the speed required to give a particular centrifugal acceleration (*g* value).** Calculate the speed required (r.p.m. value) for each of the following (express your answers to three significant figures):

(a) RCF of 1500 *g*, using a bench centrifuge with an average radius of rotation of 95 mm.

(b) RCF of 50 000 *g*, using a high-speed centrifuge with an average radius of rotation of 135 mm.

(c) RCF of 13000 *g*, using a bench-top microcentrifuge with an average radius of rotation of 5.5 cm.

58.4 **Calculate the difference in centrifugal acceleration (*g* value) between the top, middle and bottom of a centrifuge tube.** Assuming that the minimum, average and maximum radial distances of a centrifuge tube in a swing-out rotor of a bench centrifuge operating at 5000 r.p.m. are 45 mm, 70 mm and 95 mm respectively, what are the corresponding *g* values at the top, middle and bottom of the tube when the centrifuge is operating? (Express your answer to three significant figures.)

The absorption and emission of electromagnetic radiation of specific energy (wavelength) is a characteristic feature of many molecules, involving the movement of electrons between different energy states, in accordance with the laws of quantum mechanics. Spectroscopic techniques are used to measure and interpret such interactions between molecules and radiation.

UV/visible spectrophotometry

This is a widely used technique for measuring the absorption of radiation in the visible and UV regions of the spectrum. A spectrophotometer is an instrument designed to allow precise measurement at a particular wavelength, while a colorimeter is a simpler instrument, using filters to measure broader wavebands (e.g. light in the green, red or blue regions of the visible spectrum).

Principles of light absorption

Two fundamental principles govern the absorption of light by a solution:

- The absorption of light passing through a solution is exponentially related to the number of molecules of the absorbing solute, i.e. the solute concentration $[C]$.
- The absorption of light passing through a solution is exponentially related to the length of the absorbing solution, l.

These principles are combined in the Beer–Lambert relationship (sometimes referred to simply as 'Beer's Law'), usually expressed in terms of absorbance (A) – the logarithm of the ratio of incident light (I_0) to emergent light (I):

$$A = \varepsilon l [C] \qquad [59.1]$$

where A is absorbance, ε is a constant for the absorbing substance at a specific wavelength, termed the absorptivity, or absorption coefficient, and $[C]$ is expressed either as $\text{mol}\,l^{-1}$ or $\text{g}\,l^{-1}$ (see p. 138) and l is given in cm. This relationship is extremely useful, since most spectrophotometers are constructed to give a direct measurement of absorbance, sometimes also termed extinction (E), of a solution (older texts may use the outdated term optical density, OD).

 KEY POINT *The Beer–Lambert relationship states that there is a direct linear relationship between the concentration of a substance in a solution, [C], and the absorbance of that solution, A.*

Absorbance at a particular wavelength is often shown as a subscript, e.g. A_{550} represents the absorbance at 550 nm. The proportion of light passing through the solution is known as the transmittance (T), and is calculated as the ratio of the emergent and incident light intensities.

Some instruments have two scales:

- an exponential scale from zero to infinity, measuring absorbance;
- a linear scale from 0 to 100, measuring (per cent) transmittance.

For most practical purposes, the Beer–Lambert relationship (eqn [59.1]) applies and you should use the absorbance scale.

Using spectroscopy – this technique is valuable for:

- *tentatively identifying compounds, by determining their absorption or emission spectra;*
- *quantifying substances, either singly or in the presence of other compounds, by measuring the signal strength at an appropriate wavelength;*
- *determining molecular structure;*
- *following reactions, by measuring the disappearance of a substance, or the appearance of a product as a function of time.*

Definitions

Absorbance (A) – this is given by:

$A = \log_{10}(I_0/I)$, usually shown as A_x where 'x' is the wavelength in nanometres.

Transmittance (T) – this is usually expressed as a percentage often at a particular wavelength, T_x, where

$T_x = (I/I_0) \times 100\,(\%)$.

Example For incident light $(I_0) = 1.00$ and emergent light $(I) = 0.16$ (expressed in relative terms) then $A = \log_{10}(1.00 \div 0.16) = \log_{10} 6.25 = 0.796$ (to three significant figures). The corresponding transmittance, $T = (0.16 \div 1.00) \times 100 = 16\%$.

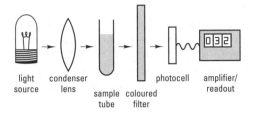

light   condenser   photocell   amplifier/
source   lens   readout
              sample   coloured
              tube   filter

Fig. 59.1 Components of a colorimeter

SAFETY NOTE *Take care not to spill water when using a colorimeter or spectrophotometer, due to the risk of electric shock (switch off at mains and seek assistance if this should happen).*

Colorimeter

This can be used with solutions where the test substance is highly coloured and present as the major constituent, e.g. haemoglobin in blood, or where a substance is assayed by adding a reagent that gives a coloured product (a chromophore), e.g. amino acid assay using ninhydrin reagent. Quantification of a particular substance requires a calibration curve, constructed using known amounts of the compound measured at the same time as the test samples, rather than using the Beer–Lambert relationship.

The light source is usually a tungsten filament bulb, focused by a condenser lens to give a parallel beam of light that passes through a glass sample tube or cuvette containing the solution, then through a coloured filter to a photocell detector, which develops an electrical potential in direct proportion to the intensity of the light falling on it (Fig. 59.1). The signal from the photocell is then amplified and passed to a galvanometer, or digital readout, calibrated on a logarithmic scale (see Box 59.1).

The broad bandwidth of most filters means that colorimetry cannot be used to identify a particular compound, nor to distinguish between two compounds with closely related absorption characteristics, e.g. in mixed solution. The photocells used in colorimeters have coefficients of variation (p. 420) of around 0.5 per cent, so they are not suitable for work requiring a high degree of precision. In the simplest instruments, the logarithmic measurement scale has arbitrary units, adjusted via sensitivity/scale zero controls, and values obtained on one instrument will not be directly comparable with other instruments, or with the same instrument on different settings. A colorimeter is not suitable for quantitative work at a particular wavelength.

Box 59.1 How to use a colorimeter

1. **Switch on and stabilise** allowing at least 5 min for the lamp to warm up before use.

2. **Choose a filter that is complementary** to the colour of the substance to be measured. Thus haemoglobin is red because it absorbs light within the blue/green regions of the spectrum and should be measured using a blue filter. Similarly, a blue substance should be measured using a red filter.

3. **Set the scale zero** using an appropriate solution blank.

4. **Adjust the sensitivity** to give a reasonable deflection on the galvanometer/readout device, e.g. for a standard solution containing a known amount of the test substance. You would normally adjust the sensitivity so that your highest standard gave a reading close to the maximum on the readout device.

5. **Analyse your samples and standard solutions**, making sure you take all the measurements you need in a single 'run', otherwise they will not be directly comparable.

6. **Use the same tube in the same orientation** in the sample holder to improve precision, since individual tubes may differ in their light-absorbing characteristics, wall thickness, etc.

7. **Rinse the tube between samples**, making sure that the rinsing solution does not dilute the next sample and that the outside of the tube is dry before it is loaded into the instrument.

8. **Make frequent checks on reproducibility** by repeat measurements of the same solution (e.g. after every six to eight samples). You should also prepare test and standard solutions in duplicate.

9. **Plot a calibration curve for your standard solutions** and draw the line of best fit through the points. Do not worry if it is a curve, rather than a straight line, or if it does not pass through the origin. Do not extrapolate the calibration curve. If a sample has a reading greater than the highest standard it should be diluted and reassayed. Chapter 50 gives further advice on preparing and using calibration curves.

Spectroscopic techniques

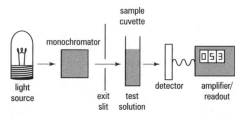

Fig. 59.2 Components of a UV/visible spectrophotometer

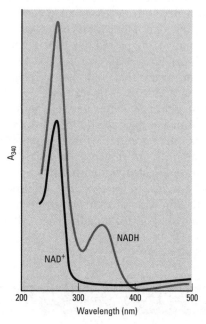

Fig. 59.3 Absorption spectra of nicotinamide adenine dinucleotide in oxidised (NAD^+) and reduced (NADH) form. Note the 340 nm absorption peak, used for quantitative work (p. 369).

Using plastic disposable cuvettes – these are adequate for work in the near-UV region, e.g. for enzyme studies using nicotinamide coenzymes, at 340 nm (p. 319), as well as the visible range. Table 21.2 (p. 126) gives details of the spectral cut-off wavelengths of different materials.

Measuring low absorbances – for accurate readings with dilute solutions, use a cuvette with a longer optical path length (e.g. 5 cm, rather than 1 cm).

UV/visible spectrophotometer

The principal components of a UV/visible spectrophotometer are shown in Fig. 59.2. High-intensity tungsten bulbs are used as the light source in basic instruments, capable of operating in the visible region (i.e. 400–700 nm). Deuterium lamps are used for UV spectrophotometry (200–400 nm); these lamps are fitted with quartz envelopes, since glass does not transmit UV radiation.

A major improvement over the simple colorimeter is the use of a diffraction grating to produce a parallel beam of monochromatic light from the (polychromatic) light source. In practice the light emerging from such a monochromator does not have a single wavelength, but is a narrow band of wavelengths. This bandwidth is an important characteristic, since it determines the wavelengths used in absorption measurements – the bandwidth of basic spectrophotometers is around 5–10 nm, while research instruments have bandwidths of less than 1 nm.

Bandwidth is affected by the width of the exit slit (the slit width), since the bandwidth will be reduced by decreasing the slit width. To obtain accurate data at a particular wavelength setting, the narrowest possible slit width should be used. However, decreasing the slit width also reduces the amount of light reaching the detector, decreasing the signal-to-noise ratio. The extent to which the slit width can be reduced depends upon the sensitivity and stability of the detection/amplification system and the presence of stray light.

Most UV/visible spectrophotometers are designed to take cuvettes with an optical path length of 1 cm. Disposable plastic cuvettes are suitable for routine work in the visible range using aqueous and alcohol-based solvents, and glass cuvettes are useful for other organic solvents. Glass cuvettes are manufactured to more exacting standards, so use optically matched glass cuvettes for accurate work, especially at low absorbances (< 0.1), where any differences in the optical properties of cuvettes for reference and test samples will be pronounced. Glass and plastic absorb UV light, so quartz cuvettes must be used at wavelengths below 300 nm.

 KEY POINT *Before taking a measurement, make sure that cuvettes are clean, unscratched, dry on the outside, filled to the correct level and located in the correct position in their sample holders.*

Proteins and nucleic acids in biological samples can accumulate on the inside faces of glass/quartz cuvettes, so remove any deposits using acetone on a cotton bud, or soak overnight in $1 \, mol \, l^{-1}$ nitric acid. Corrosive and hazardous solutions must be used in cuvettes with tightly fitting lids, to prevent damage to the instrument and to reduce the risk of accidental spillage.

Basic instruments use photocells similar to those used in colorimeters or photodiode detectors. In many cases, a different photocell must be used at wavelengths above and below 550–600 nm, due to differences in the sensitivity of such detectors over the visible waveband. The detectors used in more sophisticated instruments give increased sensitivity and stability when compared with photocells.

Digital displays are increasingly used in preference to needle-type meters, as they are not prone to parallax errors and misreading of the scale. Some digital instruments can be calibrated to give a direct readout of the concentration of the test substance.

Measuring high absorbances in color-imetric analysis – *if any final solution has an absorbance that is too high to be read with accuracy on your spectro-photometer (i.e. A > 1), it is bad practice to dilute the solution so that it can be measured. This dilutes both the sample molecules and the colour reagents to an equal extent. Instead, you should dilute the original sample and re-assay.*

Examples in quantitative analysis
The molar absorptivity of NADH is $6.22 \times 10^3 \, \mathrm{l \, mol^{-1} \, cm^{-1}}$ at 340 nm. For a test solution giving an absorbance of 0.21 in a cuvette with a light path of 5 mm, using eqn [59.1] this is equal to a concentration of:

$$0.21 = 6.22 \times 10^3 \times 0.5 \times [C]$$
$$[C] = 0.0000675 \, \mathrm{mol \, l^{-1}}$$
$$(\text{or } 67.5 \, \mathrm{\mu mol \, l^{-1}}).$$

The specific absorptivity ($10 \, \mathrm{g \, l^{-1}}$) of double-stranded DNA is 200 at 260 nm, therefore a solution containing $1 \, \mathrm{g \, l^{-1}}$ will have an absorbance of $200/10 = 20$. For a DNA solution, giving an absorbance of 0.35 in a cuvette with a light path of 1.0 cm, using eqn [59.1] this is equal to a concentration of:

$$0.35 = 20 \times 1.0 \, [C]$$
$$[C] = 0.0175 \, \mathrm{g \, l^{-1}}$$
$$(\text{equivalent to } 17.5 \, \mathrm{\mu g \, ml^{-1}}).$$

Chlorophylls a and b in vascular plants and green algae can be assayed in 90% v/v acetone/water by measuring the absorbance of the mixed solution at two wavelengths, according to the formulae:

Chlorophyll a (mg $\mathrm{l^{-1}}$) = $11.93 \, A_{664} - 1.93 \, A_{647}$

Chlorophyll b (mg $\mathrm{l^{-1}}$) = $20.36 \, A_{647} - 5.5 \, A_{664}$

Note: different equations are required for other solvents.

 SAFETY NOTE *In atomic spectro-scopy, the use of high-pressure gas cylinders can be particularly hazardous. Always consult a member of staff before using such apparatus.*

 KEY POINT *Basic spectrophotometers are most accurate within the absorbance range from 0.00 to 1.00 and your standard and test solutions should be prepared to give readings within this range.*

Types of UV/visible spectrophotometer

Basic instruments are single-beam spectrophotometers in which there is only one light path. The instrument is set to zero absorbance using a blank solution, which is then replaced by the test solution, to obtain an absorbance reading. An alternative approach is used in double-beam spectrophotometers, where the light beam from the monochromator is split into two separate beams, one beam passing through the test solution and the other through a reference blank. Absorbance is then measured by an electronic circuit that compares the output from the reference (blank) and sample cuvettes. Double-beam spectrophotometry reduces measurement errors caused by fluctuations in output from the light source or changes in the sensitivity of the detection system, since reference and test solutions are measured at the same time (Box 59.2). Recording spectrophotometers are double-beam instruments, designed to record either the difference in absorbance between reference and test solutions across a predetermined waveband to give an absorption spectrum (Fig. 59.3), or to record the change in absorbance at a particular wavelength as a function of time (e.g. in an enzyme assay, see Chapter 52).

Quantitative spectrophotometric analysis

A single substance in solution can be quantified using the Beer–Lambert relationship (eqn [59.1]), provided its absorptivity is known at a particular wavelength (usually the absorption maximum for the substance, since this will give the greatest sensitivity). The molar absorptivity is the absorbance given by a solution with a concentration of 1 mol $\mathrm{l^{-1}}$ ($= 1 \, \mathrm{kmol \, m^{-3}}$) of the compound in a light path of 1 cm. The appropriate value may be available from tabulated spectral data (e.g. Anon., 1963), or it can be determined experimentally by measuring the absorbance of known concentrations of the substance (Box 59.2) and plotting a standard curve (Chapter 50). This should confirm that the relationship is linear over the desired concentration range and the slope of the line will give the molar absorptivity.

The specific absorptivity is the absorbance given by a solution containing $10 \, \mathrm{g \, l^{-1}}$ (i.e. 1% w/v) of the compound in a light path of 1 cm. This is useful for substances of unknown molecular weight, e.g. proteins or nucleic acids, where the amount of substance in solution is expressed in terms of its mass, rather than as a molar concentration. For use in eqn [59.1], the specific absorptivity should be divided by 10 to give the solute concentration in g $\mathrm{l^{-1}}$.

This simple approach cannot be used for mixed samples. In such cases, it may be possible to estimate the amount of each substance by measuring the absorbance at several wavelengths, e.g. protein estimation in the presence of nucleic acids (p. 320). Further details on spectrophotometric methods of determining the amount of protein in an aqueous sample are given in Chapter 52.

Atomic spectroscopy

Atoms of certain metals will absorb and emit radiation of specific wavelengths when heated in a flame, in direct proportion to the number of atoms present. Atomic spectrophotometric techniques measure the

Box 59.2 How to use a spectrophotometer

1. **Switch on and select the correct lamp** for your measurements (e.g. deuterium for UV, tungsten for visible light).

2. **Allow at least 15 min for the lamp to warm up** and for the instrument to stabilise before use.

3. **Select the appropriate wavelength:** on older instruments a dial is used to adjust the monochromator, whereas newer machines have microprocessor-controlled wavelength selection.

4. **Select the appropriate detector:** some instruments choose the correct detector automatically (on the basis of the specified wavelength), and others have manual selection.

5. **Choose the correct slit width** (if available): this may be specified in the protocol you are following, or may be chosen on the manufacturer's recommendations.

6. **Insert appropriate reference blank(s):** single-beam instruments use a single cuvette, whereas double-beam instruments use two cuvettes (a matched pair for accurate work). The reference blanks should match the test solution in all respects apart from the substance under test, i.e. they should contain all reagents apart from this substance. *Make sure that the cuvettes are positioned correctly, with their polished (transparent) faces in the light path, and check that they are accurately located in the cuvette holder(s).*

7. **Check/adjust the 0% transmittance:** most instruments have a control that allows you to zero the detector output in the absence of any light (so-called dark current correction). Some microprocessor-controlled instruments carry out this step automatically.

8. **Set the absorbance reading to zero:** usually via a dial, or digital readout.

9. **Analyse your samples:** replace the appropriate reference blank with a test sample, allow the absorbance reading to stabilise (5–10 s) and read the absorbance value from the meter/readout device. For absorbance readings greater than 1 (i.e. <10% transmittance), the signal-to-noise ratio is too low for accurate results. Your analysis may require a calibration curve (Chapter 50), or you may be able to use the Beer–Lambert relationship (eqn [59.1]) to determine the concentration of test substance in your samples.

10. **Check the scale zero at regular intervals** using a reference blank, e.g. after every 10 samples.

11. **Check the reproducibility of the instrument:** measure the absorbance of a single solution several times during your analysis. It should give the same value.

Problems (and solutions): inaccurate/unstable readings are most often due to incorrect use of cuvettes, e.g. dirt, fingerprints or test solution on outside of cuvette (wipe the polished faces using a soft tissue before insertion into the cuvette holder), condensation (if cold solutions aren't allowed to reach room temperature before use), air bubbles (which scatter light and increase the absorbance; tap gently to remove), insufficient solution (causing refraction of light at the meniscus), particulate material in the solution (centrifuge before use, where necessary) or incorrect positioning in light path (locate in correct position).

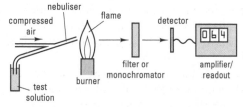

Fig. 59.4 Components of a flame photometer

> ⚠️ **SAFETY NOTE** *When carrying out acid digestion, always work within a fume hood and wear gloves and safety glasses throughout the procedure. Rinse any spillages with a large volume of water.*

absorption or emission of particular wavelengths of UV and visible light, to identify and quantify such metals.

Flame atomic-emission spectrophotometry (or flame photometry)

The principal components of a flame photometer are shown in Fig. 59.4. A liquid sample is converted into an aerosol in a nebuliser (atomiser) before being introduced into the flame, where a small proportion (typically less than 1 in 10 000) of the atoms will be raised to a higher energy level, releasing this energy as light of a particular wavelength, which is passed through a filter to a photocell detector. Flame photometry is used to measure the alkali metal ions K^+, Na^+ and Ca^{++} in biological fluids. Box 59.3 gives details on how to use a flame photometer.

Atomic-absorption spectrophotometry (or flame absorption spectrophotometry)

This technique is applicable to a broad range of metal ions, including those of Pb, Cu, Zn, etc. It relies on the absorption of light of a specific

Box 59.3　How to use a flame photometer

1. **Switch on the instrument and allow it to stabilise.** Light the flame and then wait for at least 5 min before analysing your solutions.

2. **Check for impurities in your reagents.** For example, if you are measuring K^+ in an acid digest of some biological material, e.g. plant tissue, you should check the K^+ content of a reagent blank (a solution containing everything except the soil, prepared in exactly the same way as the samples). Once converted to concentration (see below) you should then subtract this value from the concentration determined for each of your sample solutions, to obtain the true K^+ content.

3. **Quantify your samples using a calibration curve** (p. 303). Calibration standards should cover the expected concentration range for the test solutions – your calibration curve may be non-linear, especially at higher concentrations (e.g. above $1\ mmol\,l^{-1}$, i.e. $1\ mol\,m^{-3}$ in SI units).

4. **Assay all solutions in duplicate,** so that repeatability can be assessed (if duplicate readings do not agree, repeat until you are satisfied that the measurements are reliable).

5. **Check your calibration.** Make repeated measurements of a standard solution of known concentration after every six or seven samples, to confirm that the instrument calibration is still valid.

6. **Consider the possibility of interference.** Other metal atoms may emit light that is detected by the photocell, since the filters cover a wider waveband than the emission wavelength of a particular element. This can be a serious problem if you are trying to measure low concentrations of a particular metal in the presence of high concentrations of other metals (e.g. Na^+ in sea water), or if there are other substances present that form complexes with the test metal, suppressing the signal (e.g. phosphate).

Plotting calibration curves in quantitative analysis – do not force your calibration line to pass through zero if it clearly does not. There is no reason to assume that the zero value is any more accurate than any other reading you have made (see also Chapter 50).

wavelength by atoms dispersed in a flame. The appropriate wavelength is provided by a cathode lamp, coated with the element to be analysed, focused through the flame and on to the detector. When the sample is introduced into the flame, it will decrease the light detected in direct proportion to the amount of metal present. Practical advantages over flame photometry include improved sensitivity, increased precision and decreased interference. Newer variants of this method include flameless atomic absorption spectrophotometry and atomic fluorescence spectrophotometry, both of which are more sensitive than the flame-absorption technique.

Other spectroscopic methods

The range of advanced techniques used in biological research is detailed in Table 59.1 (see also Reed et al. (2007) for e.g. fluorescence spectrophotometry).

Table 59.1　Advanced spectroscopic techniques

Technique	Biological applications
Spectrofluorimetry	Quantification of fluorescent molecules, e.g. chlorophyll, porphyrins Enzyme assay, using fluorogenic substrates
Infrared (IR) spectrometry	Gas analysis Identification of molecular structure
Electron spin resonance (ESR) spectrometry	Metalloprotein analysis Free radical behaviour
Mass spectrometry	Biochemical structure/identification Used with gas chromatography (p. 375)
Nuclear magnetic resonance (NMR) spectrometry	Macromolecule structure (^1H and ^{13}C NMR) Metabolic studies (^{13}C and ^{31}P NMR) Intracellular pH (^{31}P NMR)
Circular dichroism spectroscopy	Macromolecule structure, e.g. proteins and nucleic acids

Text references

Anon. (1963) *Tables of Spectrophotometric Absorption Data for Compounds used for the Colorimetric Detection of Elements (International Union of Pure and Applied Chemistry)*. Butterworth-Heinemann, London.

Reed, R.H., Holmes, D.H., Weyers, J.D.B. and Jones, A.M. (2007) *Practical Skills in Biomolecular Sciences*, 3rd edn. Prentice Hall, Harlow.

Sources for further study

Gore, M.G. (ed.) (2000) *Spectrophotometry and Spectrofluorimetry: A Practical Approach*, 2nd edn. Oxford University Press, Oxford.

Harris, D.A. (1996) *Light Spectroscopy*. Bios Scientific Publishers, Oxford.
[Gives practical guidance on techniques, applications and interpretation of results.]

Wilson, K. and Walker, J. (eds.) (2005) *Principles and Techniques of Biochemistry and Molecular Biology*, 6th edn. Cambridge University Press, Cambridge.
[Also covers a range of other topics, including chromatography and centrifugation.]

Study exercises

59.1 **Write a protocol for using a spectrophotometer.** After reading this chapter, prepare a detailed stepwise protocol explaining how to use one of the spectrophotometers in your department. Ask another student or a tutor to evaluate your protocol and provide you with feedback.

59.2 **Use the Beer–Lambert relationship in quantitative spectrophotometric analysis.** Calculate the following (express your answer to three significant figures):

(a) The concentration of NADH ($\mu mol\,l^{-1}$) in a test solution giving an absorbance at 340 nm (A_{340}) of 0.53 in a cuvette with a path length of 1 cm, based on a molar absorptivity for NADH of $6220\,l\,mol^{-1}\,cm^{-1}$ at this wavelength.

(b) The amount of NADH (nmol) in 20 ml of a test solution where $A_{340} = 0.62$ in a cuvette with a path length of 5 cm, based on a molar absorptivity for NADH of $6220\,l\,mol^{-1}\,cm^{-1}$ at this wavelength.

(c) The mass concentration ($\mu g\,ml$) of double-stranded DNA in a test solution giving an absorbance at 260 nm (A_{260}) of 0.57 in a cuvette of path length 5 mm, based on an absorptivity of $20\,l\,g^{-1}\,cm^{-1}$.

(d) The amount (ng) of double-stranded DNA in a 50 μl subsample from of a test solution where $A_{260} = 0.31$ in a cuvette of path length 1 cm, based on an absorptivity of $20\,l\,g^{-1}\,cm^{-1}$.

(e) The chlorophyll a content, in μg (g fresh weight)$^{-1}$, of a plant leaf of fresh weight 1.56 g extracted in 50 ml of 90% acetone, where $A_{664} = 0.182$ at 664 nm and $A_{647} = 0.035$, based on the equation: Chlorophyll a $(mg\,l^{-1}) = 11.93\,A_{664} - 1.93\,A_{647}$.

59.3 **Determine the molar absorptivity of a substance in aqueous solution.** A solution of *p*-nitrophenol containing $8.8\,\mu g\,ml^{-1}$ gave an absorbance of 0.535 at 404 nm in a cuvette of path length 1 cm. What is the molar absorptivity of *p*-nitrophenol at 404 nm, expressed to three significant figures? (Note: M_r of *p*-nitrophenol is 291.27.)

59.4 **Determine the concentration of metal ions based on atomic spectroscopy of test and standard solutions.** The following data represent a set of calibration standards for K^+ in aqueous solution, measured by flame photometry:

Absorbance of standard solutions containing K^+ at up to $0.5\,mmol\,l^{-1}$

K^+ concentration ($mmol\,l^{-1}$)	Absorbance
0	0.000
0.1	0.155
0.2	0.279
0.3	0.391
0.4	0.537
0.5	0.683

Draw a calibration curve using the above data and use this to estimate the amount of K^+ in a test sample prepared by digestion of 0.482 g of tissue in a final volume of 25 ml of solution, giving an absorbance of 0.429 when measured at the same time as the standards shown above. Express your answer in $\mu mol\,K^+$ (g tissue)$^{-1}$, to three significant figures. See also Study exercise 50.1 for a similar exercise, based on atomic absorption spectrophotometry of Zn.

Chromatography is used to separate the individual constituents within a sample on the basis of differences in their physical characteristics, e.g. molecular size, shape, charge, volatility, solubility and/or adsorptivity. The essential components of a chromatographic system are:

- A stationary phase, either a solid, a gel or an immobilised liquid, held by a support matrix.
- A chromatographic bed: the stationary phase may be packed into a glass or metal column, spread as a thin layer on a sheet of glass or plastic, or adsorbed on cellulose fibres (paper).
- A mobile phase, either a liquid or a gas that acts as a solvent, carrying the sample through the stationary phase and eluting from the chromatographic bed.
- A delivery system to pass the mobile phase through the chromatographic bed.
- A detection system to monitor the test substances.

Individual substances interact with the stationary phase to different extents as they are carried through the system, enabling separation to be achieved.

 KEY POINT *In a chromatographic system, those substances that interact strongly with the stationary phase will be retarded to the greatest extent while those that show little interaction will pass through with minimal delay, leading to differences in distances travelled or elution times.*

Types of chromatographic system

Chromatographic systems can be categorised according to the form of the chromatographic bed, the nature of the mobile and stationary phases and the method of separation.

Thin-layer chromatography (TLC) and paper chromatography
Here, you apply the sample as a single spot near one end of the sheet or plate, by microsyringe or microcapillary. This sheet is allowed to dry fully, then it is transferred to a glass tank containing a shallow layer of solvent (Fig. 60.1). Remove the sheet when the solvent front has travelled across 80 – 90 per cent of its length.

You can express movement of an individual substance in terms of its relative frontal mobility, or R_F value, where:

$$R_F = \frac{\text{distance moved by substance}}{\text{distance moved by solvent}} \qquad [60.1]$$

Alternatively, you may express movement with respect to a standard of known mobility, as R_X, where:

$$R_X = \frac{\text{distance moved by test substance}}{\text{distance moved by standard}} \qquad [60.2]$$

The R_F (or R_X) value is a constant for a particular substance and solvent system (under standard conditions) and closely reflects the partitioning of

---

Making compromises in chromatography – *the process is often a three-way compromise between:*

1. *separation of analytes;*
2. *time taken for analysis;*
3. *volume of eluent.*

Thus, if you have a large sample volume and want to achieve a good separation of a mixture of analytes, the time taken for the chromatography will be lengthy.

SAFETY NOTE The solvents used as the mobile phases of chromatographic systems are often toxic and may produce noxious fumes – where necessary, work in a fume hood.

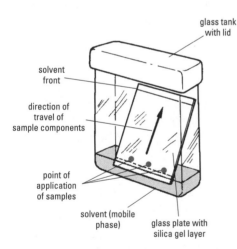

solvent front

direction of travel of sample components

point of application of samples

solvent (mobile phase)

glass tank with lid

glass plate with silica gel layer

Fig. 60.1 Components of a TLC system

Using a TLC system *– it is essential that you allow the solvent to pre-equilibrate in the chromatography tank for at least 2 hours before use, to saturate the atmosphere with vapour. Deliver drops of sample with a blunt-ended microsyringe. Make sure you know exactly where each sample is applied, so that R_F values can be calculated.*

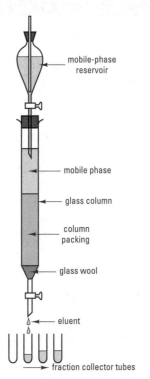

Fig. 60.2 Equipment for column chromatography (gravity-feed system)

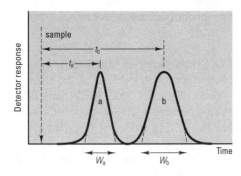

Fig. 60.3 Peak characteristics in a chromatographic separation i.e. a chromatogram

---

Using HPLC – *HPLC is a versatile form of chromatography, used with a wide variety of stationary and mobile phases, to separate individual compounds of a particular class of molecules on the basis of size, polarity, solubility or adsorption characteristics.*

---

the substance between the stationary and mobile phases. Tabulated values are available for a range of biological molecules and solvents (e.g. Stahl, 1969). However, you should analyse one or more reference compounds on the same sheet/plate as your unknown sample, to check their R_F values.

Column chromatography

Here, you pack a glass column with the appropriate stationary phase and equilibrate the mobile phase by passage through the column, either by gravity (Fig. 60.2), or using a low-pressure peristaltic pump. You can then introduce the sample to the top of the column, to form a discrete band of material. This is then flushed through the column by the mobile phase. If the individual substances have different rates of migration, they will separate within the column, eluting at different times as the mobile phase travels through the column.

You can detect eluted substances by collecting the mobile phase as it elutes from the column in a series of tubes (discontinuous monitoring), either manually or with an automatic fraction collector. Fractions of 2 – 5 per cent of the column bed volume are usually collected and analysed, e.g. by chemical assay. You can now construct an elution profile (or chromatogram) by plotting the amount of substance against either time, elution volume or fraction number, which should give a symmetrical peak for each substance (Fig. 60.3).

You can express the migration of a particular substance at a given flow rate in terms of its retention time (t), or elution volume (V_e). The separation efficiency of a column is measured by its ability to distinguish between two similar substances, assessed in terms of:

- selectivity (α), measured using the following equation, which takes into account the retention times of the two peaks (i.e. t_a and t_b), plus the column dead time (t_0):

$$\alpha = \frac{t_b - t_0}{t_a - t_0} \qquad [60.3]$$

The column dead time is the time it takes for an unretained compound to pass through the column without interacting with the stationary phase.

- resolution (R), quantified in terms of the retention time and the base width (W) of each peak:

$$R = \frac{2(t_a - t_b)}{W_a + W_b} \qquad [60.4]$$

The subscripts a and b in eqns [60.3] and [60.4] refer to substances a and b respectively (Fig. 60.3). For most practical purposes, R values of 1 or more are satisfactory, corresponding to 98 per cent peak separation for symmetrical peaks.

High-performance liquid chromatography (HPLC)

Column chromatography originally used large 'soft' stationary phases that required low-pressure flow of the mobile phase to avoid compression; separations were usually time-consuming and of low resolution ('low performance'). Subsequently, the production of small, incompressible, homogeneous particulate support materials and high-pressure pumps with reliable, steady flow rates have enabled high-performance systems to be developed. These systems operate at pressures up to 10 MPa, forcing the mobile phase through the column at a high flow rate to give rapid separation with reduced band broadening, due to smaller particle size.

HPLC columns are usually made of stainless steel, and all components, valves, etc., are manufactured from materials that can withstand the high pressures involved. The two main solvent-delivery systems are:

- Isocratic separation: a single solvent (or solvent mixture) is used throughout the analysis.
- Gradient elution separation: the composition of the mobile phase is altered using a microprocessor-controlled gradient programmer, which mixes appropriate amounts of two different substances to produce the required gradient.

Most HPLC systems are linked to a continuous monitoring detector of high sensitivity, e.g. proteins may be detected spectrophotometrically by monitoring the absorbance of the eluent at 280 nm as it passes through a flow cell (cuvette). Other detectors can be used to measure changes in fluorescence, refractive index, ionisation, radioactivity, etc. The detector is linked to a recorder or microcomputer to produce an elution profile. Most detection systems are non-destructive, which means that you can collect eluent with an automatic fraction collector for further study (Fig. 60.4).

Separation of macromolecules (especially proteins and nucleic acids) usually requires 'biocompatible' systems in which stainless steel components are replaced by titanium, glass or fluoroplastics, using lower pressures to avoid denaturation, e.g. the Pharmacia FPLC system. Such separations are carried out using ion-exchange, gel-permeation and/ or hydrophobic-interaction chromatography (p. 378).

> **Learning from experience** – *if you are unable to separate your test substance(s) using a particular method, do not regard this as a failure, but instead, think about what this tells you about either the substance(s) or your sample.*

> **Applications of high-performance liquid chromatography** – *the speed and sensitivity of HPLC makes this the method of choice for the separation of many small molecules of biological interest, normally using reverse-phase partition chromatography (p. 376).*

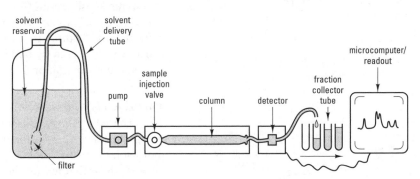

Fig. 60.4 Components of an HPLC system

Gas chromatography (GC)

Modern GC uses capillary chromatography columns (internal diameter 0.1–0.5 mm) up to 50 m in length (Fig. 60.5). The stationary phase is generally a cross-linked silicone polymer, coated as a thin film on the inner wall of the capillary: at normal operating temperatures, this behaves in a similar manner to a liquid film, but is far more robust. The mobile phase ('carrier gas') is usually nitrogen or helium. Selective separation is achieved as a result of the differential partitioning of individual compounds between the carrier gas and silicone polymer phases. The separation of most biomolecules is influenced by the temperature of the column, which may be constant during the analysis ('isothermal' – usually 50–250 °C) or, more commonly, may increase in a pre-programmed manner (e.g. from 50 °C to

> **Applications of gas chromatography** – *GC is used to separate volatile, non-polar compounds: substances with polar groups must be converted to less polar derivatives prior to analysis, to prevent adsorption on the column, leading to poor resolution and peak tailing.*

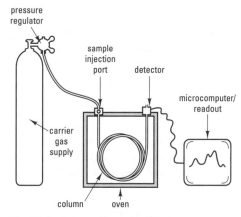

Fig. 60.5 Components of a GC system

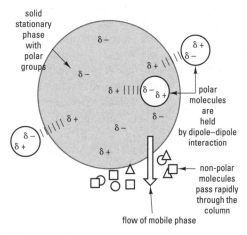

Fig. 60.6 Adsorption chromatography (polar stationary phase)

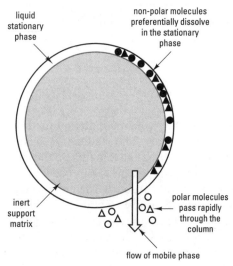

Fig. 60.7 Liquid–liquid partition chromatography, e.g. reverse-phase HPLC

250 °C at 10 °C per min). Samples are injected on to the 'top' of the column, through an injection port containing a gas-tight septum. The output from the column can be monitored by:

- Flame ionisation: the outflow gas is passed through a flame where any organic compounds will be ionised and subsequently detected by an electrode mounted near the flame tip.
- Electron capture: using a beta-emitting radioisotope as the means of ionisation. This is capable of detecting extremely small amounts (pmol) of electrophilic compounds.
- Spectrometry: including mass spectrometry (GC-MS) and infrared spectrometry (GC-IR).
- Thermal conductivity: changes in the composition of the gas at the outflow alter the resistance of a platinum wire.

GC can only be used directly with samples capable of volatilisation at the operating temperature of the column, e.g. short-chain fatty acids. Other substances need to be chemically modified to produce more volatile compounds, e.g. long-chain fatty acids are usually analysed as methyl esters while monosaccharides are converted to trimethylsilyl derivatives.

Separation methods

Adsorption chromatography

This is a form of solid–liquid chromatography. The stationary phase is a porous, finely divided solid that adsorbs molecules of the test substance on its surface due to dipole–dipole interactions, hydrogen bonding and/or van der Waals interactions (Fig. 60.6). The range of adsorbents is limited, e.g. polystyrene-based resins (for non-polar molecules), silica, aluminium oxide and calcium phosphate (for polar molecules). Most adsorbents must be activated by heating to 110–120 °C before use, since their adsorptive capacity is decreased by bound water. Adsorption chromatography can be carried out in column or thin-layer form, using a range of organic solvents.

Partition chromatography

This is based on the partitioning of a substance between two liquid phases, in this instance the stationary and mobile phases. Substances that are more soluble in the mobile phase will pass rapidly through the system, whereas those that favour the stationary phase will be retarded (Fig. 60.7). In normal phase partition chromatography the stationary phase is a polar solvent, usually water, supported by a solid matrix (e.g. cellulose fibres in paper chromatography) and the mobile phase is an immiscible, non-polar organic solvent. For reverse-phase partition chromatography the stationary phase is a non-polar solvent (e.g. a C_{18} hydrocarbon, such as octadecylsilane) that is chemically bonded to a porous support matrix (e.g. silica), while the mobile phase can be chosen from a wide range of polar solvents, usually water or an aqueous buffered solution containing one or more organic solvents, e.g. acetonitrile. Solutes interact with the stationary phase through non-polar interactions and so the *least* polar solutes elute last from the column. Solute retention and separation are controlled by changing the composition of the mobile phase (e.g. % v/v acetonitrile). Reverse-phase high-performance liquid chromatography (RP–HPLC) can separate a range of non-polar, polar and ionic biomolecules, including peptides, proteins, oligosaccharides and vitamins.

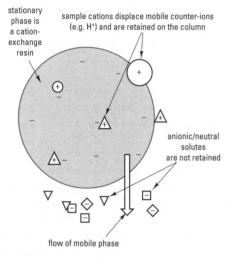

Fig. 60.8 Ion-exchange chromatography (cation exchanger)

Ion-exchange chromatography

Here, separations are carried out using a column packed with a porous matrix that has a large number of ionised groups on its surfaces, i.e. an ion-exchange resin. The groups may be cation or anion exchangers, depending upon their affinity for positive or negative ions. The net charge of a resin depends on the pK_a of the ionisable groups and the solution pH, in accordance with the Henderson–Hasselbalch equation (p. 150).

For most practical applications, you should select the ion-exchange resin and buffer pH so that the test substances are strongly bound by electrostatic attraction to the ion-exchange resin on passage through the system, while the other components of the sample are rapidly eluted (Fig. 60.8). You can then elute the bound ions by changing the pH (which will alter the affinity of the test substance for the resin) or by raising the salt concentration of the mobile phase (to displace the bound ions); for instance, if you pass a solution of increasing concentration through the system, the weakly bound ions will elute first while the strongly bound ions will be elute at a higher concentration.

Ion-exchange chromatography can be used to separate mixtures of a wide range of ions, including amino acids, peptides, proteins and nucleotides. Electrophoresis is an alternative means of separating charged molecules, as described on p. 334 for DNA.

Gel-permeation chromatography (molecular-exclusion chromatography)

Here, the stationary phase is a cross-linked gel containing a network of minute pores and channels of various sizes. Large molecules may be completely excluded from these pores, passing through the interstitial spaces and eluting rapidly from the column in the liquid mobile phase. Smaller molecules will penetrate the gel matrix to an extent that will depend upon their size and shape, retarding their progress through the column (Fig. 60.9).

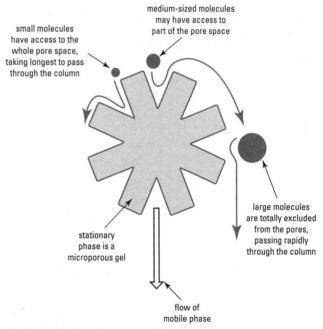

Fig. 60.9 Gel-permeation chromatography

Substances will therefore be eluted from a gel-permeation column in decreasing order of their molecular size.

Cross-linked dextrans (e.g. Sephadex), agarose (e.g. Sepharose) and polyacrylamide (e.g. Bio-Gel) are used to separate mixtures of macromolecules, particularly enzymes, antibodies and other globular proteins. Calibration of a gel-filtration column using molecules of similar shape and known molecular mass enables the molecular mass of other components to be estimated, since a plot of elution volume (V_e) against $\log_{10}$ molecular mass is approximately linear. A further application of gel-permeation chromatography is for separating low-molecular-mass and high-molecular-mass components, e.g. desalting a protein extract using a Sephadex G-25 column is faster and more efficient than dialysis.

Affinity chromatography

This is a highly specific form of adsorption chromatography, where a particular binding molecule (or ligand) is covalently attached to a solid matrix in such a way that the affinity of the ligand for its complementary molecule is unchanged. The immobilised ligand is then packed into a chromatography column where it will selectively adsorb the complementary molecule, preventing its passage through the system (Fig. 60.10). Once the sample has been applied to the column and the contaminating substances washed through with buffer, the purified complementary molecule can be eluted, e.g. by changing the pH or salt concentration, or by adding other substances with a greater affinity for the ligand. Providing a suitable ligand is available, affinity chromatography can be used for single-step purification of small quantities of a particular molecule in the presence of large amounts of contaminating substances. Examples of ligands include:

- triazine dyes, for protein purification;
- enzyme substrates and co-factors, for certain enzymes;
- antibodies, for specific antigens (see p. 309);
- protein A, for IgG antibody purification;
- single-stranded oligonucleotides, for complementary nucleic acids, e.g. mRNA, or particular single-stranded DNA sequences;
- lectins, for specific monosaccharide subunits.

Hydrophobic-interaction chromatography

This technique is used to separate proteins. The stationary phase consists of a non-polar group (e.g. an octyl or phenyl group) bound to a support matrix. This binds proteins differentially, according to the number and position of hydrophobic surface groups. The principle is similar to reverse-phase HPLC (Fig. 60.7), but samples are analysed under non-denaturing conditions, to retain biological activity.

Quantitative analysis

Most detectors and chemical assay systems give a linear response with increasing amounts of the test substance over a given 'working range' of concentration. Alternative ways of converting the measured response to an amount of substance are:

- External standardisation: this is applicable where the sample volume is sufficiently precise to give reproducible results (e.g. HPLC, column chromatography). Peak areas (or heights) of known amounts of the

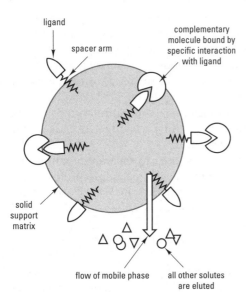

Fig. 60.10 Affinity chromatography

Elution of substances from an affinity system – make sure that your elution conditions do not affect the interaction between the ligand and the stationary phase, or you may elute the ligand from the column.

Problems with peaks – non-symmetrical peaks may result from column overloading, co-elution of solutes, poor packing of the stationary phase, or interactions between the substance and the support material.

substance give a calibration factor or calibration curve, used to calculate the amount of test substance in the sample (Chapter 50).

- Internal standardisation: where you add a known amount of a reference substance (not originally present in the sample) to the sample, to give an additional peak in the elution profile. You determine the response of the detector to the test and reference substances by analysing a standard containing known amounts of both substances, to provide a response factor (r), where:

$$r = \frac{\text{peak area (or height) of test substance}}{\text{peak area (or height) of reference substance}} \qquad [60.5]$$

Use this response factor to quantify the amount of test substance (Q_t) in a sample containing a known amount of the reference substance (Q_r), from:

Using external standardisation – samples and standards should be analysed more than once, to confirm the reproducibility of the technique.

$$Q_t = \frac{\text{peak area (or height) of test substance}}{\text{peak area (or height) of reference substance}} \times (Q_r \div r) \quad [60.6]$$

Internal standardisation should be the method of choice wherever possible, since it is unaffected by small variations in sample volume (e.g. for GC microsyringe injection). The internal standard should be chemically similar to the test substance(s) and must give a peak that is distinct from all others in the sample. An additional advantage of an internal standard that is chemically related to the test substance is that it may show up problems due to changes in detector response, etc. A disadvantage is that it may be difficult to find the space to fit an internal standard peak into a complex chromatogram.

Using an internal standard – you should add an internal standard to the sample at the first stage in the extraction procedure, so that any loss or degradation of test substance during purification is accompanied by an equivalent change in the internal standard.

Text reference

Stahl, E. (1969) *Thin Layer Chromatography – A Laboratory Handbook*, 2nd edn. Springer-Verlag, Berlin.

Sources for further study

Cazes, J. (2005) *Encyclopedia of Chromatography*. CRC Press, Boca Raton, Florida.

Gooding, K.M. and Regnier, F.E. (2002) *HPLC of Biological Macromolecules Revised and Expanded*. Marcel Dekker, New York.

Vijayalashmi, M.A. (2002) *Biochromatography*. Taylor & Francis, London.

Wilson, K. and Walker, J. (eds) (2005) *Principles and Techniques of Biochemistry and Molecular Biology*, 6th edn. Cambridge University Press, Cambridge.

Chromatography

60.1 Calculate R_F and R_X values from a chromatogram. The figure below represents the separation of three pigments by thin-layer chromatography.

(a) What is the R_F value of each pigment?

(b) What is the mobility of B and C, relative to A (R_A)?

Express both answers to three significant figures.

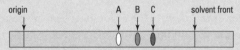

origin   A   B   C   solvent front

Thin-layer chromatographic separation of a mixture of pigments (A, B and C)

60.2 Calculate the resolution and selectivity of two components from a chromatogram. Two compounds were separated by column chromatography, giving retention times of 4 min 30 s for A and 6 min 12 s for B, while a compound that was completely excluded from the stationary phase was eluted in 1 min 35 s. The base width of peak A was 40 s and the base width of peak B was 44 s. Calculate (a) the selectivity and (b) the resolution for these two compounds (express all answers to three significant figures).

60.3 Calculate the amount of substance in a chromatographic separation using an internal standard. The chromatogram below represents the separation of three carbohydrates by GC.

Peak A corresponds to mannitol, peak B to sucrose and peak C to the internal standard, trehalose, at 1.50 mg. The table below gives retention times and areas of the peaks:

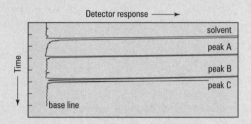

Gas–liquid chromatographic separation of three carbohydrates. Note that the plotter shows peaks A and B to be off-scale: however, the instrument still gives a valid area measurement (see table below).

Retention times and peak areas for carbohydrates A, B and C

Peak	Retention time (min)	Area (relative)
A	3.92	2060
B	5.82	1898
C	6.03	604

Given a response factor (r) of 1.26 for mannitol and 0.92 for sucrose (relative to the internal standard), determine the amount of A and B in the sample (express your answer to three significant figures).

Analysis and presentation of data

The process of discovering the meaning within your results can be fascinating. There are two main elements to this process:

- Exploratory data manipulation – this is used to investigate the nature of your results and suggest possible patterns and relationships within the data set. The aim is to generate hypotheses for further investigation. Exploratory techniques allow you to visualise the form of your data. They are ideal for examining results from pilot 'studies', but should be used throughout your investigations.
- Confirmatory analysis – this is used to test the hypotheses generated during the exploratory phase. The techniques required are generally statistical in nature and are dealt with in Chapter 66.

 KEY POINT *Spreadsheets (Chapter 11) are invaluable tools for data manipulation and transformation: complex mathematical procedures can be carried out rapidly and the results visualised almost immediately using the inbuilt graphing functions. Spreadsheets also facilitate the statistical analysis of data (see Chapters 65 and 66).*

Organising numbers

In order to organise, manipulate and summarise data, you should:

- Simplify the numbers, e.g. by rounding or taking means. This avoids the detail becoming overwhelming.
- Rearrange your data in as many ways as possible for comparison.
- Display in graphical form; this provides an immediate visual summary that is relatively easy to interpret.
- Look for an overall pattern in the data – avoid getting lost in the details at this stage.
- Look for any striking exceptions to that pattern (outliers) – they often point to special cases of particular interest or to errors in the data produced through mistakes during the acquisition, recording or copying of data.
- Move from graphical interpretations to well-chosen numerical summaries and/or verbal descriptions, including where applicable an explanatory hypothesis.

After collecting data, the first step is often to count how frequently each value occurs and to produce a frequency table. The frequency is simply the number of times a value occurs in the data set, and is, therefore, a count. The raw data could be acquired using a tally chart system to provide a simple frequency table. To construct a tally chart (e.g. Fig. 61.1):

- enter only one tally at a time;
- if working from a data list, cross out each item on the list as you enter it on to the tally chart, to prevent double entries;
- check that all values are crossed out at the end and that the totals agree.

Summarising your results – *original, unsummarised data belong only in your primary record, either in laboratory books or as computer records. You should produce summary tables to organise and condense original data.*

Colour	Tally	Total
Green	III	3
Blue	HHT III	8
Red	IIII	4
White	HHT HHT II	12
Black	I	1
Maroon	III	3
Yellow	II	2
		33

Fig. 61.1 An example of a tally chart

Producing a histogram – *a neatly constructed tally chart will double as a rough histogram.*

Table 61.1 An example of a frequency table

Size class	Frequency	Relative frequency (%)
0–4.9	7	2.6
5–9.9	23	8.6
10–14.9	56	20.9
15–19.9	98	36.7
20–24.9	50	18.7
25–29.9	30	11.2
30–34.9	3	1.1
Total	267	99.8*

* ≠ 100 due to rounding error.

Stem	Leaves
7	23
7	55
7	6
7	9
8	000
8	233
8	45555
8	77
8	888899
9	0000111111
9	2333333
9	44555555555
9	66677777
9	88888999
10	00

Fig. 61.2 A simple 'stem and leaf' plot of a data set. The 'stem' shows the common component of each number, while the 'leaves' show the individual components, e.g. the top line in this example represents the numbers 72 and 73.

Example Allometry involves a logarithmic transformation of data that reveals aspects of the relationship between the dimensions of organisms (see p. 294).

Convert the data to a formal table when complete (e.g. Table 61.1). Because proportions are easier to compare than class totals, the table may contain a column to show the relative frequency of each class. Relative frequency can be expressed in decimal form (as a proportion of 1) or as a percentage (as a proportion of 100).

Graphing data

Graphs are an effective way to investigate trends in data and can reveal features that are difficult to detect from a table, e.g. skewness of a frequency distribution. The construction and use of graphs is described in detail in Chapter 62. When investigating the nature of your data, the main points are as follows:

- Make the values stand out clearly; attention should focus on the actual data, not the labels, scale markings, etc. (contrast with the requirements for constructing a graph for data presentation, see p. 387).
- Avoid clutter in the graph; leave out grid lines and try to use the simplest graph possible for your purpose.

 KEY POINT *Use a computer spreadsheet with graphics options whenever possible: the speed and flexibility of these powerful tools should allow you to explore every aspect of your data rapidly and with relatively little effort (see Chapters 11 and 62).*

Displaying distributions

A visual display of a distribution of values is often useful for variables measured on an interval or ratio scale (p. 156). The distribution of a variable can be displayed by a frequency table for each value or, if the possible values are numerous, groups (classes) of values of the variable. Graphically, there are two main ways of viewing such data:

- histograms, (see pp. 389, 390), generally used for large samples;
- stem and leaf plots (e.g. Fig. 61.2), often used for samples of less than 100: these retain the actual values and are faster to draw by hand. The main drawback is the limitation imposed by the choice of stem values since these class boundaries may obscure some features of the distribution.

These displays allow you to look at the overall shape of a distribution and to observe any significant deviations from the idealised theoretical ones. Where necessary, you can use data transformations to investigate any departures from standard distribution patterns such as the normal distribution or the Poisson distribution.

Transforming data

Transformations are mathematical functions applied to data values. They are particularly valuable where your results are related to areas and volumes (e.g. leaf area, body mass).

The most common use of transformations is to prepare data sets so that specific statistical tests may be applied. For instance, if you find that your data distribution is unimodal but not symmetrical (p. 415), it is often useful to apply a transformation that will redistribute the data values to

Using transformations – note that if you wish to conserve the order of your data, you will need to take negative values when using a reciprocal function (i.e. $-1/n^n$). This is essential when using a box plot to compare graphically the effects of transformations on the five-number summary of a data set.

form a symmetrical distribution. The object of this exercise is often to find the function that most closely changes the data into a standard normal distribution, allowing you to apply a wide range of parametric hypothesis-testing statistics (see Chapter 66). A frequently used transformation is to take logarithms of one or more sets of values: if the data then approximate to a normal distribution, the relationship is termed 'log-normal'.

Some general points about transformations are:

- They should be made on the raw data, not on derived data values: this is simpler, mathematically valid, and more easily interpreted.
- The transformed data can be analysed like any other numbers.
- Transformed data can be examined for outliers, which may be more important if they remain after transformation.

Figure 61.3 presents a ladder of transformations that will help you decide which transformations to try (see also Table 66.1). Note that percentage and proportion data are usually arc-sine transformed, which is a more complex procedure; consult Sokal and Rohlf (1994) for details.

Figure 61.4 illustrates the following 'quick-and-easy' way to choose a transformation:

1. Calculate the 'five-number summary' for the untransformed data (p. 419).
2. Present the summary graphically as a 'box-and-whisker' plot (p. 419).
3. Decide whether you need to correct for positive or negative skew (p. 421).
4. Apply one of the 'mild' transformations in Fig. 61.3 *on the five-number summary values only.*
5. Draw a new box-and-whisker plot and see whether the skewness has been corrected.
6. If the skewness has been undercorrected, try again with a stronger transformation. If it has been overcorrected, try a milder one.
7. When the distribution appears to be acceptable, transform the full data set and recalculate the summary statistics. If necessary use a statistical test to confirm that the transformed data are normally distributed (p. 430).
8. If no simple transformation works well, you may need to use non-parametric statistics when comparing data sets.

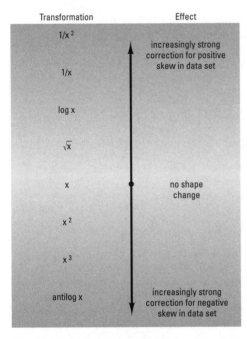

Transformation

$1/x^2$

$1/x$

$\log x$

$\sqrt{x}$

x

x^2

x^3

antilog x

Effect

increasingly strong correction for positive skew in data set

no shape change

increasingly strong correction for negative skew in data set

Fig. 61.3 Ladder of transformations (after J.W. Tukey)

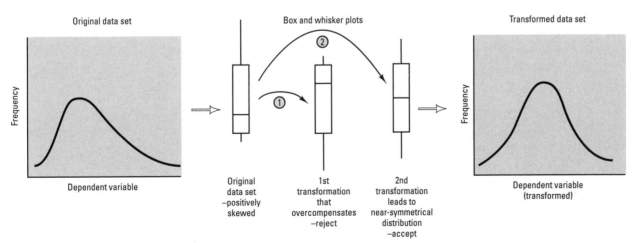

Original data set

Box and whisker plots

Transformed data set

Frequency

Dependent variable

Original data set –positively skewed

1st transformation that overcompensates –reject

2nd transformation leads to near-symmetrical distribution –accept

Frequency

Dependent variable (transformed)

Fig. 61.4 Illustration of the processes of transforming a data set

Text references and sources for further study

Heath, D. (1995) *An Introduction to Experimental Design and Statistics for Biology*. UCL Press, London.

Quinn, G.P. and Keough, M.J. (2002) *Experimental Design and Data Analysis for Biologists*. Cambridge University Press, Cambridge.

Sokal, R.R. and Rohlf, F.J. (1994) *Biometry*, 3rd edn. W.H. Freeman, San Francisco.

Study exercises

The following study exercises can be most easily carried out using a spreadsheet such as Microsoft Excel.

61.1 Compare frequency distributions for samples with different sample sizes. Compute relative frequencies for each class in the table below. Graph the data as a frequency polygon.

Frequency distribution data

Hyphal length (μm)	Treatment B	Treatment A	Hyphal length (μm)	Treatment B	Treatment A
0	0	0	13	6	30
1	1	0	14	3	26
2	2	0	15	1	22
3	4	1	16	1	18
4	8	1	17	0	14
5	13	2	18	0	12
6	22	5	19	0	8
7	36	9	20	0	5
8	48	14	21	0	4
9	46	21	22	0	2
10	32	26	23	0	2
11	18	31	24	0	1
12	11	33	25	0	1

61.2 Use a stem and leaf plot. Work out the mean and standard deviation of the data contained in the following stem and leaf plot:

6	1334
6	56788
7	013344
7	556667899
8	11224
8	569
9	03
9	6

Stem and leaf plot

61.3 Use transformations to correct skew in a data set. Find a transformation that will make the data of treatment B in Study exercise 61.1 approximately symmetrical about the mean. Demonstrate graphically that you have accomplished this.

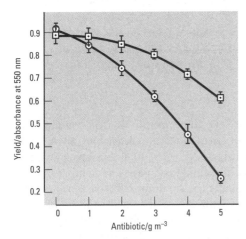

Fig. 62.1 Effect of antibiotic on yield of two bacterial isolates: ○, sensitive isolate; □, resistant isolate. Vertical bars show standard errors ($n = 6$).

Graphs can be used to show detailed results in an abbreviated form, displaying the maximum amount of information in the minimum space. Graphs and tables present findings in different ways. A graph (figure) gives a visual impression of the content and meaning of your results, while a table provides an accurate numerical record of data values. You must decide whether a graph should be used, e.g. to illustrate a pronounced trend or relationship, or whether a table (Chapter 63) is more appropriate.

A well-constructed graph will combine simplicity, accuracy and clarity. Planning of graphs is needed at the earliest stage in any write-up as your accompanying text will need to be structured so that each graph delivers the appropriate message. Therefore, it is best to decide on the final form for each of your graphs before you write your text. The text, diagrams, graphs and tables in a laboratory write-up or project report should be complementary, each contributing to the overall message. In a formal scientific communication it is rarely necessary to repeat the same data in more than one place (e.g. as a table and as a graph). However, graphical representation of data collected earlier in tabular format may be applicable in laboratory practical reports.

Practical aspects of graph drawing

The following comments apply to graphs drawn for laboratory reports. Figures for publication, or similar formal presentation, are usually prepared according to specific guidelines provided by the publisher/organiser.

> **KEY POINT** *Graphs should be self-contained – they should include all material necessary to convey the appropriate message without reference to the text. Every graph must have a concise explanatory title to establish the content. If several graphs are used, they should be numbered, so they can be quoted in the text.*

- Consider the layout and scale of the axes carefully. Most graphs are used to illustrate the relationship between two variables (x and y) and have two axes at right angles (e.g. Fig. 62.1). The horizontal axis is known as the abscissa (x axis) and the vertical axis as the ordinate (y axis).
- The axis assigned to each variable must be chosen carefully. Usually the x axis is used for the independent variable (e.g. treatment) while the dependent variable (e.g. biological response) is plotted on the y axis (p. 185). When neither variable is determined by the other, or where the variables are interdependent, the axes may be plotted either way round.
- Each axis must have a descriptive label showing what is represented, together with the appropriate units of measurement, separated from the descriptive label by a solidus or 'slash' ($/$), as in Fig. 62.1, or by brackets, as in Fig. 62.2.
- Each axis must have a scale with reference marks ('tics') on the axis to show clearly the location of all numbers used.
- A figure legend should be used to provide explanatory detail, including a key to the symbols used for each data set.

Selecting a title – it is a common fault to use titles that are grammatically incorrect: a widely applicable format is to state the relationship between the dependent and independent variables within the title, e.g. 'The relationship between enzyme activity and external pH'.

Remembering which axis is which – a way of remembering the orientation of the x axis is that x is a 'cross', and it runs 'across' the page (horizontal axis) while y is the first letter of yacht, with a large vertical mast (vertical axis).

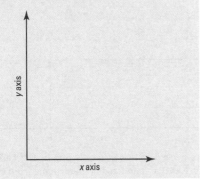

Using graphs

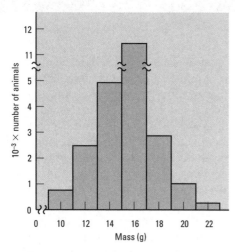

Fig. 62.2 Frequency distribution of masses for a sample of animals (sample size 24 085); the size class interval is 2 g.

Example For a data set where the smallest number on the log axis is 12 and the largest number is 9000, three-cycle log–linear paper would be used, covering the range 10–10 000 (Fig 62.3).

Handling very large or very small numbers

To simplify presentation when your experimental data consist of either very large or very small numbers, the plotted values may be the measured numbers multiplied by a power of 10: this multiplying power should be written immediately before the descriptive label on the appropriate axis (as in Fig. 62.2). However, it is often better to modify the primary unit with an appropriate prefix (p. 159) to avoid any confusion regarding negative powers of 10.

Size

Remember that the purpose of your graph is to communicate information. It must not be too small, so use at least half an A4 page and design your axes and labels to fill the available space without overcrowding any adjacent text. If using graph paper, remember that the white space around the grid is usually too small for effective labelling. The shape of a graph is determined by your choice of scale for the x and y axes which, in turn, is governed by your experimental data. It may be inappropriate to start the axes at zero (e.g. Fig. 62.1). In such instances, it is particularly important to show the scale clearly, with scale breaks where necessary, so the graph does not mislead. Note that Fig. 62.1 is drawn with 'floating axes' (i.e. the x and y axes do not meet in the lower left-hand corner), while Fig. 62.2 has clear scale breaks on both x and y axes.

Graph paper

In addition to conventional linear (squared) graph paper, you may need the following:

- Probability graph paper. This is useful when one axis is a probability scale (e.g. p. 430)
- Log–linear graph paper. This is appropriate when one of the scales shows a logarithmic progression, e.g. the exponential growth of cells in liquid culture (p. 272). Log–linear paper is defined by the number of logarithmic divisions (usually termed 'cycles') covered (e.g. Fig. 62.3), so make sure you use a paper with the appropriate number of cycles for your data. An alternative approach is to plot the log-transformed values on 'normal' graph paper.

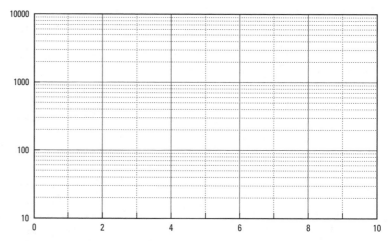

Fig. 62.3 Representation of three-cycle log–linear graph paper, marked up to show a y axis (log) scale from 10 to 10000 and an x axis (linear) scale from 0 to 10.

- Log–log graph paper. This is appropriate when both scales show a logarithmic progression, e.g. in allometry (p. 294).

Types of graph

Different graphical forms may be used for different purposes, including:

- Plotted curves – used for data where the relationship between two variables can be represented as a continuum (e.g. Fig. 62.4).
- Scatter diagrams – used to visualise the relationship between individual data values for two interdependent variables (e.g. Fig. 62.5) often as a preliminary part of a correlation analysis (p. 434).
- Three-dimensional graphs show the interrelationships of three variables, often one dependent and two independent (e.g. Fig. 62.6). A contour diagram is an alternative method of representing such data.
- Histograms represent frequency distributions of continuous variables (e.g. Fig. 62.7). An alternative is the tally chart (p. 383).
- Frequency polygons emphasise the form of a frequency distribution by joining the coordinates with straight lines, in contrast to a histogram. This is particularly useful when plotting two or more frequency distributions on the same graph (e.g. Fig. 62.8).
- Bar charts represent frequency distributions of a discrete qualitative or quantitative variable (e.g. Fig. 62.9). An alternative representation is the line chart (Fig. 66.3, p. 429).
- Pie charts illustrate portions of a whole (e.g. Fig. 62.10).
- Pictographs give a pictorial representation of data (e.g. Fig. 62.11).

The plotted curve

This is the commonest form of graphical representation used in biology. The key features are outlined below and given in checklist form in Box 62.1, while Box 62.2 gives advice on using Microsoft Excel.

Box 62.1   Checklist for the stages in drawing a graph

The following sequence can be used whenever you need to construct a plotted curve: it will need to be modified for other types of graph.

1. **Collect all of the data values and statistical values** (in tabular form, where appropriate).
2. **Decide on the most suitable form of presentation**: this may include transformation (p. 384) to convert data to linear form.
3. **Choose a concise descriptive title**, together with a reference (figure) number and date, where necessary.
4. **Determine which variable is to be plotted on the *x* axis and which on the *y* axis.**
5. **Select appropriate scales for both axes** and make sure that the numbers and their location (scale marks) are clearly shown, together with any scale breaks.
6. **Decide on appropriate descriptive labels for both axes**, with SI units of measurement, where appropriate.
7. **Choose the symbols for each set of data points** and decide on the best means of representation for statistical values.
8. **Plot the points** to show the coordinates of each value with appropriate symbols.
9. **Draw a trend line for each set of points**. Use a see-through ruler, so you can draw the line to have an equal number of points on either side of it.
10. **Write a figure legend**, to include a key that identifies all symbols and statistical values and any descriptive footnotes.

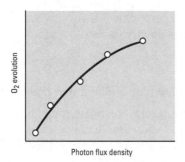

Fig. 62.4 Plotted curve: the rate of photosynthetic O_2 evolution as a function of photon flux density.

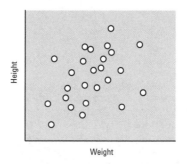

Fig. 62.5 Scatter diagram: height and weight of individual animals in a sample.

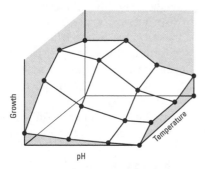

Fig. 62.6 Three-dimensional graph: growth of an organism as a function of temperature and pH.

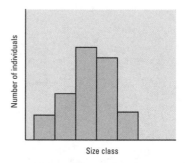

Fig. 62.7 Histogram: the number of plants within different size classes.

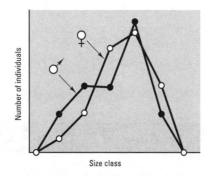

Fig. 62.8 Frequency polygon: frequency distributions of male and female animals according to size.

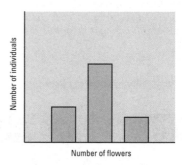

Fig. 62.9 Bar chart: number of flowers per plant.

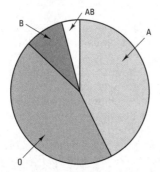

Fig. 62.10 Pie chart: relative abundance of human blood groups.

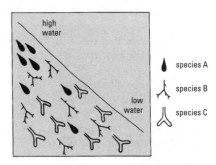

Fig. 62.11 Pictograph: distribution of plants on a rocky shore.

Box 62.2   How to create and amend graphs within a spreadsheet (Microsoft Excel) for use in coursework reports and dissertations

Microsoft Excel can used to create plotted curves, bar charts and histograms of reasonable quality, but only if you know how to amend the default settings to improve the overall effect, so that they meet the standard required for practical and project reports. As with a hand-drawn graph, the basic stages in graph drawing (Box 62.1) still apply.

Producing a plotted curve or bar chart
1. **Create the appropriate type of graph (chart) for your data.** Enter your data into the spreadsheet (e.g. in two columns), select the data array (highlight the appropriate cells by clicking and holding down the left mouse button and dragging) then use the 'Chart Wizard' icon (or the *Insert > Chart* option) to select a particular *Chart type*. For a plotted curve, use the *XY (Scatter)* option with unconnected points as the *Chart sub-type* – the line will be added at a later stage (note: never use the *Line Plot* because this will create a graph with evenly spaced x axis entries, regardless of their actual values). For a bar chart (p. 389), select the *Column* option. Work through each step of the Chart Wizard, selecting appropriate options for *Chart source*, *Chart options* (including labels for x and y axes) and *Location*. The graphs shown in Fig. 62.12(a) and Fig. 62.13(a) were producing using the default settings in Excel 2003.
2. **Change the default settings to improve the overall appearance.** Consider each element of the image

in turn, including the overall size, height and width of your graph (click on the chart using the left-hand mouse button to show the 'sizing handles' and drag these to resize). The graphs shown in Fig. 62.12(b) and Fig. 62.13(b) were produced by altering the settings, typically by moving the cursor over the feature and clicking the right mouse button to reveal an additional menu of editing/ formatting options.

The examples given below are for illustrative purposes only, and should not necessarily be regarded as a prescriptive list:

Example for a plotted curve (compare Fig. 62.12(a) with Fig. 62.12(b)):
- Grey background removed using *Clear* function, and border line around graph removed using *Format Chart Area > Patterns > Border > None*.
- Horizontal gridlines can be either removed using *Clear* function or, if desired (Fig. 62.12(b)), changed using *Chart Options > Gridlines* to show major and minor gridlines, then selecting *Format Gridlines* option to change the *Colour, Scale* and *Weight* of the gridlines, to make them more like those of conventional graph paper.
- Unnecessary legend box on right-hand side removed using the *Clear* option.
- Axes reformatted using the *Format Axis* menu (*Patterns, Scale* and *Number* options).

(a)

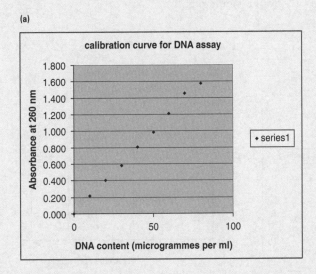

(b)

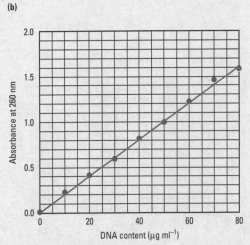

Fig. 1   **Calibration curve for DNA assay.** Performed using an A100X spectrophotometer (on 01.04.07). Values shown are averages of triplicate measurements.

Fig. 62.12   Examples of a plotted curve produced in Microsoft Excel using (a) default settings and (b) modified (improved) settings

(continued)

Box 62.2 (continued)

- Axis label changes achieved using the *Format Axis Title > Font* menu options (including superscript for selected parts – alternatively, cut and paste symbols/text from Word).
- Data point style changed using the *Format Data Series > Patterns > Marker > Custom* options for *Style, Foreground* (edge of symbol), *Background* (centre of symbol) and *Size* (note: *Line* function not used, since it joins individual data points, rather than giving a linear plot).
- Straight line of best fit added using the *Format Data Series > Add Trendline > Type > Linear* option, with the line thickness then adjusted using the *Format Trendline > Patterns > Custom > Weight* option (see also Box 50.2, p. 306).

Example for a bar chart (compare Fig. 62.13(a) with Fig. 62.13(b)):

- Grey background removed using *Clear* function, and border line around graph removed using *Format Chart Area > Patterns > Border > None*; colour of horizontal gridlines changed to light grey using *Format Gridlines > Patterns > Custom > Colour* option.
- Unnecessary legend box on right-hand side removed using the *Clear* option.
- Axes reformatted (tic marks, scales and font) using the *Format Axis* menu (*Patterns, Scale* and *Font* options).
- Axis labels changed to Times New Roman regular 12-point font size using *Format Axis Title > Font, > Font style* and *> Size* options.
- Bar colour removed using *Format Data Series > Patterns > Area*, selecting the white colour (not option *None*).

- Individual numbers shown using the *Format Data Series > Labels > Value* option; bar width altered using *Format Data Series > Options > Gap Width*.

Note that for both types of graph, it is better not to use the default *Chart Title* option within Excel, which places the title at the top of the chart (as in Fig. 62.13(a)), but to cut and paste your untitled graph into a word processor such as Microsoft Word and then type a formal figure legend below the graph itself (as in Fig. 62.12(b) and Fig. 62.13(b)). However, once your graph is embedded into Word, it is generally best not to make further amendments – go back to the original Excel file, make the required change(s) and reinsert the graph into Word.

Producing a histogram
The histogram function in Excel requires a little more effort to master, compared with the 'Chart Wizard' graph-drawing function. Essentially, a histogram is a graphical display of frequencies (counts) for a continuous quantitative variable (pp. 389, 395), where the data values are grouped into discrete classes. It is possible to select the upper limit for each class into which you want your data to be grouped (termed 'bin range values' in Excel).

An alternative approach is to let the software select the group ranges for you: Excel selects evenly distributed bins between the minimum and maximum values (this is often less effective than selecting your own class boundaries).

The following steps outline the procedure used to create the histograms shown in Fig. 62.14 for the table of data opposite (length, in mm, of 24 leaf petioles from a single plant).

(a)

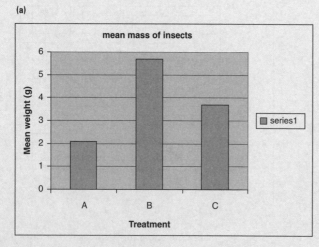

(b)

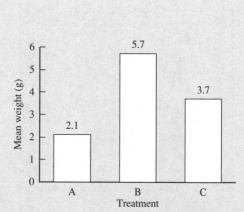

Fig. 2 Bar chart for mean mass of insects treated using three different methods. Key: A = chloroform, B = saline, C = formalin.

Fig. 62.13 Examples of a bar chart produced in Microsoft Excel using (a) default settings and (b) modified (improved) settings

(continued)

Box 62.2 (continued)

7.2	6.5	7.1	8.5	6.6	7.2
7.0	7.3	8.6	9.1	7.5	8.3
7.1	5.7	7.3	7.6	6.9	7.1
8.3	7.6	5.4	8.6	7.9	8.0

1. **Enter the raw data values,** e.g. as a single column of numbers, or as an array, as above.
2. **Decide the class intervals to be used.** Base your choice on the number of data points and the maximum and minimum values (for a small data set, you may be able to do this by quickly examining the data set, whereas for a large data set, use the Excel – *COUNT, MAX* and *MIN* functions) – a typical histogram would have 4–10 classes depending on the level of discrimination required. Enter the upper limit for each class (bin range values) in a separate array of cells close to the data, in ascending order (in the above example, 6, 7, 8 and 9 were chosen – the few data values above the final bin value will be shown on the histogram as a group labelled 'more').
3. **Select the histogram function, then input your data and class-interval values.** From the dropdown menu under *Tools* select *Data Analysis > Histogram* (if the *Data Analysis* option is not already activated on the PC you are using, select using the *Add Ins* function within this menu, but note that this addition may require the Microsoft Office software CD) then *Histogram*. Input your data into the *Input Range* box (highlight the appropriate cells by clicking and dragging while holding down the left mouse button) and similarly input the *Bin Range* values into the appropriate box (if this is left empty,

Excel will select default *Bin Range* values) – most of the remaining boxes can be left empty, though you must click the last box to get a *Chart Output*, otherwise the software will give the numerical counts for each group, without a histogram. Click *OK* and entries in a new worksheet will be created showing the upper limits of each group (column labelled *Bin*) along with the counts for each group (column labelled *Frequency*), plus a poorly constructed chart based on default settings, as shown in Fig. 62.14(a) (note that the default output is a bar chart rather than a histogram, since there are gaps between the groups).

Example for a histogram Fig. 62.14(b) shows the following modifications compared with Fig. 62.14(a):
- Chart resized to increase height (use sizing handles); grey background removed using *Clear* function and border line around graph removed using *Format Chart Area > Patterns > Border > None*.
- Title and unnecessary legend box removed using *Clear* option; y scale amended using *Format Axis > Scale* command; x labels (classes) amended by typing class limits directly into the cells containing the bin values.
- Axes reformatted using the *Format Axis* menu, and new text typed directly into the axis label box (double-click on the chart to access).
- Bar colour changed to light grey using *Format Data Series > Patterns > Area* option.
- Bar chart converted to true histogram (no gaps between bars) using the *Format Data Series > Options menu,* setting *Gap Width* = 0.
- Figure legend added below figure in Microsoft Word, following copying and pasting of the Excel histogram into a Word file.

(a)

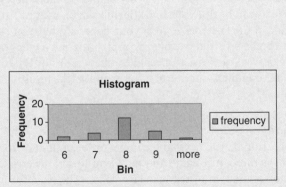

(b)

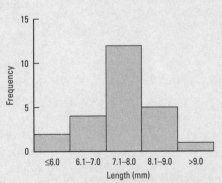

Fig. 3 Histogram of petiole lengths (mm) in a single plant (n = 24).

Fig. 62.14 Examples of histogram output from Microsoft Excel using (a) default settings and (b) modified (improved) settings

Data points

Each data point must be shown accurately, so that any reader can determine the exact values of x and y. In addition, the results of each treatment must be readily identifiable. A useful technique is to use a dot for each data point, surrounded by a hollow symbol for each treatment (see Fig. 62.1). An alternative is to use symbols only (Fig. 62.4), though the coordinates of each point are defined less accurately. Use the same symbol for the same entity if it occurs in several graphs and provide a key to all symbols.

Statistical measures

If you are plotting average values for several replicates and if you have the necessary statistical knowledge, you can calculate the standard error (p. 420, or the 95 per cent confidence limits (p. 434) for each mean value and show these on your graph as a series of vertical bars (see Fig. 62.1). Make it clear in the legend whether the bars refer to standard errors or 95 per cent confidence limits and quote the value of n (the number of replicates per data point). Another approach is to add a least significant difference bar (p. 431) to the graph.

Interpolation

Once you have plotted each point, you must decide whether to link them by straight lines or a smoothed curve. Each of these techniques conveys a different message to your reader. Joining the points by straight lines may seem the simplest option, but may give the impression that errors are very low or non-existent and that the relationship between the variables is complex. Joining points by straight lines is appropriate in certain graphs involving time sequences (e.g. the number of animals at a particular site each year), or for repeat measurements where measurement error can be assumed to be minimal (e.g. recording a patient's temperature in a hospital, to emphasise any variation from one time point to the next). However, in most plotted curves the best straight line or curved line should be drawn (according to appropriate mathematical or statistical models, or by eye), to highlight the relationship between the variables – after all, your choice of a plotted curve implies that such a relationship exists. Don't worry if some of your points do not lie on the line: this is caused by errors of measurement and by biological variation (p. 157). Most curves drawn by eye should have an equal number of points lying on either side of the line. You may be guided by 95 per cent confidence limits, in which case your curve should pass within these limits wherever possible.

Curved lines can be drawn using a flexible curve, a set of French curves, or freehand. In the latter case, turn your paper so that you can draw the curve in a single, sweeping stroke by a pivoting movement at the elbow (for larger curves) or wrist (for smaller ones). Do not try to force your hand to make complex, unnatural movements, as the resulting line will not be smooth.

Extrapolation

Be wary of extrapolation beyond the upper or lower limit of your measured values. This is rarely justifiable and may lead to serious errors. Whenever extrapolation is used, a dotted line ensures that the reader is aware of the uncertainty involved. Any assumptions behind an extrapolated curve should also be stated clearly in your text.

The histogram

Whereas a plotted curve assumes a continuous relationship between the variables by interpolating between individual data points, a histogram involves no such assumptions. Histograms are also used to represent frequency distributions (p. 383), where the y axis shows the number of times a particular value of x was obtained (e.g. Fig. 62.2). As in a plotted curve, the x axis represents a continuous quantitative variable that can take any value within a given range (e.g. plant height), so the scale must be broken down into discrete classes and the scale marks on the x axis should show either the mid-points (mid-values) of each class, or the boundaries between the classes.

The columns are contiguous (adjacent to each other) in a histogram, in contrast to a bar chart, where the columns are separate because the x axis of a bar chart represents discrete values.

Interpreting graphs

The process of analysing a graph can be split into five phases:

1. Consider the context. Look at the graph in relation to the aims of the study in which it was reported. Why were the observations made? What hypothesis was the experiment set up to test? This information can usually be found in the Introduction or Results section of a report. Also relevant are the general methods used to obtain the results. This might be obvious from the figure title and legend, or from the Materials and Methods section.

2. Recognise the graph form and examine the axes. First, what kind of graph is presented (e.g. histogram, plotted curve)? You should be able to recognise the main types summarised on pp. 389–90 and their uses. Next, what do the axes measure? You should check what quantity has been measured in each case and what units are used.

3. Look closely at the scale of each axis. What is the starting point and what is the highest value measured? For the x-axis, this will let you know the scope of the treatments or observations (e.g. whether they lasted for 5 min or 20 years; whether a concentration span was two-fold or fifty-fold). For each axis, it is especially important to note whether the values start at zero; if not, then the differences between any treatments shown may be magnified by the scale chosen (see Box 62.3).

4. Examine the symbols and curves. Information will be provided in the key or legend to allow you to determine what these refer to. If you have made your own photocopy of the figure, it may be appropriate to note this directly on it. You can now assess what appears to have happened. If, say, two conditions have been observed while a variable is altered, when exactly do they differ from each other; by how much; and for how long?

5. Evaluate errors and statistics. It is important to take account of variability in the data. For example, if mean values are presented, the underlying errors may be large, meaning that any difference between two treatments or observations at a given x-value could simply have arisen by chance. Thinking about the descriptive statistics used (Chapter 65) will allow you to determine whether apparent differences could be significant in both statistical and biological senses.

Box 62.3   How graphs can misrepresent and mislead

1. **The 'volume' or 'area' deception** – this is mainly found in histogram or bar chart presentations where the size of a symbol is used to represent the measured variable. For example, the amount of hazardous waste produced in different years might be represented on a chart by different sizes of a chemical drum, with the *y* axis (height of drum) representing the amount of waste. However, if the symbol retains its *shape* for all heights as in Fig. 62.15(a), its *volume* will increase as a cubic function of the height, rather than in direct proportion. To the casual observer, a two-fold increase may look like an eight-fold one, and so on. Strictly, the *height* of the symbol should be the measure used to represent the variable, with no change in symbol width, as in Fig. 62.15(b).

2. **Effects of a non-zero axis** – A non-zero axis acts to emphasise the differences between measures by reducing the range of values covered by the axis. For example, in Fig. 62.16(a), it looks as if there are large differences in mass between males and females; however, if the scale is adjusted to run from zero, then it can be seen that the differences are not large as a proportion of the overall mass. Always scrutinise the scale values carefully when interpreting any graph.

3. **Use of a relative rather than absolute scale** – this is similar to the above, in that data compared using relative scales (e.g. percentage or ratio) can give the wrong impression if the denominator is not the same in all cases. In Fig. 62.17(a), two treatments are shown as equal in *relative* effect, both resulting in 50 per cent relative response compared (say) to the respective controls. However, if treatment A is 50 per cent of a control value of 200 and treatment B is 50 per cent of a control value of 500, then the actual difference in *absolute* response would have been masked, as shown by Fig. 62.17(b).

4. **Effects of a non-linear scale** – when interpreting graphs with non-linear (e.g. logarithmic) scales, you may interpret any changes on an imagined linear scale. For example, the pH scale is logarithmic, and linear changes on this scale mean less in terms of absolute H^+ concentration at high (alkaline) pH than they do at low (acidic) pH. In Fig. 62.18(a), the cell density in two media is compared on a logarithmic scale, while in Fig. 62.18(b), the same data are graphed on a linear scale. Note, also, that the log *y*-axis scale in Fig. 62.18(a) cannot be shown to zero, because there is no logarithm for 0.

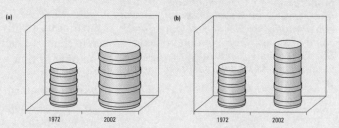

Fig. 62.15   Increase in pesticide use over a 30 year period

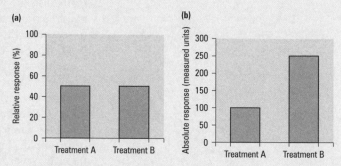

Fig. 62.16   Average mass of males and females in test group

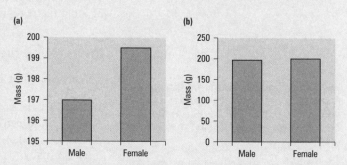

Fig. 62.17   Responses to treatments A and B

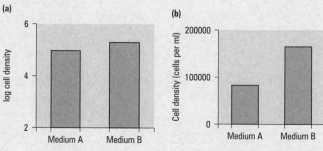

Fig. 62.18   Effect of different media on cell density

(continued)

Box 62.3 (continued)

5. **Unwarranted extrapolation** – a graph may be extrapolated to indicate what would happen if a trend continued, as in Fig. 62.19(a). However, this can only be done under certain assumptions (e.g. that certain factors will remain constant or that relationships will hold under new conditions). There may be no guarantee that this will actually be the case. Figure 62.19(b) illustrates other possible outcomes if the experiment were to be repeated with higher values for the x-axis.

6. **Failure to account for data point error** – this misrepresentation involves curves that are overly complex in relation to the scatter in the underlying data. When interpreting graphs with complex curves, consider the errors involved in the data values. It is probably unlikely that the curve would pass through all the data points unless the errors were very small. Figure 62.20(a) illustrates a curve that appears to assume zero error and is thus overly complex, while Fig. 62.20(b) shows a curve that takes possible errors of the points into account.

7. **Failure to reject outlying points** – this is a special case of the previous example. There may be many reasons for outlying data, from genuine mistakes to statistical 'freaks'. If a curve is drawn through such points on a graph, it indicates that the point carries equal weight with the other points, when in fact, it should probably be ignored. To assess this, consider the accuracy of the measurement, the number and position of adjacent points, and any special factors that might be involved on a one-off basis. Figure 62.21(a) shows a curve where an outlier (arrowed) has perhaps been given undue weight when showing the presumed relationship. If there is good reason to think that the point should be ignored, then the curve shown in Fig. 62.21(b) would probably be more valid.

8. **Inappropriate fitted line** – here, the mathematical function chosen to represent a trend in the data might be inappropriate. A straight line might be fitted to the data, when a curve would be more correct, or vice versa. These cases can be difficult to assess. You need to consider the theoretical validity of the model used to generate the curve (this is not always stated clearly). For example, if a straight line is fitted to the points, the implicit underlying model states that one factor varies in direct relation to another, when the true situation may be more complex. In Fig. 62.22(a), the relationship has been shown as a linear relationship, whereas an exponential relationship, as shown in Fig. 62.22(b), could be more correct.

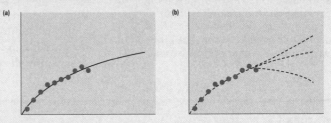

Fig. 62.19 Extrapolation of data under different assumptions

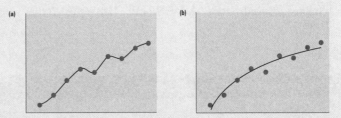

Fig. 62.20 Fitted curves under different assumptions of data error

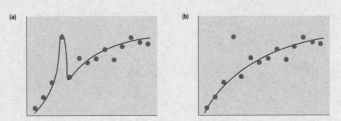

Fig. 62.21 Curves with and without outlier taken into account

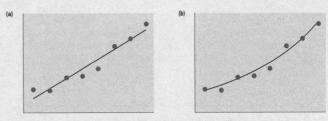

Fig. 62.22 Different mathematical model used to represent trends in data

Using graphs

Understanding graphs within scientific papers – the legend should be a succinct summary of the key information required to interpret the figure without further reference to the main text. This is a useful approach when 'skimming' a paper for relevant information (pp. 18–19).

Sometimes graphs are used to mislead. This may be unwitting, as in an unconscious favouring of a 'pet' hypothesis of the author. Graphs may be used to 'sell' a product in the field of advertising or to favouring of a viewpoint as, perhaps, in politics. Experience in drawing and interpreting graphs will help you spot these flawed presentations, and understanding how graphs can be erroneously presented (Box 62.3) will help you avoid the same pitfalls.

Sources for further study

Briscoe, M.H. (1996) *Preparing Scientific Illustrations: a Guide to Better Posters, Presentations and Publications.* Springer-Verlag, New York.

Carter, M., et al. *Graphing with Excel.* Available: http://www.ncsu.edu/labwrite/res/gt/gt-menu.html Last accessed 09/04/07.
[Online tutorial from the US-NSF LabWrite 2000 Project.]

Institute of Biology (2000) *Biological Nomenclature: Recommendations on Terms, Units and Symbols.* Institute of Biology, London.
[Includes a section on presentation of data.]

Robbins, N.B. (2005) *Creating More Effective Graphs.* Wiley, New York.

Study exercises

62.1 Select appropriate graphical presentations. Choose an appropriate graphical form for each of the following examples:
   (a) Interaction between pH and cation concentration on enzyme activity.
   (b) Proportion of different petal shapes in a countywide survey of *Primula vulgaris* flowers.
   (c) Relationship between pulse rate and age in humans.
   (d) Number of bacteria per field of view.
   (e) Effect of copper concentration on the activity of an enzyme.

62.2 Create a pie chart. Display the following information in the form of a pie chart. Do *not* use a spreadsheet for this exercise.

Behaviour of male brown bear under zoo conditions

Behaviour category	Occasions observed
Eating	37
Scratching	8
Sexual display	13
Mating	2
Sleeping	25

62.3 Create a frequency distribution histogram. The following table gives data for the haemoglobin levels of 100 people. Plot a histogram showing the frequency distribution of the data. Write a brief description of the important features of the distribution.

Haemoglobin content (g l^{-1}) in blood

11.1	14.2	13.5	9.8	12.0	13.9	14.1	14.6	11.0	12.3
13.4	12.9	12.9	10.0	13.1	11.8	12.6	10.7	8.1	11.2
13.8	12.4	12.9	11.3	12.7	12.4	14.6	15.1	11.2	9.7
11.3	14.7	10.8	13.3	11.9	11.4	12.5	13.0	11.6	13.1
9.3	13.5	14.6	11.2	11.7	10.9	12.4	12.0	12.1	12.6
10.9	12.1	13.4	9.5	12.5	11.6	12.2	8.8	10.7	11.1
10.2	11.7	10.4	14.0	14.9	11.5	12.0	13.2	12.1	13.3
12.4	9.4	13.2	12.5	10.8	11.7	12.7	14.1	10.4	10.5
13.3	10.6	10.5	13.7	11.8	14.1	10.3	13.6	10.4	13.9
11.7	12.8	10.4	11.9	11.4	10.6	12.7	11.4	12.9	12.1

62.4 Find examples of misleading graphs. Create a portfolio of examples of misleading graphs taken from newspapers. For each graph, state what aspect is misleading (see Box 62.3) and, where possible, attempt to show the data correctly in a new graph.

A table is often the most appropriate way to present numerical data in a concise, accurate and structured form. Assignments and project reports should contain tables that have been designed to condense and display results in a meaningful way and to aid numerical comparison. The preparation of tables for recording primary data is discussed on p. 192.

Decide whether you need a table, or whether a graph is more appropriate. Histograms and plotted curves can be used to give a visual impression of the relationships within your data (p. 389). On the other hand, a table gives you the opportunity to make detailed numerical comparisons.

 KEY POINT *Always remember that the primary purpose of your table is to communicate information and allow appropriate comparison, not simply to put down the results on paper.*

Preparation of tables

Title

Constructing titles – take care over titles as it is a common mistake in student practical reports to present tables without titles, or to misconstruct the title.

Every table must have a brief descriptive title. If several tables are used, number them consecutively so they can be quoted in your text. The titles within a report should be compared with one another, making sure they are logical and consistent and that they describe accurately the numerical data contained within them.

Structure

Saving space in tables – you may be able to omit a column of control data if your results can be expressed as percentages of the corresponding control values.

Display the components of each table in a way that will help the reader understand your data and grasp the significance of your results. Organise the columns so that each category of like numbers or attributes is listed vertically, while each horizontal row shows a different experimental treatment, organism, sampling site, etc. (as in Table 63.1). Where appropriate, put control values near the beginning of the table. Columns that need to be compared should be set out alongside each other. Use rulings to subdivide your table appropriately, but avoid cluttering it up with too many lines.

Table 63.1 Characteristics of selected photoautotrophic microbes

Division	Species	Optimum [NaCl]* (mol m^{-3})	Intracellular carbohydrate	
			Identity	Quantity[†] (nmol (g dry wt)$^{-1}$)
Chlorophyta	*Scenedesmus quadruplicatum*	340	Sucrose	49.7
	Chlorella emersonii	780	Sucrose	102.3
	Dunaliella salina	4700	Glycerol	910.7
Cyanobacteria	*Microcystis aeruginosa*	<20[‡]	None	0.0
	Anabaena variabilis	320	Sucrose	64.2
	Rivularia atra	380	Trehalose	ND

* Determined after 28-day incubation in modified Von Stosch medium.
[†] Individual samples, analysed by gas–liquid chromatography.
[‡] Poor growth in all media with added NaCl (minimum NaCl concentration 5 mol m^{-3}).
ND Sample lost: no quantitative data.

Headings and sub-headings

These should identify each set of data and show the units of measurement, where necessary. Make sure that each column is wide enough for the headings and for the longest data value.

Numerical data

Within the table, do not quote values to more significant figures than necessary, as this will imply spurious accuracy (pp. 157, 408). By careful choice of appropriate units for each column you should aim to present numerical data within the range 0 to 1 000. As with graphs, it is less ambiguous to use derived SI units, with the appropriate prefixes, in the headings of columns and rows, rather than quoting multiplying factors as powers of 10. Alternatively, include exponents in the main body of the table (see Table 24.1), to avoid any possible confusion regarding the use of negative powers of 10.

Other notations

Avoid using dashes in numerical tables, as their meaning is unclear; enter a zero reading as '0' and use 'NT' not tested or 'ND' if no data value was obtained, with a footnote to explain each abbreviation. Other footnotes, identified by asterisks, superscripts or other symbols in the table, may be used to provide relevant experimental detail (if not given in the text) and an explanation of column headings and individual results, where appropriate. Footnotes should be as condensed as possible. Table 63.1 provides examples.

Statistics

In tables where the dispersion of each data set is shown by an appropriate statistical parameter, you must state whether this is the (sample) standard deviation, the standard error (of the mean) or the 95 per cent confidence limits and you must give the value of n (the number of replicates). Other descriptive statistics should be quoted with similar detail, and hypothesis-testing statistics should be quoted along with the value of P (the probability). Details of any test used should be given in the legend, or in a footnote.

Text

Sometimes a table can be a useful way of presenting textual information in a condensed form (see examples on pp. 347 and 371).

When you have finished compiling your tabulated data, carefully double-check each numerical entry against the original information, to ensure that the final version of your table is free from transcriptional errors. Box 63.1 gives a checklist for the major elements of constructing a table.

Examples If you measured the width of a fungal hypha to the nearest one-tenth of a micrometre, quote the value in the form '52.6 µm'.

Quote the width of a fungal hypha as 52.6 µm, rather than 0.000 052 6 m or 52.6 10^{-6} m.

Saving further space in tables – in some instances a footnote can be used to replace a whole column of repetitive data.

Using spreadsheets and word-processing packages – these can be used to prepare high-quality versions of tables for project work (Box 63.2).

Box 63.1 Checklist for preparing a table

Every table should have the following components:

1. **A title**, plus a reference number and date where necessary.
2. **Headings for each column and row**, with appropriate units of measurement.
3. **Data values**, quoted to the nearest significant figure and with statistical parameters, according to your requirements.
4. **Footnotes** to explain abbreviations, modifications and individual details.
5. **Rulings to emphasise groupings** and distinguish items from each other.

Box 63.2 How to use a word processor (Microsoft Word) or a spreadsheet (Microsoft Excel) to create a table for use in coursework reports and dissertations

Creating tables with Microsoft Word: Word-processed tables are suitable for text-intensive or number-intensive tables, although in the second case entering data can be laborious. When working in this way, the natural way to proceed is to create the 'shell' of the table, add the data, then carry out final formatting on the table.

1. **Move the cursor to the desired position in your document.** This is where you expect the top left corner of your table to appear. Go to the *Table* command, then choose *Insert > Table*.
2. **Select the appropriate number of columns and rows.** Don't forget to add rows and columns for headings. As default, a full-width table will appear, with single rulings for all cell boundaries, with all columns of equal width and all rows of equal height.

   Example of a 4 x 3 Table:

3. **Customise the columns.** By placing the cursor over the vertical rulings then 'dragging', you can adjust their width to suit your heading text entries, which should now be added.

Heading 1	Heading 2	Heading 3	Heading 4

4. **Work through the table adding the data.** Entries can be numbers or text.

Heading 1	Heading 2	Heading 3	Heading 4
xx	xx	xx	xx
xx	xx	xx	

5. **Make further adjustments to column and row widths to suit.** For example, if text fills several rows within a cell, consider increasing the column width, and if a column contains only single or double digit numbers, consider shrinking its width. To combine cells, first highlight them, then use *Table > Merge Cells*. You may wish to reposition text within a cell using *Format > Paragraph > Spacing*.

Heading 1	Heading 2	Heading 3	Heading 4
xx	xx	xx	xx
	xx	xx	xx

6. **Finally, remove selected borders to cells.** One way to do this is using the table borders options accessed from the *Table borders* button on the toolbar, so that your table looks like the examples shown in this chapter.
7. **Add a table title.** This should be positioned *above* the table (*c.f.* a figure title and legend p. 390), legend and footnotes.

   Final version of the Table:

   Table xx. A table of some data

Heading 1	Heading 2[a]	Heading 3	Heading 4
Aaa	xx	xx	xx
	xx	xx	xx

   [a]An example of a footnote

Creating tables with Microsoft Excel:
Tables derived from spreadsheets are effective when you have lots of numerical data, especially when these are stored or created using the spreadsheet itself. When working in this way, you can design the table as part of an output or summary section of the spreadsheet, add explanatory headings, format, then possibly export to a word processor when complete.

1. **Design the output or summary section.** Plan this as if it were a table, including adding text headings within cells.

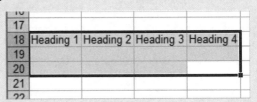

(continued)

Box 63.2 (continued)

2. **Insert appropriate formulae within cells to produce data**. If necessary, formulae should draw on the other parts of the spreadsheet.

17				
18	Heading 1	Heading 2	Heading 3	Heading 4
19	Aaa	=A1	=C3*5	=SDEV(A1:A12)
20	Bbb	=A2	=F45/G12	=SDEV(B1:B12)
21				
22				

3. **Format the cells**. This is important to control the number of decimal places presented (*Format > Cells > Number*).
4. **Adjust column width to suit**. You can do this via the column headings, by placing the cursor over the rulings between columns then 'dragging'.

17				
18	Heading 1	Heading 2	Heading 3	Heading 4
19	Aaa	=A1	=C3*5	=SDEV(A1:A12)
20	Bbb	=A2	=F45/G12	=SDEV(B1:B12)
21				
22				

5. **Add rulings as appropriate**. Use the borders menu on the toolbar as described above.

17				
18	Heading 1	Heading 2	Heading 3	Heading 4
19	Aaa	=A1	=C3*5	=SDEV(A1:A12)
20	Bbb	=A2	=F45/G12	=SDEV(B1:B12)
21				

6. **Add 'real' data values to the spreadsheet**. This should result in the summary values within the table being filled. Check that these are presented with the appropriate number of significant figures (p. 408).
7. **The table can now be copied and pasted to a Word document**. If you wish to link the spreadsheet and the word-processed document so that the latter is updated whenever changes are made to the spreadsheet values, then click *Paste* on the *Formatting* toolbar. Click *Paste Options* next to the data, and then select *Match Destination Table Style and Link to Excel* or *Keep Source Formatting and Link to Excel*.

Sources for further study

Kirkup, L. (1994) *Experimental Methods: An Introduction to the Analysis and Presentation of Data*. Wiley, New York.

Simmonds, D. and Reynolds, L. (1994) *Data Presentation and Visual Literacy in Medicine and Science*. Butterworth-Heinemann, London.

Study exercises

63.1 Redesign a table of data. Using the following example, redraft the table to improve layout and correct inconsistencies.

Concentrations of low molecular weight solutes in bacteria

Bacterium	Constituent	Concentration
Escherichia coli	Proline	21.0 mmol l^{-1}
Escherichia coli	Trehalose	1.547×10^2 kmol m^{-3}
Bacillus subtilis	Proline	39.7 mmol l^{-1}
Bacillus subtilis	Glutamate	0.0521 mmol cm^{-3}
Staphylococcus aureus*	Glutamate	15 260 mmol m^{-3}
Escherichia coli	Glutamate	0.50% w/v
Bacillus subtilis	Trehalose	<0.001% w/v

* Proline and trehalose were not measured.

(continued)

63.2 Devise a text-based table. After reading through this chapter and working from memory, draw up a table listing the principal components of a typical table in the first column, and brief comments on the major features of each component in the second column.

63.3 Interpret data from a table. The table below shows representative data for energy, protein and niacin (vitamin B3) as a function of age for human males and females. The typical average mass ('weight') of each class is also shown.

(a) In what age class does the mass of females exceed that of males?

(b) In what age class is the energy requirement of males greatest in proportion to the energy requirement of females?

(c) Over what age range do males and females have the same protein requirement?

(d) In what age class do males have their highest niacin requirement, and how does this compare with females?

Selected age-related characteristics of human males and females

Age (years)	Average mass (kg)		Energy requirement (MJ)		Protein requirement (g)		Niacin requirement (mg)	
	Male	Female	Male	Female	Male	Female	Male	Female
<1	8	8	3.4	3.2	14	14	6	6
1–3	13	13	5.2	4.9	16	16	9	9
4–6	20	20	7.2	6.5	24	24	12	12
7–10	28	28	8.2	7.3	28	28	13	13
11–14	45	46	9.3	7.7	45	46	17	15
15–18	66	55	11.5	8.8	59	44	20	15
19–50	79	63	10.6	8.1	63	50	19	15
>50	77	65	9.5	7.7	63	50	15	13

The following boxes give advice on dealing with numerical procedures:

- Box 22.2: preparing solutions (p. 130)
- Box 23.1: molar concentrations (p. 139)
- Box 46.1: cell-counting chambers (p. 275)
- Box 46.2: plate (colony) counts (p. 276)
- Box 50.1: calibration curves (p. 304)
- Box 56.1: radioactivity (p. 349)

Biology often requires a numerical or statistical approach. Not only is mathematical modelling an important aid to understanding, but computations are often needed to turn raw data into meaningful information or to compare them with other data sets. Moreover, calculations are part of laboratory routine, perhaps required for making up solutions of known concentration (see p. 130 and below) or for the calibration of a microscope (see p. 258). In research, 'trial' calculations can reveal what input data are required and where errors in their measurement might be amplified in the final result (see p. 187).

 KEY POINT *If you find numerical work difficult, practice at problem-solving is especially important.*

Practising at problem-solving:

- demystifies the procedures involved, which are normally just the elementary mathematical operations of addition, subtraction, multiplication and division (Table 64.1);
- allows you to gain confidence so that you don't become confused when confronted with an unfamiliar or apparently complex form of problem;
- helps you recognise the various forms a problem can take as, for instance, in crossing experiments in classical genetics (Box 53.1).

Table 64.1 Sets of numbers and operations

Sets of numbers	
Whole numbers:	0, 1, 2, 3, ...
Natural numbers:	1, 2, 3, ...
Integers:	... −3, −2, −1, 0, 1, 2, 3, ...
Real numbers:	integers and anything between (e.g. −5, 4.376, 3/16, π, $\sqrt{5}$)
Prime numbers:	subset of natural numbers divisible by 1 and themselves only (i.e. 2, 3, 5, 7, 11, 13, ...)
Rational numbers:	p/q where p (integer) and q (natural) have no common factor (e.g. 3/4)
Fractions:	p/q where p is an integer and q is natural (e.g. −6/8)
Irrational numbers:	real numbers with no exact value (e.g. π)
Infinity:	(symbol ∞) is larger than any number (technically not a number as it does not obey the laws of algebra)
Operations and symbols	
Basic operators:	+, −, × and ÷ will not need explanation; however, / may substitute for ÷, ∗ may substitute for × or this operator may be omitted
Powers:	a^n, i.e. 'a to the power n', means a multiplied by itself n times (e.g. $a^2 = a \times a =$ 'a squared', $a^3 = a \times a \times a =$ 'a cubed'). n is said to be the index or exponent. Note $a^0 = 1$ and $a^1 = a$
Logarithms:	the common logarithm (log) of any number x is the power to which 10 would have to be raised to give x (i.e. the log of 100 is 2; $10^2 = 100$); the antilog of x is 10^x. Note that there is no log for 0, so take this into account when drawing log axes by breaking the axis. Natural or Napierian logarithms (ln) use the base e ($= 2.71828...$) instead of 10
Reciprocals:	the reciprocal of a real number a is $1/a$ ($a \neq 0$)
Relational operators:	$a > b$ means 'a is greater (more positive) than b', $<$ means less than, $\leqslant$ means less-than-or-equal-to and $\geqslant$ means greater-than-or-equal-to
Proportionality:	$a \propto b$ means 'a is proportional to b' (i.e. $a = kb$, where k is a constant). If $a \propto 1/b$, a is inversely proportional to b ($a = k/b$)
Sums:	Σx_i is shorthand for the sum of all x values from $i = 0$ to $i = n$ (more correctly the range of the sum is specified under the symbol)
Moduli:	$|x|$ signifies modulus of x, i.e. its absolute value (e.g. $|4| = |-4| = 4$)
Factorials:	$x!$ signifies factorial x, the product of all integers from 1 to x (e.g. 3! = 6). Note 0! = 1! = 1

Steps in tackling a numerical problem

The step-by-step approach outlined below may not be the fastest method of arriving at an answer, but most mistakes occur where steps are missing, combined or not made obvious, so a logical approach is often better.

Have the right tools ready

Scientific calculators (p. 117) greatly simplify the numerical part of problem-solving. However, the seeming infallibility of the calculator may lead you to accept an absurd result that could have arisen because of faulty key-pressing or faulty logic. Make sure you know how to use all the features on your calculator, especially how the memory works; how to introduce a constant multiplier or divider; and how to obtain an exponent (note that the 'exp' button on most calculators gives you 10^x, not 1^x or y^x; so 1×10^6 would be entered as $\boxed{1}\ \boxed{\text{exp}}\ \boxed{6}$, *not* $\boxed{10}\ \boxed{\text{exp}}\ \boxed{6}$).

Approach the problem thoughtfully

If the individual steps have been laid out on a worksheet, the 'tactics' will already have been decided. It is more difficult when you have to adopt a strategy on your own, especially if the problem is presented as a story and it isn't obvious which equations or rules need to be applied.

- Read the problem carefully as the text may give clues as to how it should be tackled. Be certain of what is required as an answer before starting.
- Analyse what kind of problem it is, which effectively means deciding which equation(s) or approach will be applicable. If this is not obvious, consider the dimensions/units of the information available and think how they could be fitted to a relevant formula. In examinations, a favourite ploy of examiners is to present a problem such that the familiar form of an equation must be rearranged (see Table 64.2 and Box 64.1). Another is to make you use two or more equations in series (see Box 64.2). If you are unsure whether a recalled formula is correct, a dimensional analysis can help: write in all the units for the variables and make sure that they cancel out to give the expected answer.
- Check that you have, or can derive, all of the information required to use your chosen equation(s). It is unusual but not unknown for examiners to supply redundant information. So, if you decide not to use some of the information given, be sure why you do not require it.
- Decide in what format and units the answer should be presented. This is sometimes suggested to you. If the problem requires many changes in the prefixes to units, it is a good idea to convert all data to base SI units (multiplied by a power of 10) at the outset.
- If a problem appears complex, break it down into component parts.

Table 64.2 Simple algebra – rules for manipulating

If $a = b + c$, then $b = a - c$ and $c = a - b$
If $a = b \times c$, then $b = a \div c$ and $c = a \div b$
If $a = b^c$, then $b = a^{1/c}$ and $c = \log a \div \log b$
$a^{1/n} = \sqrt[n]{a}$
$a^{-n} = 1 \div a^n$
$a^b \times a^c = a^{(b+c)}$ and $a^b \div a^c = a^{(b-c)}$
$(a^b)^c = a^{(b \times c)}$
$a \times b = \text{antilog}(\log a + \log b)$

Present your answer clearly

The way you present your answer obviously needs to fit the individual problem. The example shown in Box 64.2 has been chosen to illustrate several important points, but this format would not fit all situations. Guidelines for presenting an answer include:

(a) Make your assumptions explicit. Most mathematical models of biological phenomena require that certain criteria are met before they can be

---

Box 64.1   Example of using the algebraic rules of Table 64.2

Problem: if $a = (b - c) \div (d + e^n)$, find e

1. Multiply both sides by $(d + e^n)$; formula becomes: $a(d + e^n) = (b - c)$

2. Divide both sides by a; formula becomes:
$$d + e^n = \frac{b - c}{a}$$

3. Subtract d from both sides; formula becomes: $e^n = \dfrac{b - c}{a} - d$

4. Raise each side to the power $1/n$; formula becomes: $e = \left\{ \dfrac{b - c}{a} - d \right\}^{1/n}$

---

legitimately applied (e.g. 'assuming the tissue is homogeneous . . .'), while some approaches involve approximations that should be clearly stated (e.g. 'to estimate the mouse's skin area, its body was approximated to a cylinder with radius x and height y . . .').

(b) Explain your strategy for answering, perhaps giving the applicable formula or definitions that suit the approach to be taken. Give details of what the symbols mean (and their units) at this point.

(c) Rearrange the formula to the required form with the desired unknown on the left-hand side (see Table 64.2).

(d) Substitute the relevant values into the right-hand side of the formula, using the units and prefixes as given (it may be convenient to convert values to SI beforehand). Convert prefixes to appropriate powers of 10 as soon as possible.

(e) Convert to the desired units step by step, i.e. taking each variable in turn.

(f) When you have the answer in the desired units, rewrite the left-hand side and underline the answer for emphasis. Make sure that the result is presented to an appropriate number of significant figures (see below).

Check your answer
Having written out your answer, you should check it methodically, answering the following questions:

- Is the answer realistic? You should be alerted to an error if a number is absurdly large or small. In repeated calculations, a result standing out from others in the same series should be double-checked.

- Do the units make sense and match up with the answer required? Don't, for example, present a volume in units of m^2.

- Do you get the same answer if you recalculate in a different way? If you have time, recalculate the answer using a different 'route', entering the numbers into your calculator in a different form and/or carrying out the operations in a different order.

Rounding: decimal places and significant figures

In many instances, the answer you produce as a result of a calculation will include more figures than is justified by the accuracy and precision of the original data. Sometimes you will be asked to produce an answer to a specified number of decimal places or significant figures and other times you will be expected to decide for yourself what would be appropriate.

Units – never write any answer without its unit(s) unless it is truly dimensionless.

Rounding off to a specific number of significant figures – do not round off numbers until you arrive at the final answer or you will introduce errors into the calculation.

Box 64.2 Model answer to a typical biological problem

Problem

Estimate the total length and surface area of the fibrous roots on a maize seedling from measurements of their total fresh weight and mean diameter. Give your answers in m and cm^2 respectively to four significant figures.

Measurements

Fresh weight[a] = 5.00 g, mean diameter[b] = 0.5 mm.

Answer

Assumptions: (1) the roots are cylinders with constant radius[c] and the 'ends' have negligible area; (2) the root system has a density of 1 000 kg m^{-3} (i.e. that of water[d]).

Strategy: from assumption (1), the applicable equations are those concerned with the volume and surface area of a cylinder (Table 64.3), namely:

$$V = \pi r^2 h \qquad\qquad [64.1]$$
$$A = 2\pi rh \text{ (ignoring ends)} \qquad [64.2]$$

where V is volume (m^3), A is surface area (m^2), $\pi \approx 3.14159$, h is height (m) and r is radius (m). The total length of the root system is given by h and its surface area by A. We can find h by rearranging eqn [64.1] and then substitute its value in eqn [64.2] to get A.

To calculate total root length: rearranging eqn [64.1], we have $h = V/\pi r^2$. From measurements[e], $r = 0.25$ mm $= 0.25 \times 10^{-3}$ m.

From density = weight/volume,

$$V = \text{fresh weight/density}$$
$$= 5\,\text{g}/1000\,\text{kg m}^{-3}$$
$$= 0.005\,\text{kg}/1000\,\text{kg m}^{-3}$$
$$= 5 \times 10^{-6}\,\text{m}^3$$

Total root length,

$$h = V/\pi r^2$$
$$5 \times 10^{-6}\,\text{m}^3/3.141\,59 \times (0.25 \times 10^{-3}\,\text{m})^2$$

$\therefore$ Total root length = 25.46 m

To calculate surface area of roots: substituting value for h obtained above into eqn [64.2], we have:

Root surface area

$$= 2 \times 3.141\,59 \times 0.25 \times 10^{-3}\,\text{m} \times 25.46\,\text{m}$$
$$= 0.04\,\text{m}^2$$
$$= 0.04 \times 10^4\,\text{cm}^2$$
$$\text{(there being } 100 \times 100 = 10^4\,\text{cm}^2\,\text{per m}^2\text{)}$$

$\therefore$ Root surface area = 400.0 cm^2

Notes

(a) The fresh weight of roots would normally be obtained by washing the roots free of soil, blotting them dry and weighing.

(b) In a real answer you might show the replicate measurements giving rise to the mean diameter.

(c) In reality, the roots will differ considerably in diameter and each root will not have a constant diameter throughout its length.

(d) This will not be wildly inaccurate as about 95 per cent of the fresh weight will be water, but the volume could also be estimated from water displacement measurements.

(e) Note conversion of measurements into base SI units at this stage and on line 3 of the root volume calculation. Forgetting to halve diameter measurements where radii are required is a common error.

 KEY POINT *Do not simply accept the numerical answer from a calculator or spreadsheet, without considering whether you need to modify this to give an appropriate number of significant figures or decimal places.*

Rounding to n *decimal places*

This is relatively easy to do.

1. Look at the number to the right of the nth decimal place.
2. If this is less than five, simply 'cut off' all numbers to the right of the nth decimal place to produce the answer (i.e. round down).
3. If the number is greater than five, 'cut off' all numbers to the right of the nth decimal place and add one to the nth decimal place to produce the answer (i.e. round up).

Examples

The number 4.123 correct to two decimal places is 4.12

The number 4.126 correct to two decimal places is 4.13

The number 4.1251 correct to two decimal places is 4.13

The number 4.1250 correct to two decimal places is 4.12

The number 4.1350 correct to two decimal places is 4.14

The number 99.99 correct to one decimal place is 100.0.

4. If the number is 5, then look at further numbers to the right to determine whether to round up or not.

5. If the number is *exactly* 5 and there are no further numbers to the right, then round to the nearest even number. *Note:* When considering a large number of calculations, this procedure will not affect the overall mean value. Some rounding systems do the opposite to this (i.e. round to the nearest odd number), while others always round up where the number is exactly 5 (which *will* affect the mean). Take advice from your tutor and stick to one system throughout a series of calculations.

Whenever you see any number quoted, you should assume that the last digit has been rounded. For example, in the number 22.4, the '.4' is assumed to be rounded and the calculated value may have been between 22.35 and 22.45.

Quoting to n *significant figures*

The number of significant figures indicates the degree of approximation in the number. For most cases, it is given by counting all the figures except zeros that occur at the beginning or end of the number. Zeros *within* the number are always counted as significant. The number of significant figures in a number like 200 is ambiguous and could be one, two or three; if you wish to specify clearly, then quote as e.g. 2×10^2 (one significant figure), 2.0×10^2 (two significant figures), etc. to avoid spurious accuracy (pp. 157, 400). When quoting a number to a specified number of significant figures, use the same rules as for rounding to a specified number of decimal places, but do not forget to keep zeros before or after the decimal point. The same principle is used if you are asked to quote a number to the 'nearest 10', 'nearest 100', etc.

When deciding for yourself how many significant figures to use, adopt the following rules of thumb:

Examples

The number of significant figures in 194 is 3

The number of significant figures in 2305 is 4

The number of significant figures in 0.003482 is 4

The number of significant figures in 210×10^8 is 3 (21×10^9 would be 2).

- Always round *after* you have done a calculation. Use *all* significant figures available in the measured data during a calculation.
- If adding or subtracting with measured data, then quote the answer to the number of decimal places in the data value with the least number of decimal places (e.g. $32.1 - 45.67 + 35.6201 = 22.1$, because 32.1 has one decimal place).
- If multiplying or dividing with measured data, keep as many significant figures as are in the number with the least number of significant places (e.g. $34901 \div 3445 \times 1.3410344 = 13.59$, because 3445 has four significant figures).
- For the purposes of significant figures, assume 'constants' (e.g. number of mm in a metre) have an infinite number of significant figures.

Examples

The number of significant figures in 3051.93 is 6

To five significant figures, this number is 3051.9

To four significant figures, this number is 3052

To three significant figures, this number is 3050

To two significant figures, this number is 3100

To one significant figure, this number is 3000

3051.93 to the nearest 10 is 3050

3051.93 to the nearest 100 is 3100

Note that in this last case you must include the zeros before the decimal point to indicate the scale of the number (even if the decimal point is not shown). For a number less than 1, the same would apply to the zeros before the decimal point. For example, 0.00305193 to three significant figures is 0.00305. Alternatively, use scientific notation (in this case, 3.05×10^{-3}).

Some reminders of basic mathematics

Errors in calculations sometimes appear because of faults in mathematics rather than computational errors. For reference purposes, Tables 64.1–64.3 give some basic mathematical principles that may be useful. Eason et al. (1992) or Stephenson (2003) should be consulted for more advanced/specific needs.

Table 64.3 Geometry and trigonometry – analysing shapes

Shape/object	Diagram	Perimeter	Area
Two-dimensional shapes			
Square		$4x$	x^2
Rectangle		$2(x + y)$	xy
Circle		$2\pi r$	πr^2
Ellipse		$\pi[1.5(a + b) - \sqrt{a} * b]$ (approx.)	πab
Triangle (general)		$x + y + z$	$0.5zh$
(right-angled)		$x + y + r$ $\sin\theta = y/r,\ \cos\theta = x/r,$ $\tan\theta = y/x;\ r^2 = x^2 + y^2$	$0.5xy$

Shape/object	Diagram	Surface area	Volume
Three-dimensional shapes			
Cube		$6x^2$	x^3
Cuboid		$2xy + 2xz + 2yz$	xyz
Sphere		$4\pi r^2$	$4\pi r^3/3$
Ellipsoid		no simple formula	πrab
Cylinder		$2\pi rh + 2\pi r^2$	$\pi r^2 h$
Cone and pyramid		$0.5PL + B$	$BL/3$

Key: x, y, z = sides a, b = half minimum and maximum axes; r = radius or hypotenuse; h = height; B = base area; L = perpendicular height; P = perimeter of base.

Percentages and proportions

A percentage is just a fraction expressed in terms of hundredths, indicated by putting the percentage sign (%) after the number of hundredths. So 35% simply means 35 hundredths. To convert a fraction to a percentage, just multiply the fraction by 100. When the fraction is in decimal form, multiplying by 100 to obtain a percentage is easily achieved just by moving the decimal point two places to the right.

> **Examples**
> 1/8 as a percentage is $1 \div 8 \times 100 = 100 \div 8 = 12.5\%$
> 0.602 as a percentage is $0.602 \times 100 = 60.2\%$.

To convert a percentage to a fraction, just remember that, since a percentage is a fraction multiplied by 100, the fraction is the percentage divided by 100. For example: $42\% = 42/100 = 0.42$. In this example, since we are dealing with a decimal fraction, the division by 100 is just a matter of moving the decimal point two places to the left (42% could be written as 42.0%). Percentages greater than 100% represent fractions greater than 1. Percentages less than 1 may cause confusion. For example, 0.5% means half of one per cent (0.005) and must not be confused with 50% (which is the decimal fraction 0.5).

> **Examples**
> 190% as a decimal fraction is $190 \div 100 = 1.9$
> 5/2 as a percentage is $5 \div 2 \times 100 = 250\%$.

To find a percentage of a given number, just express the percentage as a decimal fraction and multiply the given number. For example: 35% of 500 is given by $0.35 \times 500 = 175$. To find the percentage change in a quantity, work out the difference (= value 'after' − value 'before'), and divide this difference by the original value to give the fractional change, then multiply by 100.

> **Example** A population falls from 4 million to 3.85 million. What is the percentage change? The decrease in numbers is $4 - 3.85 = -0.15$ million. The fractional decrease is $-0.15 \div 4 = -0.0375$ and we multiply by 100 to get the percentage change = minus 3.75%.

Exponents

Exponential notation is an alternative way of expressing numbers in the form a^n ('a to the power n'), where a is multiplied by itself n times. The number a is called the base and the number n the exponent (or power or index). The exponent need not be a whole number, and it can be negative if the number being expressed is less than 1. See Table 64.2 for other mathematical relationships involving exponents.

> **Example** $2^3 = 2 \times 2 \times 2 = 8$.

Scientific notation

In scientific notation, also known as 'standard form', the base is 10 and the exponent a whole number. To express numbers that are not whole powers of 10, the form $c \times 10^n$ is used, where the coefficient c is normally between 1 and 10. Scientific notation is valuable when you are using very large numbers and wish to avoid suggesting spurious accuracy. Thus if you write 123 000, this may suggest that you know the number to ±0.5, whereas 1.23×10^5 might give a truer indication of measurement accuracy (i.e. implied to be ±500 in this case). Engineering notation is similar, but treats numbers as powers of 10 in groups of 3, i.e. $c \times 10^0$, 10^3, 10^6, 10^9, etc. This corresponds to the SI system of prefixes (p. 159).

> **Example** Avogadro's number, ≈ 602 352 000 000 000 000 000 000, is more conveniently expressed as $6.023\,52 \times 10^{23}$.

A useful property of powers when expressed to the same base is that when multiplying two numbers together, you simply add the powers, while if dividing, you subtract the powers. Thus, suppose you counted 8 bacteria in a known volume of a 10^{-7} dilution (see p. 132), there would be 8×10^7 in the same volume of undiluted solution; if you now dilute this 500-fold (5×10^2), then the number present in the same volume would be $8/5 \times 10^{(7-2)} = 1.6 \times 10^5 = 160\,000$.

Logarithms

When a number is expressed as a logarithm, this refers to the power n that the base number a must be raised to give that number. Any base could be used, but the two most common are 10, when the power is referred to as $\log_{10}$ or simply log, and the constant e (2.718282), used for mathematical convenience in certain situations, when the power is referred to as $\log_e$ or ln (natural logarithm). Note that (a) logs need not be whole numbers; (b) there is no log value for the number zero; and (c) that $\log_{10} = 0$ for the number 1.

To obtain logs, you will need to use the log key on your calculator, or special log tables (now largely redundant). To convert back (antilog), use

- the $\boxed{10^x}$ key, with $x = $ log value;
- the $\boxed{\text{inverse}}$ then the $\boxed{\log}$ key; or
- the $\boxed{y^x}$ key, with $y = 10$ and $x = $ log value.

If you have used log tables, you will find complementary antilogarithm tables to do this.

There are many uses of logarithms in biology, including pH ($= -\log[H^+]$), where $[H^+]$ is expressed in $\text{mol}\,l^{-1}$ (see p. 146); the exponential growth of micro-organisms, where if log(cell number) is plotted against time, a straight-line relationship is obtained (see p. 272); and allometric studies of growth and development, where if data are plotted on log axes, a series of straight-line relationships may be found (see p. 294).

> **Examples** The logarithm to the base 10 ($\log_{10}$) of 1000 is 3, since $10^3 = 1000$. The logarithm to the base e ($\log_e$ or ln) of 1000 is 6.907755 (to six decimal places).

> **Examples** (use to check the correct use of your own calculator)
> 102963 as a $\log_{10} = 5.012681$ (to six decimal places)
> $10^{5.012681} = 102962.96$
> (Note loss of accuracy due to loss of decimal places.)
> 102963 as a natural logarithm (ln) = 11.542125 (to six decimal places), thus $2.718282^{11.542125} = 102963$.

Linear functions and straight lines

One of the most straightforward and widely used relationships between two variables x and y is that represented by a straight-line graph, where the corresponding mathematical function is known as the equation of a straight line, where:

$$y = a + bx \qquad \qquad [64.3]$$

In this relationship, a represents the intercept of the line on the y (vertical) axis, i.e. where $x = 0$, and therefore $bx = 0$, while b is equivalent to the slope (gradient) of the line, i.e. the change in y for a change in x of 1. The constants a and b are sometimes given alternative symbols, but the mathematics remains unchanged, e.g. in the equivalent expression for the slope of a straight line, $y = mx + c$. Figure 64.1 shows what happens when these two constants are changed, in terms of the resultant straight lines.

The two main applications of the straight-line relationship are:

1. Function fitting. Here, you determine the mathematical form of the function, i.e. you estimate the constants a and b from a data set for x and y, either by drawing a straight line by eye (p. 394) and then working out the slope and y intercept, or by using linear regression (p. 435) to obtain the most probable values for both constants. When putting a straight line of best fit by eye on a hand-drawn graph, note the following:

- Always use a *transparent* ruler, so you can see data points on either side of the line.
- For a data series where the points do not fit a perfect straight line, try to have an equal number of points on either side of the line, as in

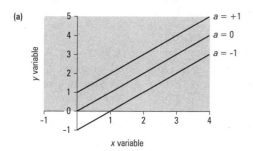

(a)

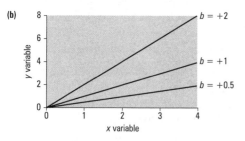

(b)

Fig. 64.1. Straight-line relationships ($y = a + bx$), showing the effects of (a) changing the intercept at constant slope, and (b) changing the slope at constant intercept.

Fig. 64.2(a), and try to minimise the average distance of these points from the line.

- Once you have drawn the line of best fit use this line, rather than your data values, in all subsequent procedures (e.g. in a calibration curve, Chapter 50).
- Tangents drawn to a curve give the slope (gradient) at a particular point, e.g. in an enzyme reaction progress curve, Fig. 52.2. These are best drawn by bringing your ruler up to the curve at the exact point where you wish to estimate the slope and then trying to make the two angles immediately on either side of this point approximately the same, by eye (Fig. 64.2(b)).
- Once you have drawn the straight line or tangent, choose two points reasonably far apart at either end of your line and then draw construction lines to represent the change in y and the change in x between these two points: make sure that your construction lines are perpendicular to each other. Determine the slope as the change in y divided by the change in x (Fig. 64.2).

2. Prediction. Where a and b are known, or have been estimated, you can use eqn [64.3] to predict any value of y for a specified value of x, e.g. during exponential growth of a cell culture (p. 272), where $\log_{10}$ cell number (y) increases as a linear function of time (x): note that in this example the dependent variable has been transformed to give a linear relationship. You will need to rearrange eqn [64.3] in cases where a prediction of x is required for a particular value of y (e.g. in calibration curves, p. 303, or bioassays, p. 295), as follows:

$$x = (y - a) \div b \qquad [64.4]$$

This equation can also be used to determine the intercept on the x (horizontal) axis, i.e. where $y = 0$.

Hints for some typical problems

Calculations involving proportions or ratios
The 'unitary method' is a useful way of approaching calculations involving proportions or ratios, such as those required when making up solutions from stocks (see also Chapter 23) or as a subsidiary part of longer calculations.

1. If given a value for a multiple, work out the corresponding value for a single item or unit.
2. Use this 'unitary value' to calculate the required new value.

Calculations involving series
Series (used in e.g. dilutions, see also p. 131) can be of three main forms:

1. Arithmetic, where the *difference* between two successive numbers in the series is a constant, e.g. 2, 4, 6, 8, 10, ...
2. Geometric, where the *ratio* between two successive numbers in the series is a constant, e.g. 1, 10, 100, 1000, 10000, ...
3. Harmonic, where the values are reciprocals of successive whole numbers, e.g. $1, \frac{1}{2}, \frac{1}{3}, \frac{1}{4}, \ldots$

Note that the logs of the numbers in a geometric series will form an arithmetic series (e.g. 0, 1, 2, 3, 4, ... in the above case). Thus, if a

---

Examples
Using eqn [64.3], the predicted value for y for a linear function where $a = 2$ and $b = 0.5$, where $x = 8$ is: $y = 2 + (0.5 \times 8) = 6$.
Using eqn [64.4], the predicted value for x for a linear function where $a = 1.5$ and $b = 2.5$, where $y = -8.5$ is:
$x = (-8.5 - 1.5) \div 2.5 = -4$.
Using eqn [64.4] the predicted x intercept for a linear function where $a = 0.8$ and $b = 3.2$ is: $x = (0 - 0.8) \div 3.2 = -0.25$.

Example   A lab schedule states that 5 g of a compound with a relative molecular mass of 220 are dissolved in 400 ml of solvent. For writing up your Materials and Methods, you wish to express this as $mol\,l^{-1}$.
1. If there are 5 g in 400 ml, then there are $5 \div 400$ g in 1 ml.
2. Hence, 1000 ml will contain $5 \div 400 \times 1000$ g = 12.5 g.
3. 12.5 g = $12.5 \div 220$ mol = 0.0568 mol, so [solution] = 56.8 mmol l^{-1} ($= 56.8$ mol m^{-3}).

Examples   For a geometric dilution series involving ten-fold dilution steps, calculation of concentrations is straightforward, e.g. two serial decimal dilutions (= 100-fold dilution) of a solution of NaCl of 250 mmol l^{-1} will produce a dilute solution of $250 \div 100 = 2.5$ mmol l^{-1}. Similarly, for an arithmetic dilution series, divide by the overall dilution to give the final concentration, e.g. a sixteen-fold dilution of a solution of NaCl of 200 mg ml^{-1} will produce a dilute solution of $200 \div 16 = 12.5$ mg ml^{-1}.

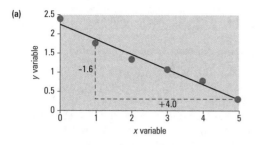

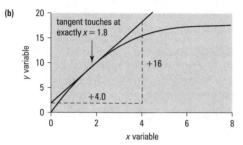

(a)

(b)

tangent touches at exactly $x = 1.8$

Fig. 64.2. Drawing straight lines. (a) Simple linear relationship, giving a straight line with an intersect of 2.3 and a slope of $-1.6 \div 4.0 = 0.4$. (b) Tangent drawn to a curve at $x = 1.8$, giving a slope of $16 \div 4 = 4$.

quantity y varies with a quantity x such that the rate of change in y is proportional to the value of y (i.e. it varies in an exponential manner), a semi-log plot of such data will form a straight line. This form of relationship is relevant for exponentially growing cell cultures (p. 272) and radioactive decay (p. 373).

Statistical calculations

The need for long, complex calculations in statistics has largely been removed because of the widespread use of spreadsheets with statistical functions (Chapter 11) and specialised programs such as SPSS and Minitab. It is, however, important to understand the principles behind what you are trying to do (see Chapters 65 and 66) and interpret the program's output correctly, either using the 'help' function or a reference manual.

Problems in Mendelian genetics

These cause difficulties for many students. The key is to recognise the different types of problems and to practise so you are familiar with the techniques for solving them. Chapter 53 deals with the different types of crosses you will come across and methods of analysing them, including the use of the χ^2 (Chi^2) test.

Text references

Eason, G., Coles, C.W. and Gettinby, G. (1992) *Mathematics and Statistics for the Bio-Sciences*. Ellis Horwood, Chichester.

Stephenson, F.H. (2003) *Calculations in Molecular Biology and Biotechnology*. Academic Press, London.

Sources for further study

Anon. *S.O.S. Mathematics*. Available: http://www.sosmath.com/
Last accessed 09/04/07.
[A basic Web-based guide with very wide coverage.]

Britton, N.F. (2003) *Essential Mathematical Biology*. Springer-Verlag. New York.
[Describes a range of mathematical applications in population dynamics, epidemiology, genetics, biochemistry and medicine.]

Cann, A.J. (2002) *Maths from Scratch for Biologists*. Wiley, Chichester.
[Deals with basic manipulations, formulae, units, molarity, logs and exponents as well as basic statistical procedures.]

Causton, D.R. (1992) *A Biologist's Basic Mathematics*. Cambridge University Press, Cambridge.

Forster, P.C. (1999) *Easy Mathematics for Biologists*. Harwood, Amsterdam.

[Covers basic principles, units, logarithms, ratios and proportions, concentrations, equations, rates and graphs.]

Harris, M., Taylor, G. and Taylor, J. (2005) *Catch Up Maths and Stats for the Life and Medical Sciences*. Scion, Bloxham.
[Covers a range of basic mathematical operations and statistical procedures.]

Koehler, K.R. *College Physics for Students of Biology and Chemistry*. Available: http://www.rwc.uc.edu/koehler/biophys/text.html.
Last accessed 09/04/07.
[A 'hypertextbook' written for first-year undergraduates. Assumes that you have a working knowledge of algebra.]

Lawler, G. (2006) *Understanding Maths. Basic Mathematics Explained*. Studymates, Abergele.

Hints for solving numerical problems

64.1 Rearrange the following formulae.

(a) If $y = ax + b$, find b
(b) If $y = ax + b$, find x
(c) If $x = y^3$, find y
(d) If $x = 3^y$, find y
(e) If $x = (1 - y)(z^p + 3)$, find z
(f) If $x = (y - z)^{1/n}/pq$, find n

64.2 Work with decimal places or significant figures. Give the following numbers to the accuracy indicated:

(a) 214.51 to three significant figures
(b) 107 029 to three significant figures
(c) 0.0450 to one significant figure
(d) 99.817 to two decimal places
(e) 99.897 to two decimal places
(f) 99.997 to two decimal places
(g) 6255 to the nearest 10
(h) 134 903 to the nearest ten thousand

State the following:

(i) the number of significant figures in 3400
(j) the number of significant figures in 3400.3
(k) the number of significant figures in 0.00167
(l) the number of significant figures in 1.00167
(m) the number of decimal places in 34.46
(n) the number of decimal places in 0.00167

64.3 Carry out calculations involving percentages. Answer the following questions, giving your answers to two decimal places:

(a) What is 6/35ths expressed as a percentage?
(b) What is 0.0755 expressed as a percentage?
(c) What is 4.35% of 322?
(d) A rat's weight increased from 55.23 g to 75.02 g. What is the percentage increase in its weight?

64.4 Practise using exponents and logarithms.

(a) Write out (i) the charge on an electron and (ii) the speed of light *in vacuo* in longhand (i.e. without using powers of 10). See Table 26.4 for values given in scientific notation.
(b) Compute the following values: $10^{3.624}$; $\log(6.37)$; $e^{-2.32}$; $\ln(1123)$; $6^{3.2}$. A calculator should be used, but round the output to give five significant figures.

64.5 Practise using geometric formulae. Select appropriate formulae from Table 64.3, and use them to compare the surface-area-to-volume ratios of the following:

(a) a spherical unicellular alga of diameter 30 μm;
(b) a cylindrical root of radius 0.5 mm (ignoring ends).

64.6 Practise working with linear functions (note also that Chapter 50 includes study exercises based on linear functions and plotting straight lines). Assuming a linear relationship between x and y, calculate the following (give your answers to three significant figures):

(a) x, where $y = 7.0$, $a = 4.5$ and $b = 0.02$;
(b) x, where $y = 15.2$, $a = -2.6$ and $b = -4.46$;
(c) y, where $x = 10.5$, $a = 0.2$ and $b = -0.63$;
(d) y, where $x = 4.5$, $a = -1.8$ and $b = 4.1$.

Whether obtained from observation or experimentation, most data in biology exhibit variability. This can be displayed as a frequency distribution (e.g. Fig. 62.7). Descriptive (or summary) statistics quantify aspects of the frequency distribution of a sample. You can use them to:

- condense a large data set for presentation in figures or tables;
- provide estimates of parameters of the frequency distribution of the population being sampled (p. 427).

 KEY POINT *The appropriate descriptive statistics to choose depend on both the type of data, i.e. quantitative, ranked, or qualitative, and the nature of the underlying population frequency distribution.*

If you have no clear theoretical grounds for assuming what the underlying frequency distribution is like, graph one or more sample frequency distributions, ideally with a sample size >100. The tally system for recording data (see p. 383) can give an immediate visual indication of the frequency distribution as data are collected.

The methods used to calculate descriptive statistics depend on whether data have been grouped into classes. You should use the original data set if it is still available, because grouping into classes loses information and accuracy. However, large data sets may make calculations unwieldy, and are best handled using computer programs.

Three important features of a frequency distribution that can be summarised by descriptive statistics are:

- the sample's location, i.e. its position along a given dimension representing the dependent (measured) variable (Fig. 65.1);
- the dispersion of the data, i.e. how spread out the values are (Fig. 65.2);
- the shape of the distribution, i.e. whether symmetrical, skewed, U-shaped, etc. (Fig. 65.3).

Measuring location

Here, the objective is to pinpoint the 'centre' of the frequency distribution, i.e. the value about which most of the data are grouped. The chief measures of location are the mean, median and mode. Figure 65.4 shows how to choose among these for a given data set.

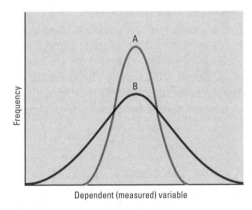

Fig. 65.1 Two distributions with different locations but the same dispersion. The data set labelled B could have been obtained by adding a constant to each datum in the data set labelled A.

Fig. 65.2 Two distributions with different dispersions but the same location. The data set labelled A covers a relatively narrow range of values of the dependent (measured) variable, while that labelled B covers a wider range.

Example Box 65.1 shows a set of data and the calculated values of the measures of location, dispersion and shape for which methods of calculation are outlined here. Check your understanding by calculating the statistics yourself and confirming that you arrive at the same answers.

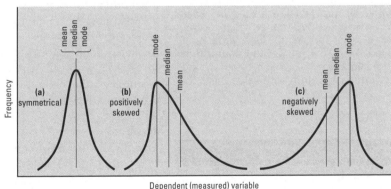

Fig. 65.3 Symmetrical and skewed frequency distributions, showing relative positions of mean, median and mode.

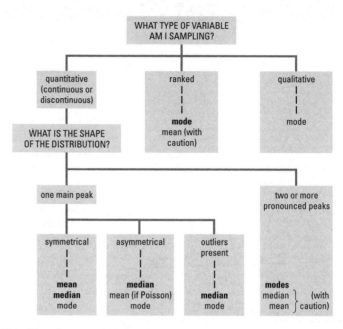

Fig. 65.4 Choosing a statistic for characterising a distribution's location. Statistics written in bold are the preferred option(s).

Use of symbols – *Y is used in Chapters 65 and 66 to signify the dependent variable in statistical calculations (following the example of Sokal and Rohlf, 1994, Heath, 1995 and Wardlaw, 2000). Note, however, that some authors use X or x in analogous formulae and many calculators refer to e.g. $\bar{x}$, Σx^2, etc., for their statistical functions.*

Definition

Rank – the position of a data value when all the data are placed in order of ascending magnitude. If ties occur, an average rank of the tied variates is used. Thus, the rank of the datum 6 in the sequence 1,3,5,6,8,8,10 is 4; the rank of each datum with value 8 is 5.5.

Definition

An outlier – any datum that has a value much smaller or bigger than most of the data.

Mean

The mean (denoted $\bar{Y}$ and also referred to as the arithmetic mean) is the average value of the data. It is obtained from the sum of all the data values divided by the number of observations (in symbolic terms, $\Sigma Y/n$). The mean is a good measure of the centre of symmetrical frequency distributions. It uses all of the numerical values of the sample and therefore incorporates all of the information content of the data. However, the value of a mean is greatly affected by the presence of outliers (extreme values). The arithmetic mean is a widely used statistic in biology, but there are situations when you should be careful about using it (see Box 65.2 for examples).

Median

The median is the mid-point of the observations when ranked in increasing order. For odd-sized samples, the median is the middle observation; for even-sized samples it is the mean of the middle pair of observations. Where data are grouped into classes, the median can only be estimated. This is most simply done from a graph of the cumulative frequency distribution, but can also be worked out by assuming the data to be evenly spread within the class. The median may represent the location of the main body of data better than the mean when the distribution is asymmetric or when there are outliers in the sample.

Mode

The mode is the most common value in the sample. The mode is easily found from a tabulated frequency distribution as the most frequent value. If data have been grouped into classes then the term modal class is used for the class containing most values. The mode provides a rapidly and easily found estimate of sample location and is unaffected by outliers. However,

Box 65.1 Descriptive statistics for an illustrative sample of data

Value (Y)	Frequency (f)	Cumulative frequency	fY	fY²
1	0	0	0	0
2	1	1	2	4
3	2	3	6	18
4	3	6	12	48
5	8	14	40	200
6	5	19	30	180
7	2	21	14	98
8	0	21	0	0
Totals	$21 = \Sigma f (=n)$		$104 = \Sigma fY$	$548 = \Sigma fY^2$

In this example, for simplicity and ease of calculation, integer values of Y are used. In many practical exercises, where continuous variables are measured to several significant figures and where the number of data values is small, giving frequencies of 1 for most of the values of Y, it may be simpler to omit the column dealing with frequency and list all the individual values of Y and Y^2 in the appropriate columns. To gauge the underlying frequency distribution of such data sets, you would need to group individual data into broader classes (e.g. all values between 1.0 and 1.9, all values between 2.0 and 2.9, etc.) and then draw a histogram (p. 395). Calculation of certain statistics for data sets that have been grouped in this way (e.g. median, quartiles, extremes) can be tricky and a statistical text should be consulted.

Statistic	Value*	How calculated
Mean	4.95	$\Sigma fY/n$, i.e. 104/21
Median	5	Value of the $(n+1)/2$ variate, i.e. the value ranked $(21+1)/2 = 11$th (obtained from the cumulative frequency column)
Mode	5	The most common value (Y value with highest frequency)
Upper quartile	6	The upper quartile is between the 16th and 17th values, i.e. the value exceeded by 25% of the data values
Lower quartile	4	The lower quartile is between the 5th and 6th values, i.e. the value exceeded by 75% of the data values
Semi-interquartile range	1.0	Half the difference between the upper and lower quartiles, i.e. $(6-4)/2$
Upper extreme	7	Highest Y value in data set
Lower extreme	2	Lowest Y value in data set
Range	5	Difference between upper and lower extremes
Variance (s^2)	1.65	$s^2 = \dfrac{\Sigma fY^2 - (\Sigma fY)^2/n}{n-1}$ $= \dfrac{548 - (104)^2/21}{20}$
Standard deviation (s)	1.28	$\sqrt{s^2}$
Standard error (SE)	0.280	$s/\sqrt{n}$
95% confidence limits	4.36 – 5.54	$\bar{Y} \mp t_{0.05}[20] \times \text{SE}$, (where $t_{0.05}[20] = 2.09$, Table 66.2)
Coefficient of variation (CoV)	25.9%	$100s/\bar{Y}$

*Rounded to three significant figures (see p. 408), except when it is an exact number.

the mode is affected by chance variation in the shape of a sample's distribution and it may lie distant from the obvious centre of the distribution.

The mean, median and mode have the same units as the variable under discussion. However, whether these statistics of location have the same or similar values for a given frequency distribution depends on the symmetry and shape of the distribution. If it is near-symmetrical with a single peak, all three will be very similar; if it is skewed or has more than one peak, their values will differ to a greater degree (see Fig. 65.3).

Measuring dispersion

Here, the objective is to quantify the spread of the data about the centre of the distribution. Figure 65.5 indicates how to decide which measure of dispersion to use.

Range

The range is the difference between the largest and smallest data values in the sample (the extremes) and has the same units as the measured variable. The range is easy to determine, but is greatly affected by outliers. Its value may also depend on sample size: in general, the larger this is, the greater will be the range. These features make the range a poor measure of dispersion for many practical purposes.

Semi-interquartile range

The semi-interquartile range is an appropriate measure of dispersion when a median is the appropriate statistic to describe location. For this, you need to determine the first and third quartiles, i.e. the medians for those

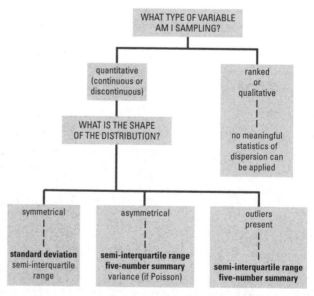

Fig. 65.5 Choosing a statistic for characterising a distribution's dispersion. Statistics written in bold are the preferred option(s). Note that you should match statistics describing dispersion with those you have used to describe location, i.e. standard deviation with mean, semi-interquartile range with median.

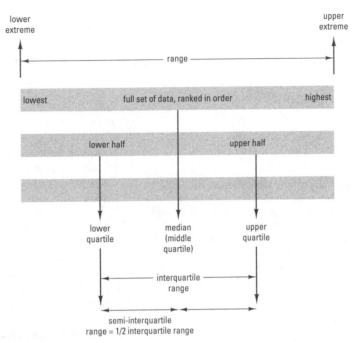

Fig. 65.6 Illustration of median, quartiles, range and semi-interquartile range

data values ranked below and above the median of the whole data set (see Fig. 65.6). To calculate a semi-interquartile range for a data set:

1. Rank the observations in ascending order.
2. Find the values of the first and third quartiles.
3. Subtract the value of the first quartile from the value of the third.
4. Halve this number.

For data grouped in classes, the semi-interquartile range can only be estimated. Another disadvantage is that it takes no account of the shape of the distribution at its edges. This objection can be countered by using the so-called 'five-number summary' of a data set, which consists of the three quartiles and the two extreme values; this can be presented on graphs as a box and whisker plot (see Fig. 65.7) and is particularly useful for summarising skewed frequency distributions. The corresponding 'six-number summary' includes the sample's size.

Variance and standard deviation

For symmetrical frequency distributions, an ideal measure of dispersion would take into account each value's deviation from the mean and provide a measure of the average deviation from the mean. Two such statistics are the sample variance, which is the sum of squares $(\Sigma(Y - \bar{Y})^2)$ divided by $n - 1$ (where n is the sample size), and the sample standard deviation, which is the positive square root of the sample variance.

The variance (s^2) has units that are the square of the original units, while the standard deviation (s, or SD) is expressed in the original units, one reason s is often preferred as a measure of dispersion. Calculating s or s^2 longhand is a tedious job and is best done with the help of a calculator or

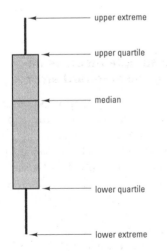

Fig. 65.7 A box and whisker plot, showing the 'five-number summary' of a sample as it might be used on a graph.

computer. If you don't have a calculator that calculates s for you, an alternative formula that simplifies calculations is:

$$s = +\sqrt{\frac{\Sigma Y^2 - (\Sigma Y)^2/n}{n-1}}$$ [65.1]

To calculate s using a calculator:

1. Obtain ΣY, square it, divide by n and store in memory.
2. Square Y values, obtain ΣY^2, subtract memory value from this.
3. Divide this answer by $n - 1$.
4. Take the positive square root of this value.

Take care to retain significant figures, or errors in the final value of s will result. If continuous data have been grouped into classes, the class mid-values or their squares must be multiplied by the appropriate frequencies before summation (see example in Box 65.1). When data values are large, longhand calculations can be simplified by coding the data, e.g. by subtracting a constant from each datum, and decoding when the simplified calculations are complete (see Sokal and Rohlf, 1994).

Coefficient of variation

The coefficient of variation (CoV) is a dimensionless measure of variability relative to location that expresses the sample standard deviation, usually as a percentage of the sample mean, i.e.

$$\text{CoV} = 100s/\bar{Y} \, (\%)$$ [65.2]

This statistic is useful when comparing the relative dispersion of data sets with widely differing means or where different units have been used for the same or similar quantities.

A useful application of the CoV is to compare different analytical methods or procedures, so that you can decide which involves the least proportional error – create a standard stock solution, then base your comparison on the results from several subsamples analysed by each method. You may find it useful to use the CoV to compare the precision of your own results with those of a manufacturer, e.g. for an autopippettor (p. 123). The smaller the CoV, the more precise (repeatable) is the apparatus or technique (note: this does not mean that it is necessarily more *accurate*, see p. 157).

Measuring the precision of the sample mean as an estimate of the true value using the standard error

Most practical exercises are based on a limited number of individual data values (a sample) that are used to make inferences about the population from which they were drawn. For example, the haemoglobin content might be measured in blood samples from 100 adult females and used as an estimate of the adult female haemoglobin content, with the sample mean ($\bar{Y}$) and sample standard deviation (s) providing estimates of the true values of the underlying population mean (μ) and the population standard deviation (σ). The reliability of the sample mean as an estimate of the true (population) mean can be assessed by calculating a statistic termed the standard error of the sample mean (often abbreviated to standard error or SE), from:

$$\text{SE} = s/\sqrt{n}$$ [65.3]

Using a calculator for statistics – make sure you understand how to enter individual data values and which keys will give the sample mean (usually shown as $\bar{X}$ or $\bar{x}$) and sample standard deviation (often shown as σ_{n-1}). In general, you should not use the population standard deviation (usually shown as σ_n).

Example Consider two methods of bioassay for a toxin in fresh water. For a given standard, Method A gives a mean result of $= 50$ 'response units' with $s = 8$, while Method B gives a mean result of $= 160$ 'response units' with $s = 18$. Which bioassay gives the more reproducible results? The answer can be found by calculating the CoV values, which are 16 per cent and 11.25 per cent respectively. Hence, Method B is the more precise, even though the absolute value of s is larger.

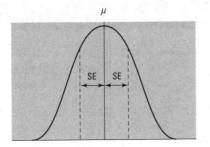

Fig. 65.8 Frequency distribution of sample means around the population mean (μ). Note that SE is equivalent to the standard deviation of the sample means, for sample size = n.

Strictly, the standard error is an estimate of the dispersion of repeated sample means around the true (population) value: if several samples were taken, each with the same number of data values (n), then their means would cluster around the population mean (μ) with a standard deviation equal to SE, as shown in Fig. 65.8. Therefore, the *smaller* the SE, the more reliable the sample mean is likely to be as an estimate of the true value, since the underlying frequency distribution would be more tightly clustered around μ. At a practical level, eqn [65.3] shows that SE is directly affected by the dispersion of individual data values within the sample, as represented by the sample standard deviation (s). Perhaps more importantly, SE is inversely related to the *square root* of the number of data values (n). Therefore, if you wanted to increase the precision of a sample mean by a factor of 2 (i.e. to reduce SE by half), you would have to increase n by a factor of 2^2 (i.e. four-fold).

Summary descriptive statistics for the sample mean are often quoted as $\bar{Y} \pm \text{SE}(n)$, with the SE being given to one significant figure more than the mean. For example, summary statistics for the sample mean and standard error for the data shown in Box 65.1 would be quoted as 4.95 ± 0.280 ($n = 21$). You can use such information to carry out a t-test between two sample means (Box 66.1); the SE is also useful because it allows calculation of confidence limits for the sample mean (p. 434).

Describing the 'shape' of frequency distributions

Frequency distributions may differ in the following characteristics:

- number of peaks;
- skewness or asymmetry;
- kurtosis or pointedness.

The shape of a frequency distribution of a small sample is affected by chance variation and may not be a fair reflection of the underlying population frequency distribution: check this by comparing repeated samples from the same population or by increasing the sample size. If the original shape were due to random events, it should not appear consistently in repeated samples and should become less obvious as sample size increases.

Genuinely bimodal or polymodal distributions may result from the combination of two or more unimodal distributions, indicating that more than one underlying population is being sampled (Fig. 65.9). An example of a bimodal distribution is the height of adult humans (females and males combined).

A distribution is skewed if it is not symmetrical, a symptom being that the mean, median and mode are not equal (Fig. 65.3). Positive skewness is

Fig. 65.9 Frequency distributions with different numbers of peaks. A unimodal distribution (a) may be symmetrical or asymmetrical. The dotted lines in (b) indicate how a bimodal distribution could arise from a combination of two underlying unimodal distributions. Note here how the term 'bimodal' is applied to any distribution with two major peaks – their frequencies do not have to be exactly the same.

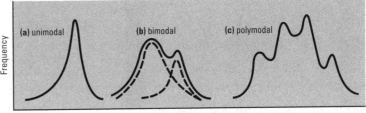

Box 65.2 Three examples where simple arithmetic means are inappropriate

Mean	n
6	4
7	7
8	1

1. **If means of samples are themselves meaned, an error can arise if the samples are of different size.** For example, the arithmetic mean of the means in the table shown left is 7, but this does not take account of the different 'reliabilities' of each mean due to their sample sizes. The correct weighted mean is obtained by multiplying each mean by its sample size (n) (a 'weight') and dividing the sum of these values by the total number of observations, i.e. in the case shown, $(24 + 49 + 8)/12 = 6.75$.

2. **When making a mean of ratios (e.g. percentages) for several groups of different sizes, the ratio for the combined total of all the groups is not the mean of the proportions for the individual groups.** For example, if 20 rats from a batch of 50 are male, this implies 40% are male. If 60 rats from a batch of 120 are male, this implies 50% are male. The mean percentage of males $(50 + 40)/2 = 45\%$ is *not* the percentage of males in the two groups combined, because there are $20 + 60 = 80$ males in a total of 170 rats $= 47.1\%$ approx.

pH value	$[H^+]$ (mol l^{-1})
6	1×10^{-6}
7	1×10^{-7}
8	1×10^{-8}
mean	3.7×10^{-7}
$-\log_{10}$ mean	6.43

3. **If the measurement scale is not linear, arithmetic means may give a false value.** For example, if three media had pH values 6, 7 and 8, the appropriate mean pH is not 7 because the pH scale is logarithmic. The definition of pH is $-\log_{10}[H^+]$, where $[H^+]$ is expressed in mol l^{-1} ('molar'); therefore, to obtain the true mean, convert data into $[H^+]$ values (i.e. put them on a linear scale) by calculating $10^{(-pH\ value)}$ as shown. Now calculate the mean of these values and convert the answer back into pH units. Thus, the appropriate answer is pH 6.43 rather than 7. Note that a similar procedure is necessary when calculating statistics of dispersion in such cases, so you will find these almost certainly asymmetric about the mean.

Mean values of log-transformed data are often termed geometric means – they are sometimes used in microbiology and in cell culture studies, where log-transformed values for cell density counts are averaged and plotted (p. 272), rather than using the raw data values. The use of geometric means in such circumstances serves to reduce the effects of outliers on the mean.

where the longer 'tail' of the distribution occurs for higher values of the measured variable; negative skewness where the longer tail occurs for lower values. Some biological examples of characteristics distributed in a skewed fashion are volumes of plant protoplasts, insulin levels in human plasma and bacterial colony counts.

Kurtosis is the name given to the 'pointedness' of a frequency distribution. A platykurtic frequency distribution is one with a flattened peak, while a leptokurtic frequency distribution is one with a pointed peak (Fig. 65.10). While descriptive terms can be used, based on visual observation of the shape and direction of skew, the degree of skewness and kurtosis can be quantified and statistical tests exist to test the 'significance' of observed values (see Sokal and Rohlf, 1994), but the calculations required are complex and best done with the aid of a computer.

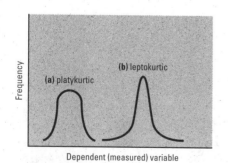

Fig. 65.10 Examples of the two types of kurtosis

Using computers to calculate descriptive statistics

There are many specialist statistical packages (e.g. SPSS) that can be used to simplify the process of calculation of statistics. Note that correct interpretation of the output requires an understanding of the terminology used and the underlying process of calculation, and this may best be obtained by working through one or more examples by hand before using these tools. Spreadsheets offer increasingly sophisticated statistical analysis functions, some examples of which are provided in Box 65.3 for Microsoft Excel.

Box 65.3 How to use a spreadsheet (Microsoft Excel) to calculate descriptive statistics

Method 1: Using spreadsheet functions to generate the required statistics.

Suppose you had obtained the following set of data, stored within an array (block of columns and rows) of cells (A2:L6) within a spreadsheet:

	A	B	C	D	E	F	G	H	I	J	K	L
1	My data set											
2	4	4	3	3	5	4	3	7	7	3	5	3
3	6	2	9	7	3	4	5	6	6	9	4	8
4	5	3	2	5	4	5	7	2	8	3	6	3
5	11	3	5	2	4	3	7	8	4	4	4	3
6	3	6	8	5	6	4	3	4	3	6	10	5

The following functions could be used to extract descriptive statistics from this data set:

Descriptive statistic	Example of use of function[a,b]	Result for the above data set
Sample size n	=COUNT((A2:L6)	60
Mean	=AVERAGE(A2:L6)[c]	4.9
Median	=MEDIAN(A2:L6)	4.0
Mode	=MODE(A2:L6)	3
Upper quartile	=QUARTILE(A2:L6,3)[d]	6.0
Lower quartile	=QUARTILE(A2:L6,1)	3.0
Semi-interquartile range	=QUARTILE(A2:L6,3)-QUARTILE(A2:L6,1)	3.0
Upper extreme	=QUARTILE(A2:L6,4) *or* =MAX(A2:L6)	11
Lower extreme	=QUARTILE(A2:L6,0) *or* =MIN(A2:L6)	2
Range	=MAX(A2:L6)- MIN(A2:L6)[e]	9.0
Variance	=VAR(A2:L6)	4.464
Standard deviation	=STDEV(A2:L6)	2.113
Standard error	=STDEV(A2:L6)/(SQRT(COUNT(A2:L6)))[f]	0.273
Coefficient of variation	=100*STDEV(A2:L6)/AVERAGE(A2:L6)	43.12%

Notes:
[a] Typically, in an appropriate cell, you would *Insert > Function > COUNT*, then select the input range and press return.
[b] Other descriptive statistics can be calculated – these mirror those shown in Box 65.1, but for this specific data set.
[c] There is no 'MEAN' function in Microsoft Excel.
[d] The first argument within the brackets relates to the array of data, the second relates to the quartile required (consult the *Help* feature for further information).
[e] There is no direct 'RANGE' function in Microsoft Excel.
[f] There is no direct 'STANDARD ERROR' function in Microsoft Excel. The SQRT function returns a square root and the COUNT function determines the number of filled data cells in the array.

(continued)

Box 65.3   (continued)

Method 2: Using the *Tools > Data Analysis* option – This can automatically generate a table of descriptive statistics for the data array selected, although the data must be presented as a single row or column. This option might need to be installed for your network or personal computer before it is available to you (in the latter case use the *Add Ins > Analysis ToolPak* option from the *Tools* menu – consult the *Help* feature for details). Having entered or rearranged your data into a row or column, the steps involved are as follows:

1. Select *Tools > Data Analysis*.
2. From the *Data Analysis box,* select *Descriptive Statistics.*
3. Input your data location into the *Input Range* (left-click and hold down to highlight the column of data).
4. From the menu options, select *Summary Statistics* and *Confidence Level for Mean: 95%*.
5. When you click *OK* you should get a new worksheet, with descriptive statistics and confidence limits shown. Alternatively, at step 3, you can select an area of your current worksheet as a data output range (select an area away from any existing content as these cells would otherwise be overwritten by the descriptive statistics output table).
6. Change the format of the cells to show each number to an appropriate number of decimal places. You may also wish to make the columns wider so you can read their content.

7. For the data set shown above, the final output table should look as shown in Table 65.1.

Table 65.1   Descriptive statistics for a data set.

Column1[a,b]

Mean	4.9
Standard error	0.27
Median	4.0
Mode	3
Standard deviation	2.113
Sample variance	4.464
Kurtosis	0.22
Skewness	0.86
Range	9.00
Minimum	2.0
Maximum	11.0
Sum	294
Count	60
Confidence level (95.0%)	0.55

Notes:
[a] These descriptive statistics are specified (and are automatically presented in this order) – any others required can be generated using Method 1.
[b] A more descriptive heading can be added if desired - this is the default.

Text references and sources for further study

Heath, D. (1995) *An Introduction to Experimental Design and Statistics for Biology*. UCL Press, London.

Schmuller, J. (2005) *Statistical Analysis with Excel for Dummies*. Wiley, Hoboken, New Jersey.

Sokal, R.R. and Rohlf, F.J. (1994) *Biometry*, 3rd edn. W.H. Freeman, San Francisco.

Wardlaw, A.C. (2000) *Practical Statistics for Experimental Biologists*, 2nd edn. Wiley, New York.

Study exercises

65.1 Choose appropriate descriptive statistics. State for the following data and distribution types what statistics of location and dispersion (Figs. 65.4 and 65.5) you should choose.

(a) Colour of each egg in nests as judged against a colour chart.

(b) Clutch size (number of eggs) in bird nests visited.

(c) Beak lengths for chicks hatched in nests, where there is a pronounced difference in beak length between the sexes.

65.2 Practise calculating descriptive statistics. Using the data set given below, calculate the following statistics:

(a) range
(b) variance
(c) standard deviation
(d) coefficient of variation
(e) standard error.

Answers (b) to (e) should be given to four significant figures.

65.3 Calculate and interpret standard errors. Two samples, A and B, gave the following descriptive statistics (measured in the same units): Sample A, mean $= 16.2$, standard deviation $= 12.7$, number of data values $= 12$; Sample B, mean $= 13.2$, standard deviation 14.4, number of data values $= 20$. Which has the lower standard error in absolute terms and in proportion to the sample mean? (Express answers to three significant figures.)

65.4 Compute a mean value correctly. A researcher finds that the mean diameter of limpets on three seashore stones designated A, B and C is 3.0, 2.5 and 2.0 mm respectively. He computes the mean limpet diameter as 2.5 mm, but forgets that the sample sizes were 24, 37 and 6 respectively. What is the true mean diameter of the limpets in this sample? (Answer to three significant figures.)

Set of data

9	6	7	5	7	7	8	6	5	5	7	8
5	8	7	7	6	7	8	6	5	7	7	6
3	6	8	9	9	6	7	8	5	6	5	5
8	8	7	5	6	5	8	6	7	5	7	6
5	6	7	8	7	6	7	7	8	8	9	4

This chapter outlines the philosophy of hypothesis-testing statistics, indicates the steps to be taken when choosing a test, and discusses features and assumptions of some important tests. For details of the mechanics of tests, consult appropriate texts (e.g. Sokal and Rohlf, 1994; Heath, 1995; Wardlaw, 2000). Most tests are now available in statistical packages for computers (see p. 78) and many in spreadsheets (p. 70).

To carry out a statistical test:

1. Decide what it is you wish to test (create a null hypothesis and its alternative).
2. Determine whether your data fit a standard distribution pattern.
3. Select a test and apply it to your data.

Setting up a null hypothesis

Hypothesis-testing statistics are used to compare the properties of samples either with other samples or with some theory about them. For instance, you may be interested in whether two samples can be regarded as having different means, whether the counts of an organism in different quadrats can be regarded as randomly distributed, or whether property A of an organism is linearly related to property B.

 KEY POINT *You can't use statistics to* prove *any hypothesis, but they can be used to assess* how likely *it is to be wrong.*

Statistical testing operates in what at first seems a rather perverse manner. Suppose you think a treatment has an effect. The theory you actually test is that it has no effect; the test tells you how improbable your data would be if this theory were true. This 'no effect' theory is the null hypothesis (NH). If your data are very improbable under the NH, then you may suppose it to be wrong, and this would support your original idea (the 'alternative hypothesis'). The concept can be illustrated by an example. Suppose two groups of subjects were treated in different ways, and you observed a difference in the mean value of the measured variable for the two groups. Can this be regarded as a 'true' difference? As Fig. 66.1 shows, it could have arisen in two ways:

- Because of the way the subjects were allocated to treatments, i.e. all the subjects liable to have high values might, by chance, have been assigned to one group and those with low values to the other (Fig. 66.1(a)).
- Because of a genuine effect of the treatments, i.e. each group came from a distinct frequency distribution (Fig. 66.1(b)).

A statistical test will indicate the probabilities of these options. The NH states that the two groups come from the same population (i.e. the treatment effects are negligible in the context of random variation). To test this, you calculate a test statistic from the data, and compare it with tabulated critical values giving the probability of obtaining the observed or a more extreme result by chance (see Boxes 66.1 and 66.2). This probability is sometimes called the significance of the test.

Note that you must take into account the degrees of freedom (d.f.) when looking up critical values of most test statistics. The d.f. is related to the

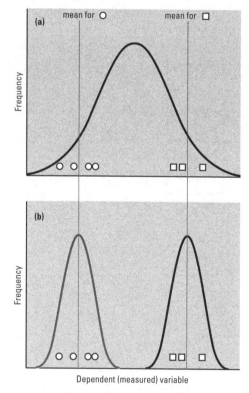

Fig. 66.1 Two explanations for the difference between two means. In case (a) the two samples happen by chance to have come from opposite ends of the same frequency distribution, i.e. there is no true difference between the samples. In case (b) the two samples come from different frequency distributions, i.e. there is a true difference between the samples. In both cases, the means of the two samples are the same.

size(s) of the samples studied; formulae for calculating it depend on the test being used. Biologists normally use two-tailed tests, i.e. we have no expectation beforehand that the treatment will have a positive or negative effect compared with the control (in a one-tailed test we expect one particular treatment to be bigger than the other). Be sure to use critical values for the correct type of test.

By convention, the critical probability for rejecting the NH is 5 per cent (i.e. $P = 0.05$). This means we reject the NH if the observed result would have come up by chance a maximum of one time in twenty. If the modulus of the test statistic is less than or equal to the tabulated critical value for $P = 0.05$, then we accept the NH and the result is said to be 'not significant' (NS for short). If the modulus of the test statistic is greater than the tabulated value for $P = 0.05$, then we reject the NH in favour of the alternative hypothesis that the treatments had different effects and the result is 'statistically significant'.

Two types of error are possible when making a conclusion on the basis of a statistical test. The first occurs if you reject the NH when it is true and the second if you accept the NH when it is false. To limit the chance of the first type of error, choose a lower probability, e.g. $P = 0.01$, but note that the critical value of the test statistic increases when you do this and results in the probability of the second error increasing. The conventional significance levels given in statistical tables (usually 0.05, 0.01, 0.001) are arbitrary. Increasing use of statistical computer programs now allows the actual probability of obtaining the calculated value of the test statistic to be quoted (e.g. $P = 0.037$).

Note that if the NH is rejected, this does not tell you which of many alternative explanations are true. Also, it is important to distinguish between statistical significance and biological relevance: identifying a statistically significant difference between two samples doesn't mean that this will carry any biological importance.

Comparing data with parametric distributions

A parametric test is one that makes particular assumptions about the mathematical nature of the population distribution from which the samples were taken. If these assumptions are not true, then the test is obviously invalid, even though it might give the answer we expect. A non-parametric test does not assume that the data fit a particular pattern, but it may assume some things about the distributions. Used in appropriate circumstances, parametric tests are better able to distinguish between true but marginal differences between samples than their non-parametric equivalents (i.e. they have greater 'power').

The distribution pattern of a set of data values may be biologically relevant, but it is also of practical importance because it defines the type of statistical tests that can be used. The properties of the main distribution types found in biology are given below with both rules of thumb and more rigorous tests for deciding whether data fit these distributions.

Binomial distributions

These apply to samples of any size from populations when data values occur independently in only two mutually exclusive classes (e.g. type A or type B). They describe the probability of finding the different possible combinations of the attribute for a specified sample size k

---

Definition

Modulus – the absolute value of a number, e.g. modulus $-3.385 = 3.385$.

Quoting significance – the convention for quoting significance levels in text, tables and figures is as follows:
$P > 0.05 =$ 'not significant' (or NS)
$P \leqslant 0.05 =$ 'significant' (or *)
$P \leqslant 0.01 =$ 'highly significant' (or **)
$P \leqslant 0.001 =$ 'very highly significant' (or ***)

Thus, you might refer to a difference in means as being 'highly significant ($P \leqslant 0.01$)'. For this reason, the word 'significant' in its everyday meaning of 'important' or 'notable' should be used with care in scientific writing.

Choosing between parametric and non-parametric tests – always plot your data graphically when determining whether they are suitable for parametric tests as this may save a lot of unnecessary effort later.

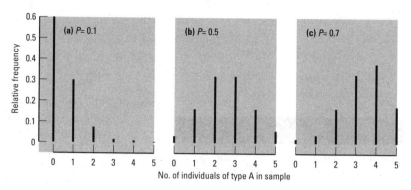

Fig. 66.2 Examples of binomial frequency distributions with different probabilities. The distributions show the expected frequency of obtaining n individuals of type A in a sample of 5. Here P is the probability of an individual being type A rather than type B.

(e.g. out of 10 specimens, what is the chance of 8 being type A?). If p is the probability of the attribute being of type A and q the probability of it being type B, then the expected mean sample number of type A is kp and the standard deviation is $\sqrt{kpq}$. Expected frequencies can be calculated using mathematical expressions (see Sokal and Rohlf, 1994). Examples of the shapes of some binomial distributions are shown in Fig. 66.2. Note that they are symmetrical in shape for the special case $p = q = 0.5$ and the greater the disparity between p and q, the more skewed the distribution.

Some biological examples of data likely to be distributed in binomial fashion are: possession of two alleles for seed-coat morphology (e.g. smooth and wrinkly); whether an organism is infected with a microbe or not; whether an animal is male or female. Binomial distributions are particularly useful for predicting gene segregation in Mendelian genetics and can be used for testing whether combinations of events have occurred more frequently than predicted (e.g. more siblings being of the same sex than expected). To establish whether a set of data is distributed in binomial fashion: calculate expected frequencies from probability values obtained from theory or observation, then test against observed frequencies using a χ^2 test (p. 330) or a G test (see Wardlaw, 2000).

Poisson distributions

These apply to discrete characteristics that can assume low whole-number values, such as counts of events occurring in area, volume or time. The events should be 'rare' in that the mean number observed should be a small proportion of the total that could possibly be found. Also, finding one count should not influence the probability of finding another. The shape of Poisson distributions is described by only one parameter, the mean number of events observed, and has the special characteristic that the variance is equal to the mean. The shape has a pronounced positive skewness at low mean counts, but becomes more and more symmetrical as the mean number of counts increases (Fig. 66.3).

Some examples of characteristics distributed in a Poisson fashion are: number of plants in a quadrat; number of microbes per unit volume of medium; number of animals parasitised per unit time; number of radioactive disintegrations per unit time. One of the main uses for the

Tendency towards the normal distribution – under certain conditions, binomial and Poisson distributions can be treated as normally distributed:

- where samples from a binomial distribution are large (i.e. > 15) and p and q are close to 0.5;
- for Poisson distributions, if the number of counts recorded in each outcome is greater than about 15.

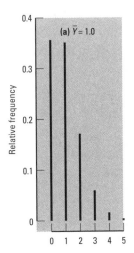
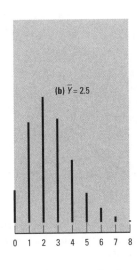
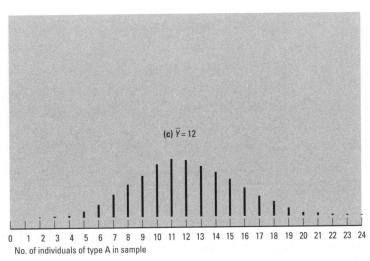

Fig. 66.3 Examples of Poisson frequency distributions differing in mean. The distributions are shown as line charts because the independent variable (events per sample) is discrete.

Poisson distribution is to quantify errors in count data such as estimates of cell densities in dilute suspensions (see p. 275). To decide whether data are Poisson distributed:

- Use the rule of thumb that if the coefficient of dispersion ≈ 1, the distribution is likely to be Poisson.
- Calculate 'expected' frequencies from the equation for the Poisson distribution and compare with actual values using a χ^2 test or a G-test.

It is sometimes of interest to show that data are *not* distributed in a Poisson fashion, e.g. the distribution of parasite larvae in hosts. If $s^2/\bar{Y} > 1$, the data are 'clumped' and occur together more than would be expected by chance; if $s^2/\bar{Y} < 1$, the data are 'repulsed' and occur together less frequently than would be expected by chance.

Normal distributions (Gaussian distributions)

These occur when random events act to produce variability in a continuous characteristic (quantitative variable). This situation occurs frequently in biology, so normal distributions are very useful and much used. The bell-like shape of normal distributions is specified by the population mean and standard deviation (Fig. 66.4): it is symmetrical and configured such that 68.27 per cent of the data will lie within ± 1 standard deviation of the mean, 95.45 per cent within ± 2 standard deviations of the mean, and 99.73 per cent within ± 3 standard deviations of the mean.

Some biological examples of data likely to be distributed in a normal fashion are: fresh weight of plants of the same age; linear dimensions of bacterial cells; height of either adult female or male humans. To check whether data come from a normal distribution, you can:

- Use the rule of thumb that the distribution should be symmetrical and that nearly all the data should fall within $\pm 3s$ of the mean and about two-thirds within $\pm 1s$ of the mean.
- Plot the distribution on normal probability graph paper. If the distribution is normal, the data will tend to follow a straight line (see Fig. 66.5). Deviations from linearity reveal skewness and/or kurtosis

Definition

Coefficient of dispersion $= s^2/\bar{Y}$. This is an alternative measure of dispersion to the coefficient of variation (p. 420).

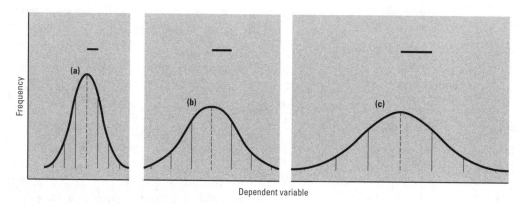

Fig. 66.4 Examples of normal frequency distributions differing in mean and standard deviation. The horizontal bars represent population standard deviations for the curves, increasing from (a) to (c). Vertical dashed lines are population means, while vertical solid lines show positions of values ±1, 2 and 3 standard deviations from the means.

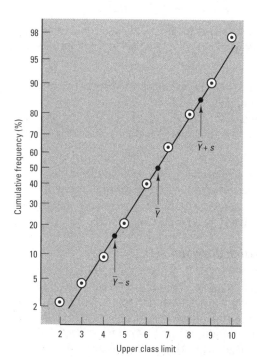

Fig. 66.5 Example of a normal probability plot. The plotted points are from a small data set where the mean $\bar{Y} = 6.93$ and the standard deviation $s = 1.895$. Note that values corresponding to 0% and 100% cumulative frequency cannot be used. The straight line is that predicted for a normal distribution with $\bar{Y} = 6.93$ and $s = 1.895$. This is plotted by calculating the expected positions of points for $\bar{Y} \pm s$. Since 68.3% of the distribution falls within these bounds, the relevant points on the cumulative frequency scale are $50 \pm 34.15\%$; thus this line was drawn using the points (4.495, 15.85) and (8.285, 84.15) as indicated on the plot.

(see p. 422), the significance of which can be tested statistically (see Sokal and Rohlf, 1994).

● Use a suitable statistical computer program to generate predicted normal curves from the $\bar{Y}$ and s values of your sample(s). These can be compared visually with the actual distribution of data and can be used to give 'expected' values for a χ^2 test or a G-test.

The wide availability of tests based on the normal distribution and their relative simplicity means you may wish to transform your data to make them more like a normal distribution. Table 66.1 provides transformations that can be applied (see also Fig. 61.3). The transformed data should be tested for normality as described above before proceeding – don't forget that you may need to check that transformed variances are homogeneous for certain tests (see below).

A very important theorem in statistics, the Central Limit Theorem, states that as sample size increases, the distribution of a series of means from any frequency distribution will become normally distributed. This fact can be used to devise an experimental or sampling strategy that ensures that data are normally distributed, i.e. using means of samples as if they were primary data.

Table 66.1 Suggested transformations altering different types of frequency distributions to the normal type. To use, modify data by the formula shown; then examine effects with the tests described on pp. 427–30.

Type of data; distribution suspected	Suggested transformation(s)
Proportions (including percentages); binomial	arcsine $\sqrt{x}$ (also called the angular transformation)
Scores; Poisson	$\sqrt{x}$ or $\sqrt{(x + 1/2)}$ if zero values present
Measurements; negatively skewed	x^2, x^3, x^4, etc. (in order of increasing strength)
Measurements; positively skewed	$1/\sqrt{x}$, $\sqrt{x}$, ln x, $1/x$ (in order of increasing strength)

Definition

Homogeneous variance – uniform (but not necessarily identical) variance of the dependent variable across the range of the independent variable. The term homoscedastic is also used in this sense. The opposite of homogeneous is heterogeneous (= hereoscedastic).

Understanding 'degrees of freedom' – this depends on the number of values in the data set analysed, and the method of calculation depends on the statistical test being used. It relates to the number of observations that are free to vary before the remaining quantities for a data set can be determined.

Checking the assumptions of a test – always acquaint yourself with the assumptions of a test. If necessary, test them before using the test.

Choosing a suitable statistical test

Comparing location (e.g. means)

If you can assume that your data are normally distributed, the main test for comparing two means from independent samples is Student's t-test (see Boxes 66.1 and 66.2, and Table 66.2). This assumes that the variances of the data sets are homogeneous. Tests based on the t-distribution are also available for comparing means of paired data or for comparing a sample mean with a chosen value.

When comparing means of two or more samples, analysis of variance (ANOVA) is a very useful technique. This method also assumes data are normally distributed and that the variances of the samples are homogeneous. The samples must also be independent (e.g. not subsamples). The test statistic calculated is denoted F and it has two different degrees of freedom related to the number of means tested and the pooled number of replicates per mean. The nested types of ANOVA are useful for letting you know the relative importance of different sources of variability in your data. Two-way and multi-way ANOVAs are useful for studying interactions between treatments.

For data satisfying the ANOVA requirements, the least significant difference (LSD) is useful for making planned comparisons among several means (see Sokal and Rohlf, 1994). Any two means that differ by more than the LSD will be significantly different. The LSD is useful for showing on graphs.

The chief non-parametric tests for comparing the locations of two samples are the Mann–Whitney U-test and the Kolmogorov–Smirnov test. The former assumes that the frequency distributions of the samples are similar, whereas the latter makes no such assumption. In both cases the sample's size must be $\geqslant 4$ and for the Kolmogorov–Smirnov test the samples must have equal sizes. In the Kolmogorov–Smirnov test, significant differences found with the test could be due to differences in location or shape of the distribution, or both.

Table 66.2 Critical values of Student's t statistic (for two-tailed tests). Reject the null hypothesis at probability P if your calculated t value equals or exceeds the value shown for the appropriate degrees of freedom = $(n_1 - 1) + (n_2 - 1)$.

Degrees of freedom	Critical values for $P = 0.05$	Critical values for $P = 0.01$	Critical values for $P = 0.001$
1	12.71	63.66	636.62
2	4.30	9.92	31.60
3	3.18	5.84	12.94
4	2.78	4.60	8.61
5	2.57	4.03	6.86
6	2.45	3.71	5.96
7	2.36	3.50	5.40
8	2.31	3.36	5.04
9	2.26	3.25	4.78
10	2.23	3.17	4.59
12	2.18	3.06	4.32
14	2.14	2.98	4.14
16	2.12	2.92	4.02
20	2.09	2.85	3.85
25	2.06	2.79	3.72
30	2.04	2.75	3.65
40	2.02	2.70	3.55
60	2.00	2.66	3.46
120	1.98	2.62	3.37
∞	1.96	2.58	3.29

Choosing and using statistical tests

Box 66.1 How to carry out a *t*-test

The *t*-test was devised by a statistician who used the pen name 'Student', so you may see it referred to as Student's *t*-test. It is used when you wish to decide whether two samples come from the same population or from different ones (Fig. 66.1). The samples might have been obtained by selective observation (Chapter 27) or by applying two different treatments to an originally homogeneous population (Chapter 31).

The null hypothesis (NH) is that the two groups can be represented as samples from the same overlying population (Fig. 66.1(a)). If, as a result of the test, you accept this hypothesis, you can say that there is no significant difference between the group means.

The alternative hypothesis is that the two groups come from different populations (Fig. 66.1(b)). By rejecting the NH as a result of the test, you can accept the alternative hypothesis and say that there is a significant difference between the sample means, or, if an experiment were carried out, that the two treatments affected the samples differently.

How can you decide between these two hypotheses? On the basis of certain assumptions (see below), and some relatively simple calculations, you can work out the probability that the samples came from the same population. If this probability is very low, then you can reasonably reject the NH in favour of the alternative hypothesis, and if it is high, you will accept the NH.

To find out the probability that the observed difference between sample means arose by chance, you must first calculate a '*t* value' for the two samples in question. Some computer programs (e.g. Minitab) provide this probability as part of the output, otherwise you can look up statistical tables (e.g. Table 66.2). These tables show 'critical values' – the borders between probability levels. If your value of *t* equals or exceeds the critical value for probability *P*, you can reject the NH at this probability ('level of significance'). Note that:

- for a given difference in the means of the two samples, the value of *t* will get larger the smaller the scatter within each data set; and
- for a given scatter of the data, the value of *t* will get larger, the greater the difference between the means.

So, at what probability should you reject the NH? Normally, the threshold is arbitrarily set at 5 per cent – you quite often see descriptions like 'the sample means were significantly different ($P < 0.05$)'. At this 'significance level' there is still up to a 5 per cent chance of the *t* value arising by chance, so about one in twenty times, on average, the conclusion will be wrong. If *P* turns out to be lower, then this kind of error is much less likely.

Tabulated probability levels are generally given for 5, 1 and 0.1 per cent significance levels (see Table 66.2). Note that this table is designed for 'two-tailed' tests, i.e. where the treatment or sampling strategy could have resulted in either an increase or a decrease in the measured values. These are the most likely situations you will deal with in biology.

Examine Table 66.2 and note the following:

- The larger the size of the samples (i.e. the greater the 'degrees of freedom'), the smaller *t* needs to be to exceed the critical value at a given significance level.
- The lower the probability, the greater *t* needs to be to exceed the critical value.

The mechanics of the test

A calculator that can work out means and standard deviations is helpful.

1. **Work out the sample means $\bar{Y}_1$ and $\bar{Y}_2$ and calculate the difference between them.**

2. **Work out the sample standard deviations s_1 and s_2.** (NB if your calculator offers a choice, chose the '$n-1$' option for calculating s – see p. 420).

3. **Work out the sample standard errors $SE_1 = s_1/\sqrt{n_1}$ and $SE_2 = s_2/\sqrt{n_2}$; now square each, add the squares together, then take the positive square root of this (n_1 and n_2 are the respective sample sizes, which may, or may not, not be equal).**

4. **Calculate *t* from the formula:**

$$t = \frac{\bar{Y}_1 - \bar{Y}_2}{\sqrt{((SE_1)^2) + ((SE_2)^2)}} \qquad [66.1]$$

   The value of *t* can be negative or positive, depending on the values of the means; this does not matter and you should compare the modulus (absolute value) of *t* with the values in tables.

5. **Work out the degrees of freedom $= (n_1 - 1) + (n_2 - 1)$.**

6. **Compare the *t* value with the appropriate critical value (see e.g. Table 66.2) and decide on the significance of your findings (see p. 427).**

Box 66.2 provides a worked example – use this to check that you understand the above procedures.

Assumptions that must be met before using the test

The most important assumptions are:

- The two samples are independent and randomly drawn (or if not, drawn in a way that does not create bias). The test assumes that the samples are quite large.
- The underlying distribution of each sample is normal. This can be tested with a special statistical test, but a rule of thumb is that a frequency distribution of the data should be (a) symmetrical about the mean and (b) nearly all of the data should be within 3 standard deviations of the mean and about two-thirds within 1 standard deviation of the mean (see p. 429).
- The two samples should have uniform variances. This again can be tested (by an *F*-test), but may be obvious from inspection of the two standard deviations.

Box 66.2 Worked example of a *t*-test

Suppose the following data were obtained in an experiment (the units are not relevant):

Control: 6.6, 5.5, 6.8, 5.8, 6.1, 5.9
Treatment: 6.3, 7.2, 6.5, 7.1, 7.5, 7.3

Using the steps outlined in Box 66.1, the following values are obtained (denoting control with subscript 1, treatment with subscript 2):

1. $\bar{Y}_1 = 6.1167$; $\bar{Y}_2 = 6.9833$: difference between means $= \bar{Y}_1 - \bar{Y}_2 = -0.8666$

2. $s_1 = 0.49565$; $s_2 = 0.47504$

3. $SE_1 = 0.49565/2.44949 = 0.202348$
   $SE_2 = 0.47504/2.44949 = 0.193934$

4. $t = \dfrac{-0.8666}{\sqrt{(0.202348^2 + 0.193934^2)}} = \dfrac{-0.8666}{0.280277} = -3.09$

5. d.f. $= (5 + 5) = 10$

6. Looking at Table 66.2, we see that the modulus of this t value exceeds the tabulated value for $P = 0.05$ at 10 degrees of freedom $(= 2.23)$. We therefore reject the NH, and conclude that the means are different at the 5% level of significance. If the modulus of t had been $\leqslant 2.23$, we would have accepted the NH. If the modulus of t had been > 3.17, we could have concluded that the means are different at the 1% level of significance.

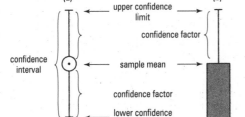

Fig. 66.6 Graphical representation of confidence limits as 'error bars' for (a) a sample mean in a plotted curve, where both upper and lower limits are shown; and (b) a sample mean in a histogram, where, by convention, only the upper value is shown. For data that are assumed to be symmetrically distributed, such representations are often used in preference to the 'box and whisker' plot shown on p. 419. Note that SE is an alternative way of representing sample imprecision/error (e.g. Fig. 62.1).

Confidence limits for statistics other than the mean – consult an advanced statistical text (e.g. Sokal and Rohlf, 1994) if you wish to indicate the reliability of estimates of e.g. population variances.

Suitable non-parametric comparisons of location for paired data (sample size $\geqslant 6$) include Wilcoxon's signed rank test, which is used for quantitative data and assumes that the distributions have similar shape. Dixon and Mood's sign test can be used for paired data scores where one variable is recorded as 'greater than' or 'better than' the other.

Non-parametric comparisons of location for three or more samples include the Kruskal–Wallis H-test. Here, the number of samples is without limit and they can be unequal in size, but again the underlying distributions are assumed to be similar. The Friedman S-test operates with a maximum of five samples and data must conform to a randomised block design. The underlying distributions of the samples are assumed to be similar.

Comparing dispersions (e.g. variances)

If you wish to compare the variances of two sets of data that are normally distributed, use the F-test. For comparing more than two samples, it may be sufficient to use the F_{max}-test, on the highest and lowest variances. The Scheffé–Box (log-anova) test is recommended for testing the significance of differences between several variances. Non-parametric tests exist but are not widely available: you may need to transform the data and use a test based on the normal distribution.

Determining whether frequency observations fit theoretical expectation

The χ^2 test (Box 53.2) is useful for tests of 'goodness of fit', e.g. comparing expected and observed progeny frequencies in genetical experiments or comparing observed frequency distributions with some theoretical function. One limitation is that simple formulae for calculating χ^2 assume that no expected number is less than 5. The G-test ($2I$-test) is used in similar circumstances.

Comparing proportion data

When comparing proportions between two small groups (e.g. whether 3/10 is significantly different from 5/10), you can use probability tables such as those of Finney et al. (1963) or calculate probabilities from formulae; however, this can be tedious for large sample sizes. Certain proportions can be transformed so that their distribution becomes normal.

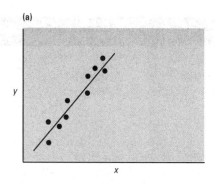

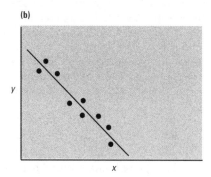

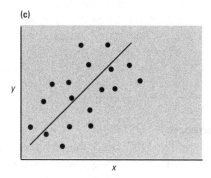

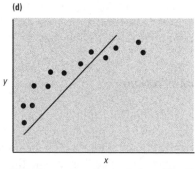

Fig. 66.7 Examples of correlation. The linear regression line is shown. In (a) and (b), the correlation between x and y is good: for (a) there is a positive correlation and the correlation coefficient, r, would be close to 1; for (b) there is a negative correlation and the correlation coefficient would be close to −1. In (c) there is a weak positive correlation and r would be close to 0. In (d) the correlation coefficient may be quite large, but the choice of linear regression is clearly inappropriate.

Placing confidence limits on an estimate of a population parameter

On many occasions, a sample statistic is used to provide an estimate of a population parameter, and it is often useful to indicate the reliability of such an estimate. This can be done by putting confidence limits on the sample statistic, i.e. by specifying an interval around the statistic within which you are confident that the true value (the population parameter) is likely to fall, at a specified level of probability. The most common application is to place confidence limits on the mean of a sample taken from a population of normally distributed data values. In practice, you determine a confidence factor for a particular level of probability that is added to and subtracted from the sample mean ($\overline{Y}$) to give the upper confidence limit and lower confidence limit respectively. These are calculated as:

$$\overline{Y} + (t_{P[n-1]} \times \text{SE}) \text{ for the upper limit and}$$
$$\overline{Y} - (t_{P[n-1]} \times \text{SE}) \text{ for the lower limit} \qquad [66.2]$$

where $t_{P[n-1]}$ is the tabulated critical value of Student's t statistic for a two-tailed test with $n - 1$ degrees of freedom at a specified probability level (P) and SE is the standard error of the sample mean (p. 420). The 95 per cent confidence limits (i.e. $P = 0.05$) tells you that, on average, 95 times out of 100 the interval between the upper and lower limits will contain the true (population) value. Confidence limits are often shown as 'error bars' for individual sample means plotted in graphical form. Figure 66.6 illustrates how this is applied to plotted curves and histograms (note that this can be carried out for data series within a Microsoft Excel graph (chart) using the *Format data series* and *Y error bars* commands).

Correlation and regression

These methods are used when testing the relationship between data values for two variables. Correlation is used to measure the extent to which changes in the two sets of data values occur together in a linear manner. If one variable can be assumed to be dependent upon the other (i.e. a change in X causes a particular change in Y), then regression techniques can be used to provide a mathematical description of the underlying relationship between the variables, e.g. to find a line of best fit for a data series. If there is no a priori reason to assume dependency, then correlation methods alone are appropriate.

A correlation coefficient measures the strength of the linear relationship between two variables, but does not describe the relationship. The coefficient is expressed as a number between −1 and +1: a positive coefficient indicates a direct relationship, where the two variables change in the same direction, while a negative coefficient indicates an inverse relationship, where one variable decreases as the other increases (Fig. 66.7). The nearer the coefficient is to −1 or +1, the stronger the linear relationship between the variables, i.e. the less 'scatter' there would be about a straight line of best fit (note that this does *not* imply that one variable is dependent upon the other). A coefficient of 0 implies that the two variables show no linear association and therefore the closer the correlation coefficient is to zero, the weaker the linear relationship. The importance of graphing data is shown by the case illustrated in Fig. 66.7(d).

Pearson's product moment correlation coefficient (r) is the most commonly used statistic for testing correlations. The test is valid only if both variables are normally distributed. Statistical tests can be used to decide whether the correlation is significant (e.g. using a one-sample t-test to see whether r is significantly different from zero, based on the equation:

$$t = r \div \sqrt{[(1 - r^2) \div (n - 2)]} \text{ at } n - 2 \text{ degrees of freedom,} \qquad [66.3]$$

where n is the number of paired observations. If one or both variables are not normally distributed, then you should calculate an alternative non-parametric coeffcient, e.g. Spearman's coefficient of rank correlation (r_s) or Kendall's coefficient of rank correlation (τ). These require the two sets of data to be ranked separately, and the calculation can be complex if there are tied (equal) ranks. Spearman's coefficient is said to be better if there is any uncertainty about the reliability of closely ranked data values.

If underlying theory or empirical graphical analysis indicate a linear relationship between a dependent and an independent variable, then linear regression can be used to estimate the mathematical equation that links the two variables. Model I linear regression is the standard approach, and is available within general-purpose software programs such as Microsoft Excel (Box 66.3), and on some scientific calculators. It is suitable for experiments where a dependent variable Y varies with an *error-free* independent variable X in accordance with the relationship $Y = a + bX + e_Y$, where e_Y represents the residual (error) variability in the Y variable. For example, this relationship might apply in a laboratory procedure where you have carefully controlled the independent variable and the X values can be assumed to have zero error (e.g. in a calibration curve, see Chapter 50, or in a time course experiment where measurements are made at exact time points). The regression analysis gives estimates for a and b (equivalent to the slope and intercept of the line of best fit, p. 411): computer-based programs usually provide additional features, e.g. residual values for Y (e_Y), estimated errors for a and b, predicted values of Y along with graphical plots of the line of best fit (the trend line) and the residual values. In order for the model to be valid, the residual (error) values should be normally distributed around the trend line and their variance should be uniform (homogeneous), i.e. there should be a similar scatter of data points around the trend line along the x-axis (independent variable).

If the relationship is not linear, try a transformation (see p. 385). For example, this is commonly done in analysis of enzyme kinetics (see Fig. 52.4). However, you should be aware that the transformation of data to give a straight line can lead to errors when carrying out linear regression analysis: take care to ensure that (a) the assumptions listed in the previous paragraph are valid for the transformed data set and (b) the data points are evenly distributed throughout the range of the independent variable. If these criteria cannot be met, non-linear regression may be a better approach, but for this you will require a suitable computer program, e.g. GraphPad Prism.

The strength of the relationship between Y and X in Model I linear regression is best estimated by the coefficient of determination (r^2 or R^2), which is equivalent to the square of the Pearson correlation coefficient. The coefficient of determination varies between 0 and $+1$ and provides a measure of the goodness of fit of the Y data to the regression line: the closer the value is to 1, the better the fit. In effect, r^2 represents the fraction of the variance in Y that can be accounted for by the regression equation. Conversely, if you subtract this value from 1, you will obtain the residual

Using more advanced types of regression – these include:

- Model II linear regression, which applies to situations where a dependent variable Y varies with an independent variable X, and where both variables may have error terms associated with them.
- Multiple regression, which applies when there is a relationship between a dependent variable and two or more independent variables.
- Non-linear regression, which extends the principles of linear regression to a wide range of functions. Technically, this method is more appropriate than transforming data to allow linear regression.

Advanced statistics books should be consulted for details of these methods, which may be offered by some statistical computer programs.

Example If a regression analysis gives a value for r^2 of 0.75 (i.e. $r = 0.84$), then 75% of the variance in Y can be explained by the trend line, with $1 - r^2 = 0.25$ (25%) remaining as unexplained (residual) variation.

Box 66.3   Using a spreadsheet (Microsoft Excel) to calculate hypothesis-testing statistics

Presented below are three examples of the use of Microsoft Excel to investigate hypotheses about specific data sets. In each case, there is a brief description of the problem; a table showing the data analysed; an outline of the Microsoft Excel commands used to carry out the analysis and an annotated table of results from the spreadsheet.

Example 1: A t-test

As part of a project, a student applied a chemical treatment to a series of flasks containing fungal cultures with nutrient solution. An otherwise similar set of control flasks received no chemical treatment. After three weeks' growth, she measured the wet mass of the filtered cultures:

Wet mass of samples (g)

Replicate	1	2	3	4	5	6	7	8	Mean	Variance
Treated with ZH52	2.342	2.256	2.521	2.523	2.943	2.481	2.601	2.449	2.515	0.042
Control	2.658	2.791	2.731	2.402	3.041	2.668	2.823	2.509	2.703	0.038

The student proposed the null hypothesis that there was no difference between the two means and tested this using a t-test, as she had evidence from other studies that the fungal masses of replicate flasks were normally distributed. She also established, by calculation, that the assumption that the populations had homogeneous variances was likely to be valid. Using the *Tools > Data Analysis > t-Test: Two-Sample Assuming Equal Variance* option, with *Hypothesized Mean Difference* = 0 and *Alpha* $(=P) = 0.05$, and adjusting the number of significant figures displayed, the following table was obtained:

t-test: Two-sample assuming equal variances

	Variable 1	Variable 2
Mean	2.515	2.703
Variance	0.042	0.038
Observations	8	8
Pooled variance	0.040	
Hypothesized mean difference	0	
d.f.	14	
t Stat	−1.881	
P(T <= t) one-tail	0.040	
t Critical one-tail	1.761	
P(T <= t) two-tail	0.081	
t Critical two-tail	2.145	

The value of t obtained was −1.881 (row 7, 't stat') and the probability of obtaining this value for a two-tailed test (row 10) was 0.081 (or 8.1%), so the student was able to accept the null hypothesis and conclude that ZH52 had no significant effect on fungal growth in these circumstances.

Example 2: An ANOVA test

A biochemist made six replicate measurements of four different batches (A–D) of alcohol dehydrogenanse, obtaining the following data:

Alcohol dehydrogenase activity (U I^{-1})

Batch/Replicate	1	2	3	4	5	6	Mean	Variance
A	0.562	0.541	0.576	0.545	0.542	0.551	0.552833	0.000189
B	0.531	0.557	0.537	0.521	0.559	0.538	0.540500	0.000221
C	0.572	0.568	0.551	0.549	0.564	0.559	0.560500	0.000085
D	0.532	0.548	0.541	0.538	0.547	0.536	0.540333	0.000039

The biochemist wanted to know whether the observed differences were statistically significant, so he carried out an ANOVA test, assuming the samples were normally distributed and the variances in the three populations were homogeneous. Using the *Tools > Data Analysis > Anova: Single Factor* option, with *Alpha* $(=P) = 0.05$, and adjusting the number of significant figures displayed, the following table was obtained:

ANOVA: Single Factor

SUMMARY

Groups	Count	Sum	Average	Variance
A	6	3.317	0.552833	0.000189
B	6	3.243	0.5405	0.000221
C	6	3.363	0.5605	8.51E-05
D	6	3.242	0.540333	3.95E-05

ANOVA

Source of Variation	SS	d.f.	MS	F	P-value	F crit
Between groups	0.001761	3	0.000587	4.397856	0.015669	3.098391
Within groups	0.002669	20	0.000133			
Total	0.00443	23				

The F value calculated was 4.397856. This comfortably exceeds the stated critical value (F_{crit}) of 3.098391, and the probability of obtaining this result by chance (P-value) was calculated as 0.015669 (1.57% to three significant figures); hence the biochemist was able to reject the null hypothesis and conclude that there was a significant difference in average enzyme activity between the four batches, since $P < 0.05$. Such a finding might lead on to an investigation into why there was batch variation, e.g. had they been stored differently?

(continued)

Box 66.3 (continued)

Example 3: Testing the significance of a correlation

A researcher wanted to know whether his observations of earthworm casts on the surface of closely mown grass were related to how wet the soil was. He took weekly measurements of precipitation using a rain gauge and counted the mean number of casts per m^2 taking mean results from nine quadrats per weekly observation:

Observation	Precipitation in previous week (mm)	Mean density of casts (m^{-2})
1	11	4.4
2	1	3.1
3	0	2.3
4	5	4.6
5	8	4.5
6	2	3.3
7	4	3.5
8	15	6.4

The researcher used the Microsoft Excel function PEARSON(array1, array2) to obtain a value of +0.927 857 674 for Pearson's product moment correlation coefficient r, specifying the precipitation data as array1 and the cast density as array2. He then used a spreadsheet to calculate the t statistic (p. 435) for this r value, using eqn. [66.3]. The value of t was 7.037, with six degrees of freedom. The critical value from tables (e.g. Table 66.2) at $P = 0.001$ is 5.96, so he concluded that there was a very highly significant positive correlation between his two sets of observations. The investigator moved on from this observation and next investigated the effect of artificial hosepipe rainfall in a sheltered grass plot, to test whether there was a causal relationship involved.

(error) component, i.e. the fraction of the variance in Y that cannot be explained by the line of best fit. Multiplying the values by 100 allows you to express these fractions in percentage terms.

Using computers to calculate hypothesis-testing statistics

As with the calculation of descriptive statistics (p. 423), specialist statistical packages such as SPSS and MINITAB can be used to simplify the calculation of hypothesis-testing statistics. The correct use of the software and interpretation of the output requires an understanding of relevant terminology and of the fundamental principles governing the test, which is probably best obtained by working through one or more examples by hand before using these tools (e.g. Box 53.2, Box 66.2). Spreadsheets offer increasingly sophisticated statistical analysis functions, three examples of which are provided in Box 66.3.

Text references and sources for further study

Finney, D.J., Latscha, R., Bennett, B.M. and Hsu, P. (1963) *Tables for Testing Significance in a 2 × 2 Table*. Cambridge University Press, Cambridge.

Heath, D. (1995) *An Introduction to Experimental Design and Statistics for Biology*. UCL Press, London.

Schmuller, J. (2005) *Statistical Analysis with Excel for Dummies*. Wiley, Hoboken, New Jersey.

Sokal, R.R. and Rohlf, F.J. (1994) *Biometry*, 3rd edn. W.H. Freeman, San Francisco.

Wardlaw, A.C. (2000) *Practical Statistics for Experimental Biologists*, 2nd edn. Wiley, New York.

Study exercises

66.1 Calculate 95 per cent confidence limits. What are the 95 per cent confidence limits of a sample with a mean = 24.7, standard deviation = 6.8 and number of data values = 16? (Express your answer to three significant figures.)

66.2 Use the Poisson distribution. In a sample of 15 snails, a researcher finds the following number of parasite larvae per snail: 0, 0, 0, 0, 1, 1, 1, 3, 4, 5, 5, 7, 7, 9, 9. Using the rule of thumb on p. 429, decide whether the parasites are 'clumped' or 'repulsed' in distribution on their host. What might this mean in biological terms?

66.3 Practise using a *t*-test. A biology student examines the effect of adding a plant hormone to pea plants. She dissolves an appropriate amount of the compound in ethanol and applies 25 µl of this to the uppermost stipules of the treated plants. With the controls, she applies the same amount of pure ethanol. After three days, she measures the distance between the 2nd and 3rd internodes of the plants and obtains the results shown at the top of the next column. Carry out a *t*-test on the data and draw appropriate conclusions.

Internode distance in cm

Control	7.5	8.1	7.6	6.2	7.5	7.8	8.9
Treatment	5.6	7.5	8.2	6.7	3.5	6.5	5.9

66.4 Interpret the output from Excel linear regression analysis. The following output represents a regression analysis for an experiment measuring the uptake of an amino acid by a cell suspension (in $pmol\,cell^{-1}$) against time (in minutes). Based on this output, what is the form and strength of the underlying linear relationship? (Express the coefficients to three significant figures.)

Summary output from Microsoft Excel spreadsheet linear regression analysis

Regression statistics

Multiple R	0.985335951
R square	0.970886937
Adjusted R square	0.963608672
Standard error	2.133876419
Observations	6

ANOVA

	d.f.	SS	MS	F	Significance F
Regression	1	607.4062857	607.4063	133.3954	0.000320975
Residual	4	18.21371429	4.553429		
Total	5	625.62			

	Coefficients	Standard error	t Stat	P-value	Lower 95%	Upper 95%	Lower 95.0%	Upper 95.0%
Intercept	1.171428571	1.544386367	0.758507	0.490383	−3.11648428	5.459341423	−3.11648428	5.459341423
X variable 1	2.945714286	0.255047014	11.54969	0.000321	2.237588784	3.653839787	2.237588784	3.653839787

Answers to study exercises

We have attempted to provide an 'answer' to all of the study exercises. Where the question is open-ended and no 'correct' answer can be given, we have provided a *Tip* or *Tips*, which should either help with your general approach, indicate which resources are worth consulting, or provide a pointer to relevant material within the book. Where a non-numerical question has a 'correct' answer, we have provided a model text-based answer. If a calculation is involved, we have shown the steps involved and have indicated the **correct answer underlined and in bold**.

1   The importance of transferable skills

1.1   *Tips:* Where you feel confident about skills you have learned at school, take care to consider whether they might need to be upgraded for university/college use. When thinking about opportunities for developing skills, remember that these may occur outside university, perhaps in work or social contexts.

1.2   *Tip:* Possible keywords/phrases for a Web search include: biology skills, C&IT skills, numeracy skills, study skills, time management.

1.3   *Tip:* You could create a grid to record your thoughts, with column headings 'University', 'Work' and 'Social', and rows for 'Short term', 'Medium term' and 'Long term'. Short term and medium term could refer to your time at university, while long term could refer to after graduation.

2   Managing your time

2.1   *Tips:* Model your spreadsheet on Fig. 2.1. It isn't too difficult to fill in details at the end of each day either directly into the spreadsheet or on a printed version. You will need to collect information for at least a week and possibly longer before you see patterns emerging.

2.2   *Tips:* It is important to include both social and study-related activities on your lists. Short-term lists deal with issues of the day or week; medium-term for the term or semester and longer-term for one year onwards. Relate each of these lists to appropriate goals you may have over these timescales.

2.3   *Tips:* This task could be related to (a) your self-analysis of transferable skills (Chapter 1), or (b) to some large task such as carrying out a final-year project. In the latter case, for example, there are small jobs that can be done as you proceed with the larger task, such as writing parts of the Introduction or compiling a list of references, and these will reduce time pressure towards the end of the task.

3   Working with others

3.1   *Tips:* Think about your contributions to past group activities while doing this exercise, and, if necessary, get feedback from friends and colleagues. You may feel that you fit into more than one 'natural' role: this is quite often the case.

3.2   *Tip:* This exercise might best be associated with a specific teamwork activity, but this need not be restricted to study – group contributions with clubs, societies or in employment are equally valid. Strategies for improvement will depend on your identified weaknesses. For example, if you felt too shy to contribute effectively, you might want to practise public speaking (see Chapter 14) or attend an assertiveness workshop.

3.3   *Tip:* Columns 4 and 5 of Table 3.1 are relevant, particularly if you are able to identify a 'natural' role or roles for yourself in group situations (as in exercise 3.1). Can you relate your strengths and weaknesses to specific events that have occurred during teamwork, and might you handle the situation differently in future?

4   Taking notes from lectures and texts

4.1   *Tips:* If you are used to using a particular note-taking method that seems to work well, you may feel reluctant to experiment. Try out the new method by taking notes at a seminar, tutorial or meeting where it will not matter so much if the technique does not work well at the first attempt.

4.2   *Tip:* Make a list of any missing notes or handouts and ask a colleague if you can work from their notes to fill in these gaps.

4.3   *Tip:* To compare methods, test your recall some time afterwards. Take a blank sheet of paper and see what you can write about each topic.

5   Learning and revising

5.1   *Tips:* When constructing your revision timetable, remember the following important points:
- Break large topics into 'bite-sized' units.
- Mix up hard or less interesting topics with those you enjoy revising (don't ignore them).
- Keep an appropriate balance between work and relaxation.
- Offer yourself rewards if work targets are completed.

5.2   *Tips:* You may wish to keep one of the past papers (preferably a recent one) aside to use in a mock exam. Remember that learning objectives can be just as important as exam papers in helping you to revise.

5.3   *Tips:* Read through the list of active revision tips on pp. 26–7 and decide which might work best for you. Carry out a trial of the method with one of your topics where the material has seemed difficult to learn. Experiment until you arrive at a solution that works, remembering (a) that different techniques might work best in different subjects and (b) a little variety might make the revision process more interesting.

6   Curriculum options, assessments and exams

6.1   *Tip:* Some questions to ask yourself about past exam performances... Did you prepare well enough? Did you run out of time during the exam? Did you spend too long on a particular question or section, and have to rush another? Were your answers direct and at the appropriate depth? Did you misinterpret any questions, or miss out part of any answers? Were your writing skills, including planning, up to the standard required?

6.2 *Tip:* When revising, try to find a topic you find hard to understand and see whether your colleague has approached it in the same way as you. If not, you may be able to learn from this.

6.3 *Tip:* Discuss your tactics with a colleague sitting the same exam. What topics do you agree are likely to come up?

7 Preparing your curriculum vitae

7.1 *Tip:* This exercise might be attempted with a friend you can trust to give you a frank opinion. Remember that your personal qualities can be developed in both curricular and extra-curricular activities.

7.2 *Tip:* Working with a friend or group from your class, compare notes on how you have organised your CVs and then modify your own CV, taking on board good ideas from your peers. Create a physical file for your revised CV and add hand-written notes, updating it as and when required.

7.3 *Tip:* Use the 'Prospects' website noted in the Sources for further study to gain ideas before your appointment with the careers service. Their library or resource centre may also have useful information.

8 Finding and citing published information

8.1
(a) The Dewey Decimal Classification system; the US Library of Congress system; or other, as appropriate.
(b) *Nature*: Dewey: 505.0942; Library of Congress: Q1.N3; (or other, as appropriate). Note: the 'Per' or 'per' that may precede these numbers refers to the fact that this is a periodical, rather than a book.
(c) *Tip:* Answer as appropriate – this will depend on the layout of your particular library.
(d) (i) 'A Structure for Deoxyribose Nucleic Acid'; (ii) 'Analysis of the genome sequence of the flowering plant *Arabidopsis thaliana*'.

8.2 *Tips:* The main point here is that methods of writing down references are diverse. You should also find that the details recorded are the same, although they may be in a different order or type style. You must pay attention to the precise instructions given to ensure that you get this right when writing up a project report or dissertation.

8.3 *Tips:* Three possible sites are:
(a) SI units: http://www.ex.ac.uk/cimt/dictunit/dictunit.htm (accessed 09/04/07)
(b) Bat Pollination: http://habitatnews.nus.edu.sg/pub/naturewatch/text/a062a.htm (accessed 09/04/07)
(c) Blood cells: http://users.rcn.com/jkimball.ma.ultranet/BiologyPages/B/Blood.html (accessed 09/04/07)

8.4 *Tips:* When writing a handwritten essay, the Vancouver method in which references are listed in the order cited might be the most convenient to use, because you could write down all the references cited on a separate sheet as you wrote out the essay. However, if you felt that it was important for the reader to see the author(s) and publication dates of the papers, then the Harvard method might be more appropriate – however, you would need to wait until the essay was fully written before organising the literature cited section in author and date order. In a word-processed review, you might prefer the Vancouver method as this might allow the text to flow more smoothly without being interrupted by lists of author names and dates. On the other hand, the Harvard system is easier to use as you write with a word processor, because you can simply 'slot' each new reference into the correct place in the list. In an academic journal, it might be assumed that the readership would want to know the authors and dates of the references cited, so the Harvard system would probably be the most appropriate. The exception might be a journal like *Nature*, which uses the Vancouver system, in accord with its aim of presenting important new scientific advances to a wide audience. *Nature* readers want to be able to assimilate each paper's findings rapidly without getting distracted by citation details.

9 Evaluating information

9.1 (a) P; (b) P; (c) S; (d) S; (e) P; (f) S; (g) P; (h) P; (i) S; (j) S; (where P = primary, S = secondary). Where journals contain both original papers and reviews, we have classed them as primary.

9.2

Some potential debating points on whether *'Genetic modification of crops is a good thing'*

For	Against
Large increases in yield could be engineered, helping to alleviate hunger in developing countries	Transfer of new genes to other crops might occur, with unknown and unpredictable effects
The range of possible modifications to crops is much greater than in conventional breeding – entirely new forms of crops are possible	Super-weeds derived from genetically modified plants might spread uncontrollably and disrupt the ecological balance
It is possible to modify crops much faster than with conventional breeding	The long-term effects of e.g. modified proteins on human health are untested
Very specific changes are possible, e.g. adding a particular vitamin or specific amino acid to a crop, effectively improving nutrition rather than calorific intake	Our effort (and money) would be better applied to birth control, basic medicine, and better implementation of cropping practice

(continued)

Table continued

For	Against
Genetic modification can be used to deter pests, reducing pesticide use and thereby benefiting the environment	The public are so concerned about the safety of foods derived from genetically modified crops that they will fail economically anyway, since consumers will avoid food labelled as containing genetically modified material

9.3 *Tip:* Use Box 62.3 as a source to indicate ways in which graphs may be used to misrepresent information. You may find this exercise easier to accomplish with printed graphs. Those on television may not be shown for long enough to allow you to carry out a detailed analysis. Nevertheless, they represent a rich source of material to criticise.

10 Using online resources
10.1

    (a) Norman Borlaug is credited with being the architect of the Green Revolution. He received the Nobel peace prize for 1970.

    (b) *Conolophus subcristatus* is the Galapagos land iguana.

    (c) Smithsonian Information
        PO Box 37012
        SI Building, Room 153, MRC 010
        Washington, DC 20013-7012, USA

10.2 *Tips:* Note the number of hits found for each search engine. Make a note or printout of the top ten sites located by each search engine. Compare these site listings to reveal the quality of data obtained. Repeat this exercise for meta-search engines. Make lists of the component search engines used by the main meta-search engines to help you devise a strategy for the best coverage of the Internet.

10.3 *Tips:* Within the 'bookmark' editor for your browser, create folders for each of your study courses and then create subfolders for each module that you are doing. Now use some searches to put relevant bookmarks in each folder. Don't forget to include a folder for transferable skills sites; there are a lot of supportive sites available that merit frequent visits.

11 Using spreadsheets
11.1 See figure below. *Tips:* Depending on the precise method you use, you may need to provide the *Name*, *Values* and *Category Labels* for the *Chart Wizard* or adjust those assumed. When creating the graph, select the 'as new sheet' option when prompted for a *Chart Location*. For printing in black and white, select black and

white hatching/shading options after click-selecting chart segments. Ensure the segments of the chart are labelled (i.e. not the default legend option) using the *Chart/Chart Options* menu.

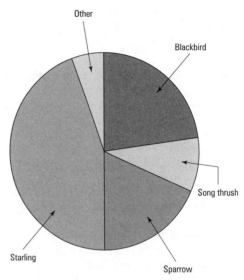

Visits of birds to feeding area as a proportion of total time

11.2 See figure below. *Tips:* When setting up the spreadsheet, create appropriate formulae in new columns for the transformed data; graph using the '*XY (Scatter)*' option in the Microsoft Excel *Chart Wizard* to make sure the data are correctly spaced on the *x*-axis. Plot time on the *x*-axis and the transformed cell count data on the *y*-axis. You will need to create a separate graph to examine the effects of each transformation, because the *y*-axis scales will be different. Select the *Line None* option for the relevant (log transformed) series on the *Data Series* Menu, then use the *Add Trendline* option (linear model) within the *Chart* menu in order to add a linear regression line (note: this can also be done using the *Regression* command in the *Data Analysis* option on the *Tools* menu, but the procedures required are much more complex). Adjust the chart output to suit, e.g. by changing gridlines, background and legends, before copying and pasting to the Microsoft Word document.

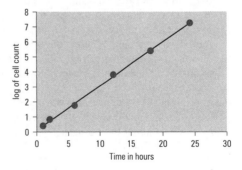

Log-linear graph of data

Answers to study exercises

11.3 *Tips:* Take care when entering the date data in the spreadsheet. You may need to specify the format of the cells for the date column (*Format* menu, *Cells* option). Highlight all parts of the spreadsheet table including column headers, then use the *Data* menu, *AZ↓ Sort* option to rearrange data. You do not need to fill in all the response boxes and can select 'none' if a box contains criteria from a previous sort.

12 Word processors, databases and other packages

12.1 *Tips:*
(a) Ensure list components are in separate paragraphs; highlight list; from the *Table* menu select *AZ↓ Sort* function; ensure options are set to *paragraphs*, *text*, *ascending*.
(b) From the *Edit* menu, select *Replace* option, then fill in the relevant boxes.
(c) From *Edit* menu, select *Replace* option, then fill in the relevant boxes; select *More ▾* then *Format*, *Font*, *Italic*.
(d) From the *View* menu, select *Header and Footer* and use the small on-screen menu to add text and numbering as desired. Formatting commands (e.g. centre) will work in the header and footer boxes. As an alternative for adding page numbers, this can be done from the *Insert* menu, *Page Numbers…* option.
(e) Margins can be adjusted from *File* menu, *Page Setup*, *Margins* option.
(f) Select paragraphed text, then use the *Format* menu, *Bullets and Numbering* option, using *Customise* if the option you desire is not displayed within the panel.
(g) Type 'alternative' within a word page. Ensure cursor is within word (or highlight word); from *Tools* menu select *Language*, *Thesaurus* options. Perhaps 'option' might be a useful synonym?
(h) From *Tools* menu, select *Spelling and Grammar* to start a spell check. Take care if the checker queries UK spellings (it may be set to accept US spellings). Note: this is *not* a substitute for reading your document – the spellchecker will miss typing errors like 'form' instead of 'from'.
(i) For whole document, triple click in left margin to select all, then use *Tools* menu, *Word Count* option. For selected text, highlight (select) the part you wish to count, then carry out same instructions.
(j) While in first document, use *File* menu, *Open* option to open another. You can switch between them using the *Window* menu, where you will see all open documents listed. Note: remember to save changes to all documents when closing down your editing session.

12.2 *Tips:* Although there are several ways to modify tables, you may find it useful to try the following commands (think carefully about their order).
- *Table, Insert Table*
- *Table, Merge Cells*
- *Table, Distribute Columns Evenly*

- *Tables and borders*, modify borders commands (various) Consult Box 63.2 or use the *Help* feature if you run into problems.

12.3 *Tips:* You might start your search on the networked 'desktop' and icons visible to you when you log on. Use 'help' menus and 'wizards' to investigate how to use the programs, or consult any manuals available in the computer suite.

13 Organising a poster display

13.1 *Tip:* Use Fig. 13.1 and the sub-headings of Chapter 13 as a starting point for assessing the pros and cons of your designs.

13.2 Ten-point checklist – what makes a good poster presentation?
- ☐ *Title* – is this concise, specific and interesting?
- ☐ *Author(s)* – does the poster show details of the names and addresses of all contributors?
- ☐ *Structure* – is the overall structure clear to the observer, with good use of sub-headings?
- ☐ *Flow* – is the layout of the poster clear, in terms of how the audience is expected to 'read' each section?
- ☐ *Content* – does the presentation have a clear introduction, main message and conclusion?
- ☐ *Text* – does the poster contain an appropriate amount of written material in a 'readable' font?
- ☐ *Graphics* – does the poster make good use of diagrams, figures and images?
- ☐ *Size and scale* – does the poster make effective use of the space available?
- ☐ *Colour* – does the presentation use colour effectively?
- ☐ *Conclusion* – is the 'take-home message' clear to the observer?

13.3 *Tip:* You might use the ten-point checklist given above, or your own version from Study exercise 13.2 to assess selected posters by giving a mark out of 10 for each item in the list. In a group exercise, you can see how your 'scores' compare with those of other students, and discuss items where scores are different.

14 Giving a spoken presentation

14.1 Ten-point checklist – what makes a good spoken presentation?
- ☐ *Introduction* – does the speaker explain the structure and rationale of the topic at the outset?
- ☐ *Audience* – is the material presented at an appropriate level?
- ☐ *Content* – does the presentation cover the topic effectively?
- ☐ *Speed and structure* – are the main elements of the presentation presented clearly, and at an appropriate pace?
- ☐ *Voice* – is the presentation delivered in an engaging, active vocal style?
- ☐ *Audio-visual aids* – does the speaker make good use of visual images (e.g. graphs, photographs, etc.)?
- ☐ *Image* – does the presenter make good use of body language?

☐ *Notes* – does the speaker use notes to structure the talk, rather than reading from a written text?

☐ *Conclusion* – is a clear 'take-home message' delivered in the final stages of the presentation?

☐ *Questions* – does the presenter deal with questions effectively?

14.2 *Tips:* You could try marking speakers out of 10 for each of the points listed in Study exercise 14.1 (using either the checklist provided above or your own version), or you could simply prepare a list of the good and bad points of the presentations you have evaluated.

14.3 *Tips:* When evaluating your own performance: (i) remember that most of us are overcritical when viewing ourselves on video, tending to focus on relatively minor points, e.g. facial expression, rather than broader aspects, e.g. clarity of delivery and content; (ii) when giving a talk to a group of fellow students you might ask them to use all or part of the checklist from Study exercise 14.1 as a way of providing feedback.

15 General aspects of scientific writing

15.1 *Tips:* When brainstorming, think of all aspects of the topic that could be relevant (even if you could not give precise details at this stage). When you meet together, discuss which of your combined ideas would be most important in crafting an answer and what order you might place them in.

15.2 *Tip:* Try to apply what you have learned in the next assignment you write.

15.3 *Tips:* Use the spelling list to help you when writing and make a conscious effort to memorise the spelling each time you use it. The act of writing the vocabulary list will help you to memorise new words and their meanings. If you have trouble learning definitions in biology, you could extend these ideas to create a personal glossary of terms.

16 Writing essays

16.1 *Tip:* Can you determine what lies behind each question? Some question titles are quite direct, but others ask for information in a more subtle way. Use your imagination to think beyond the immediate question, linking to learning objectives or to other relevant material covered in lectures.

16.2 *Tips:* Compare your ideas with those of a partner, or ask tutors or lecturers what they think of your efforts. A useful adjunct to this exercise is to write the first sentence of your answer to avoid 'writer's block'.

16.3 *Tips:* Note that each of the stages (b)–(d) requires a separate 'review' – it is difficult, if not impossible, to make a good job of all aspects in a single 'pass' through the text. Ideally, you would reprint the essay between reviews, having made corrections.

17 Reporting practical and project work

17.1 *Tips:* To assimilate the required style for Materials and Methods, you may find it appropriate to read some research papers from the journal section of your library. Note particularly that the Materials and Methods are a description of 'what was done', not a set of instructions.

17.2 *Tips:* Again, you may find it appropriate to read some research papers from the journal section of your library to assimilate the normal style for describing research results.

17.3 *Tips:* Abstracts are difficult to write: try to include a synopsis of all the main sections of the paper, including its aims, methods, key results and conclusions. Ignore the word-length restriction until you have noted the main points, then try to reduce (or, less likely, expand) your effort to match the requirement.

18 Writing literature surveys and reviews

18.1 Five differences between reviews and papers are:

1. Reviews deal with material in general terms, taking an overview, whereas papers tend to look at a focused, single aspect of a topic.
2. The subdivisions are generally not the same.
3. There is little or no use of original data in reviews.
4. There is little detail of methods in reviews.
5. They tend to use different reference-citation systems.

18.2 *Tip:* It should be possible to do the second part of this exercise online at your library – you may need to enlist the help of library staff if unsure of how to do this.

18.3 *Tip:* The Abstract and Conclusions sections will provide information about the key points addressed. It might be helpful to use a highlighter on a photocopied version of review to note important statements within the main body of the review.

19 Your approach to practical work

19.1 Possible reasons why practical work might be of value in a university course include:

- practicals reinforce lecture material, in agreement with the adage: 'I hear and I forget, I see and I remember, I do and I understand';
- practical procedures provide students with an opportunity to develop their manual skills and laboratory competences;
- investigative procedures and problem-solving practical exercises enable students to improve their skills in experimental design and the application of scientific method;
- practicals enable students to develop their abilities to observe and measure biological systems, and to record the outcome;
- students are able to analyse and interpret the data from practical exercises and to consider the validity of their results, including sources of error, etc.;
- the results of practical procedures give students an opportunity to develop skills in reporting and presenting 'experiments' in written format.

19.2 *Tip:* Possible items for a snail dissection: specimen (snail); scalpel; blunt seekers; coarse and fine scissors; wax dish and pins; pointed seeker; dissection microscope; paper tissues; water; Pasteur pipette; first aid kit nearby – in case of accident.

19.3

(a) **40**

(b) 44.26666, expressed to 4 significant figures = **44.27**

(c) 0.01953125, expressed to 3 significant figures = **0.0195**

(d) 1.6×10^5

(e) 0.0313, expressed to 3 decimal places = **0.031**

20   Health and safety

20.1

(a) p. 119.

(b) Fig. 20.1 (p. 119).

(c) Fig. 20.2 (p. 120).

(d) Fig. 45.1 (p. 264).

(e) Fig. 56.4 (p. 353).

20.2   *Tip:* The locations of each item will vary, according to the layout of your chosen laboratory. You should ask someone in charge (e.g. demonstrator, lecturer, technician or laboratory manager) if you are unable to find any of the items listed.

20.3

(a) *Tip:* The answer will vary, according to your university, but is likely to cover the main points listed on p. 120.

(b) In the UK, water-filled fire extinguishers are red – use for wood, paper and textiles; CO_2 extinguishers are black, or have a black panel – use for flammable liquids and electrical fires; dry powder extinguishers are blue, or have a blue panel – use for all types of fires (wood, paper, textiles, gaseous fires, flammable liquids and electrical fires); foam extinguishers are beige, or have a beige panel – use for flammable liquids (note that other colour conventions may apply outside the UK).

(c) *Tip:* This will vary, depending upon your department's procedure. However, it is likely to include a specific written record book for accidents.

(d) *Tip:* There will often be separate codes of practice for work involving non-pathogenic microbes (e.g. general microbiology classes) and pathogenic microbes (e.g. in medical microbiology courses). The latter requires a higher level of safe working practice, and may include the use of disposable aprons, etc.

20.4   *Tip:* Use either your department's chemical hazard information system or a Web database (e.g. the National Institute for Occupational Safety and Health database at: http://www.cdc.gov/niosh/database.html [accessed 09/04/07], or the Agency for Toxic Substances and Disease Registry ToxFAQs database at: http://www.atsdr.cdc.gov/toxfaq.html [accessed 09/04/07]. Alternatively, you could use a 'portal' site such as the ChemPortal site at: http://webnet3.oecd.org/echemportal/contents.aspx?ContentName=PortalInfo [accessed 09/04/07]; or the Material Safety Data Sheet (MSDS) site at http://www.ilpi.com/msds/index.html [accessed 09/04/07]).

(a) Formaldehyde is toxic and potentially carcinogenic (and also highly flammable in pure form): protective gloves and eye/face protection should be used during preparation of stock solutions. Formaldehyde can also cause allergic skin reactions and breathing the vapour should be avoided – working solutions should be used in a well-ventilated place or fume hood.

(b) Acetone is highly flammable and can form explosive peroxides on contact with strong oxidants. It should be used in a well-ventilated room, away from naked flames or spark-prone electrical equipment. Disposable gloves can be used to prevent skin contact.

(c) Sodium hydroxide is highly corrosive and an irritant. Protective gloves and safety glasses should be worn when preparing stock solutions using solid NaOH. Contact with moisture generates considerable heat, so it should always be added in small amounts to a large volume of water.

21   Working with liquids

21.1

(a) Use measuring cylinders – measure out 700 ml of ethanol using a 1000 ml measuring cylinder and measure out 300 ml of water using a 500 ml measuring cylinder: mix together in a large conical flask (e.g. 1000 ml capacity) or beaker.

(b) Use a pipettor (e.g. P20 Gilson Pipetman) set to deliver 10.0 μl and sterile RNAse-free tips, manufactured for molecular biology (note also that some disposable tips are designed to have extended, fine tips, to aid gel loading).

(c) Weigh out 20.0 mg (0.0200 g) of DNA standard using a four-place balance, add this to a 100 ml volumetric flask and make up to 100 ml with distilled, deionised water.

(d) Use a burette (e.g. 50 ml burette) and beaker or flask (e.g. 250 ml capacity), plus a magnetic stirrer and 'flea' (p. 129) for mixing.

21.2   *Tip:* Box 21.1 gives step-wise details of how to use a pipettor.

21.3   *Tip:* You may find this type of calculation easier using a spreadsheet, rather than a calculator (Box 65.3 gives further details). The model A pipettor gave a mean delivery of 0.9795 ml, with a standard deviation of 0.010 40 ml (to 5 decimal places). The model B pipettor delivered, on average, 1.0107 ml, with a standard deviation of 0.001 25 ml. The model C pipettor gave an average delivery of 1.0009 ml, with a standard deviation of 0.058 82 ml. Therefore, model C was the most accurate pipettor (with a mean value closest to the true value of 1.0000 ml) and model B was most precise (giving the most reproducible results, as shown by the lowest standard deviation). The model A pipettor was the least accurate (mean value furthest from the true value) while model C was the least precise (having the highest standard deviation).

22   Basic laboratory procedures

22.1   *Tip:* Box 22.1 and Box 23.1 give advice on preparing solutions.

(a) $50 \div 1000$ (number of moles required for 1 litre) $\times 0.1$ (volume required, expressed in litres) $\times 58.44$ (relative molecular mass of NaCl) = 0.2922 = **0.292 g**

(b) 0.1 (number of moles required for 1 litre) $\times 0.25$ (volume required, expressed in litres) $\times 182.17$ (relative molecular mass of mannitol) = 4.55425 = **4.55 g**

(c) 800 (number of μg in 1 ml) × 200 (volume, in ml) = 160 000 μg = **160 mg**

(d) 22.5 ÷ 1000 (number of moles required for 1 litre) × 0.5 (volume required, in litres) × 203.30 (relative molecular mass of $MgCl_2.6H_2O$) = 2.287125 = **2.29 g**

(e) 20 (number of ng in 1 μl) × 1000 (number of μl in 1 ml) × 400 (volume required, in ml) = 8 000 000 ng = **8.00 mg**

22.2 *Tip:* All of these can be calculated using the relationship $[C_1]V_1 = [C_2]V_2$ (eqn [23.2]). Do not forget to include the volumes of *both* solutions when calculating the final volume – e.g. in example (a) below, the final volume is 10 ml (1 + 9) and *not* 9 ml.

(a) $1 \times 0.4 = [C_2] \times 10$, therefore $[C_2] = (1 \times 0.4) \div 10 = 0.04 \, mol\,l^{-1}$ and therefore $0.04 \times 1000 = $ **40.0 mmol l^{-1}**.

(b) $10 \times 25 = [C_2] \times 500$, therefore $[C_2] = (10 \times 25) \div 500 = 0.5 \, \mu g\,ml$ and therefore $0.5 \times 1000 = $ **500 ng ml^{-1}**.

(c) $200 \times 0.01 = [C_2] \times 250$ (note that both volumes are expressed in ml), therefore $[C_2] = (200 \times 0.01) \div 250 = 0.008 \, mmol\,l^{-1}$ and therefore $0.008 \times 1\,000\,000 = 8000 \, nmol\,l^{-1}$ and therefore $8000 \div 1000 = $ **8.00 nmol ml^{-1}**

(d) $0.2 \times V_1 = 0.02 \times 250$ (note that both concentrations are expressed in $mol\,l^{-1}$), therefore $V_1 = (0.02 \times 250) \div 0.2 = $ **25.0 ml** of KCl solution, made up to a final volume of 250 ml with water.

(e) First express the concentration of the glucose solution in $mol\,m^{-3}$, by dividing by M_r, therefore $20 \div 180.16 = 0.1110124 \, mol\,m^{-3}$. Then use the relationship $[C_1]V_1 = [C_2]V_2$. Thus $0.1110124 \times V_1 = 50 \times 10^{-6} \times 1 \times 10^{-3}$, therefore $V_1 = (50 \times 10^{-6} \times 1 \times 10^{-3}) \div 0.1110124 = 4.504 \times 10^{-7} \, m^3 = $ **450 μl**. Make up 450 μl of glucose solution to $1 \times 10^{-3} \, m^3$ (1000 ml) with water.

22.3 *Tip:* Either use the relationship $[C_1]V_1 = [C_2]V_2$ (eqn [23.2]) to calculate the volumes required, or calculate the dilution factor from the ratio of the concentration in the stock solution to that required in the final solution.

(a) The dilution factor for KCl is 20-fold. To prepare the solution, use **0.050 ml ($\equiv$ 50.0 μl) of KCl stock solution, plus 0.950 ml ($\equiv$ 950 μl) of water**.

(b) The dilution factor is 40-fold for NaCl and 10-fold for glucose. To prepare the solution, use **1.25 ml of NaCl stock solution, 5.00 ml of glucose stock solution and 43.75 ml of water**.

(c) The dilution factor for NaCl is 20-fold, for KCl is 80-fold, for $CaCl_2$ is 4-fold and for glucose is 20-fold. To prepare the solution, use **5.00 ml of NaCl stock solution, 1.25 ml of KCl stock solution, 25.00 ml of $CaCl_2$ stock solution, 5.00 ml of glucose stock solution and 63.75 ml of water**.

(d) First, convert the required concentrations to the same units as those of the stock solutions. In this case, $\mu mol\,ml^{-1}$ and $mmol\,l^{-1}$ are directly equivalent (divide by 10^3 to convert from μmol to mmol, then multiply by 10^3 to convert from ml to l) so the numerical values remain the same. The dilution factor

for NaCl is 5-fold, for $CaCl_2$ is 8-fold and for KCl is 10-fold. To prepare the solution, use **2.00 ml of NaCl stock solution, 1.25 ml of $CaCl_2$ stock solution, 1.00 ml of KCl stock solution and 5.75 ml of water**.

(e) First, convert the glucose concentration from $nmol\,ml^{-1}$ to $mmol\,l^{-1}$ (divide by 10^6 to convert nmol to mmol and multiply by 10^3 to convert ml to l), giving glucose at $0.2 \, mmol\,l^{-1}$ (all other concentrations are in the same units as those of the stock solutions). The dilution factor for $CaCl_2$ is 20-fold, for KCl is 4-fold and for glucose is 25-fold. To prepare the solution, use **1.25 ml of $CaCl_2$ stock solution, 6.25 ml of KCl stock solution, 1.00 ml of glucose stock solution and 16.50 ml of water**.

23 Principles of solution chemistry

23.1 *Tip:* Box 22.1 and Box 23.1 give useful procedures and example calculations for molar concentrations.

(a) 1 molar = $1 \, mol\,l^{-1}$ = 58.44 g in 1 litre of solution = **58.440 g**

(b) $110.99 \times 100 \div 1000$ (number of g required for $100 \, mmol\,l^{-1}$) × 250 ÷ 1000 (number of g in 250 ml) = 2.77475 = **2.775 g**.

(c) $10 \, nmol\,\mu l^{-1}$ is equivalent to $10 \, mmol\,l^{-1} \div 1000$ (convert to $mol\,l^{-1}$) × 182.17 (number of $g\,l^{-1}$) × 2.5 (number of g in 2.5 l) = 4.554 25 = **4.554 g**.

(d) 5% w/v is equivalent to 5 g in 100 ml × 4 (number of g in 400 ml) = **20.000 g**.

(e) 2.50×180.16 (convert from $mol\,m^{-3}$ to $g\,m^{-3}$) × 250 ÷ 1 000 000 (number of g in 250 ml) = 0.1126 = **0.113 g**.

23.2 *Tip:* Make stepwise changes and always show your working out in any written report or assessment including calculations involving the interconversion of concentrations.

(a) $5 \div 342.3$ (convert $g\,l^{-1}$ to $mol\,l^{-1}$) = 0.0146071 = **0.0146 mol l^{-1}**. Note that this can also be expressed as $14.6 \, mmol\,l^{-1}$.

(b) 1×58.44 (convert $mol\,m^{-3}$ to $g\,m^{-3}$) ÷ 1000 (convert $g\,m^{-3}$ to $g\,l^{-1}$) = 0.05844 = **0.0584 g l^{-1}**. Note that this could also be expressed as $58.4 \, mg\,l^{-1}$.

(c) 5×0.789 (convert % v/v to % w/v) × 10 (convert % w/v to $g\,l^{-1}$) ÷ 46.06 (convert $g\,l^{-1}$ to $mol\,l^{-1}$) = 0.8564915 = **0.856 mol l^{-1}**. Note that this could also be expressed as $856 \, mmol\,l^{-1}$.

(d) $150 \div 1000$ (convert from $mmol\,l^{-1}$ to $mol\,l^{-1}$) × 180.16 (convert from $mol\,l^{-1}$ to $g\,l^{-1}$) ÷ 10 (convert from $g\,l^{-1}$ to % w/v) = **2.70 % w/v**.

(e) 1×74.55 (convert from $mol\,kg^{-1}$ to $g\,kg^{-1}$) ÷ 10 (convert from $g\,kg^{-1}$ to g per 100 g, i.e. % w/w) = 7.455 = **7.46% w/w**.

23.3

(a) *Tip:* Use eqn [23.6] and set $\phi = 1$ (ideal behaviour) to calculate the osmolality of individual solutes in each solution. (i) $NaCl = 50 \times 2 = $ **100 mosmol kg^{-1}**; (ii) $KCl = 200 \times 2 = 400 \, mosmol\,kg^{-1}$; $CaCl_2 = 40 \times 3 = 120 \, mosmol\,kg^{-1}$, giving a total osmolality of

Answers to study exercises

$400 + 120 = \textbf{520 mosmol kg}^{-1}$. (iii) $NaCl = 100 \times 2 = 200$ mosmol kg^{-1}; $KCl = 60 \times 2 = 120$ mosmol kg^{-1}; $CaCl_2 = 75 \times 3 = 225$ mosmol kg^{-1}, giving a total osmolality of $200 + 120 + 225 = \textbf{545 mosmol kg}^{-1}$.

(b) *Tip:* Use the relationship 1 osmol kg^{-1} = 2.479 MPa at 25 °C (p. 142) to calculate osmotic pressure, as follows: $100 \div 1000$ (convert mosmol kg^{-1} to osmol kg^{-1} $\times 2.479 = 0.2479 = \textbf{0.248 MPa}$. (i) $520 \div 1000$ (convert mosmol kg^{-1} to osmol kg^{-1}) $\times 2.479 = 1.28908 = \textbf{1.29 MPa}$. (ii) $545 \div 1000$ (convert mosmol kg^{-1} to osmol kg^{-1}) $\times 2.479 = 1.351055 = \textbf{1.35MPa}$.

Tip: You can use eqn [23.8] to calculate the osmotic potential of each solution. In each case, the numerical value stays the same as the value for osmotic pressure, as shown above, but there is a negative sign for osmotic potential, compared with osmotic pressure: (i) $\textbf{-0.248 MPa}$. (ii) $\textbf{-1.29 MPa}$. (iii) $\textbf{-1.35 MPa}$.

24 pH and buffer solutions

24.1 *Tip:* Use eqn [24.5] to interconvert between pH and [H$^+$].
(a) $7.4 = -\log_{10}[H^+]$, therefore $[H^+] = \textbf{3.98} \times \textbf{10}^{-8} \textbf{ mol l}^{-1}$.
(b) $4.1 = -\log_{10}[H^+]$, therefore $[H^+] = 7.94 \times 10^{-5}$ mol l^{-1} and therefore $\textbf{7.94} \times \textbf{10}^{-2} \textbf{ mol m}^{-3}$.
(c) pH $= -\log_{10}[2 \times 10^{-5}]$, therefore pH $= \textbf{4.70}$.
(d) pH $= -\log_{10}[10^{-12.5}]$, therefore pH $= \textbf{12.5}$.
(e) First convert [H$^+$] from mol m^{-3} to mol l^{-1}. $[H^+] = 2.8 \times 10^{-8}$ mol l^{-1}. Then use eqn 24.5. pH $= -\log_{10}[2.8 \times 10^{-8}]$, therefore pH $= \textbf{7.55}$

24.2

(a) Since the assay does not involve whole cells or organelles (where toxicity would be a factor to consider), TRIS buffer would be suitable. It has a pK$_a$ of 8.3 which is close to the required pH. If the enzymic reaction involves no major production of acids or bases a buffer concentration of 50 mmol l^{-1} would be sufficient, otherwise a higher concentration (e.g. of 200 mmol l^{-1}) might be more appropriate.
(b) Here, you could use one of the zwitterionic 'Good buffers' such as MES, which has a pK$_a$ of 6.1, and this is in the direction of any expected pH change during the experiment (*E. coli* cells normally show a decrease in pH during growth, as substrates are metabolised to acidic end products). A reasonably high concentration of buffer would be appropriate (e.g. 200 mmol l^{-1}) if the experiment is to run for any length of time.
(c) TRIS buffer would *not* be suitable, as it shows a large change in pH with temperature (p. 150) and can also inhibit photosynthetic activity – either HEPES or PIPES would be suitable alternatives.
(d) Zwitterionic 'Good buffers' would *not* be suitable, as they interfere with some protein assays (p. 151) – citrate or acetate buffer would be suitable alternatives.

24.3 *Tip:* Rearrange eqn [24.7] to give $\log_{10}[A^-]/[HA] = $ pH $-$ pK$_a$.
(a) $\log_{10}[A^-]/[HA] = 3.8 - 4.8 = -1$, so $\textbf{[A}^-\textbf{]/[HA]} = \textbf{0.1}$ (1 to 10);

(b) $\log_{10}[A^-]/[HA] = 9.5 - 9.2 = 0.3$ so $\textbf{[A}^-\textbf{]/[HA]} = \textbf{1.995}$ (approximately 2 to 1);
(c) $\log_{10}[A^-]/[HA] = 8.1 - 7.5 = 0.6$ so $\textbf{[A}^-\textbf{]/[HA]} = \textbf{3.98}$ (approximately 4 to 1).

25 The principles of measurement

25.1

(a) quantitative, discontinuous, ratio
(b) quantitative, continuous, ratio
(c) qualitative, discontinuous, nominal
(d) quantitative, discontinuous, ratio
(e) qualitative, discontinuous, nominal
(f) qualitative, discontinuous, ordinal

25.2 *Tip:* Sources of error are likely to include:
- misreading of the scales;
- imprecisely defined axes of measurement;
- curvature of the base when measuring height;
- irregularities of the base of the shell when measuring length or breadth.

The precision of the ruler is ± 0.5 mm, while that of the callipers is ± 0.05 mm. Your protocol should include factors such as defining the axes to be measured; a strategy for dealing with irregular edges; a strategy for measuring height consistently and the precision of recording required.

25.3 The results indicate that balance A has a bias of +0.05 g across the weighing range, while balance B has a consistent bias of about +0.04 per cent. Both balances may be precise, but neither is accurate.

26 SI units and their use

26.1

(a) $\underline{\textbf{0.997 atmospheres}}$
(b) $\underline{\textbf{836 000 mm}^{-2}}$
(c) $\underline{\textbf{568 ml}}$
(d) $\underline{\textbf{310 K}}$
(e) $\underline{\textbf{72.6 kg}}$

26.2

(a) $\underline{\textbf{10 m}}$
(b) $\underline{\textbf{15 μl}}$
(c) $\underline{\textbf{5 GJ}}$
(d) $\underline{\textbf{65 km s}^{-1}}$
(e) $\underline{\textbf{100 pg}}$

26.3 *Tip:* The units can be put in place of the symbols and cancelled out as follows:

$$\frac{m^2 \times Pa}{Pa \, s \times m} = \frac{m^2}{m \, s} = m \, s^{-1}$$

27 Making observations

27.1 *Tips:* Having made your initial, unguided counts, repeat using methods to guide your counting such as drawing radial lines, concentric circles or a grid. You may need a protocol for deciding which subdivision the colonies on dividing lines are in. The actual number of points is 143, and you should find that counts using subdivisions are both more accurate and more precise (p. 157) than whole-area counts.

27.2 *Tips:* Start by joining equivalent points on the sections to give you an indication of the broad structure before attempting to interpret any further detail. Try not to rush to a judgement as to the end result, and remain objective.

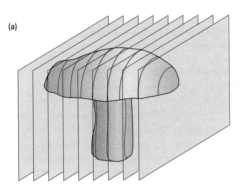

(a)

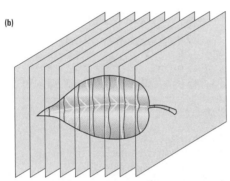

(b)

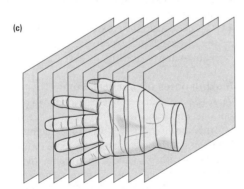

(c)

3-D representations of the organisms or parts of organism shown as sections in the question

27.3 *Tips:* When doing this exercise concentrate solely on doing a line diagram without any attempt at shading. Compare line lengths and the shapes of spaces and avoid thinking of the names of component parts.

28 Drawing and diagrams

28.1 *Tip:* Use construction lines as described in the text to ensure that the diagram is well proportioned. Ask a colleague or tutor to comment on the diagram and any labelling.

28.2 *Tip:* Use any of the flowcharts in the book (e.g. Fig. 33.1) as a model for style.

28.3

Comparison of the main types of biological diagram

| Feature | Type of diagram | | |
	Cell diagram	Tissue diagram	Morphological diagram
Should the diagram carry a scale?	Yes	Yes	Yes
What might be an appropriate scale?	<1 mm	1–10 mm	1 mm–1 m
Approximate number of cells visible	1 complete cell of each type, plus part of neighbours	No cells visible, only outlines of tissues	No cells visible
Are organelles visible?	Yes	No	No
Should shading be used?	No	No	Rarely
Should labelling be used?	Yes	Yes	Yes

29 Basic fieldwork procedures

29.1 Items omitted are: indelible marker; first aid kit equipment; two-way radio or mobile telephone; sampling equipment.

29.2 *Tip:* Your listings will vary in detail with the precise objectives of the study but some examples are shown in the table below. The reasons for including each item should be relatively easy to note, and problems could include, for example, carrying heavy and unwieldy items; taking care of easily damaged items; and keeping water-sensitive items dry.

Items required for ecological investigations

Rocky shore	Meadow
Generic items from Table 29.1	Generic items from Table 29.1
Quadrats of appropriate shape/size	Quadrats of appropriate shape/size
Measuring tape	Measuring tape

(continued)

Answers to study exercises

Table continued

Rocky Shore	Meadow
Surveying equipment (for levelling)	Pegs and twine for laying out grids
Sample bags, ties and labels	Point-quadrat frame
Compass	Sample bags, labels and vasculum
Knife/scalpel for removing specimens	Trowel for digging specimens
Tide tables	

29.3 *Tip:* Some examples would include: traffic; traffic pollution; trespass in relation to access; antisocial and unfriendly behaviour from local people; equipment removal/theft; over-interested people, especially children.

30 Samples and sampling

30.1

(a) The mean of the ten 1% sub-samples, is 288.7, so the original 5 litre sample contained 28 870 copepods. The density per litre is therefore, $28\ 870 \div 5 = \underline{\mathbf{5774}}$.

(b) A cubic metre of water contains 1000 litres and taking the result from (a) this would contain **5 774 000** copepods.

30.2 *Tip:* You will need to decide on the optimum shape of the quadrat: a circle has the smallest edge length and that minimises edge effects. You will have to decide on a strategy as to when an animal at the edge is in or out of the counting area, e.g. if the peak of the shell is within the area it is counted but if it outside, it is not. The area of the quadrat must be related to the density of the contained organisms since too small a quadrat would make the sample size too small, while too large a quadrat makes counting more difficult (how will you prevent the possibility of double-counting or the possibility of missing individuals?); a quadrat of approximately $0.25\ m^2$ might satisfy relevant criteria.

30.3 *Tip:* Use Microsoft Excel to create a graph of the data and then use it to work out how many samples are required to provide results within the range 45–55 ($\pm10\%$ of 50). The minimum required is about 20.

31 Scientific method and design of experiments

31.1 A possible answer (one of many) is given in the following figure.

A possible layout of treatments in the glasshouse experiment

The Latin square design ensures that treatments appear with equal probability in each row and column; hence, if there are gradients of potentially confounding variables, such as temperature and light in this case, their effects are spread evenly among treatments.

31.2 *Tip:* For Microsoft Excel, try the formula $=INT(5*(RAND())+1)$.

31.3 *Tip:* It might be easier to lay out your answer like the table below:

Interactions between treatments in the cell proliferation experiment

Treatment	Mean result	Observed effect of treatment(s)*	Expected effect of treatments	Classification
Control	5.0	0.0	–	–
A	7.0	2.0	–	–
B	8.0	3.0	–	–
C	10.0	5.0	–	–
A + B	7.0	2.0	$2 + 3 = 5$	Antagonism
A + C	12.0	7.0	$2 + 5 = 7$	No interaction
B + C	17.0	12.0	$3 + 5 = 8$	Synergism
A + B + C	13.0	8.0	$2 + 3 + 5 = 10$	Antagonism

* Control value subtracted

Notes:
- The observed effect is the mean result minus the control (baseline value).
- The expected result is the additive effect of individual treatments.
- The classification is based on the difference between the observed and expected effects, shown in columns 3 and 4 above (see Fig. 31.5).

32 Making notes of practical work

32.1 *Tips:* Use a spreadsheet to create a table to record your data but include some blank columns to record additional information that perhaps was not thought of beforehand. You will need to record, e.g., the time of male and female bird visits with or without food items, the nature of any food items (this will probably require the use of binoculars), and the day of the observation. Serendipitous observations of e.g. disposal of waste from the nest should also be included. This study could usefully involve use of a tape recorder so that observation is continuous, although the use of a voice might itself disturb the bird activity. If you have to use a written record a predetermined code system will minimise writing.

32.2 Using a tape recorder has obvious advantages when data can be recorded orally, especially when there is a need to avoid taking your eyes off a situation as in Study exercise 32.1. However, the data are potentially insecure and there is a real risk of losing data due to failure of the machine to perform as expected. If using this method, you should check the quality of the recording at regular intervals.

Disadvantages include the fact that data have to be transcribed after the period of observation is complete, requiring more time and introducing further possibility of error in the transcription process.

32.3 *Tip:* Use a spreadsheet to lay out the table once you have decided what it needs to contain and what statistics you might wish to calculate. You can use spreadsheet functions to help you calculate some of the statistics.

33 Project work

33.1 *Tips:*

- Divide the time available for your project into relevant parts concerned with different aspects of your work (note some of these could run in parallel, such as experimental work and research for your Introduction).
- A Gantt chart might be an appropriate method of presenting this (see e.g. http://freegroups.net/gantt/, [accessed 09/04/07]).
- Set yourself realistic target dates for completion of each section.
- Identify important potential 'sticking points' (e.g. time taken for delivery of reagents and other supplies) and work around these (e.g. by doing library work in the meantime).
- Make sure you allow some flexibility (say 5 per cent) to allow for slippage or unforeseen problems.

33.2 *Tip:* Even if you prefer to write up your work on paper first, create appropriate computer files soon afterwards. Devise a simple and logical system for naming the files.

33.3 *Tip:* Remember that a word processor or a spreadsheet may have sufficient functions for your database needs for storing reference details (see p. 47).

34 Collecting animals and plants

34.1

(a) Use a plankton net (100 μm mesh) or a water bottle. Transport in a Thermos flask if required live.

(b) Collect at night from the surface of grassland using a torch (live) or use dilute formalin solution (or its equivalent) sprayed on the surface (causes them to rise to the surface), although this may adversely affect the worms and should not be used if required live. Transport live worms in large plastic bags with a layer of dead leaf material to retain moisture.

(c) Use a trowel to dig them up or a pair of scissors to collect the aerial parts only. Place in vasculum or plastic bags to prevent dehydration during transport.

(d) Collect ants using a pooter (Fig. 34.1(e)). Place in a sealed (plastic) container for transport. Be alert to the fact that many are able to sting or spray acidic fluids; you should wear gloves and appropriate protective clothing.

34.2

(a) Kill using alcohol (to produce a 70% solution) or formalin (to produce a 4% solution) and decant excess fluid once the dead animals have settled out.

(b) Preserve earthworms in a fixative solution such as 70% alcohol or 4% formalin, depending upon the intended use of the material.

(c) Preservation is usually by drying between sheets of blotting paper, but entire plants could be preserved in alcohol solution.

(d) Use 70% alcohol both to kill and to preserve.

34.3 Labels should contain at least the following information:

- date
- time
- location
- habitat details
- method of collection
- preservation method.

35 Fixing and preserving animals and plants

35.1 *Tip:* Determine which fixative is appropriate for the particular objective and then decide how to ensure adequate penetration of the fixative. Regarding earthworms, see answer to Study exercise 34.2(b) above, but note that they are too large to allow adequate fixation of internal organs when the animal is entire.

35.2 *Tip:* Many of the more delicate seaweeds need to be laid out under water before preservation. Pick a typical specimen from the middle of the available size range and tease it out on to the sheet of paper while under water. Larger seaweeds may be able to be treated more like vascular plants and can often be dried between sheets of blotting paper.

35.3 *Tip:* When selecting your five fixatives, select those with clearly different compositions, so that you can find out about the range of problems encountered (see Table 35.1).

36 Collecting and isolating microbes

36.1

(a) A range of approaches is possible, including taking swabs, which are then surface spread on to an agar-based medium (e.g. a general-purpose medium such as nutrient agar, or blood agar), or direct 'imprints' from the skin surface on to agar-based 'contact' plates.

(b) Important factors to consider here are: firstly, limiting the exposure of the sample to oxygen – for example, by taking a core through the mud and putting the core into an anaerobic jar (p. 213) immediately after sampling; and secondly, keeping the sample cool, since the organisms are psychrotrophs – for example, by putting the anaerobic jar into an ice bucket during transportation to the laboratory.

(c) You will need to consider (i) where to sample; (ii) how much water to collect (typically, faecal bacteria are often counted in volumes of up to 100 ml of sea water); and (iii) how many samples to collect (typically, at least three separate volumes of 100 ml). The samples should not be exposed to sunlight, as this may inactivate the bacteria; they should be kept at 1–4 °C during transit to the laboratory and processed within 6 hours of collection.

Answers to study exercises

(d) One approach is to take 'contact' plates by pressing the selected surface of the leaf on to a suitable agar-based medium (e.g. a non-selective medium such as place count agar for heterotrophic bacteria). This should be done as soon as possible after collecting the leaves, and is feasible under field conditions. Incubation at different temperatures would give counts for psychrophiles, mesophiles and thermophiles.

36.2

(a) Use positive phototaxis towards a unidirectional light source. 'Sloppy' agar (e.g. 0.5% w/v agar) can help reduce contamination due to other micro-organisms.

(b) Rinse free of other microbes using sterile water, then use sonication to disrupt the biofilm and create a suspension for further investigation (e.g. culture). Alternatively, you might try an enzyme mixture formulated to hydrolyse the biofilm polymers, if their composition is known.

(c) Use membrane filtration (e.g. pore size 0.45 μm) to 'concentrate' the enterococci, then transfer the membrane (face up) to a suitable selective medium (e.g. glucose azide medium) for enumeration after incubation at an appropriate temperature, e.g. at 44 °C.

(d) Use a micromanipulator to separate the target bacterium from the other microbes. Alternatively, you might try dilution to extinction (p. 212), but you would be likely to look through a large number of cultures before finding the distinctively shaped bacterium.

36.3

(a) NaCl at 7.5 per cent is selective for halotolerant bacteria, while mannitol is present as a specific carbohydrate nutrient source for *S. aureus*. Phenol red in the medium turns yellow when mannitol is metabolised to acidic end products. *S. aureus* grows as larger colonies with a yellow 'halo' in an otherwise red medium, due to mannitol breakdown. Other *Staphylococcus* spp. grow as smaller colonies, with no yellow 'halo'.

(b) The surfactant sodium lauryl sulphate is selective for coliform bacteria. Lactose is present as a specific carbohydrate nutrient source, plus phenol red as a pH indicator. Coliforms, including *E. coli*, grow in the presence of sodium lauryl sulphate and metabolise lactose to acidic end products, giving large yellow colonies. A selective temperature of 44 °C can be used to grow thermotolerant (faecal) coliforms, most of which are *E. coli*.

(c) Sodium azide is the selective ingredient, inhibiting the growth of most obligate aerobes and facultative anaerobes but leaving *Enterococcus* spp. unaffected, as they are aerotolerant anaerobes (see Fig. 36.1). The medium also contains a tetrazolium salt as a general indicator of metabolic activity, used to help visualise the small blue-purple colonies.

(d) Polymyxin is present as a selective agent for *Bacillus* spp. Mannitol and pyruvate are present as specific carbohydrate nutrient sources. *B. cereus*, in contrast to most other species of *Bacillus*, can utilise pyruvate but not mannitol, causing the pH to rise (most other *Bacillus* spp. use both mannitol and pyruvate, causing the pH to drop or remain stable). Bromophenol blue changes from green (neutral) to peacock blue (alkaline) for *B. cereus*. The egg yolk lecithin is broken down by a lecithinase produced by this bacterium, giving an opaque 'halo' of precipitated degradation products around the colonies.

37 Naming and classifying organisms

37.1 The correct classifications are as in the table below:

Hierarchical classifications for four species

	Limpet	Great White Shark	Giant Redwood tree	Earthworm
Kingdom	Animalia	Animalia	Plantae	Animalia
Phylum/Division	Mollusca	Vertebrata	Pinophyta	Annelida
Class	Gastropoda	Elasmobranchiata	Pinopsida	Oligochaeta
Order	Archaeogastropoda	Lamniformes	Pinales	Haplotaxida
Family	Patellidae	Lamnidae	Taxodiaceae	Lumbricidae
Genus	*Patella*	*Carcharodon*	*Sequoiadendron*	*Lumbricus*
species	*vulgata*	*carcharias*	*giganteum*	*terrestris*

37.2 Each listing has one pair of taxa transposed with an adjacent one.
- For the nautilus, look up Cephbase at http://www.cephbase.utmb.edu/(accessed 09/04/07).

- For the edible mushroom look up sites such as Natural Perspective: The Fungus Kingdom http://www.perspective.com/nature/fungi/index.html (accessed 09/04/07).

- For the bacterium, look up the NCBI Taxonomy browser at: http://www.ncbi.nlm.nih.gov/Taxonomy/tax.html/(accessed 09/04/07).

37.3 *Tip:* Use both textbooks and Internet sites (e.g. Maddison, D.R. (2001) *Tree of Life Project.* Available: http://tolweb.org/tree/phylogeny.html. Last accessed 09/04/07) for this activity. Look in particular for modern developments at the molecular taxonomy level that inform latest opinions.

38 Identifying plants and animals

38.1 A possible key:
1. Animal with soft body and hard rigid shell 2
   Animal without a shell. 3
2. Phylum Mollusca
       Shell in two parts . . . . Class Bivalvia
       Shell in one part . . . . Class Gastropoda
3. Animal with jointed legs 4
   Animal without jointed legs 6
4. 3 pairs of jointed legs
       . . . . Phylum Arthropoda, Subphylum Insecta
   More than 3 pairs of jointed legs 5
5. 4 pairs of jointed legs
       . . . . . . Phylum Arthropoda, Subphylum Arachnida
   More than 4 or more pairs of jointed legs
       . . . . Phylum Arthropoda, Subphylum Crustacea
6. Animal segmented
       . . . . Phylum Annelida, Class Oligochaetae
   Animal not segmented 7
7. Animal attached . . . . Phylum Coelenterata
   Animal free-living 8
8. Animal thread-like . . . . Phylum Nematoda
   Animal with flattened body . Phylum Platyhelminthes

38.2 *Tip:* For plants, it might be easier to choose a specimen that is in flower. You may need to learn some relevant morphological terminology (usually given at the front of keys).

38.3 *Tip:* Over 1 000 000 keys were found in a general search using Google, so there should be plenty of choice, and you may wish to specify a type of organism as well as 'identification key' in your search.

39 Identifying microbes

39.1 Colony appearance is typically as follows (note that the growth medium can modify the appearance from that noted below):
(a) Shiny, off-white, mucoid, slightly raised circular colonies with an entire (smooth) margin, 2–3 mm in diameter.
(b) Mucoid, flat spreading colonies with an undulate margin, often with green pigmentation, 4–5 mm in diameter.
(c) Circular 'pinpoint' colonies, convex and often with yellow or orange pigmentation, 1–2 mm in diameter.
(d) Circular off-white colonies, with a depressed centre ('draughtsman' colonies), α-haemolytic on blood agar, 1–3 mm in diameter.
(e) Flat irregular off-white colonies with a matt surface texture and lobed margin, 2–4 mm in diameter.
(f) Raised circular white colonies with an entire (smooth) margin, 3–4 mm in diameter.

39.2
(a) N′-N′-N′-N′-tetramethyl-*p*-phenylenediamine acts as an electron donor to cytochrome oxidase and is converted from the colourless, reduced form to the oxidised form, which has a strong purple-blue colour (a positive result).
(b) Indole, produced as a result of the breakdown of tryptophan, can be detected by a variety of methods including Kovac's reagent (*p*-dimethylamino-benzaldehyde in alcohol).
(c) β-galactosidase can be detected using a suitable chromogenic substrate (a synthetic substrate that liberates a coloured product when cleaved by the target enzyme), such as *o*-nitrophenyl-β-D-galactoside (ONP-GAL), or a fluorogenic substrate (a synthetic substrate that liberates a fluorescent product when cleaved by the target enzyme), such as 4-methylumbelliferyl-β-D-galactoside (MU-GAL).
(d) Hydrolysis of urea to ammonia and CO_2 is most easily demonstrated by including a pH indicator such as phenol red in the medium, to visualise the alkalinisation that accompanies this reaction. In this instance, the medium turns from yellow to red as the pH rises.
(e) The decarboxylation of lysine under anaerobic conditions can be detected by an alkalinisation of the medium, usually visualised using bromophenol purple, which changes from yellow to purple as the pH rises due to the action of lysine decaroboxylase.
(f) H_2S production from S-containing amino acids is most easily demonstrated by incorporating an iron-containing salt in the medium (e.g. ferrous ammonium sulphate). The H_2S produced by the growing bacteria combines with the ferrous ions to form a black deposit of iron sulphide. In tube tests, a filter paper strip containing lead acetate turns black on exposure to H_2S.

39.3
(a) *Proteus mirabilis*
(b) *Escherichia coli*
(c) *Salmonella* spp
(d) *Morganella morganii*
(e) *Escherichia coli*

40 The purpose and practice of dissection

40.1 *Tip:* This question requires considerable thought as even cell biology lines originally started from organs dissected out of animals. Chemical analysis of tissue samples will likewise require some form of dissection to remove the sample. Don't forget the dissection of plant organs – grafting is a form of dissection technique.

40.2
(a) Blunt seeker that allows separation of tissues with minimal risk of damaging the nerve or blood vessel.
(b) Fine scissors that allow incisions to be finely controlled so as not to damage the tissues within the cavity.

(c) Scalpel as the easiest and cleanest tool for removing tissues in relative bulk.

40.3 The usual rule is to dissect from the side opposite the central nervous system, i.e. from the dorsal surface of most invertebrates and the ventral surface of vertebrates.

41 Introduction to microscopy

41.1 (a) T; (b) F; (c) F; (d) T; (e) T; (f) T; (g) T; (h) F; (i) F; (j) T.

41.2 Missing words are: diffracted; near-transparent; resolution; interference; contrast; 3-D; focus; optical sectioning; polarised; laser.

41.3

(a) Embed; stain

(b) Air dry; stain

(c) Dehydrate

42 Preparing specimens for light microscopy

42.1

(a) Chlorazol black

(b) Phloroglucinol/HCl or toluidine blue

(c) Lactophenol cotton blue

(d) Neutral red

(e) Mallory

(f) Gram

42.2

(a) To preserve material and prevent degradation.

(b) To assist further processing of the specimen by removing water.

(c) To assist sectioning.

(d) To provide a thin slice of specimen suitable for viewing at best resolution.

(e) To enhance the contrast of the specimen and to identify chemical constituents.

(f) To protect specimens and allow them to be stored.

42.3 *Tip:* Use a Web database to find out about chemical hazards. The following answers were compiled using the Biochemical Safety website at: http://biochemlinks.com/bclinks/safety.cfm (accessed 09/04/07).

In general, all of the stains cause irritation to skin and eyes, and can be harmful if ingested in large amounts. Safety glasses and disposable gloves should be worn to reduce the risk of exposure during use. Specific additional comments for each stain include:

(a) Chlorazol black is toxic, is known to be carcinogenic and may cause foetal abnormalities.

(b) Neutral red is known to be mutagenic and to cause chromosomal abnormalities in animals.

(c) Malachite green is toxic and has caused reproductive and foetal abnormalities in laboratory animals.

(d) Safranin can cause permanent eye damage in humans and animals, and its toxicological properties remain to be established.

43 Setting up and using a light microscope

43.1 *Tip:* To memorise the parts, it may help to imagine yourself setting up a microscope.

43.2

(a) Provides optimal resolution by ensuring light source is focused on slide.

(b) Provides magnified image to the eyepiece lenses.

(c) Enhances image contrast by cutting down on stray light reaching the objective lens.

(d) Sets up the microscope for the distance between your pupils, so you can use both eyepieces.

(e) Allows you to return to same place on a slide.

43.3 Answer (c) is correct (see pp. 252–3).

44 Interpreting microscope images

44.1 One eyepiece micrometer unit $= 100~\mu m/11 = 9.09~\mu m$; 25 eyepiece units $= 25 \times 9.09~\mu m = \textbf{227}~\boldsymbol{\mu m}$.

44.2 Area of field $= \pi r^2 = 3.14159(0.2125^2) = 0.1419~mm^2$. Area density $= 25/0.1419 = \textbf{176 mm}^{-2}$.

44.3 Volume of field $=$ area $\times$ thickness $= 0.1419 \times 0.1 = 0.01419~mm^3$. Volume density $= 32/0.01419 = \textbf{2255 mm}^{-3}$.

44.4

(a) $100 \times (34.0 \div 519.1) = \underline{\textbf{6.55\%}}$

(b) Mean count for cytoplasm $=$ counts for cytosol + nucleus + mitochondria + chloroplasts + other cytoplasmic bodies $= (35.0 + 16.9 + 5.1 + 160.8 + 2.6) = 220.4$. Expressed as percentage $= 100 \times (220.4 \div 519.1) = \underline{\textbf{42.46\%}}$

(c) From (b) mean cytoplasmic count $= 220.4$, hence cytoplasm : vacuole ratio $= 220.4 : 264.7 = \underline{\textbf{0.83}}$.

(d) $160.8 \div 220.4 = \underline{\textbf{72.96\%}}$

45 Sterile technique

45.1 *Tip:* Consider the suitability of the various types of sterilisation procedures listed in Chapter 45, and whether they might be applicable for each item.

(a) Tips for pipettors are often sterilised by irradiation during the manufacturing process and may be supplied in sterile form. Autoclaving (p. 265) is the easiest approach to routine laboratory sterilisation of these items.

(b) Blood is collected and handled using sterile technique, to avoid contamination, and therefore it does not require sterilisation – it is added to sterilised agar medium after autoclaving, once the medium has cooled to below 50 °C.

(c) Autoclave, e.g. at 121 °C for 15 minutes.

(d) Either dry-heat sterilisation in a metal pipette can within a hot-air oven at 160 °C for $\geqslant 2$ h, or autoclaving at 121 °C for 15 min in the same type of pipette can.

(e) Red-heat sterilisation in a Bunsen flame (Fig. 45.2).

(f) Filter-sterilisation through a pre-sterilised filter unit, typically with a pore size of 0.45 µm or 0.2 µm. The filter-sterilised antibiotic solution is mixed thoroughly with the molten agar medium at 50 °C, then poured immediately. The filter unit may be purchased ready-sterilised, or prepared by autoclaving.

45.2 *Tip:* You will need to consult UK ACDP guidelines and categorisation (Anon., 1995, p. 270) – some information on categorisation and safe working with specific pathogens is available on the Web (note that other countries may use a slightly different categorisation, e.g. biosafety levels, BSL 1–4 in the USA):

(a) Category 2

(b) Category 2 (gloves should also be worn)

(c) Category 3

(d) Category 1

(e) Category 3

(f) Category 3

(g) Category 2 (a microbiological safety cabinet should be used)

(h) Category 3

(i) Category 1

(j) Category 4

45.3 The following list covers six important points, and is based on material in Chapter 45. However, you and your fellow students may think of other valid points:

- Spread plates show the typical colonial form of the microbe, whereas most of the colonies in a pour plate are contained within the agar and typical colony morphology cannot be observed.

- Pour plates are more suitable for counting higher numbers of microbes, since the colonies are smaller and are contained within the agar medium, in contrast to spread plates.

- Heat-sensitive microbial cells may be damaged by the temperature of the molten agar used in pour plating (45–50 °C) and may not grow on subsequent incubation.

- Some liquid will adhere to the spreader, and this may give a lowered count compared with a pour plate. Similarly, when the sample and molten agar are pre-mixed before pouring (see Fig. 45.6), some of the cells may not reach the pour plate, leading to a reduced count (note that this is one reason why some workers carry out the mixing of sample and molten agar within the Petri plate, rather than in the bottle).

- Each type of plating procedure requires different apparatus – spread plating requires a sterile spreader and pre-poured plates, while pour plating requires molten agar at 50 °C (usually kept in a water bath until needed).

- There is a fire risk associated with using ethanol to flame-sterilise a glass spreader. This risk is absent in pour plates, and can be removed in spread plates by using a disposable plastic spreader.

46 Cell culture

46.1 *Tip:* Use eqns [46.1] and [46.2] for these calculations.

(a) Substituting the appropriate values into eqn [46.1], $\mu = [2.303 \ (6.5314789 - 4.7160033)] \div 9 = 0.46456 = $ **0.465 h^{-1}**. Similarly, using eqn [46.2], $g = (0.301 \times 9) \div (6.5314789 - 4.7160033) = 1.492\,171 = $ **1.49 h** (approximately 1 h 29 min, *not* 1 h 49 min).

(b) Substituting the appropriate values into eqn [46.1], $\mu = [2.303 \ (7.8260748 - 4.4771213) \div 200 = 0.0385632 = $ **0.0386 min^{-1}**. Similarly, using eqn [46.2], $g = (0.301 \times 200) \div (7.8260748 - 4.4771213) = 17.975765 = $ **18.0 min**.

(c) Substituting the appropriate values into eqn [46.1], $\mu = [2.303 \ (2.39794 - 1.50515)] \div 2.5 = 0.8224382 = $ **0.822 h^{-1}**. Similarly, using eqn [46.2], $g = (0.301 \times 2.5) \div (2.39794 - 1.50515) = 0.8428634 = $ **0.843 h** (approximately 51 min). Note that the time difference in this example is two and a half hours, not 2.3 hours.

46.2 *Tip:* Use eqn [46.6] (Box 46.1) to carry out these calculations.

(a) $6.42 \div 1 \div (2.5 \times 10^{-7}) \times 1 = 2.568 \times 10^{-7} = $ **2.57 $\times$ 10^7 ml^{-1}**

(b) Total cell count $= 698$, so $698 \div 20 \div (8 \times 10^{-7}) \times 1 = 4.3625 \times 10^7 = $ **4.36 $\times$ 10^7 ml^{-1}**.

(c) $78 \div 25 \div (2.5 \times 10^{-7}) \times 10^2 = 1.248 \times 10^9 = $ **1.25 $\times$ 10^9 ml^{-1}** (note the correction factor of 10^2, to account for the dilution factor of 10^{-2}).

(d) $6.8 \div 1 \div (2 \times 10^{-5}) \times 0.1 = $ **3.40 $\times$ 10^4 ml^{-1}** (note the correction factor of 0.1, because the suspension was concentrated by a factor of 10).

46.3 *Tip:* Use eqn [46.7] (Box 46.2) to carry out the calculations.

(a) $(54.4 \div 0.1) \times 10^5 = $ **5.44 $\times$ 10^7 CFU ml^{-1}**.

(b) The mean count per spread plate is $(34 + 40 + 37) \div 3 = 37.0$ CFU. So $(37.0 \div 0.2) \times 10^3 = $ **1.85 $\times$ 10^5 CFU ml^{-1}**.

(c) The mean count per pour plate is $(211 + 186 + 194 + 202) \div 4 = 198.25$ CFU. So $(198.25 \div 0.5) \times 20 = $ **7.93 $\times$ 10^3 CFU ml^{-1}**.

(d) The mean count per plate is $(35 + 41 + 32) \div 3 = 36$ CFU. So $(36 \div 1) \times 50 = 1800$ per g and therefore $1800 \times 100 = $ **1.80 $\times$ 10^5 CFU per 100 g**. (Note that the multiplication factor to correct for dilution was 50 in this case, since the seafood had been diluted to 2% w/v.)

(e) This example does not require eqn [46.7] in full. The mean count for 250 ml was $(45 + 32) \div 2 = 38.5$ CFU. Therefore, the count per 100 ml is $38.5 \div 250 \times 100 = $ **15.4 CFU per 100 ml**.

(f) The mean count per plate is $(28 + 32 + 39) \div 3 = 33.0$, so using eqn [46.7], $(33.0 \div 0.5) \times 10^1 = $ **660 CFU ml^{-1}**. This value is higher than the EU regulations (maximum 100 CFU ml^{-1}), so this water sample does not meet the specification required of mineral water.

47 Working with animal and plant tissues and cells

47.1 Assuming that each solute behaves in an ideal manner, then the osmolarity of each constituent will be equivalent to the amount in grams per litre divided by the relative molecular mass and multiplied by the number of osmotically active particles produced from each solute molecule when it dissolves in water. Thus $NaCl = 6 \div 58.44 \times 2 = 0.2053388$ osmol l^{-1}; $KCl = 0.075 \div 74.55 \times 2 = 0.0020121$ osmol l^{-1}; $CaCl_2 = 0.1 \div 110.99 \times 3 = 0.0027029$ osmol l^{-1}; $NaHCO_3 = 0.1 \div 84.01 \times 2$ (note that $NaHCO_3$ gives only two osmotically active entities in

Answers to study exercises

water) $= 0.0023807$ osmoll^{-1}. Added together, this gives an osmolarity of 0.2124345 osmol$l^{-1} = \underline{\mathbf{212.4\ mosmol\,l^{-1}}}$. Note that you should use the full numerical values until the last step, to avoid introducing 'rounding' errors.

47.2 *Tip:* The UK National Culture Collection website at http://www.ukncc.co.uk/ (accessed 09/04/07) provides coordinated information for a number of individual UK collections. The UK National Collection of Type Cultures (NCTC) can be found at http://www.hpa.org.uk/nctc/default.htm/ (accessed 09/04/07); the American Collection of Type Cultures (ATCC) is at http://www.atcc.org/; the World Federation of Culture Collections has a website at http://wdcm.nig.ac.jp/wfcc/index.xml (accessed 09/04/07). Box 10.5 gives further details on the evaluation of Web-based resources.

47.3

(a) HeLa cells are named after their source, Ms Henrietta Lacks, of Baltimore, USA.

(b) Vero cells are derived from the kidney of an African Green Monkey (*Cercopithecus aethiops*), known as the vervet monkey, hence 'vero'.

(c) BHK stands for 'baby hamster kidney', the original source of this cell line.

(d) CHO stands for 'Chinese hamster ovary', the original source of this cell line.

(e) TBY-2 stands for tobacco bright yellow strain 2.

48 Photography and imaging

48.1 *Tip:* The key issues here relate to:

- The ability to take a good image in the first place; film still provides the highest-quality image but you do not know what it is like until it has been processed – by this time you might not be able to repeat the photograph. Digital cameras provide instant information as to the success of the image creation, allowing repetition until it is adequate.

- The ability to subsequently manipulate the image to improve it. Digital images are easily manipulated but, of course, film material can also be digitised once it has been processed.

48.2 *Tip:* The key point about making an image is that it provides a record of an event or an action and this can be particularly important when things cannot be repeated. In addition, there are many modern microscopic methods that depend on image creation for their use, e.g. scanning electron microscopy.

48.3 *Tip:* Decide first on the available lighting options available for the imaging equipment. This will to a large extent determine what type of film will be required. Poor lighting requires fast films, but these generally produce poorer image quality, so maximum effort should be made to provide good lighting conditions. This might require the use of artificial aids such as electronic flashes.

49 Measuring growth and responses

49.1

(a) $(38.5 - 35)/7 = \underline{\mathbf{500\ mg\,day^{-1}}}$.

(b) $((38.5 - 35)/35) \times 100 = \underline{\mathbf{10\%}}$.

49.2 *Tip:* You need to graph heart rate against log concentration values; this should give a sigmoidal curve (see below). Having obtained log concentration estimates graphically from the heart rate values for the two loch samples, convert back to mmoll^{-1} concentration by raising 10 to the power of the estimated log concentration (see p. 411).

(a) The estimate for (i) is **approximately 1.58×10^{-5} mol m^{-3}**; and for (ii) **approximately 6.31×10^{-5} mol m^{-3}**.

(b) Estimate (ii) is the least reliable owing to the shallowness of the log concentration versus heart rate curve at this point: a small error in heart rate will lead to a large difference in estimated concentration.

(c) If sample (ii) were diluted approximately 4-fold, then tested again, this would shift the 'expected' heart rate downwards to the most sensitive part of the graph. The dilution factor would need to be accounted for when estimating the true concentration.

(d) There is no guarantee that the only substance affecting the *Daphnia* sp. heart rate in the loch samples is the toxin tested. Other substances present might increase or decrease heart rate.

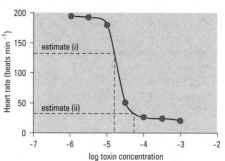

Responses of *Daphnia* to toxin

49.3 *Tip:* You will first need to convert the nisin concentrations of the standard solutions to log$_{10}$ values, before plotting zone diameter against log$_{10}$ nisin concentration (see figure below). Based on the calibration curve, solution 5 contains antilog$_{10}$ $1.24 = \underline{\mathbf{17.4\ \mu g\ ml^{-1}}}$.

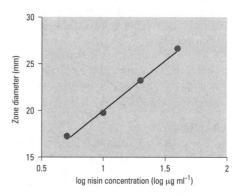

Effect of nisin on inhibitory zone diameter for *B. subtilis* colonies

50 Calibration and its application to quantitative analysis

50.1 See figure below – note that this calibration line is approximately linear up to $\approx 4.0\ \mu g\,ml^{-1}$ but it becomes increasingly curved above this point. Water sample (a) contains Zn at **1.70 $\mu g\,ml^{-1}$**; water sample (b) contains Zn at $3.5 \times 20 = \overline{\textbf{70.0 μg\,ml}^{-1}}$; water sample (c) contains $8.3 \times 5 = \textbf{41.5 μg\,ml}^{-1}$ (note that the graph curves at higher concentrations, making the estimate for water sample (c) less reliable than the other two – a better approach might have been to dilute this sample further and re-assay).

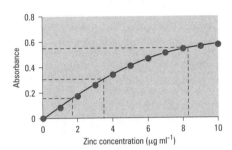

Calibration curve for a series of solutions containing zinc

50.2 Note that this data set is a very good fit to a straight line– linear regression analysis (p. 435) carried out using Microsoft Excel, which gives a value for r^2 of 0.996829051, with an intercept of 0.001714286 and a slope of 7.424285714. The protein content of each solution could be determined directly from a printout of the graph (e.g. figure below). However, the calculations below show how you can substitute values into the equation for a straight line, $Y = a + bX$ (p. 411), to give the required answers.

(a) $0.225 = 0.001714286 + 7.424285714X$, so $X = (0.225 - 0.001714286) \div 7.424285714 = 0.030075043 = \textbf{0.0301 mg protein}$.

(b) $0.465 = 0.001714286 + 7.424285714X$, so $X = (0.465 - 0.001714286) \div 7.424285714 = 0.062409385 = \textbf{0.0624 mg protein}$.

(c) $0.682 = 0.001714286 + 7.424285714X$, so $X = (0.682 - 0.001714286) \div 7.424285714 = 0.091629786 = \textbf{0.0916 mg protein}$.

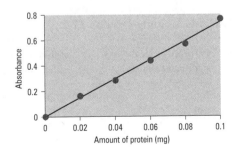

Calibration curve for a series of protein standards

50.3 *Tip:* Check that you understand the general requirements for good graph drawing, detailed in Chapter 62.

(a) The title of the figure does not give sufficient information to enable the reader to understand the content of the graph.

(b) No details are given as to the nature of each set of data values – are these curves for two separate substances?

(c) The x axis has a poor choice of units – it would be better to use μg, then all of the values would be 1000 times smaller.

(d) The x axis runs beyond the highest value (8 μg), wasting around 20 per cent of the horizontal space.

(e) The scale marks (tic marks) on the x axis are missing.

(f) The x axis has no zero shown.

(g) The y axis has too many numbers, and too many minor scale marks – it is probably better to have a tic mark every 0.1 absorbance units. (Note that absorbance is a dimensionless term, so no units are required – this is *not* an error.)

(h) The data values represented by the square symbols show a reasonable fit to the lower linear trend line, but the data values represented by the round symbols do not – the latter set of data shows a clear curvilinear relationship and therefore a linear trend line is not valid.

(i) Lines should not run through the symbols, as is the case for several of the square symbols.

51 Immunological methods

51.1 *Tip:* Take care to make sure that your construction lines are carefully drawn by hand – they must be parallel to the appropriate axis. The test sample contains 7.5 μg of antigen, approximately.

51.2 *Tip:* Remember that you need to plot the *square* of the ring diameter versus the amount of antigen in each standard. The test sample has a squared ring diameter of 42.25 mm^2 and contains 11 μg of antigen (see figure overleaf).

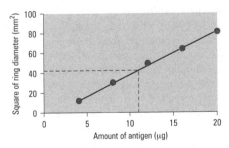

Calibration curve for a series of standards assayed by single radial immunodiffusion

51.3

(a) Wells 1 and 2 in the figure (p. 317) show the 'prozone' phenomenon, with a negative haemagglutination test result (a small 'button' of cells), due to an excess of antibody. Wells 3 to 10 show strong positive haemagglutination, with an even 'carpet' of red blood cells rather than a 'button'. Wells 11 and 12 show a negative reaction, due to the further dilution of the test serum.

(b) The lowest dilution giving a positive result is well 10, which is 9 doubling dilutions of the original 10-fold dilution, i.e. $10 \times 2^9 = 5120$-fold. Therefore the titre is 1 in 5120.

(c) The test serum is positive, since the titre is greater than 1 in 5000.

52 Enzyme studies

52.1

(a) *Tip:* Tackle this problem by converting the absorbance change per minute to amount of product formed, in nmol: $0.25 \div 0.122 \times 10 = 20.491803$ nmol min^{-1} = **20.5 nmol min^{-1}**.

(b) *Tip:* Express the activity in the appropriate units (mol s^{-1}) to give the number of katals. From part (a), enzyme activity = 20.491803 nmol min$^{-1} \div 60 =$ 0.3415301 nmol s^{-1} = 0.3415301 $\times 10^{-9}$ mol s^{-1} = 0.3415301 $\times 10^{-9}$ kat = **0.342 $\times 10^{-9}$ kat** (or 0.342 nkat). Note that you should use the full numerical values in these calculations and not the answers expressed to 3 significant figures, to avoid introducing 'rounding' errors.

(c) *Tip:* Divide the enzyme activity by the amount of protein present in the assay to give the specific activity. Thus, in terms of nmol: $20.491803 \div 2.5 =$ 8.1967212 = **8.20 nmol min^{-1} (mg protein)$^{-1}$**. In terms of kat: $0.3415301 \times 10^{-9} \div 2.5 = 0.136612 \times 10^{-9}$ = **0.137 $\times 10^{-9}$ kat (mg protein)$^{-1}$**.

52.2 *Tip:* Prepare a table, with columns for each of the additional steps required, i.e. (i) total enzyme activity (activity per ml × total volume of extract), (ii) total protein content (protein concentration per ml × total volume of extract), and (iii) specific activity (total enzyme activity ÷ total protein content). Then, use eqns [52.4] and [52.5] to calculate yield and *n*-fold purification and add these to two further columns. The table below shows how this can be applied in this case (note that the values are not rounded off until the final stage).

52.3 *Tip:* If you do not have specialised software for determining kinetic constants, you can use Microsoft Excel to prepare a suitable plot, e.g. a Lineweaver–Burk plot. Enter the data from the table in the question into two columns, then prepare a further two columns to contain

the reciprocals of each value (note that you cannot plot the zero values, since there is no reciprocal). Then use the chart 'wizard' to prepare a double reciprocal plot: note that Microsoft Excel will only draw a trend line between the points – you will need either to extend this line by hand, as shown in the figure below, or use linear regression analysis (p. 435) to give the *y* intercept and slope (if you divide the *y* intercept by the slope, this will give you the *x* intercept, but without the negative sign). In the example shown in the figure below, the *y* intercept ($1/V_{max}$) is 0.013443707, so $V_{max} = 74.38424731 = $ **74.4 U**. Similarly, the *x* intercept ($-1/K_m$) is -0.055498669, so $K_m = 18.01845 = $ **18.0 µmol l^{-1}**.

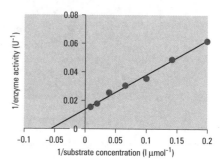

Lineweaver–Burk double reciprocal plot.

53 Mendelian genetics

53.1 The table below shows the layout of the Punnett square.

Punnett square for cross RrOO × RrOo

Gametes	Parent 1	
	RO	**rO**
RO	RROO	RrOO
Ro	RROo	RrOo
rO	RrOO	rrOO
ro	RrOo	rrOo

(Parent 2 labels the left column: RO, Ro, rO, ro)

Genotype ratios: RROO : RROo : RrOo : RrOO : rrOO : rrOo in ratio 1 : 1 : 2 : 2 : 1 : 1

Phenotype ratio: red closed : pink closed : white closed in ratio 1 : 2 : 1

Calculation of the yield and purification of an enzyme using data from table in Study exercise 52.2

Stage:	Total enzyme activity (U)	Total protein content (mg)	Specific activity (U mg^{-1})	Percentage yield (to 3 significant figures)	*n*-fold purification (to 3 significant figures)
1	93.6	374.4	0.25	100	1.00
2	84.0	119.0	0.705882352	89.7	2.82
3	64.0	25.0	2.56	68.4	10.3
4	54.0	7.0	7.714285714	57.7	30.9
5	39.42	2.04	19.32352941	42.1	77.3

53.2 Probability of blue-eyed $= 0.25$
Probability of blonde $= 0.5$
Probability of girl $= 0.5$
Overall probability $= 0.25 \times 0.5 \times 0.5 = \underline{\mathbf{0.0625}}$

53.3 Assuming normal mice are homozygous for tail presence, the approximately $1 : 1$ phenotype ratio in the F_1 indicates that the tailless phenotype is heterozygous and possibly dominant. The approximately $1 : 2 : 1$ ratio of phenotypes in the F_2 generation (counting 'dead on birth' as a phenotype) also indicates that tailless might be dominant over normal. The results could be explained if the tailless gene is fatal when double recessive. There is no sex linkage because the ratio of males to females is roughly equivalent in all progeny phenotypes.

53.4 Use notation $X =$ female, $Y =$ male, $X^R =$ red allele, $X^r =$ white allele.

(a) Assume progeny ratio is $2 : 1 : 1$, then cross must be $X^R X^r$ (red-eyed heterozygous female) $\times X^R Y$ (red-eyed male), giving progeny genotypes $X^R X^R$, $X^R X^r$, $X^R Y$ and $X^r Y$ in the ratio $1 : 1 : 1 : 1$ and phenotypes as observed.

(b) Assume progeny ratio is $1 : 1$, then cross must be *either* $X^R X^R$ (red-eyed homozygous female) $\times X^r Y$ (white-eyed male), giving progeny genotypes $X^R X^r$, $X^R Y$ in the ratio $1 : 1$ and the phenotypes as observed; *or* $X^R X^R$ (red-eyed homozygous female) $\times X^R Y$ (red-eyed male) giving progeny genotypes $X^R X^R$, $X^R Y$ in the ratio $1 : 1$ and the phenotypes as observed.

(c) Assume progeny ratio is $1 : 1 : 1 : 1$, then cross must be $X^R X^r$ (red-eyed heterozygous female) $\times X^r Y$ (white-eyed male), giving progeny genotypes $X^R X^r$, $X^r X^r$, $X^R Y$ and $X^r Y$ in the ratio $1 : 1 : 1 : 1$ and the phenotypes as observed.

53.5 The expected values are $137 : 274 : 137$, based on a $1 : 2 : 1$ ratio. The χ^2 value is $1.05 + 3.07 + 2.11 = 6.23$ (see Box 53.2 for method of working). Since this value is greater than the critical value for Chi^2 for 2 degrees of freedom ($n - 1$ categories $= 3 - 1$) at $P = 0.05$ (which is 5.99, Table 53.1), the geneticist should reject the null hypothesis of a $1 : 2 : 1$ outcome.

54 Molecular biology techniques

54.1

(a) Reduces the level of contaminating proteins.

(b) Removes salt contamination and precipitates DNA.

(c) Degrades contaminating RNA (ribonuclease) and proteins (protease).

(d) Lysozyme digests the peptidoglycan of the bacterial cell wall and EDTA aids its access through the Gram-negative bacterial outer membrane (p. 229) to reach the peptidoglycan layer.

(e) Sodium dodecyl sulphate at high concentration solubilises membranes and causes cell lysis.

(f) NaOH aids cell lysis and disrupts macromolecules. Neutralisation with potassium acetate allows plasmid DNA to return to a soluble form, while leaving

chromosomal DNA and other macromolecules as an insoluble mass.

The order of use of these reagents is: (d); (e); (f); (c); (a); (b).

54.2 First, determine the fragment sizes (in kbp) for each digest:

Eco RI: 7 kbp;

Hin dIII: 7 kbp;

Pvu II: 7 kbp.

The results for the individual enzymes confirm that the plasmid is cut once, creating a single linear molecule of 7 kbp (note that an intact, circular, plasmid would run further on the gel, since it would be more compact).

Eco RI and *Hin* dIII: 1 and 6 kbp;

Eco RI and *Pvu* II: 3 and 4 kbp;

Hin dIII and *Pvu* II: 2 and 5 kbp;

Eco RI, *Hin* dIII and *Pvu* II: 1, 2 and 4 kbp.

The approach to working out the relative locations of each restriction site is to match up the sizes of the band, or bands, that no longer appear on the gel when the enzymes are combined and then try to piece together the fragments, to see how they were combined in the intact molecule by sketching out possible configurations. The exercise is not unlike a jig-saw puzzle. The results for the paired enzyme combinations confirm that each cuts the plasmid once, since two fragments are obtained. The minimum distance between two enzymes is given by the smaller band in each case: 1 kbp between *Eco* RI and *Hin* dIII, 3 kbp between *Eco* RI and *Pvu* II, and 2 kbp between *Hin* dIII and *Pvu* II. In this instance, the results when all three enzymes are combined can then be compared with those for any pair of enzymes (e.g. *Eco* RI and *Pvu* II), showing that the *Hin* dIII site must lie between *Eco* RI and *Pvu* II sites, since the 3 kbp fragment is cleaved into a 1 kbp and a 2 kbp fragment, giving the map shown in the figure below. (Note that the results cannot tell us the *absolute* orientation of the sites, but only their relative position: a mirror image of the map shown in the figure below would give the same results. Complex restriction digests are best analysed using an appropriate software package: alternatively, restriction sites can be predicted on the basis of DNA sequence information.)

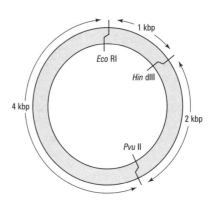

Restriction map of plasmid pARJ

54.3 *Tip:* Work through the calculation in small steps, taking into account the various volumes and dilutions used, to arrive at the correct answer. First, calculate the number of transformants in the entire sample: 180×10 (to correct for ten-fold dilution) $\div 200 \times 500$ (to convert from 200 µl to 500 µl) $= 4500$ CFU. Next, calculate the amount of plasmid DNA (in µg) used in the procedure: $20 \times 0.1 = 2.0$ ng $= 0.002$ µg. Finally, divide the number of transformants by the amount of DNA, to give the answer: $4500 \div 0.002 = 2\,250\,000 = \mathbf{2.25 \times 10^6}$ **CFU (µg DNA)$^{-1}$.** An alternative approach is to calculate the amount of plasmid DNA (in µg) present in the sub-sample used on the spread plate, as 0.1×20 (ng added) $\div 10$ (dilution factor) $\times 200 \div 500$ (fractional volume spread) $= 0.08$ ng $= 0.00008$ µg. Next, divide the number of transformants on the spread plate by this value to give the answer: $180 \div 0.00008 = \mathbf{2.25 \times 10^6}$ **CFU (µg DNA)$^{-1}$.**

55 Light measurement

55.1 First convert PFFD to photosynthetic irradiance (PI) using Table 55.1: $1610 \div 4.6 = 350$ W m^{-2} ($\equiv$ J m^{-2} s^{-1}). Then, calculate the PI for the leaf area of 45 cm^2 ($\equiv 0.0045$ m^2): $350 \times 0.0045 = 1.575$ J s^{-1}. Finally, calculate the total energy supplied over the 30 minute (1800 s) period: $1.575 \times 1800 = 2835$ J $= \mathbf{2840}$ **J.**

55.2 *Tip:* Convert values to PPFD expressed in µmol photons m^{-2} s^{-1} using the conversion factors given in Table 55.2, where necessary.
(a) 300×4.6 (to convert to PPFD) $= \mathbf{1380}$ **µmol photons m^{-2}s^{-1}**;
(b) $6.5 \div 1000$ (to convert to µmol) $\times 10\,000$ (to convert to m^{-2}) $= \mathbf{65.0}$ **µmol photons m^{-2}s^{-1}**;
(c) this value is already expressed in the appropriate units, **275 µmol photons m^{-2}s^{-1}**;
(d) 35.2×4.6 (to convert to PPFD) $= 161.92 = \mathbf{162}$ **µmol photons m^{-2}s^{-1}**;
(e) $1.15 \div 1000$ (to convert to W) $\times 10\,000$ (to convert to m^{-2}) $\times 5.0$ (to convert to PPFD) $= \mathbf{57.5}$ **µmol photons m^{-1}s^{-1}**;
The locations ranked in order of decreasing PPFD are therefore as follows: (a); (c); (d); (b); (e).

55.3 *Tip:* If you simply use the term 'light meter' in a general-purpose search engine (e.g. Google), you will find out mostly about photographic light meters, and these are not suitable for scientific purposes – you'll need to add other terms, such as photosynthesis, quantum, irradiance, photon flux density, etc. Relevant features might include: measuring scale(s); range(s); detector/receiver type; cost; portability/weight; power supply; the type of sensor (e.g. 3-D, flat, or rod-shaped); whether the sensor can be used in water; whether the meter can be used with a data logger.

56 Radioactive isotopes and their uses

56.1 Assume a ^{14}C half-life of 5715 years (Table 56.1). Using formula 56.1, $e^x = 0.5725$, and so $x = \ln(0.5725) = -0.5578$. If $-0.5578 = -0.693(t/5715)$, then $t = (-0.5578/-0.693) \times 5715 = 4600$. **The rat visited about 4600 years ago.**

56.2
(a) **1200 Bq**
(b) **4.44 × 10^7 dpm**
(c) **1.20 µCi**
(d) **30.0 Bq g^{-1}**
(e) **28.0 pmol**

56.3
(a) The solution contains $250 \times 10^{-6} \times 5$ mol $= 1.25 \times 10^{-3}$ mol. The specific activity is therefore $55 \times 10^6 / 1.25 \times 10^{-3}$ Bq mol$^{-1} = \mathbf{4.4 \times 10^{10}}$ **Bq mol^{-1}.**
(b) $79.2 \times 10^5 / 4.4 \times 10^{10} = 1.8 \times 10^{-4}$ mol per 2 h per 10^7 cells is equivalent to **2.5 fmol s^{-1} cell^{-1}.**

57 Measuring oxygen content

57.1
(a) 0.285 (value in mmol l^{-1}m from Table 57.1) $\div 1000 \times 4$ (to calculate mmol in 4 ml) $\times 1000$ (to convert from mmol to µmol) $= \mathbf{1.14}$ **µmol.**
(b) $0.273 \div 1000 \times 20$ (to calculate mmol in 20 ml) $\times 1000$ (to convert from mmol to µmol) $= \mathbf{5.46}$ **µmol.**
(c) 0.317 (estimated as the mid-point between the values at 14 °C and 16 °C) $\div 1000 \times 10$ (to calculate mmol in 10 ml) $\times 1000$ (to convert from mmol to µmol) $= \mathbf{3.17}$ **µmol.**
(d) $6.75 \div 1000 \times 250$ (to calculate mg in 250 ml) $= 1.6875 = \mathbf{1.69}$ **mg.**
(e) 4.73 (estimated as the mid-point between the values at 24 °C and 26 °C) $\div 1000 \times 200$ (to calculate ml in 200 ml) $= \mathbf{0.946}$ **ml.**

57.2 First, calculate the oxygen content of the electrode chamber as $0.235 \div 1000 \times 5 = 0.001175$ mmol ($\equiv 1.175$ µmol). This is equal to 200 divisions on the chart recorder paper, so therefore each division will correspond to $1.175 \div 200 = 0.005875$ µmol ($\equiv 5.875$ nmol). The rate of oxygen consumption was -25 units (-146.875 nmol) per 2 min, equivalent to $-146.875 \div 2 = -73.44$ nmol min^{-1}. Since this was the rate for 4.1×10^9 cells, the rate per 10^9 cells is $-73.44 \div 4.1 = \mathbf{-17.9}$ **nmol min^{-1} (10^9 cells)$^{-1}$.**

57.3 *Tip:* Use a Microsoft Excel spreadsheet (i) to convert the data to µmol; (ii) to plot the converted data and (iii) to estimate the slope of the line (see Box 50.2). The calculations are as follows: first calculate the amount of oxygen in the chamber, as 0.285 (value in mmol l^{-1} from Table 57.1) $\div 1000$ (amount in 1 ml) $\times 5$ (amount in 5 ml) $\times 1000$ (convert from µmol to (mol) $= 1.425$ µmol. Next, convert the data from the table in the question from percentage to µmol by dividing by 100 and multiplying by 1.425, to give the values shown in the table opposite.

Oxygen content of illuminated electrode chamber containing a cyanobacterial suspension, based on data in the table in the question

Time	Oxygen content of chamber (μmol)
0	1.0032
1	1.0488
2	1.073025
3	1.118625
4	1.1913
5	1.224075
6	1.25685
7	1.30245
8	1.340925

These values are then plotted as a graph (see below), with the slope giving the rate of net photosynthesis: in this instance, the slope (rate of net oxygen production) is 0.043083 $\mu mol\,min^{-1}$ (based on linear regression analysis, in Microsoft Excel). Next, calculate the amount of chlorophyll a in the electrode chamber, as 2.13 ($\mu g\,ml^{-1}$) $\times\,5$ (volume of chamber, in ml) $= 10.65$ μg. Therefore the rate of net photosynthesis is $0.043083 \div 10.65 = 0.0040453521$ μmol min^{-1} (μg chlorophyll a)$^{-1}$ and therefore **4.05 $\mu mol\,min^{-1}$ (mg chlorophyll a)$^{-1}$**.

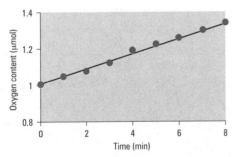

Oxygen content of an illuminated electrode chamber containing a cyanobacterial suspension

58 Centrifugation

58.1

(a) Use a high-speed centrifuge and differential centrifugation.

(b) Use silicone oil microcentrifugation (e.g. using an Eppendorf-type microfuge).

(c) Use continuous-flow centrifugation.

58.2 *Tip:* Substitute values into eqn [58.1].

(a) RCF $= 1.118 \times 125\ (4000 \div 1000)^2 = $ **2240 g**.

(b) RCF $= 1.118 \times 60\ (12000 \div 1000)^2 = $ **9660 g**.

(c) RCF $= 1.118 \times 186\ (30000 \div 1000)^2 = $ **187 000 g**.

58.3 *Tip:* Substitute values into eqn [58.2].

(a) $945.7 \sqrt{(1500 \div 95)} = $ **3760 r.p.m.**

(b) $945.7 \sqrt{(50\,000 \div 135)} = $ **18 200 r.p.m.**

(c) First convert the radius of rotation from cm to mm before substituting into eqn [58.2].
$945.7 \sqrt{(13\,000 \div 55)} = $ **14 5000 r.p.m.**

58.4 *Tip:* Use eqn [58.1] with each of the values in turn:
for $r_{min} = 45$ mm, $1.118 \times 45\ (5)^2 = $ **1260 g**;
for $r_{av} = 70$ mm, $1.118 \times 70\ (5)^2 = $ **1960 g**;
for $r_{max} = 95$ mm, $1.118 \times 95\ (5)^2 = $ **2660 g**.
This calculation illustrates the range of g values possible within a single centrifuge tube, with the maximum value being more than double the minimum value in this instance.

59 Spectroscopic techniques

59.1 *Tip:* Box 59.2 gives stepwise instructions on using a spectrophotometer – check that you have covered all of the major points. However, the exact details will vary from instrument to instrument.

59.2 *Tip:* Substitute values into eqn [59.1] and solve for [C] (margin examples are given on p. 369).

(a) Using eqn [59.1]: $0.53 = 6220 \times 1 \times [C]$, therefore $[C] = 0.000085209$ $mol\,l^{-1} = $ **85.2 $\mu mol\,l^{-1}$**.

(b) First calculate the concentration, using eqn [59.1]: $0.62 = 6220 \times 5 \times [C] = 0.0000199357$ $mol\,l^{-1} = 19.9357$ $\mu mol\,l^{-1}$. Next, calculate the amount in 20 ml: $19.9357 \div 1000$ (amount in 1 ml) $\times 20$ (amount in 20 ml) $= 0.398\,714$ μmol $= $ **398 nmol**.

(c) Using eqn [59.1] (note that you must convert the path length from mm to cm beforehand): $0.57 = 20 \times 0.5 \times [C] = 0.057$ $g\,l^{-1} = $ **57.0 $\mu g\,ml^{-1}$**.

(d) First calculate the concentration, using eqn [59.1]: $0.31 = 20 \times 1 \times [C] = 0.0155$ $g\,l^{-1}$. Next, express the concentration in the required units: 0.0155×10^9 (convert to μg) $\div 10^6$ (convert to μl) $= 15.5$ $ng\,\mu l^{-1}$. Then, calculate the amount in 50 μl: $15.5 \times 50 = $ **775 ng**.

(e) Substitute the values into the equation given: $(11.93 \times 0.182) - (1.93 \times 0.035) = 2.103\,71$ $mg\,l^{-1}$. Next, calculate the amount in 50 ml: $2.103\,71 \div 1000$ (amount in 1 ml) $\times 50$ (amount in 50 ml) $= 0.1051855$ mg. Since this is the amount of chlorophyll in 1.56 g fresh weight of leaf, divide by the fresh weight to give the content: $0.1051855 \div 1.56 = 0.0674266$ mg per g fresh weight. Finally, covert to the specified units: 0.0674266×1000 (convert from mg to μg) $= 67.4266 = $ **67.4 μg (g fresh weight)$^{-1}$**.

59.3 First determine the concentration of *p*-aminophenol, in $mol\,l^{-1}$. Thus 8.8 $\mu g\,ml^{-1} \div 10^6$ (convert to $g\,ml^{-1}$) $\times 10^3$ (convert to $g\,l^{-1}$) (291.27 (convert to mol l^{-1})) $= 0.0000302125$ $mol\,l^{-1}$. Next, substitute values into eqn [59.1] and solve for ε: $0.535 = \varepsilon \times 1 \times 0.0000302125 = $ **17 707.90236 $= 17\,700\ l\,mol^{-1}\,cm^{-1}$**. Note that you should use the full numerical value until the final stage, to avoid introducing 'rounding' errors.

59.4 The calibration graph is shown in the figure below. The test solution has a K^+ concentration of 0.32 mmol l^{-1}. This is equivalent to $0.32 \div 1000 \times 25 = 0.008$ mmol in 25 ml (the test sample volume). Next, divide by the weight of tissue used in grams: $0.008 \div 0.482 = 0.016598$ mmol (g tissue)$^{-1}$ = **16.6 µmol (g tissue)$^{-1}$**.

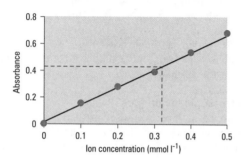

Calibration curve for a series of standards, assayed for K^+

60 Chromatography

60.1 *Tip:* Use eqn [60.1] to determine R_F values and eqn [60.2] for R_A values, based on the measured distances travelled by each component, relative to the solvent front.

(a) Using eqn [60.1], the corresponding R_F values are:
pigment A = $26 \div 47$ = **0.553**;
pigment B = $31 \div 47$ = **0.667**;
pigment C = $35.5 \div 47$ = **0.755**.

(b) Using eqn [60.2], the corresponding R_A values are:
pigment B = $31 \div 26 = 1.1923076$ = **1.19.**
pigment C = $35.5 \div 26 = 1.3653846$ = **1.37.**

60.2

(a) *Tip:* Calculate the selectivity using eqn [60.3], using the elution time for the compound excluded from the stationary phase as the column dead time (t_0), with all times converted to seconds. Thus $\alpha = (372 - 95) \div (270 - 95) = 1.5828571$ = **1.58.**

(b) *Tip:* Calculate the resolution using eqn [60.4]. Thus $R = 2(372 - 270) \div (40 + 44) = 2.4285714$ = **2.43**. This is a good level of resolution for quantitative work, giving clear separation of the two peaks.

60.3 *Tip:* Substitute relative area values for each peak in eqn [60.6], as follows:
for mannitol (peak A): $Q_A = [2060 \div 604] \times [1.50 \div 1.26] = 4.0602334$ = **4.06 mg**;
for sucrose (peak B): $Q_B = [1898 \div 604] \times [1.50 \div 0.92] = 5.1234523$ = **5.12 mg**.

61 Manipulating and transforming raw data

61.1 You first need to work out relative frequencies (see table below), then graph the data appropriately (see figure below).

Relative frequencies of hyphal lengths in treatments A and B

Hyphal length µ(m)	Relative frequency (% total)	
	Treatment A	Treatment B
0	0.0	0.0
1	0.4	0.0
2	0.8	0.0
3	1.6	0.3
4	3.2	0.3
5	5.2	0.7
6	8.7	1.7
7	14.3	3.1
8	19.0	4.9
9	18.3	7.3
10	12.7	9.0
11	7.1	10.8
12	4.4	11.5
13	2.4	10.4
14	1.2	9.0
15	0.4	7.6
16	0.4	6.3
17	0.0	4.9
18	0.0	4.2
19	0.0	2.8
20	0.0	1.7
21	0.0	1.4
22	0.0	0.7
23	0.0	0.7
24	0.0	0.3
25	0.0	0.3

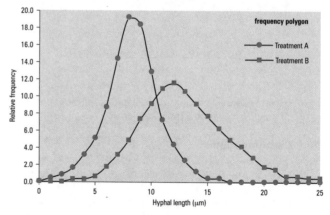

Frequency polygons for data in above table

61.2 Mean: **7.60**; standard deviation: **0.886**.

61.3 *Tip:* The frequency distribution of the untransformed data indicates that they are positively skewed (see figure opposite). The ladder of transformations shown in the figure suggests initial corrections in increasing strength being square root x and $\log(x)$. Figure (b) indicates that the square root transformation is approximately symmetrical, whereas figure (c) indicates that the log transformation results in a slight negative skew, in effect 'over-transforming' the data.

(a)

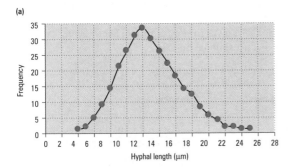

(b)

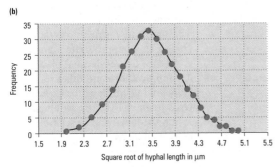

(c)

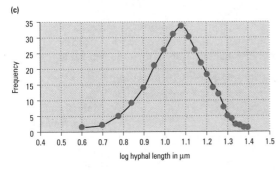

Effect of different transformations on distribution of data from Treatment B in hyphal growth experiment. (a) Untransformed data; (b) square root transformation of (x); (c) log transformation of (x)

62 Using graphs
62.1

    (a) Three-dimensional graph (Fig. 62.6).
    (b) Pie chart (Fig. 62.10).
    (c) Scatter diagram (Fig. 62.5).
    (d) Bar chart (Fig. 62.9).
    (e) Plotted curve (Fig. 62.4).

62.2 *Tip:* The data must first be calculated as a proportion of the total, and then the degrees if the relevant 'slice' worked out by multiplying by 360 (the number of degrees in a circle), as shown in the table below. Use a compass to make a circle for your graph, then use a protractor to mark out the number of degrees required per slice. The resulting graph should look like the figure below.

Behaviour of male brown bear as a proportion of total time

Behaviour category	Occasions observed	Proportion of total	Degrees for 'slice'
Eating	37	0.435	156.7
Sleeping	25	0.294	105.9
Sexual display	13	0.153	55.1
Scratching	8	0.094	33.9
Mating	2	0.024	8.5
Total	85	1	360

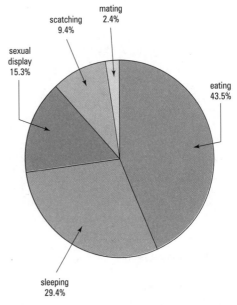

Pie chart showing proportion of time spent by male brown bears in different activities

62.3 *Tip:* First decide on the class boundaries and number of classes – in this case, the values range from 8.1 to 15.1. The subdivisions shown in the table below provide eight

Data of table on p. 398 presented as a frequency distribution

Class	Frequency
8.0–8.9	2
9.0–9.9	5
10.0–10.9	17
11.0–11.9	22
12.0–12.9	26
13.0–13.9	17
14.0–14.9	10
15.0–15.9	1

classes, which give a frequency distribution with a reasonable number of classes to aggregate the data sufficiently, but also provide enough detail to indicate the overall shape of the distribution. Note that Microsoft Excel also has a useful histogram function, but you will need to learn how to set the class boundaries (termed 'bin values') to make effective use of this function (see Box 62.2).

A histogram for these data is shown in the figure below. Your brief description should include something about location, dispersion and shape of the distribution (see Chapter 65): for example, 'The modal class of the distribution is 12.0–12.9 g l⁻¹, with a range of 7.0 g l⁻¹ (15.5–8.5 g l⁻¹). The data show a slight negative skew, with 28 values in the three classes above the mode and 46 in the four classes below it'.

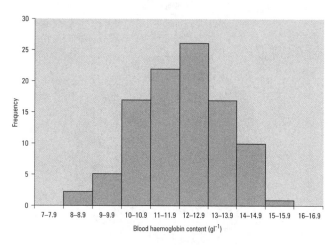

Frequency distribution of haemoglobin data from p. 398

62.4 *Tip:* The Sunday newspaper supplements often analyse news stories with the aid of graphs, and might be a good source for material, as their articles are often slanted towards a particular viewpoint, and you may find that the graphical presentation has been chosen to support this.

63 Presenting data in tables

63.1 *Tip:* To convert all of the concentrations to the same units, you will need to know the relative molecular mass of each constituent (glutamate $M_r = 147.1$; proline $M_r = 115.1$;

Concentrations of low molecular weight solutes in bacteria

Bacterium	Solute concentration (mmol l⁻¹)*		
	Glutamate	Proline	Trehalose
Bacillus subtilis	52.1	39.7	<0.1
Escherichia coli	34.0	21.0	154.7
Staphylococcus aureus	15.3	NT	NT

NT = not tested.
* All values expressed to 1 decimal place.

trehalose $M_r = 342.3$). We have used mmol l⁻¹ in this example, but other units would be equally valid, as long as you are consistent. Data expressed as per cent (w/v) were first converted to g l⁻¹ (multiplied by 10), then to mmol l⁻¹.

63.2 *Tip:* Box 63.1 gives the key aspects. The following example gives one possible layout.

The principal components of a typical table

Component	Comment
Title	Concise, self-contained description of table contents, with numbering where appropriate
Column/row headings	Identifies the content of each column and row, showing units of measurement, where appropriate
Data values	Quoted to an appropriate number of significant figures/decimal places
Footnotes	All abbreviations and other individual details must be explained
Rulings	Used to emphasise any groupings within the table, and to separate individual items

63.3

(a) **11–14 years**.
(b) *Tip:* Calculate the ratio of male : female energy requirement, in order to establish which is greatest in proportional terms, as in the following table.

Male:female energy requirements

Age	Male : female energy requirement
<1	1.0625
1–3	1.0612
4–6	1.1077
7–10	1.1233
11–14	1.2078
15–18	1.3068
19–50	1.3086
>50	1.2338

This shows that the age class where the energy requirement of males is greatest in proportion to that of females is **19–50 years**.
(c) **Birth to 10 years**.
(d) **15–18 years for males**, **11–50 years for females**.

64 Hints for solving numerical problems

64.1

(a) Subtract ax from both sides and then swap sides, so:
$b = y - ax$
(b) 1. Subtract b from both sides and then swap sides, so:
$ax = y - b$

2. Divide both sides by a, so: $\underline{x=(y-b)/a}$

(c) Raise each side to the power $\frac{1}{3}$, then swap sides, so: $\underline{y=x^{1/3}}$

(d) 1. Take logs of both sides, so: $\log x = \log(3^y)$

2. Noting $\log(3^y) = y\log 3$, substitute on right-hand side, so $\log x = y\log 3$

3. Divide both sides by $\log 3$, then swap sides, so: $\underline{y=\log x/\log 3}$

(e) 1. Divide both sides by $1-y$, so: $z^p + 3 = x/(1-y)$

2. Subtract 3 from both sides, so: $z^p = x/(1-y) - 3$

3. Raise both sides to the power $1/p$, so: $\underline{z = ((x/(1-y))3)^{\frac{1}{p}}}$

(f) 1. Multiply both sides by pq, so: $zpq = (y-z)1/n$

2. Raise each side to the power n, so: $zpqn = (y-z)$

3. Take logs of both sides, so: $n\log(zpq) = \log(y-z)$

4. Hence $\underline{n = \log(zpq)/\log(y-z)}$

64.2

(a) **215**

(b) **107 000**

(c) **0.04**

(d) **99.82**

(e) **99.90**

(f) **100.00**

(g) **6260**

(h) **130 000**

(i) **2, 3, or 4, depending on circumstances (see p. 408)**

(j) **5**

(k) **3**

(l) **6**

(m) **2**

(n) **5**

64.3

(a) **17.14%**

(b) **7.55%**

(c) **14.01**

(d) $100 \times (75.02 - 55.23) \div 55.23 = \underline{\textbf{35.83\%}}$

64.4

(a) (i) **0.00000000000000001602192 C;**

(ii) **299 792 400 m s^{-1}**

(b) **4207.3**; **0.80414**; **0.09827**; **7.0238**; **309.09**

64.5 *Tip:* Present formulae as ratios and cancel terms before calculation.

(a) Surface area of sphere $= 4\pi r^2$; volume $= 4\pi^3/3$, hence SA:V ratio $= 4\pi r^2/(4\pi r^3/3)$, which simplifies to $3r^2/r^3 = 3/r$. $r = 15 \times 10^{-6}$ m (diameter divided by 2), hence SA:V ratio $= \textbf{200 000 m}^{-1}$.

(b) Surface area of cylinder without ends $= 2\pi rh$; volume $= \pi r^2 h$, hence SA:V ratio $= 2\pi rh/\pi r^2 h$ which simplifies to $2r/r^2 = 2/r$. $r = 0.5 \times 10^{-3}$ m, hence SA:V ratio $= \textbf{4000 m}^{-1}$.

64.6 *Tip:* To calculate x, substitute values into eqn [64.4].

(a) $x = (7.0 - 4.5) \div 0.02 = \underline{\textbf{125}}$;

(b) $x = (15.4 - -2.6) \div -4.46 = -4.035874439 = \underline{\textbf{-4.04}}$.

Tip: To calculate y, substitute values into eqn [64.3].

(c) $y = 0.2 + (-0.63 \times 10.5) = -6.415 = \underline{\textbf{-6.42}}$;

(d) $y = -1.8 + (4.1 \times 4.5) = 16.65 = \underline{\textbf{16.6}}$.

65 Descriptive statistics

65.1

(a) As these are qualitative data, the mode should be used to describe location. There will be no suitable measure of dispersion.

(b) This is a quantitative but discontinuous distribution and you might choose mean or median (the latter especially for cases of asymmetrical distribution), with standard deviation or semi-interquartile range, respectively, depending on statistic of location chosen.

(c) For this bimodal continuous distribution, it may be best to quote two modes. If you can identify the sex of the bird in each case, then perhaps two distributions should be used, each with its own mean. There is unlikely to be a suitable measure of dispersion if the data for both sexes are combined (perhaps range). If you can identify the sex of the birds, separate standard deviations for males and females may be suitable, if the distributions are symmetrical.

65.2

(a) **Range = 6**

(b) **Variance = 1.769**

(c) **Standard deviation = 1.330**

(d) **Coefficient of variation = 20.15%** (CoV = $100 \times 1.330 \div 6.6$)

(e) **Standard error = 0.1717**.

65.3 *Tip:* Substitute values into eqn [65.3] to calculate the SE of each sample:

for sample A: $12.7 \div 3.464101615 = 3.666174209 = \underline{\textbf{3.67}}$;

for sample B: $14.4 \div 4.472135955 = 3.219937888 = \underline{\textbf{3.22}}$.

Thus, in absolute terms, sample B has the lower standard error (note that the higher standard deviation of sample B is more than offset by the increased number of data values). To calculate the proportional standard error, divide the SE by the mean, as follows:

for sample A: $3.666174209 \div 16.2 = 0.22630705 = \underline{\textbf{0.226}}$;

for sample B: $3.219937888 \div 13.2 = 0.243934688 = \underline{\textbf{0.244}}$.

Thus, in proportion to the mean, sample A has the lower standard error.

65.4

Limpet data (recalculated)

Stone	Mean diameter (mm)	Sample size	Mean diameter × sample size
A	3.0	24	72.0
B	2.5	37	92.5
C	2.0	6	12.0
Total		67	176.5

True mean $= 176.5/67 = \underline{\textbf{2.63 mm}}$

66 Choosing and using statistical tests

66.1 *Tip:* First calculate the standard error using eqn [65.3], as $6.8 \div 4 = 1.7$. Then, substitute values into eqn [66.2] to calculate the 95 per cent confidence limits (note that you also

Answers to study exercises

need the value for $t_{0.05[15]} = 2.131$). The upper limit is calculated as: $24.7 + (2.131 \times 1.7) = 28.3227 = \underline{\textbf{28.3}}$. The lower limit is calculated as $24.7 - (2.131 \times 1.7) = 21.0773 = \underline{\textbf{21.1}}$.

66.2 *Tip:* The variance of the data, s^2, computes as 10.543 and the mean as 3.4. Hence $s^2/\text{mean} = \underline{\textbf{3.1}}$, hence the data appear clumped. In biological terms this might mean that if a snail is infected once, then it has a greater chance of multiple infection. Perhaps parasite eggs are clumped in the snails' environment.

66.3 *Tip:* The calculated $t = \underline{\textbf{2.136}}$ with 12 degrees of freedom; the critical value at $P = 5\%$ is 2.18 (Table 65.2); hence you would accept the null hypothesis that the samples came from the same population and hence conclude that the plant hormone had no significant effect.

66.4 *Tip:* Amino acid uptake will be the dependent (y) variable and time the independent (x) variable. From the Microsoft Excel output, the coefficients are: intercept $= \underline{\textbf{1.17}}$ pmol cell^{-1} (to 3 significant figures) and slope (labelled as 'x variable 1') $= \underline{\textbf{2.95}}$ pmol cell^{-1} min^{-1} (to 3 significant figures), so the linear relationship has the following form: $y = 1.17 + 2.95x$. The strength of the relationship is measured by the coefficient of determination, termed 'R square' in the Excel output. In this case, a value of $\underline{\textbf{0.971}}$ (to 3 significant figures) shows that there is a very strong fit of the y data to the trend line.

Index

Index